DATE DUE

Tectonic Faults
Agents of Change on a Dynamic Earth

Goals for this Dahlem Workshop:

- To assess the intrinsic and extrinsic factors controlling fault evolution, from nucleation through growth to maturity,
- To evaluate the competing processes and feedback mechanisms of faulting on different time and length scales, from the surface down to the asthenosphere,
- To consider new strategies for predicting fault behavior and its impact on the rock record and on the human environment.

Report of the 95th Dahlem Workshop on
The Dynamics of Fault Zones
Berlin, January 16–21, 2005

Held and published on behalf of the
President, Freie Universität Berlin: Dieter Lenzen

Scientific Advisory Board: H. Keupp and R. Tauber, Chairpersons
                               N. S. Baer, G. Braun, P. J. Crutzen,
                               E. Fischer-Lichte, F. Hucho, K. Labitzke,
                               R. Menzel, J. Renn, H.-H. Ropers,
                               E. Sandschneider, L. Wöste

Executive Director: W. de Vivanco

Assistant Editors: G. Custance, C. Rued-Engel

Funded by: Deutsche Forschungsgemeinschaft

Tectonic Faults

Agents of Change on a Dynamic Earth

Edited by

Mark R. Handy, Greg Hirth, and Niels Hovius

Program Advisory Committee:
Mark R. Handy, Chairperson
Lukas P. Baumgartner, Anke M. Friedrich, Greg Hirth,
Walter D. Mooney, and James R. Rice

The MIT Press
Cambridge, Massachusetts
London, U.K.

in cooperation with the Freie Universität Berlin

MIT Press books may be purchased at special quantity discounts for business or sales promotional use. For information, please email special_sales@mitpress.mit.edu or write to Special Sales Department, The MIT Press, 55 Hayward Street, Cambridge, MA 02142.

This book was set in TimesNewRoman by Stasch · Verlagsservice, Bayreuth.

Printed and bound in China.

Library of Congress Cataloging-in-Publication Data

Tectonic Faults : agents of change on a dynamic Earth / edited by Mark R. Handy, Greg Hirth, and Niels Hovius.
　　p.　cm. — (Dahlem workshop reports ; 95)
"Report of the 95th Dahlem Workshop on the dynamics of fault zones, Berlin, January 16–21, 2005"
Includes bibliographical references and index.
ISBN 978-0-262-08362-1 (hardcover : alk. paper)
1. Faults (Geology)—Congresses. 2. Geodynamics—Congresses. I. Handy, Mark R. II. Hirth, Greg. III. Hovius, Niels.
QE606.T44 2007
551.8'72—dc22

　　　　　　　　　　　　　　　　　　　　　　　　　　　　　　2006033362

10　9　8　7　6　5　4　3　2　1

Contents

Dahlem Konferenzen®   vii

List of Participants   ix

1   Tectonic Faults: Agents of Change on a Dynamic Earth   1
    Mark R. Handy, Greg Hirth, and Niels Hovius

2   Fault Zones from Top to Bottom: A Geophysical Perspective   9
    Walter D. Mooney, Gregory C. Beroza, and Rainer Kind

3   Strain Localization within Fault Arrays over Timescales
    of 10^0–10^7 Years: Observations, Explanations, and Debates   47
    Patience A. Cowie, Gerald P. Roberts, and Estelle Mortimer

4   Group Report: Nucleation and Growth of Fault Systems   79
    Kevin Furlong, Rapporteur
    Gregory C. Beroza, Jean-Pierre Brun, Patience A. Cowie,
    Mark R. Handy, Walter D. Mooney, Tuncay Taymaz,
    Christian Teyssier, Alain Vauchez, and Brian Wernicke

5   Seismic Fault Rheology and Earthquake Dynamics   99
    James R. Rice and Massimo Cocco

6   Continental Fault Structure and Rheology
    from the Frictional-to-Viscous Transition Downward   139
    Mark R. Handy, Greg Hirth, and Roland Bürgmann

7   Group Report: Rheology of Fault Rocks
    and Their Surroundings   183
    Terry E. Tullis, Rapporteur
    Roland Bürgmann, Massimo Cocco, Greg Hirth, Geoffrey C. P. King,
    Onno Oncken, Kenshiro Otsuki, James R. Rice, Allan Rubin,
    Paul Segall, Sergei A. Shapiro, and Christopher A. J. Wibberley

**8   Topography, Denudation, and Deformation:
      The Role of Surface Processes in Fault Evolution   205**
      Peter O. Koons and Eric Kirby

9   Constraining the Denudational Response to Faulting   231
      Niels Hovius and Friedhelm von Blanckenburg

**10   Group Report: Surface Environmental Effects
      on and of Faulting   273**
      W. Roger Buck, Rapporteur
      Alexander L. Densmore, Anke M. Friedrich, Niels Hovius, Eric Kirby,
      Peter O. Koons, Thorsten J. Nagel, Fritz Schlunegger,
      Manfred R. Strecker, and Friedhelm von Blanckenburg

11   Fluid Processes in Deep Crustal Fault Zones   295
      Bruce W. D. Yardley and Lukas P. Baumgartner

**12   Deformation in the Presence of Fluids and Mineral Reactions:
      Effect of Fracturing and Fluid–Rock Interaction
      on Seismic Cycles   319**
      Jean-Pierre Gratier and Frédéric Gueydan

**13   Effects of Melting on Faulting and
      Continental Deformation   357**
      Claudio L. Rosenberg, Sergei Medvedev, and Mark R. Handy

**14   Group Report: Fluids, Geochemical Cycles, and
      Mass Transport in Fault Zones   403**
      Mark Person, Rapporteur
      Lukas P. Baumgartner, Bart Bos, James A. D. Connolly,
      Jean-Pierre Gratier, Frédéric Gueydan, Stephen A. Miller,
      Claudio L. Rosenberg, Janos L. Urai, and Bruce W. D. Yardley

Author Index   427

Subject Index   429

Dahlem Konferenzen®

Prof. Dr. WERNER REUTTER, Scientific Director

Arnimallee 22, 14195 Berlin-Dahlem, Germany

Purpose

The Dahlem Konferenzen are held to promote the exchange of scientific ideas and information, to stimulate cooperation between scientists, and to define avenues of future research.

Concept

Progress in understanding complex systems—whether in science or in society—requires interdisciplinary research. Yet, specialists must understand each other across disciplinary lines if they want to collaborate.

The Dahlem Konferenzen offer a unique possibility for researchers from various disciplines to approach topics from their own perspective while combining their experience. The aim of the Konferenzen is not necessarily to reach a consensus, but to identify gaps in knowledge, to find new ways of approaching contentious points, and to indicate the direction of future research.

Themes

Leading scientists submit workshop proposals on themes that

- are directed toward innovative, interdisciplinary research
- are of high-priority interest to the disciplines involved.

The proposals are submitted to the Scientific Advisory Board of the Konferenzen for consideration.

® Dahlem Konferenzen is a registered trademark in the EU.

Program Advisory Committee

A Program Advisory Committee is formed for each workshop based on the recommendations of the Workshop initiator(s). Approximately one year before the workshop, this committee convenes to decide on the scientific program, define the goals of the workshop, and select the themes for debate. Approximately 40 participants are invited on the basis of their expertise and international reputation in the relevant research topics. In addition, a young German scientist can be invited who has demonstrated outstanding potential in field(s) related to the Workshop theme.

The Dahlem Workshop Model

The Dahlem Konferenzen employ a unique format for scientific deliberation (*the Dahlem Workshop Model*) in which the invited participants meet in four interdisciplinary working groups to illuminate the workshop theme from a variety of perspectives. The basis for the group discussions are background papers written by selected participants before the workshop. These papers review particular areas of the workshop theme and pose fundamental questions for the future of research on that theme. During the workshop, each group prepares a report summarizing the results of its deliberations. Two to three workshops a year are held with this format.

Dahlem Workshop Reports

The group reports are published together with the revised background papers as a Dahlem Workshop Report. The reports are published as books by MIT Press.

History

In 1974, the Dahlem Konferenzen were established by the *Stifterverband for die Deutsche Wissenschaft* in cooperation with the *Deutsche Forschungsgemeinschaft* to promote communication and cooperation between scientific disciplines and individual scientists. Since 1990, the Dahlem Konferenzen have been a part of the Freie Universität Berlin. To date, ninety-five Dahlem Workshops have been organized with over 4000 participants. Basic costs are covered by the Freie Universität Berlin.

Name

Dahlem Konferenzen are named after the Berlin district of Dahlem, which has a rich tradition as a scientific location. Today, several Max Planck Institutes, the Freie Universität Berlin, and the Wissenschaftskolleg are located there.

List of Participants

LUKAS P. BAUMGARTNER Institut de Minéralogie et Géochimie, Université de Lausanne, BFSH 2, 1015 Lausanne, Switzerland

Metamorphic petrology, fluid–rock interaction, kinetics of mineral reactions, texture development

GREGORY C. BEROZA Department of Geophysics, Stanford University, 397 Panama Mall, Stanford, CA 94305–2215, U.S.A.

Earthquake and engineering seismology: precise earthquake locations, tomography, dynamic rupture modeling

BART BOS Materials Technology, TNO Science and Industry, P.O. Box 595, Eindhoven 5600 AN, The Netherlands

Deformation mechanics, fracture mechanics, fluid–rock interaction, experimental rock deformation

JEAN-PIERRE BRUN Geoscience Rennes, University of Rennes 1, Campus de Beaulieu, Bat. 15, Avenue du Général Leclerc, 35042 Rennes cedex, France

Continental tectonics, thrusting and extension; mechanics of brittle-ductile systems; lithosphere deformation

W. ROGER BUCK Lamont-Doherty Earth Observatory of Columbia University, Oceanography 108A, Rt. 9W, Palisades, NY 10964, U.S.A.

Continental rifting and the generation of parallel sets of normal faults; dike intrusion in rifts and along mid-ocean ridges

ROLAND BÜRGMANN Department of Earth and Planetary Science, University of California, Berkeley, 389 McCone Hall, Berkeley, CA 94720, U.S.A.

Active tectonics, crustal deformation, and space geodesy

MASSIMO COCCO Istituto Nazionale di Geofisica e Vulcanologia, Via di Vigna Murata 605, 00143 Rome, Italy

Earthquake and fault mechanics; rheology of fault zones; frictional models and dynamic simulations of earthquake ruptures; frictional heating and fluid flow

JAMES A. D. CONNOLLY　Institute for Mineralogy and Petrology, ETH Zürich, Clausiusstrasse 25, 8092 Zürich, Switzerland

Fluid flow in deformable media; metamorphic/igneous petrology

PATIENCE A. COWIE　School of GeoSciences, Grant Institute of Earth Sciences, University of Edinburgh, West Mains Road, Edinburgh EH9 3JW, Scotland, U.K.

Strain localization and variations in fault slip rates in space and time

ALEXANDER L. DENSMORE　Department of Geography, Durham University, South Road, Durham DH1 3LE, U.K.

Development of topography above active structures; evolution of catchment-fan systems; patterns of erosion associated with fault growth

ANKE M. FRIEDRICH　Institut für Geologie, Universität Hannover, Callinstr. 30, 30167 Hannover, Germany

Surface deformation and kinematics of active continental plate boundary regions; geologic context of geodetic data and landscape evolution

KEVIN FURLONG　Department of Geosciences, Pennsylvania State University, 542 Deike Building, University Park, PA 16802, U.S.A.

Lithospheric geodynamics and modeling thermal-deformational processes along plate boundaries

JEAN-PIERRE GRATIER　L.G.I.T. CNRS-Observatoire, Geosciences, Université Joseph Fourier, Rue de la Piscine, 38041 Grenoble cedex 9, France

Mechanisms of deformation in the presence of fluids; creep and compaction by pressure solution; experimental approach and observations of natural deformation; faulting and folding compatibility; 3D restoration and balancing methods

FRÉDÉRIC GUEYDAN　Géosciences Rennes, Université Rennes 1, Bat 15, Campus de Beaulieu, 35042 Rennes cedex, France

Field geology (ductile shear zones); ductile rheology (strain localization and mineral reaction); numerical modeling (lithospheric extension)

MARK R. HANDY　Geowissenschaften, Freie Universität Berlin, Malteserstr. 74–100, 12249 Berlin, Germany

Tectonics, structural geology, rock mechanics, faulting

GREG HIRTH　Department of Geology and Geophysics, Woods Hole Oceanographic Institution, MS#8, WH01, Woods Hole, MA 02543, U.S.A.

Rock mechanics, structural geology, geophysics

NIELS HOVIUS   Department of Earth Sciences, University of Cambridge, Downing Street, Cambridge CB2 3EQ, U.K.

Feedbacks between tectonics, climate, and erosion; controls on erosional landscape evolution, onshore and offshore; erosional fluxes from continents to the oceans; mechanisms of hillslope mass wasting and fluvial bedrock incision

GEOFFREY C. P. KING   Laboratoire de Tectonique, Mecanique de la Lithosphère, Institut de Physique du Globe, 4, place Jussieu, 75252 Paris cedex 05, France

The mechanics of lithospheric deformation

ERIC KIRBY   Department of Geosciences, Pennsylvania State University, 336 Deike Bldg., University Park PA 16802, U.S.A.

Interaction between surface processes and tectonics; landscape response to differential rock uplift; tectonics in Asia

PETER O. KOONS   Department of Earth Sciences, University of Maine, Bryand Global Science Center, Orono, ME 04469–5790, U.S.A.

Mechanics of atmospheric/tectonic cooperation

STEPHEN A. MILLER   Geodynamics/Geophysics, University of Bonn, Nussallee 8, 53115 Bonn, Germany

Earthquake mechanics; crustal fluid flow; fracture networks; fault zone processes

WALTER D. MOONEY   U.S. Geological Survey, 345 Middlefield Rd., MS 977, Menlo Park, CA 94025, U.S.A.

Structure, composition and evolution of the continental crust; internal physical properties of fault zones; continental tectonics; intraplate earthquakes; lithospheric structure

THORSTEN J. NAGEL   Geologisches Institut Bonn, Nussallee 8, 53115 Bonn, Germany

Structural geology and tectonics

ONNO ONCKEN   GeoForschungsZentrum Potsdam, Telegrafenberg A17, 14473 Potsdam, Germany

Structural analysis of orogens, analogue modeling, subduction zones, deformation partitioning and quantification

KENSHIRO OTSUKI   Department of Geoenvironmental Sciences, Graduate School of Sciences, Tohoku University, Aobayama, Aramaki, Aoba-ku, Sendai 980–8578, Japan

Fault rocks and fault dynamics, fractal geometry of fault zones and fault populations, water–rock interaction, earthquake prediction

MARK PERSON   Department of Geological Sciences, Indiana University, 1001 East 10th Street, Bloomington, IN 47405–1405, U.S.A.

Numerical modeling of hydrothermal fluid flow in continental rift basins and fault permeability evolution

JAMES R. RICE   Harvard University, 224 Pierce Hall, DEAS–EPS, 29 Oxford St., Cambridge, MA 02138, U.S.A.

Mechanics and physics of fault processes

CLAUDIO L. ROSENBERG   Geowissenschaften, Freie Universität Berlin, Malteserstr. 74–100, 12249 Berlin, Germany

Faults and magmatism, rheology of partially melted crust, alpine tectonics, indentation tectonics

ALLAN RUBIN   Department of Geosciences, Princeton University, 319 Guyot Hall, Princeton, NJ 08544, U.S.A.

Earthquake mechanics, using both observation and theory; dike propagation

FRITZ SCHLUNEGGER   Institute of Geological Sciences, University of Bern, Baltzerstraße 1, 3012 Bern, Switzerland

Surface processes, tectonic geomorphology, climate and surface erosion, process sedimentology, Andes, Alps

PAUL SEGALL   Department of Geophysics, Stanford University, Mitchell Earth Sciences Building, 397 Panama Mall, Stanford, CA 94305–2215, U.S.A.

Active crustal deformation, physics of faulting and magma transport

SERGEI A. SHAPIRO   Geophysik, Freie Universität Berlin, Malteserstr. 74–100, 12249 Berlin, Germany

Seismogenic processes, fluid-induced faulting, rock physics, seismic imaging, and forward and inverse scattering

MANFRED R. STRECKER   Institut für Geowissenschaften, Universität Potsdam, Postfach 60 15 53, Potsdam 14415, Germany

Neotectonics, relationship between tectonics and climate

TUNCAY TAYMAZ   Department of Geophysical Engineering, Seismology Section, Faculty of Mines, Istanbul Technical University (ITU), Maslak-TR 34390, Istanbul, Turkey

Active tectonics, regional tectonics, geodynamics, seismotectonic processes, rheology of fault zones, earthquake and fault mechanisms, and source rupture histories

CHRISTIAN TEYSSIER   Institut de Géologie et de Paléontologie, Université de Lausanne, Anthropole, 1015 Lausanne, Switzerland

Role of partial melting in evolution of orogens; rheology of lithosphere (crust–mantle coupling); deformation at obliquely convergent/divergent plate margins

TERRY E. TULLIS   Department of Geological Sciences, Brown University, 324 Brook Street, Box 1846, Providence, RI 02912–1846, U.S.A

Experimental rock deformation, rock friction, earthquake mechanics, numerical modeling

JANOS L. URAI   RWTH Aachen University, Geologie–Endogene Dynamik, Lochnerstrasse 4–20, 52056 Aachen, Germany

Deformation mechanisms, fluid–rock interaction, rock rheology, patterns of deformation at different scales

ALAIN VAUCHEZ   Geosciences Montpellier, Université de Montpellier II et CNRS, Place E. Bataillon – cc049, 34095 Montpellier cedex 05, France

Geodynamics; strain localization/distribution in middle/lower crust and mantle; crust mantle coupling/uncoupling; deformation, texture and physical properties of mantle rocks; seismic, mechanical anisotropy in the lithosphere

FRIEDHELM VON BLANCKENBURG   Institut für Mineralogie, Universität Hannover, Callinstraße 3, 30167 Hannover, Germany

Geochemistry, geochemical and isotopic expressions of Earth surface processes

BRIAN WERNICKE   Division of Geological and Planetary Sciences, California Institute of Technology, MC 170–25, 1200 E. California Blvd., Pasadena, CA 91125, U.S.A.

Continental rifting, active tectonics of intraplate fault zones

CHRISTOPHER A. J. WIBBERLEY   Laboratoire Géosciences Azur CNRS, Université de Nice-Sophia Antipolis, 250, rue A. Einstein, 06560 Valbonne, France

Fault zone structure and hydromechanical properties; mechanics of fault growth and array evolution; interdependence of fluid–rock interactions and fault zone rheology

BRUCE W. D. YARDLEY   School of Earth and Environment, Earth Sciences, University of Leeds, Leeds LS2 9JT, U.K.

Fluid–rock interactions in the crust, including metamorphic and ore-forming processes

1

Tectonic Faults

Agents of Change on a Dynamic Earth

MARK R. HANDY[1], GREG HIRTH[2], and NIELS HOVIUS[3]

[1]Department of Earth Sciences, Freie Universität Berlin, Malteserstr. 74–100, 12249 Berlin, Germany
[2]Department of Geology and Geophysics, Woods Hole Oceanographic Institution, MS#8, WH01, Woods Hole, MA 02543, U.S.A.
[3]Department of Earth Sciences, University of Cambridge, Downing Street, Cambridge CB2 3EQ, U.K.

WHAT ARE FAULTS AND WHY SHOULD WE STUDY THEM?

Movements within the Earth and at its surface are accommodated in domains of localized displacement referred to as faults or shear zones. Since the advent of the plate tectonic paradigm, faults have been recognized as primary agents of change at the Earth's surface. Faults delimit tectonic plate boundaries, accommodate plate motion, and guide stress and strain to plate interiors. In extending and contracting lithosphere, faults are the locus of burial and exhumation of large rock bodies.

Active faults are zones of enhanced seismicity with associated surface rupture, ground shaking, and mass wasting. The risk associated with seismic hazard is particularly high in densely populated areas with complex infrastructure. Because faults create morphologies that are in many ways favorable for human settlement (e.g., valleys, harbors), many large population centers are situated near active faults. Prediction of the magnitude, timing, and location of earthquakes is important to the safety and development of these centers.

Faults are also channels for the advection of fluids within the lithosphere. As such, they link the biosphere and atmosphere with the asthenosphere. In particular, faults are conduits for water, which is essential for maintaining life.

They are sites of enhanced dissolution and precipitation, and therefore often contain hydrothermal deposits rich in metal oxides, sulfides, and other minerals of value to industrial society. In addition, faults bound sedimentary basins that contain hydrocarbon resources.

Faults affect the composition of the hydrosphere and atmosphere by exposing fresh rock to weathering. In this sense, faults are a potential factor in long-term climate change. The topography created by faulting provides ecological niches that favor the evolution and migration of mammals, notably hominids. Human evolution has been facilitated by faulting.

Faults are high-permeability pathways for molten rock that ascends from source regions at depth to sinks higher in the lithosphere. Faults are also sites of melt extraction, magma–wall rock interaction, and differentiation. These processes modify both the thermal structure and composition of the Earth's crust and mantle.

Clearly, understanding faults and their underlying processes is a scientific challenge with lasting social and economic relevance. Driven by extensive research in all of these areas, our understanding of faults and faulting has developed rapidly over the past thirty years. Yet many of the factors and feedback mechanisms involved in faulting have still to be constrained. Other notions of fault evolution that have long been accepted are now being called into question. Traditional avenues of research have lost their potential to yield surprising insights. New concepts and initiatives are necessary if we are to augment our knowledge of faulting and harness this knowledge to develop models with predictive capability. This book reports on the findings of the 95th Dahlem Workshop that was devoted to this endeavor.

THE WORKSHOP

The week-long Dahlem Workshop brought together 41 scientists with backgrounds in the natural and engineering sciences, all engaged in various aspects of basic and applied research on fault systems. Prior to the meeting, the program advisory committee had agreed on three main goals for advancing fault research:

- to assess the intrinsic and extrinsic factors controlling fault evolution, from nucleation through growth to maturity and termination;
- to evaluate processes and feedback mechanisms of faulting on different time and length scales, from the surface down to the asthenosphere;
- to advance strategies for predicting fault behavior, for understanding the interaction of faulting with topography and climate, and for interpreting its impact on the rock record.

In accordance with the Dahlem Workshop format, participants were divided into four discussion groups charged with developing the following themes:

1. Nucleation and growth of fault systems
2. Rheology of fault rocks and their surroundings
3. Climatic and surficial controls on and of faulting
4. Fluids, geochemical cycles and mass transport in fault zones.

These themes encompass numerous challenges for basic research in the Earth Sciences, many of them with implications for assessing hazard and mitigating fault-induced risk. To be met, these challenges demand a broad approach in which specialized research is combined with cross-disciplinary studies to develop a new generation of models with predictive capability. The groups' deliberations were facilitated by background papers that had been written on selected aspects of these themes in the months leading up to the meeting. These papers were made available to all participants before the meeting and constitute the bulk of this book. They are complemented by the reports of the four workshop groups, which were drafted by designated rapporteurs by the end of the meeting. In the ensuing months, the authors and other participants were able to revise their papers and reports in light of the discussions and reviews of colleagues who are acknowledged below. This book is therefore the result of a week of well-informed, intensive debate and learning.

WHAT WAS LEARNED?

To answer this question, it helps to begin with some general, long-standing observations. The structure of faults in the Earth's lithosphere varies with depth and displacement: In shallow levels, initial displacement over short times (10^{-2}–10^0 s) on a complex system of fault segments (10^{-2}–10^3 m) eventually concentrates or localizes on one or more long faults (10^3–10^6 m), which remain active intermittently over extended periods of time (10^5–10^7 yr). Superposed on this long-term evolution is short-term transient behavior, exemplified by the recurrence of earthquakes (10^2–10^5 yr). The dynamic range of length and timescales of fault-related processes far exceeds the human dimension (see Figure 4.1 in Furlong et al., Chapter 4). The localization of motion on faults implies a weakening of faulted rock with respect to its surrounding host rocks. Accordingly, motion on fault surfaces and systems can be partitioned in different directions relative to the trend of a fault system. Taken together, these general characteristics reflect the interaction of fault motion history (kinematics) with fault mechanics (rheology), the ambient physical conditions of faulting (e.g., temperature, pressure, fluid properties), the physical and chemical properties of rock (mineralogy, porosity, permeability), and the rates and amounts of denudation at Earth's surface. Understanding the processes and feedbacks that govern the impact of faults at Earth's surface is destined to advance along many parallel and intertwined lines of investigation.

The geometry and internal structure of fault zones has been imaged from the surface down to the base of the lithosphere with a variety of geological and geophysical methods, as reviewed by Mooney et al. (Chapter 2). At shallow levels in the Earth's crust, active faults are discrete features, with microseismicity (M_L 1–3) concentrated on strands no more than several tens of meters wide. Damage zones on either side of this core show time-dependent changes in seismic velocity, presumably due to mineral dissolution–precipitation on the grain scale in the fractured rock. The role of fluids in healing and sealing upper crustal fault systems is considered in the context of the earthquake cycle by Gratier and Gueydan (Chapter 12). The lower depth limit of the damage zone is not well known, and reflects the need to develop imaging methods with better resolution at depth (see Furlong et al., Chapter 4, and Tullis et al., Chapter 7).

Inroads in understanding the full three-dimensional evolution of upper crustal fault systems have come from the study of rifted margins with fault activity documented by sediments in fault-bounded basins (Cowie et al., Chapter 3). The temporal resolution of fault motion at Earth's surface is obviously limited by gaps in the stratigraphic record and the inherent difficulty of discerning all length and timescales of fault activity in a large faulted domain (Buck et al., Chapter 10). Fortunately, recent advances in geochronology (e.g., surface exposure dating with cosmogenic nuclides) already allow us to constrain more precisely not only the age of sediments, but also time- and area-integrated rates of denudation (Hovius and von Blanckenburg, Chapter 9). This has facilitated the calculation of short-term slip rates on faults active over the last ca. 10^5 yr. Many of these new techniques await application, especially in regions where numerical modeling predicts that surface mass flux can perturb the mechanical stability of rocks at depth (Koons and Kirby, Chapter 8). Erosion potentially triggers a positive feedback between rock uplift (exhumation), further denudation, and the generation of topography on timescales of the earthquake cycle.

Much knowledge of fault processes at depths beneath 5 km comes from inactive (fossil), exhumed fault systems, for example, in mountain belts. Marked changes in structure are noted at the transition from brittle, frictional sliding and frictional granular flow (cataclasis) to thermally activated, viscous creep (mylonitization), as reviewed by Handy et al. in Chapter 6. The authors illustrate the dynamic nature of this transition and emphasize its significance for decoupling within the lithosphere as well as for short-term, episodic changes in fluid flux and strength. These changes are triggered by frictional or viscous instabilities and may be measurable as transient motion of the Earth's surface, especially after large earthquakes. Geophysical images and geo-electric studies support the idea of high pore-fluid pressures along thrusts and low-angle normal faults; they also indicate that faults can act as fluid conduits, barriers or both depending on the evolving properties of the fault rocks (see Mooney et al., Chapter 2). Yardley and Baumgarter (Chapter 11) underscore the impact of fluid and fluid composition, both on the structural style and on rheology of the crust. This pertains especially to the escape of volatiles during burial and prograde metamorphism, which is expected to dry out and strengthen the crust. On the other hand, the presence of fluids can weaken fault rocks in several ways; in the case of melt, even modest quantities (<5–7 vol.-%) can reduce viscosity by an order of magnitude, possibly more (Rosenberg et al., Chapter 13). Melt-induced weakening within the base of the continental crust can induce lateral crustal flow, a key process for supporting broad topographic loads like the orogenic plateaus of Tibet and the Andean Altiplano. Faults in the Earth's upper mantle, imaged by measurements of seismic anisotropy, are interpreted to be planar zones of distributed shear some 20–100 km wide (Mooney et al., Chapter 2), although more localized shearing is likely based on rare observations in ex-humed mantle shear zones. Looking even deeper, the lithosphere–asthenosphere boundary is also a major shear zone that accommodates tectonic plate motion with respect to the convecting asthenosphere. The future ability to image fault structure at these depths is contingent on improving spatial resolution even beyond that achieved by recently developed seismic receiver function methods (Furlong et al., Chapter 4).

The mechanical behavior of fault rocks is considered from different per-spectives, depending on the depth interval and conditions of faulting. Regarding the seismic response of upper crustal fault zones, Rice and Cocco (Chapter 5) point out that while rate and state friction laws are adequate descriptions of fault rock behavior at earthquake nucleation and at slow, interseismic rates, new concepts are needed to understand why faults weaken so rapidly during the rupture (growth) stage of large earthquakes. Together with these authors, Tullis et al. (Chapter 7) and Person et al. (Chapter 14) propose several testable hy-potheses for fault weakening that call for a new generation of seismic and labo-ratory experiments, as well as observations of natural fault rocks. In particular, Person et al. (Chapter 14) examine the role of metamorphic reactions and reac-tion rates in the context of upwardly and downwardly mobile fluids as a pos-sible key for the rheology of upper crustal faults during the earthquake cycle. Osmotic effects of clay minerals in faults are expected to affect pore fluid pres-sure and frictional properties of fault zones. In contrast, the viscous lower crust contains mechanical anisotropies (e.g., foliations, minerals), which play a prin-cipal role in localizing strain within shear zones on all length scales (Handy et al., Chapter 6). Scaling these inherited structures is a necessary step toward incorporating the effect of mechanical anisotropy into constitutive rheological models. This may help to constrain the response time of fault geometry and structure to changes in regional deformation rate associated with changing plate-scale kinematics.

The Earth's dynamic surface, especially in faulted areas, is the product of coupled climatic, erosional, and tectonic processes. Progress in understand-ing this coupling has been made, but quantitative, predictive models for the environmental effects on and of faulting are still far from mature. The models of Koons and Kirby (Chapter 8) demonstrate the viability of feedbacks between dynamic topography, stress distribution, and uplift rates. However, identifying limits on the time and length scales at which different surface processes can influence faulting (and vice versa) remains a principal challenge, as discussed by Buck et al. (Chapter 10). These limits are expected to depend on a host of climatic factors, as well as on the erodibility of rocks in the faulted area. Hovius and von Blanckenburg (Chapter 9) review the available geomorphological and geochemical techniques for measuring erosion and weathering on timescales relevant to faulting. These are shown to be key to understanding feedbacks between tectonics and climate, especially isostatic effects related to shifting topographic loads and climatic effects associated with CO_2 drawdown in freshly eroded areas of active faulting. The authors argue that although climatic variability and change are evident in the pattern of erosion and weathering, this pattern almost always reflects a stronger tectonic signal.

RECOMMENDATIONS FOR FUTURE RESEARCH

Rather than summarize the wealth of ideas generated by the four group reports, we end this introduction with an attempt to formulate the participants' consensus opinion on recommendations for future work in fault studies.

There was broad agreement that research should develop along both interdisciplinary and multidisciplinary lines. Faults have immediate impacts on society, but understanding them to the point where we can improve predictions of fault behavior is only possible if the underlying processes can be studied on all relevant time and length scales.

Studies should focus on *natural laboratories* and on *interacting processes*. Natural laboratories are regions of the Earth where geological and climatic processes can be characterized and quantified in a geo-historical context. For fault studies, ideal natural laboratories contain both active and fossil (exhumed) fault systems in a well-defined plate tectonic setting (orogenesis, continental transform faulting, back-arc spreading, intraplate faulting). The fault images—whether mapped from space by satellite, at the surface by eye, or resolved at great depth by geophysical methods—can yield insight into coupled processes during prolonged periods of faulting. Several natural laboratories were mentioned at the conference (e.g., Furlong et al., Chapter 4): the European Alps, the Southern Alps of New Zealand, the Aegean trench-backarc system, the North Anatolian and San Andreas faults, the Cordilleran orogens, and the Himalayan–Tibetan orogen–plateau system. The laboratory chosen obviously depends on the nature of the process(es) studied, so comparing the role of a specific process in more than one setting yields better insight into feedbacks. The best natural laboratories would have an in-depth geological, geophysical, and climatological information base. New natural laboratories can only be developed if funding agencies are willing to support prolonged campaigns whose primary objective is to collect, interpret, and assimilate large and diverse datasets. Much of this basic work is perforce interdisciplinary. Some of the technologies applied are new.

Experimental laboratory studies are needed to understand processes under controlled conditions. Specific examples of experiments pertain to fault weakening and the role of gels and fluids, as outlined, respectively, by Tullis et al. (Chapter 7) and Person et al. (Chapter 14). In some cases, these studies will require the development of new deformation apparati to better approach natural conditions in the laboratory. *Improved data acquisition and processing techniques* are needed to augment the resolution of structures and material flux in Earth and at its surface. The improvement of seismic imaging methods remains a high-priority goal of the geophysical community. Advances are also desirable in geochemical techniques, for example, to improve the precision of surface exposure ages obtained by analyzing trace amounts of cosmogenic nuclides.

Modeling is necessary to test hypotheses and to make predictions in coupled Earth systems that are too complex to understand intuitively. This effort includes both physical modeling (i.e., scaled models using analogue, Earth-like materials) and numerical/analytical modeling. Although both forms of modeling are not new to the Earth Science community, the solid Earth community should take more advantage of recent advances in computing technology to study coupled, fault-related processes. For example, fault studies should employ high-power computing facilities (supercomputing, massive parallel arrays) to test theoretical concepts on the nucleation and growth of slip surfaces at the onset of large earthquakes (Tullis et al., Chapter 7). Likewise, climate models could be adapted to test the long-term effects of faulting and weathering on atmospheric and oceanic CO_2 budgets, and therefore on climate. As in any study of complex phenomena, true progress will come from a pragmatic combination of new and existing approaches and technologies.

Outreach, i.e., public information, is not a form of research, but sharing specialized knowledge is a public duty of the scientific community. Under the fresh impression of the devastating M_w 9.3 Sumatra-Andaman earthquake and tsunami of December 26, 2004, the members of Group 3 formulated a strategy of how Earth scientists could better prepare the public for such events and how public officials might be informed of the risks associated with active faulting (*Buck* et al., Chapter 10). The mechanisms by which information flows in societies under existential stress and duress of time (e.g., in advance of short-term predictions of natural calamities, like large earthquakes) may be a field of interdisciplinary research with potential for another Dahlem Workshop.

In this Introduction, we are only able to provide a glimpse of the wealth of new ideas generated at the workshop. It is left to readers to engage each contribution in this book on its own terms.

ACKNOWLEDGMENTS

Prime thanks are due to the Dahlem staff, which did a fine job of ensuring that every phase of the conference, from the months of planning to the logistical support of the workshops, ran smoothly. Their efforts made possible a productive atmosphere in which science always came first. We wish to thank especially the core members of the staff (Julia Lupp, Caroline Rued-Engel, Gloria Custance, and Angela Daberkow) for their spirited support, especially in the face of unexpected personnel changes made by the Freie Universität just prior to the meeting. They were aided by Barbara Borek and Myriam Nauerz, who did an admirable job of filling in temporarily for Julia Lupp, the director of the Dahlem Conferences, who was unfortunately hindered from attending.

The editors would like to thank the following colleagues for their thoughtful reviews of the background papers: Brian Wernicke, Michael Weber, Charles Sammis, Mark Behn, Paul Segall, Christian Teyssier, Martyn Drury, Torgeir Andersen, Alexander Densmore, Jean-Phillipe Avouac, Guy Simpson, Arjun Heimsath, Chris Wibberley, Chris Spiers, James Connolly, Rainer Abart, Mike Brown, and two anonymous reviewers.

Finally, we acknowledge the Freie Universität and Deutsche Forschungsgemeinschaft for their financial support of this conference, which covered the conference costs, as well as the travel costs and creature comforts of all participants. The participants join the editors in hoping that the Freie Universität will continue to honor its commitment to the scientific integrity of the Dahlem Conferences.

2

Fault Zones from Top to Bottom

A Geophysical Perspective

WALTER D. MOONEY[1], GREGORY C. BEROZA[2], and RAINER KIND[3]

[1]U.S. Geological Survey, 345 Middlefield Road, MS 977,
Menlo Park, CA 94025, U.S.A.
[2]Department of Geophysics, Stanford University, Mitchell Building,
Stanford, CA, 94305–2215, U.S.A.
[3]GeoForschungsZentrum (GFZ) Potsdam, Telegrafenberg, 14473 Potsdam, Germany

ABSTRACT

We review recent geophysical insights into the physical properties of fault zones at all depths in the crust and subcrustal lithosphere. The fault core zone, where slip occurs, is thin (tens of centimeters) and can mainly be studied in trenches and in borehole well logs. The fault damage zone is wider (tens to hundred of meters) and can be measured by the analysis of fault zone-trapped waves. Such studies indicate that the damage zone extends to a depth of at least 3–5 km, but there is no agreement on the maximum depth limit. The damage zone exhibits a seismic velocity reduction (with respect to the neighboring country rock) as high as 20–50%. Significantly, this velocity reduction appears to have a temporal component, with a maximum reduction after a large rupture. The fault damage zone then undergoes a slow healing process that appears to be related to fluid–rock interactions that leads to dissolution of grain contacts and recrystallization. Deep seismic reflection profiles and teleseismic receiver functions provide excellent images of faults throughout the crust. In extensional environments these profiles show normal faulting in the upper crust and ductile extension in the lower crust. In compressional environments, large-scale low-angle nappes are evident. These are commonly multiply faulted. The very thin damage zones for these low angle faults are indicative of high pore-fluid pressures that appear to counteract the normal stresses, thereby facilitating thrusting. The presence of fluids within fault zones is also evidenced by geo-electrical studies in such diverse environments as the Himalayan and Andean orogens, the San Andreas fault, and the Dead Sea Transform. Such studies show that the fault can act as a fluid conduit, barrier, or combined conduit–barrier system depending on the physical properties of the fault core zone and damage zone. The geometry of active fault zones at depth is revealed by precise microearthquake hypocentral locations. There is considerable geometric diversity, with some strike-slip faults showing a very thin (less then 75 m wide) fault plane and others showing wider, segmented planes and/or parallel strands of faulting. A new discovery is slip-parallel, subhorizontal streaks of seismicity that have been identified on some faults. Such streaks may be due to boundaries between locked and slipping parts of the fault or lithologic variations on the fault surface. Measurements of seismic anisotropy across strike-slip faults are consistent with localized fault-parallel shear deformation in the uppermost mantle, with a width that varies between 20 and 100 km. In addition to shear deformation zones, seismic reflection profiles have imaged discrete faults in the uppermost mantle, mainly associated with paleo-continent/continent collisions. Looking deeper, the lithosphere–asthenosphere boundary may be considered as a major shear zone, considering the horizontal movement of lithospheric plates. This shear zone can be imaged with newly developed seismic receiver function methods.

INTRODUCTION

Geophysical studies of Earth's crust, including its fault zones, have developed steadily over the past 80 years. At present, an impressive array of seismic and nonseismic techniques is available to investigate the crust and uppermost mantle. These techniques include active-source refraction and reflection profiles, seismic tomography, measurements of seismic anisotropy and tele-seismic converted waves, seismicity patterns and fault zone-guided waves, borehole surveys, Global Position System (GPS) measurements of crustal deformation, and geo-electrical, magnetic, and gravity methods. In this paper, we briefly review recent geophysical progress in the study of the structure and internal properties of faults zones, from their surface exposures to their lower limit. We focus on the structure of faults within continental crystalline and competent sedimentary rock, rather than within the overlying, poorly consolidated sedimentary rocks (cf. Catchings et al. 1998; Stephenson et al. 2002). A significant body of literature exists for oceanic fracture zones (e.g., Whitmarsh and Calvert 1986; Minshull et al. 1991). Due to space limitations, this review is restricted to faults within and at the margins of the continents.

GEOLOGIC AND BOREHOLE OBSERVATIONS OF FAULTS

Geological studies show that faults are characterized by two dominant features, the core zone and the damage zone (Figure 2.1). The core zone is a thin (tens of cm) plane on which the majority of displacement along a fault is accommodated. It is defined by Chester et al. (1993) as a foliated central ultra-cataclastite layer. Examples are the 10–20 cm thick core zone of the Punchbowl and San Gabriel faults in Southern California (Chester et al. 1993) and the <5 cm thick

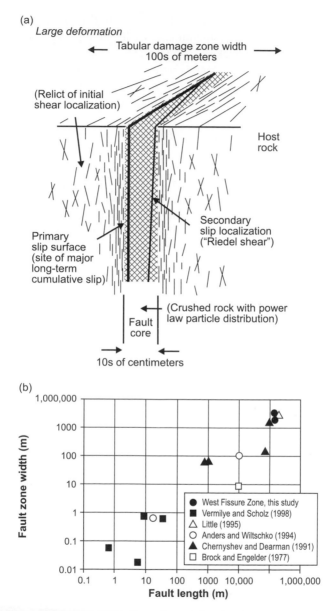

Figure 2.1 (a) Conceptual model of a fault zone showing the core zone and the broader damage zone. Geologic mapping and borehole data can identify the core zone, but most surficial geophysical methods only detect the wider damage zone. Deformation is dominated by strain weakening such that the overall evolution progresses towards geometric simplicity. Modified from Ben-Zion and Sammis (2003). (b) Empirically determined relationship between fault length and fault zone width (damage zone). Modified from Janssen et al. (2002).

core zone of the Chelungpu fault of the 1999, Chi-Chi, Taiwan earthquake. Observations made in underground mines have identified a principal slip zone, or core zone, that is 1 cm thick or less, traceable to hundreds of meters depth, suggesting that the core zone extends much deeper (Holdsworth et al. 2001b; Sibson 2003). The thickness of the core zone varies greatly along each fault, and individual studies describe particular outcrops rather than the characteristics of the entire fault.

The core zone is bounded on either side by a zone of damaged host rock that may be hundreds of meters thick for faults with large displacements (Figure 2.1a; Chester et al. 1993; Schultz and Evans 2000; Ben-Zion and Sammis 2003). The damage zone is interpreted by Ben-Zion and Sammis (2003) as the remnants of failed or abandoned fault surfaces.

Several studies indicate that the width of the damaged zone is roughly proportional to the fault length and/or the magnitude of displacement along a fault, and is controlled by several characteristics of the fault zone at depth, i.e., rheology, lithology, and stress level (Figure 2.1b; Janssen et al. 2002; Faulkner et al. 2003; Sibson 2003; Collettini and Holdsworth 2004; Famin et al. 2004). In contrast with strike-slip faults, many brittle foreland thrust faults, with up to 100 km of displacement, display a sharp "knife-edge" fault contact, with a damage zone of less than a meter or so. This remarkable slip localization is attributed to the presence of fluid at near-lithostatic pressure. This fluid pressure counteracts the normal stress on the fault surface, thereby lowering the shear strength (Sibson 2003).

Although the width of the damage zone for major strike-slip faults can amount to hundreds of meters or more (Holdsworth et al. 2001a; Braathen et al. 2004), trench investigations of strike-slip faults around the world have shown that, overwhelmingly, the bulk of the displacement occurs through successive ruptures localized within a core zone that is only a few centimeters thick (Sibson 2003). For example, this is observed along the slipping portions of the Hayward fault, where the width of surficial deformation averages tens of meters within sediments, but becomes only centimeters wide within deeper basement rocks (Sibson 2003). Determining the width and extent of the fault, as well as the degree to which fluids can penetrate along the fault plane, is extremely important for sites such as proposed nuclear repositories (e.g., Yucca Mountain, Nevada; Potter et al. 2004).

The thin core zone of the fault can be identified in boreholes, but most surface-based geophysical measurements generally cannot identify crustal features that are this thin (i.e., measured in cm). In contrast, the much wider damage zone can easily be identified with surface geophysical measurements because it is characterized by a strong (20–30%) reduction in P- and S-wave seismic velocities (i.e., a seismic low-velocity zone) and reduced electrical resistivity. Below, we discuss these measurements and their implications for the physical properties of fault zones.

FAULT STRUCTURE WITHIN THE SEISMOGENIC ZONE: FAULT ZONE-GUIDED WAVES

As noted above, the existence of a damage zone along the fault leads to strong variations in material properties within and across the fault. The variations have strong effects on seismic wave propagation. Waves that are trapped in the seismic low-velocity zone, which is typically one to several hundred meters wide in active fault zones, are known as fault zone-guided waves (Figure 2.2). Fault zone-guided waves are said to be "trapped" because they propagate within the confines of the low-velocity damaged zone, much like an organ pipe guides sound waves, or like the "SOFAR" channel in the ocean guides long-distance sound waves (Ewing and Worzel 1948). These can also be thought of as analogous to Love waves in vertically layered media in that they consist of critically reflected waves within the low-velocity material. Fault zone-guided waves have been observed in settings as diverse as the subduction zone in Japan (Fukao

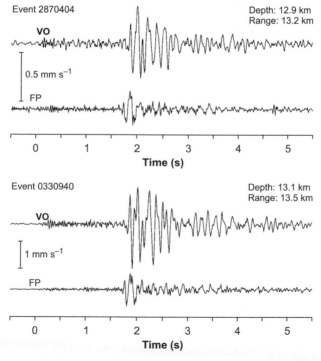

Figure 2.2 Two examples of seismograms showing fault zone-guided waves for aftershocks of the Duzce, Turkey, earthquake. In each case, the seismogram above (station VO) was recorded within the fault zone (higher amplitudes) whereas the seismogram below (station FP) was recorded outside of the fault zone (lower amplitudes). The fault zone-guided waves are the reverberations directly after the large amplitude S waves (time = 2 s).

et al. 1983), a normal fault in the Sierra Nevada foothills near Oroville, California (Leary et al. 1987), and, most commonly, in continental strike-slip environments (e.g., Li et al. 1990; Ben-Zion 1998; Ben-Zion et al. 2003; Haberland et al. 2003; Fohrmann et al. 2004). Waves that refract horizontally due to the large contrast in seismic velocity across a fault zone are known as fault zone head waves. These have also been observed in diverse environments, including both subduction zones, where low-velocity crust descends into the upper mantle (Fukao et al. 1983; Abers 2000), and continental strike-slip faults (e.g., McNally and McEvilly 1977). These two types of waves are naturally very sensitive to the detailed structure (i.e., width, depth, lateral continuity, and seismic velocity) of fault zones, and hence have the potential to reveal the properties of faults at length scales on the order of tens of meters.

The excitation and propagation of fault zone-guided waves depends critically on the geometry and extent of the seismic low-velocity zone that acts to trap the waves. If the fault zone structure contains discontinuities, then such waves will not propagate. Thus, fault zone-trapped waves have tremendous potential to define fault segmentation (Li et al. 2003). One of the outstanding questions not yet fully addressed by studies of fault zone-guided waves is how deep the low-velocity zone extends. There is some evidence (Li et al. 2000) that the low-velocity zone may extend throughout the entire depth of the seismogenic zone, defined as extending from the surface to the maximum depth of microearthquakes (10–14 km in California). However, recent results from aftershocks of the 1999 Duzce, Turkey, earthquake indicate that a significant low-velocity zone only extends to ~3 km depth (Ben-Zion et al. 2003). Similar results for the Landers, California, aftershocks suggest that the low-velocity zone extends to a depth of 2–4 km, with velocity reductions on the order of 30–40% (Peng et al. 2003).

Seismic measurements show that the lateral dimensions of some seismic low-velocity zones responsible for fault zone-guided waves are on the order of one hundred meters. This is comparable to the width of the damage zone observed on exhumed faults. However, Haberland et al. (2003) report a seismic low-velocity zone width of only 3–12 m for the Dead Sea Transform fault, Jordan, despite the more than 100 km of lateral offset on this fault. Thus, both the width and depth of the low-velocity zone are highly variable.

Another interesting issue concerning fault zone-guided waves is to what extent the seismic low velocity zone is a permanent feature, and how much it changes during the earthquake cycle. Field data obtained following the 1992 Landers, California, earthquake suggests that at least some of the decrease in fault zone velocity arises from damage to shallow materials induced by the mainshock (Li et al. 1998). The slow temporal increase in seismic velocity and fault strength after a mainshock is referred to as fault zone healing and is discussed in detailed by Gratier and Gueydan (Chapter 12). This process is difficult to define using seismic data but may consist of crack closure by dissolution of grain contacts and filling of voids by re-crystallization. Fluid–rock interactions are therefore very important in this process. The M 7.1, 1999 Hector Mine earthquake disrupted healing of the nearby Landers fault zone (Vidale and Li 2003), suggesting that strong ground motion from nearby faults can delay the healing process. Finally, the link between the observed low-velocity zone and the mechanical properties of the fault is interesting. InSar (satellite radar) (Fialko et al. 2002) imaging of faults near the Hector Mine earthquake indicate that a kilometer-wide zone reacted compliantly to the static stress change induced by that earthquake. This indicates that the rigidity of the faults is significantly lower than that of the surrounding crust.

Fault zone-guided waves provide important insight into the internal properties of fault zones with depth. Although fault zone-guided waves illuminate the internal structure of the upper few kilometers of fault zones, the internal fault properties throughout the crust, and in particular the lower depth limit of the damage zone, are not yet well known.

SEISMIC REFLECTION AND REFRACTION IMAGING OF FAULTS

Surface-based seismic methods are highly effective at imaging near-horizontal layers within the Earth. However, seismic imaging of steep structures is more difficult (Mooney and Ginzburg 1986; Storti et al. 2003; Weber et al. 2004). Likewise, due to the attenuation of high-frequency seismic energy with depth, the imaging of very thin structures, such as the core zones of faults, is best achieved with borehole geophysical methods rather than surface seismic methods. However, it is possible to extract evidence regarding near-vertical faults from observations such as travel time and amplitude delays, time offset of crustal reflectors, the observation of scattered waves from faults (Maercklin et al. 2004), and a change in the strength and coherence of crustal reflectivity (e.g., Weber et al. 2004).

Deep reflection profiles recorded around the British Isles provide excellent images of crustal and subcrustal faults and shear zones (Matthews 1986; Klemperer and Hobbs 1991). This region has undergone rifting, and these profiles show normal faulting in the upper crust and ductile extension in the lower crust, as expressed by a dense zone of reflections (Figure 2.3). Upper mantle faults have also been imaged, albeit only on a few deep seismic reflection profiles. One of the clearest examples is the Flannan reflector, offshore Scotland, which is believed to be a Caledonian thrust reactivated as an extensional shear zone (Brewer and Smythe 1986; Figure 2.3). This mantle reflector has a dip of about 30° and can be followed to a depth of 80 km. Snyder and Flack (1990) suggest that the Flannan reflector may consist of sheared mafic rocks or eclogite, or may contain hydrous minerals, such as serpentine. Layered seismic anisotropy of sheared peridotite cannot, by itself, explain the strength of the Flannan reflector (Warner and McGeary 1987). The discovery of the Flannan reflector

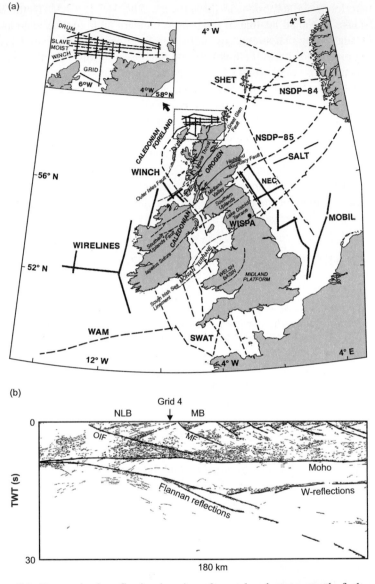

Figure 2.3 Deep seismic reflection imaging of crustal and upper mantle fault zones. (a) Location map for the British Isles with marine seismic reflection profile lines indicated (modified from Matthews et al. 1990). (b) Seismic profile DRUM located off the north coast of Scotland; see inset panel in (a). This profile shows brittle normal faults within the upper crust that merge into a zone of diffuse ductile deformation in the lower crust. The Moho is labeled at a two-way time of 10 s (30 km depth). The uppermost mantle shows two zones of reflections, labeled Flannan and W. The Flannan reflections are interpreted as a Caledonian suture that was reactivated as a lithospheric extensional fault (Flack and Warner 1990).

appeared to confirm the "jelly sandwich" model for lithospheric rheology, in which the ductile lower crust is tectonically decoupled from the brittle upper crust and uppermost mantle (e.g., Ranalli and Murphy 1987).

The European Alps provide some of the most important data regarding deep fault geometries and provide a rare opportunity to compare detailed images of crustal structure with well-determined focal depths. Seismic images of the crust are available from numerous profiles, as summarized by Pfiffner et al. (1997), Waldhauser et al. (1998), and Schmid and Kissling (2000). The NRP-20 profile across the western part of the central European Alps (Figure 2.4a) illustrates several important features: south-directed subduction of the European lithosphere resulted in the formation of large-scale nappes that are multiply folded and are the site of a high level of seismic activity (Figure 2.4b). Intra-crustal decoupling appears to have occurred at the base of the hydrous, quartz-rich intermediate crust rather than within the mafic lower crust (Schmid et al. 1996). The thickness of the seismogenic zone varies widely from over 40 km beneath the Penninic realm to less than 20 km beneath the central Alps (Figure 2.4b), where present-day seismicity is restricted to the nappe pile and is rare in the subducted crust and mantle at depth. As summarized by Schmid and Kissling (2000), the coincidence of the lower limit of seismicity with the 500° isotherm (Okaya et al. 1996) suggests that temperature is the dominant parameter controlling the brittle–ductile transition. Three possible heat sources have been considered to explain the anomalous temperature field: (a) frictional heating; (b) radiogenic heat production within accreted upper crustal material; and (c) ascent of asthenospheric magmas due to slab breakoff (Okaya et al. 1996; Bousquet et al. 1997; von Blanckenburg and Davis 1995; Wortel and Spakman 2000). The third of these sources is significant in that it would also contribute fluids and melts, as observed along the Oligo–Miocene Periadriatic fault system in the Alps.

Handy and Brun (2004) provide a critical review of lithospheric structure (as imaged in seismic reflection profiles), lithospheric strength, grain-scale deformation mechanisms, and crustal seismicity. These authors draw a distinction between the long-term (10^6–10^7 yr) rheology of the lithosphere and short-term seismicity patterns imaged today. The latter are an ambiguous indicator of long-term strength because most earthquakes are most reasonably viewed as manifestations of transient instability within shear zones. Seismicity patterns are therefore more an indication of the location of current zones of episodic decoupling than an indication of lithospheric strength.

The San Andreas fault is one of the most studied faults in the world. Shallow, high-resolution seismic surveys have produced very accurate definition of the sedimentary section and upper crust of the fault zone. Refraction/wide-angle reflection profiles show that the major strike-slip faults associated with the San Andreas fault zone are at a near-vertical orientation and cut through the entire crust, in places even offsetting the Moho (Figure 2.5; Beaudoin et al. 1996; Hole et al. 1998; Henstock and Levander 2000). The vertical Moho offsets are observed in a

(a)

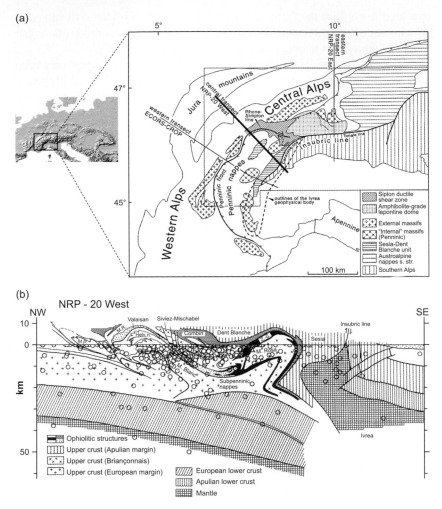

(b)

Figure 2.4 Synthesis of the deep structure and seismicity of the western part of the central Alps along the transect NRP-20 West (Schmid and Kissling 2000). (a) Location map showing the trace of the seismic profile. (b) Crustal cross section showing basement nappes, seismicity, and deep structure. Surficial faults are rooted in the middle crust. Seismicity (open circles) decreases dramatically below a depth of 10–15 km, but some earthquakes are located in the lower crust and even in the upper mantle.

highly reflective mafic layer above the Moho that is interpreted to be the remnant subducted Juan de Fuca oceanic slab (Figure 2.5). These results strongly support the concept that Californian strike-slip faults penetrate the entire crust (Figure 2.6a).

Active low-angle faults associated with the San Andreas fault system have also been imaged in seismic data. For example, the Los Angeles Area Regional Seismic Experiment (LARSE) data yielded impressive images of the hidden faults

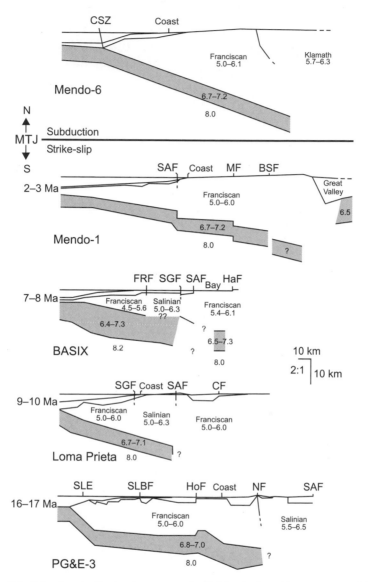

Figure 2.5 Seismic velocity structure across the California Coast Ranges, arranged from north to south (from Hole et al. 1998). The dates indicate the time of passage of the northward migrating Mendocino Triple Junction (MTJ). Seismic velocities are given in km s^{-1}. Highly reflective mafic rocks are shown in gray; the surface locations of faults are shown. Scale of vertical exaggeration is 2:1. SAF: San Andreas Fault; MF: Moho Fault; BSF: Barlett Spring Fault; FRF: Farallon Ridge Fault; SGF: San Gregorio Fault; HaF: Hayward Fault; CF: Calaveras Fault; SLE: Santa Lucia Embankment; SLBF: Santa Lucia Bank Fault; HoF: Hosgri Fault; NF: Nacimiento Fault; BASIX: (San Francisco) Bay Area Seismic Experiment; PG&E-3: Pacific Gas and Electric (Seismic Line) 3.

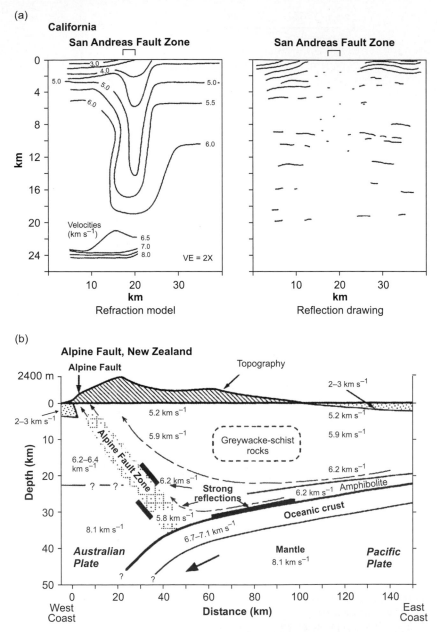

Figure 2.6 Summary of velocity structure within and adjacent to two strike-slip zones (modified from Stern and McBride 1998). (a) Refraction and reflection profiles cross the San Andreas fault in central California near the city of San Luis Obispo (after Mooney and Brocher 1987). (b) Seismic constraints on crustal structure and composition across the Alpine fault of New Zealand (Stern and McBride 1998).

within the Los Angeles basin and beneath the San Gabriel Mountains (Fuis et al. 2001, 2003). There, the high-angle Sierra Madre fault zone (within the Los Angeles basin located west of the San Andreas fault) appears to sole into a master décollement that terminates at the San Andreas fault. The San Andreas fault at this location is near-vertical and appears to extend at least to the Moho, if not deeper (Zhu 2000). This result is consistent with other examples of regional-scale strike-slip faults that appear to cut through the entire crust (often offsetting the Moho) and penetrate deep into the mantle (Storti et al. 2003). Examples include the Dead Sea Transform (ten Brink et al. 1990; Weber et al. 2004), the Great Glen fault (McBride 1995), and the Alpine (New Zealand) fault (Stern and McBride 1998; Figure 2.6b). Moho offsets are also reported in southern Tibet (Hirn et al. 1984a, b) and in northern and western Tibet (Wittlinger et al. 1998, 2001; Zhu and Helmberger 1998). However, recently obtained seismic refraction profiles have failed to confirm these results in northern Tibet (Wang et al. 2006; Zhao et al. 2006), and the size, frequency, and precise geometry of Moho offsets in Tibet should be viewed as an open question.

The geometry of faulting beneath the Himalayan orogen in central Nepal is shown in Figure 2.7 (Zhao et al. 1993; Brown et al. 1996). This is a zone of active convergence, with the Indian crust and mantle lithosphere underthrusting the Asian crust. Seismic reflection data clearly image low-angle faults to depths as great as 30 km. Thrust faults within the crust sole into a main detachment fault that appears to coincide with the top of the Indian Plate. Within the Asian crust (above a

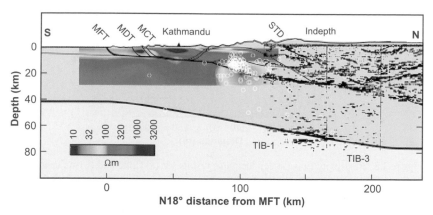

Figure 2.7 Geophysical constraints on the crustal structure across northern India and central Nepal (Avouac 2003). The conductivity section was obtained from a magnetotelluric experiment carried out across central Nepal (Lemonnier et al. 1999). High seismicity correlates with enhanced conductivity. Also shown are the seismic data from the INDEPTH Project located about 300 km east of this section (Zhao et al. 1993; Brown et al. 1996; Nelson et al. 1996). All of the thrust faults are inferred to terminate within prominent midcrustal reflectors, interpreted to be a subhorizontal ductile shear zone. MFT: Main Frontal Thrust; MDT: Main Detachment Fault; MCT: Main Central Thrust.

depth of 20–30 km), the image shows complex, interactive low-angle faulting. This seismic image confirms geological field studies indicating that syn- and post-orogenic normal faults are ubiquitous in collisional mountain belts (Chen and Chen 2004; Victor et al. 2004). It is also noteworthy that enhanced seismicity correlates with a zone of high conductivity that may contain fluids (Figure 2.7).

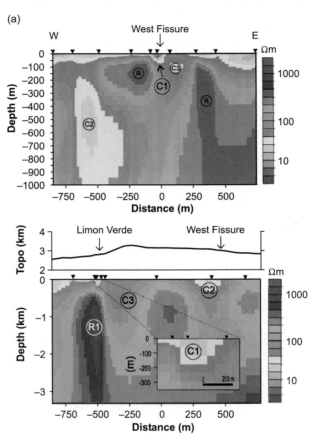

Figure 2.8 Resistivity (inverse conductivity) results for (a) Chile and (b) the central San Andreas fault, California. (a) Magnetotelluric data from two profiles crossing the West Fissure Zone in northern Chile. The conductivity model shown is without vertical exaggeration. Along the West Fissure fault zone surface trace (C1 and C2), a shallow conductive anomaly is visible down to 50–200 m, flanked by two resistive zones (R). At a greater depth beneath Limon Verde (LV) (e), the LV zone is underlain by a resistive zone (R1); modified from Janssen et al. (2002). (b) Resistivity structure of the San Andreas fault near Parkfield, California (Unsworth et al. 1997). High conductivity (here shown as low resistivity) correlates with the fault zone and has a width consistent with the expected width of the damage zone (500–800 m at this location). Solid red dots are earthquake hypocenters. Tcr: Tertiary cover; Kg: Cretaceous granite; DZ: Damage Zone; Tgv: Tertiary gravel; Kjf: Cretaceous Franciscan Assemblage.

GEO-ELECTRICAL IMAGING OF FAULTS

Magnetotelluric (MT) studies of the electrical resistivity (or, equivalently, conductivity) have been used to determine subsurface structure of shallow fracture and damage zones, as well as deeper fault zones. Geo-electrical data have identified fissure zones within the Chilean Precordilleran fault system where resistivity is reduced by fluid transport within fractured rock (Janssen et al. 2002). As seen in Figure 2.8a, the fault zone is interpreted to correlate with a shallow low-resistivity zone (<200 Ωm). Elsewhere, Bai and Meju (2003) found low resistivity correlated with normal faults that define the edges of the Ruili Basin, eastern China.

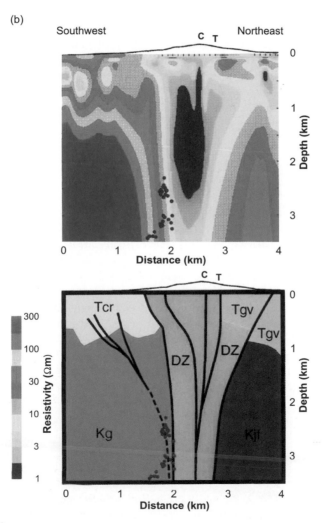

Figure 2.8 (continued)

A classic geo-electrical image is that of the shallow San Andreas fault at the deep drill hole site (SAFOD) near Parkfield, California. This image shows a near-vertical zone of low resistivity that is 500–800 m wide at the top of the seismogenic zone at a depth of 3–4 km (Unsworth et al. 1997; Figure 2.8b). A thinner zone of low resistivity may exist below 4 km but is not resolveable with surface MT measurements. Thurber et al. (2003, 2004) report that the low-resistivity zone coincides with a seismic low-velocity zone determined from seismic tomographic analysis. Unsworth et al. (1997) conclude that the low-resistivity zone consists of clay and saline fluids and that elevated levels of seismicity along this section of the San Andreas fault are correlated with the presence of fluids. As noted above, a similar correlation between elevated seismicity rates and low resistivity is evident beneath central Nepal (Figure 2.7). However, the Dead Sea Transform fault, Jordan, shows a resistivity structure that is remarkably different from the San Andreas fault. Ritter et al. (2003) report that the Dead Sea Transform fault acts as an impermeable barrier to fluid flow rather than acting as a fluid conduit. These contrasting results indicate that a fault can act as a conduit, barrier, or combined conduit–barrier system depending on the physical properties of the fault's core zone and damage zone (Caine et al. 1996; Ritter et al. 2003).

FAULTS AS ILLUMINATED BY SEISMICITY

The geometry of active fault zones at depth is revealed primarily by seismicity. Recently, precise earthquake location methods (e.g., Waldhauser and Ellsworth 2000), coupled with the use of waveform cross-correlation to reduce measurement error (e.g., Schaff et al. 2004), have greatly increased our ability to resolve the fine structure of fault zones, at least to the extent that they are illuminated by seismicity (Figure 2.9). Faults are often idealized as being planar, but geologically mapped surface fault traces are more complex than a simple plane. Since complexities in fault structure may exert a strong control on earthquake behavior, an important question is whether or not the structural complexities observed in fault zones at the Earth's surface extend through the seismogenic crust.

The 1992 Landers, California, earthquake provides a clear example of complex faulting. This event ruptured along three major fault segments: the Johnson Valley fault, the Homestead Valley fault, and the Emerson fault. Felzer and Beroza (1999) studied the complexity of the Homestead Valley–Emerson fault intersection using precise earthquake locations and concluded that the fault at depth was at least as complicated as it was along the surface rupture. More generally, earthquake relocations using the double-difference method and waveform cross-correlation (Schaff et al. 2004) suggest considerable complexity throughout the entire depth of the Landers rupture.

Schaff et al. (2002) studied the depth distribution of earthquakes on the Calaveras fault, central California, and found the fault zone to be extremely thin (75 m or less). They also found that the complex left-step in the surface trace of the fault is geometrically simpler at depth than the surface mapping would indicate. There is, however, some evidence that the base of the seismogenic zone may have some geometrical complexity. Shearer (2002) used precise earthquake relocations to discern parallel strands of seismicity at 9 km depth near the base of the Imperial fault, southern California (Figure 2.10).

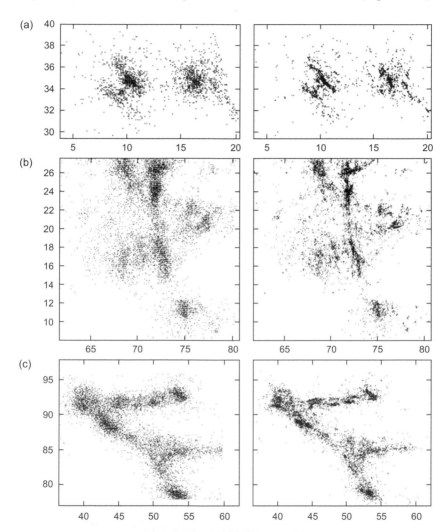

Figure 2.9 Map view of microearthquake catalog locations (left panels) and precise relo-cations (right panels) obtained from a combination of waveform cross-correlation arrival time measurements and the double-difference location method: (a) Results for a part of the 1992 Big Bear sequence; (b) results from the 1992 Joshua Tree earthquake; (c) results for the north end of the 1992 Landers earthquake. In each case, the "cloud" of earthquakes on the left was resolved into more compact, often planar, structures (after Zanzerkia 2003).

The different strands span an approximately 2 km wide zone near the base of the seismogenic zone, indicating that the width of the actively deforming zone is considerable (Figure 2.10).

One of the more striking aspects of twentieth-century seismicity of California is the degree to which the vast stretches of the San Andreas fault that ruptured in the 1857 Fort Tejon and 1906 San Francisco earthquakes have been

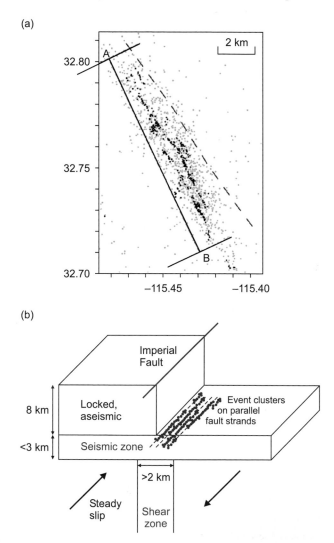

Figure 2.10 Map view of relocated earthquakes (left) and schematic interpretation (right) of the geometry of seismicity strands near the base of the seismogenic zone on the Imperial fault, California (after Shearer 2002). These strands are interpreted to define a 2 km wide shear zone.

devoid of earthquake activity—not only of large events, but down to the detectability threshold of local seismic networks. On a smaller scale, many studies have found a similar anti-correlation of large earthquake slip and small earthquake occurrence (Hill et al. 1990; Ellsworth 1990). For example, Oppenheimer et al. (1990) found a strong correspondence between the areas that slipped in moderate earthquakes on the Calaveras fault (central California) and areas that were relatively devoid of microearthquake activity. They also found that small earthquakes had a very similar spatial distribution before and after moderate earthquakes. They proposed that the areas devoid of seismicity were locked

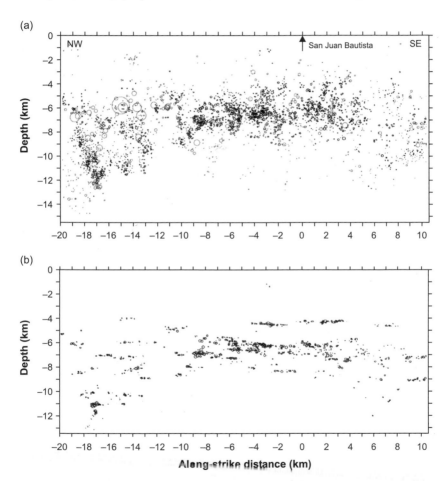

Figure 2.11 Catalog locations of microearthquakes on a vertical cross-section along the San Andreas fault near San Juan Bautista, central California (upper panel). Lower panel shows the same events after relocation. These events define subhorizontal streaks that were previously obscured by location errors. Red circles in upper panel represent large events that were not relocated and therefore do not appear in the lower panel (after Rubin et al. 1999).

portions of the fault and used this assumption to identify two likely source zones for future moderate earthquakes on the Calaveras fault.

The discovery of slip-parallel, subhorizontal "streaks" of seismicity (Rubin et al. 1999; Figure 2.11) is one of the more interesting results to come from precise earthquake relocation in recent years. These streaks have been observed on faults associated with the Southeast Rift Zone at Kilauea Volcano, Hawaii, as well as on the San Andreas, Calaveras, and Hayward faults in California. Waldhauser et al. (2004) examined several such streaks along the Parkfield segment of the San Andreas fault and concluded that in one case they appear to demarcate the boundary between locked and slipping parts of the fault, whereas

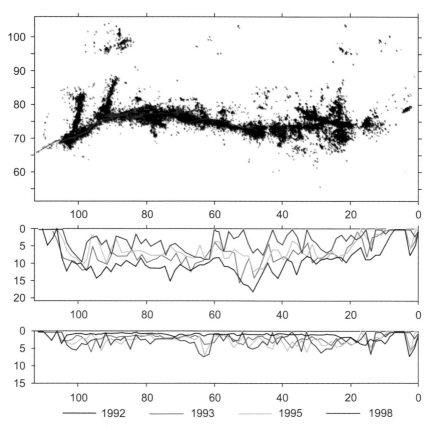

Figure 2.12 Top panel shows rotated map view of the relocated Landers aftershock sequence. Horizontal axis is distance along the fault in km. The middle panel shows the time-dependent depth of the deepest 5% of aftershocks for 1-year time intervals (key indicating color for each year at bottom: 1992 (black), 1993 (blue), 1995 (green), and 1998 (red)). There is a clear tendency for the depth of the deepest earthquakes to become more shallow with time (i.e., from black to red line), as expected for a strain rate-dependent seismic–aseismic transition.

in another case they are more easily explained as occurring at a lithologic discontinuity. An alternative model is that the seismicity streaks reflect temporal migration of slip; that is, the locus of dislocation may migrate up or down the fault.

The depth distribution of seismicity appears to be time dependent. Schaff et al. (2002) noted a temporary increase in the depth of the deepest earthquakes on the Calaveras fault in the year immediately following the 1984 Morgan Hill earthquake. This time dependence is consistent with the depth of the deepest earthquakes being governed by the transition from frictional failure to strain rate-dependent viscous creep. Similar behavior, even much more extensive and dramatic, was observed after the 1992 Landers earthquake, California (Figure 2.12).

SEISMIC ANISOTROPY AND DEFORMATION WITHIN THE MANTLE

Nearly all rock-forming minerals are seismically anisotropic (Babuska 1981; Gebrande 1982). Consequently, all rocks exhibiting a certain degree of textural ordering can be expected to be anisotropic. Chastel et al. (1993) show that pure shear and simple shear regimes can cause different patterns of mineral alignment within ultramafic rocks. Thus, seismic anisotropy is a powerful tool for investigating mechanisms of crustal and upper mantle deformation, particularly in the vicinity of fault zones (Kind et al. 1985; Vinnik et al. 1992; Silver and Chan 1991; Savage and Silver 1993; Rabbel and Mooney 1996; Meissner et al. 2002; Savage 2003; Grocott et al. 2004; Savage et al. 2004). Seismic anisotropy is manifested by the splitting of teleseismic shear wave (S-wave) arrivals. The polarized (split) S-wave arrivals correspond to fast and slow directions of seismic velocity, respectively.

A model in which fault zone deformation is laterally distributed within the plastically deforming upper mantle is illustrated in Figure 2.13 (Teyssier and Tikoff 1998). This model shows the tectonic fabric (flow plane and direction) of the upper mantle curving into parallelism with a strike-slip shear zone (Vauchez et al. 1998; Storti et al. 2003; Vauchez and Tommasi 2003).

The seismic anisotropy of the upper mantle beneath California in the vicinity of the San Andreas fault has been studied using shear-wave splitting by several investigators (Savage and Silver 1993; Hearne 1996; Polet and Kanamori 2002; Savage 2003). The fast directions in the uppermost mantle are generally subparallel to the trend of the fault and orthogonal to the maximum horizontal compressive stress directions, as determined from shallow crustal stress indicators (i.e., the World Stress Map; Zoback 1992). For seismic stations located very close to the San Andreas fault, an optimal model of the anisotropy consists of a thin (10–20 km) sub-Moho layer with the fast direction parallel to the San Andreas fault, underlain by a layer with an fast EW-oriented direction that is parallel to North American Plate motion (Polet and Kanamori 2002). Such a model is consistent with localized, fault-parallel shear deformation within the

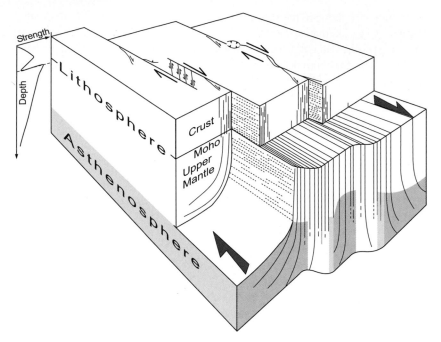

Figure 2.13   (a) Illustration showing how major continental interplate strike-slip deformation belts may ultimately root within the asthenosphere (after Teyssier and Tikoff
1998). Strike-slip faults in the upper crust pass down into increasingly broad shear zones
in the lower crust and lithospheric mantle.

uppermost mantle adjacent to the San Andreas fault. This implies that the mantle
shear depicted in Figure 2.13 is too uniform: the actual deformation may be
both more localized and two-layered, as described above.

Seismic anisotropy measured across the Dead Sea Transform fault (Ruempker
et al. 2003) shows a ~20 km wide zone in the subcrustal mantle. This is interpreted to indicate that the fault plane becomes a broad zone of distributed shear
deformation within the lower crust and mantle lithosphere. There are also reflectors in the lowermost crust (25–32 km) that dip away from the fault zone,
interpreted as contributing to an anisotropic fabric (Figure 2.13). The seismic
anisotropy measurements and asymmetric topography on the Moho discontinuity also indicate that the Dead Sea Transform fault cuts through the entire
crust (Tikoff et al. 2004; Weber et al. 2004).

In New Zealand, the relative motion of the Australian and Pacific Plates
produces mantle anisotropy that reveals shear deformation not only within a
localized inverted flower structure, but in a zone a hundred km wide within the
upper mantle (Figure 2.14; Klosko et al. 1999; Savage et al. 2004). Similar
interpretations of anisotropy have been correlated to crust–mantle coupling for
other transcurrent faults (Vauchez and Tommasi 2003).

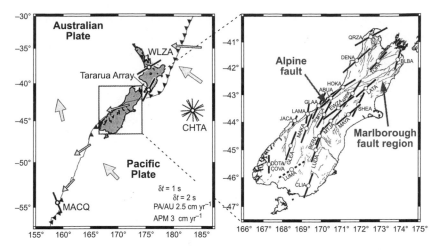

Figure 2.14   Stacked shear wave splitting measurements plotted for stations in the New Zealand SAPSE network and the Tararua Array. The dashed line at Luma represents a measurement that is not stacked. For the South Island stations, the Alpine and Marlborough fault traces are also plotted. Large arrows indicate the absolute plate motion and thinner arrows are relative motion (modified from Klosko et al. 1999).

STRUCTURE OF FAULTS AND SHEAR ZONES WITHIN THE LOWER CRUST AND MANTLE LITHOSPHERE

We distinguish faults, which contain discrete fracture planes, from shear zones, which are regions of distributed deformation. Lithospheric shear zones are large-scale features that penetrate the entire crust and mantle lithosphere (e.g., Figure 2.13). In view of the relative motion of lithospheric plates over the asthenosphere, we may consider the lithosphere–asthenosphere boundary (LAB) itself to be a major shear zone. A new technology is required for imaging such deep structures with better resolution than is currently available. Controlled source seismic experiments only rarely obtain reflected energy from such great depths. Therefore, experiments using natural seismic sources are required for this type of study.

The receiver function technique has been developed for the specific purpose of detailed imaging of the Earth's interior. The idea underlying this method is that strong teleseismic arrivals carry useful information about the Earth's structure below the recording station (the so-called "receiver function," where the recording station is the "receiver" and the Earth's structure is the "function" beneath the receiver). Receiver functions are wavelets caused by seismic discontinuities and, therefore, they mainly carry information about sharp discontinuities rather than smooth seismic velocity variations, as is determined by seismic tomography. Langston (1979) and Owens et al. (1987) used this technique to invert waveform data into velocity-depth distributions of the crust.

Ammon et al. (1990) discuss the non-uniqueness and uncertainties of the waveform inversion method.

The pioneering receiver function experiment was carried out by Nabelek et al. (1993), who mapped the continental Moho and the subducting slab beneath northern Oregon in a region where the slab is not accompanied by seismicity. Since that time, the P-wave receiver function technique (meaning application of P- to S-converted waves in the coda of the P phase) has been used in many field experiments. Deep seismic discontinuities—such as the Moho beneath high orogenic plateaus, the top and bottom of subducting oceanic crust, or detached mantle lithosphere—have been mapped in many parts of the world (Andes: Yuan et al. 2000; Bohm et al. 2002; Tibet: Yuan et al. 1997; Kosarev et al. 2001; Kind et al. 2002; Shi et al. 2004; Wittlinger et al. 2001, 2004; Canada: Bostock 1998, 1999; Alps: Kummerow et al. 2004; Phanerozoic Europe and Scandinavian shield: Alinaghi et al. 2003). In the P-wave receiver function technique, all scattered phases arrive in the coda of the main phase (P, also PP is possible). It is thus often hard to distinguish between originally converted phases and multiply reflected phases. This situation is different in the S-wave receiver function technique. The S to P conversions are precursors of S (or SKS) and the multiples arrive after the S-wave arrival. This separation of the multiples makes the S-wave receiver functions especially useful for studying the LAB. Li et al. (2004) used this approach to study the interaction of the Hawaii plume with the overriding lithosphere.

P-wave receiver functions from an east–west profile along about 26° S in the southern Puna in the central Andes are displayed in Figure 2.15 (Bohm et al. 2002; Yuan et al. 2000). The data displayed in Figure 2.15a are in the time domain as originally recorded. Zero time corresponds to the arrival time of the P-wave. Figure 2.15b shows the same data after depth migration using the IASP91 global reference model. Dark structures are positive amplitudes that correspond to conversions at discontinuities with downwardly increasing velocities. The interpretation of such a receiver function profile is similar to the interpretation of a seismic reflection profile. The first and most important step is the identification of structures that can be followed coherently over some distance. The clearest such structure is the Moho at 8 s delay time (64 km depth) beneath most of the Puna High Plateau. The Moho deepens towards the east starting at about 67° W longitude below the eastern Cordillera. East of 67° W longitude, two shallower structures are visible: one east-dipping and the other west-dipping, which meet in the middle crust at 66.2° W. These structures therefore penetrate the entire crust and form a triangular zone. This triangular zone is interpreted as part of a large thrust fault system that accommodated shortening of the Puna Plateau by westward motion of the Brazilian shield. The Puna crust indents almost 100 km into the central part of the South American crust. This triangular zone is connected to the west with another west-dipping fault between 67.1° W and about 67.7° W, again associated with thickening of the

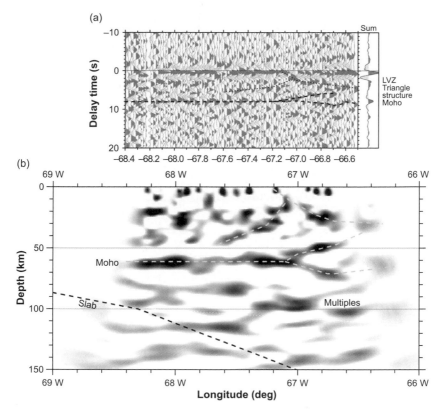

Figure 2.15 (a) Time domain receiver function profile across the southern Puna High Plateau at about 26° S latitude. The varying epicentral distances of each record are taken into account by applying a distance moveout correction (reference distance is 67°). All traces are binned over in a piercing point window of 2 km at 70 km depth without overlapping. The summation trace of all traces is also shown for emphasizing the averaged structure over the entire profile. The Moho and several significant faults that penetrate large parts of the crust beneath the eastern Cordillera are visible (Bohm et al. 2002). (b) Same data as in (a) after migration into depth domain. The subducting slab is not visible in this data. It is visible at shallower depth further west in other data (Yuan et al. 2000). LVZ: low-velocity zone

Puna crust. Thus, these data show a system of thrust faults which penetrates to the base of the crust and which probably contributed to uplift of the Puna Plateau. No continuation of this fault in the mantle lithosphere was observed.

The deep structure of Tibet has been investigated by Project INDEPTH in several multinational experiments (Nelson et al. 1996; Kind et al. 2002). Figures 2.16 and 2.17 show migrated P-wave receiver function sections of the INDEPTH II and III data from about 27° N to 37° N, along the highway between Lhasa and Golmud. The profile crosses the Yarlung Zangbo suture (YZS),

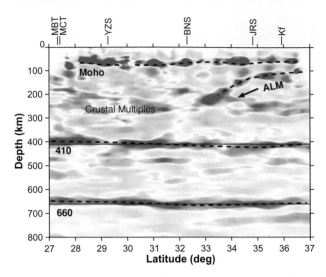

Figure 2.16 Migrated receiver function image across Tibet in a north–south transect (INDEPTH II and III data; Zhao et al. 2001; Tilmann et al. 2003; Shi et al. 2004). Moho, 410 and 660 discontinuities, and crustal multiples are seen at their expected locations. The structure ALM is interpreted as detached Asian lithospheric mantle. MBT: Main Boundary Thrust; MCT: Main Central Thrust; YZS: Yarlung Zangbo suture; BNS: Bangong suture; JRS: Jinsha River suture; KF: Kun-Lun fault. Modified from Kind et al. (2002).

Bangong suture (BNS), Jinsha River suture (JRS), and the Kun-Lun fault (KF). The Moho and the 410 and 660 km discontinuities are clearly visible (Figure 2.16). The 410 and 660 discontinuities are close to their global average depths, which indicates that the surface tectonics do not penetrate as deeply as the mantle transition zone. The Moho is at a depth of 60–80 km. There is a clear additional phase, south-dipping from the Kun-Lun fault and reaching a depth of more than 200 km below the Bangong suture (termed ALM, Asian Lithospheric Mantle). This unexpected structure is assumed to be the detached, sinking Asian mantle lithosphere. This sinking lithosphere probably does not include a significant amount of crust, since this material would undergo phase changes at large depths which presumably would give rise to earthquakes. However, there are no earthquakes in Tibet (except in the westernmost parts) with depths greater than 100 km. The sinking Asian mantle lithosphere presumably decoupled from the crust during the collision between India and Tibet. Without the low-density crust, the mantle lithosphere is negatively buoyant and sinks.

Figure 2.17 shows results that are similar to Figure 2.16 but employs a shorter-period filter and is confined to the upper 100 km. Figure 2.17a shows the direct P to S conversions from the Moho. The Moho depth shallows from about 80 km near the YZS to about 60 km north of the BNS. In the southernmost 200 km of the profile we see a doublet structure of the Moho, which is interpreted to be an

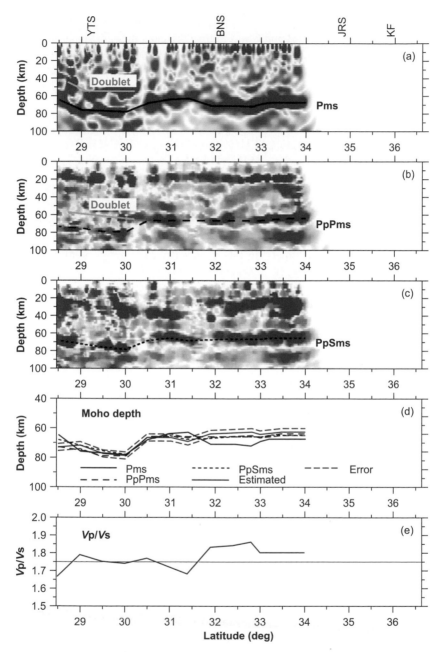

Figure 2.17 Higher resolution image of the same data as in Figure 2.16, confined to the upper 100 km (modified from Kind et al. 2002). Panels A, B, and C show migration for direct conversions and PP and PS multiples, respectively, and may be used for determining the crustal thickness and the average crustal *V*p / *V*s ratio (panels D and E).

imbrication of the Indian and Asian crust, thereby increasing the thickness of the Tibetan crust. Panels B and C in Figure 2.17 show the same data migrated for crustal multiples. The doublet is also clearly seen in the PP crustal multiples confirming the doublet observation. These multiples may be used to determine accurately the crustal thickness and the average crustal Vp/Vs ratio (lower panels in Figure 2.17).

MAPPING OF THE LITHOSPHERE–ASTHENOSPHERE BOUNDARY

The concept of a high-viscosity lithospheric layer overlying a low-viscosity asthenospheric layer was first introduced to explain the postglacial uplift of Scandinavia. Later, with the rise of the plate tectonic theory, this concept was expanded to explain the large lateral displacements of entire continents on rigid lithospheric plates that "float" on the weaker asthenospheric mantle. In this sense, the LAB is the largest shear zone on Earth. However, direct seismic observations of the LAB remain a challenging task. Indeed, the question remains whether we possess an adequate definition of the seismic lithosphere. A major problem is that the asthenosphere is a low-velocity zone, which is difficult to observe seismically. The Moho, in contrast, is marked by a sharp, easily observable increase in velocity. Therefore, the Moho is relatively well constrained on a global scale.

Few constraints on the depth of the LAB have been obtained from seismic body waves. Instead, we rely mostly on surface wave dispersion to measure the depth of the LAB. From these relatively low-resolution observations (low compared to the resolution of Moho depth) the LAB can only be imaged as a broad transition zone and is therefore difficult to locate. Fortunately, the S-wave receiver function technique can improve our ability to image the LAB. In contrast to the P-wave receiver function technique, S-wave receiver functions do not have crustal multiples that mask exactly the interesting depth range. The processing technique for S-wave and P-wave receiver functions is practically the same. The main difference is that P-wave forerunners of the S-wave phase are used, which are on a different component of the seismic record. Figure 2.18a shows summed S-wave receiver functions from Iceland and Greenland (Kumar et al. 2004). Two clear phases are visible in all seismic traces: the Moho and the LAB. The Moho delay time varies between 3–4 s in most regions, which corresponds to 24–32 km depth. More details about crustal thickness can be resolved from the shorter period P-wave receiver functions. The LAB delay times vary between 7 s in eastern Greenland (region F) and 12 s in western Greenland. These values correspond, respectively, to 70 km and 120 km depths for the LAB. Figure 2.18b shows the depth distribution of the LAB in the entire region. An unexpected result is that the LAB is about 80 km deep beneath most parts of Iceland and Greenland. The LAB is shallowest in region F, beneath the

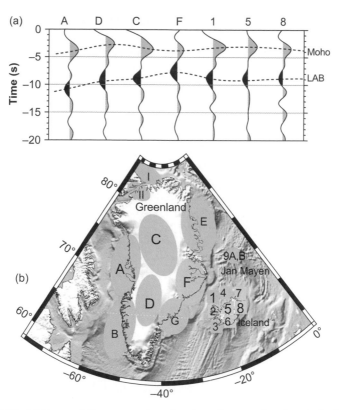

Figure 2.18 (a) Summed S-wave receiver functions for several regions in Iceland and Greenland. The traces are moveout-corrected before summation, with a reference slowness of 6.4 s deg^{-1}. This is the same reference slowness used in P receiver functions, resulting in comparable P-wave and S-wave delay times. The traces are identified by characters on top of the figure, which relate to marked regions in Figure 2.18b. The Moho and lithosphere–asthenosphere boundary (LAB) is clearly visible in each region (Kind et al. 2002). (b) Depth to the LAB marked by a color code (Kind et al. 2002). Previously, it had been assumed that the LAB could not be detected using such seismic data due to the existence of a wide transition zone.

east coast of Greenland. This is where flood basalts suggest a connection to the Iceland plume. GPS uplift data near the biggest volcano on Iceland can best be modeled with a 10–20 km thick elastic lithosphere, which is much thinner than the seismic lithosphere. However, we do not expect that seismic and mechanical (elastic) definitions of the lithosphere should agree. A definition of the lithosphere solely on the basis of temperature is inconsistent with the occurrence of thick lithosphere above a plume and near a mid-ocean spreading center. Comparison of these results suggests that physical attributes other than temperature are necessary to define the lithosphere.

OUTSTANDING QUESTIONS

How far down into the crust does the fault damage zone extend? Does the low-velocity zone associated with fault zone-trapped waves extend to the base of the seismogenic zone? Does crustal deformation become narrower in the lower crust, display an inverted flower structure, or become a wider zone of deformation?

What is the physical relationship between (a) the existence of a pronounced seismic low velocity-zone and high-conductivity zone, and (b) locked versus creeping behavior of strike-slip faults?

Are zones of enhanced electrical conductivity commonly associated with high levels of seismic activity, as suggested for the San Andreas fault at Parkfield, California, and beneath central Nepal? If so, does this imply that high fluid pressure commonly triggers faulting?

Is there always a time dependence of motion at the base of the seismogenic zone? If so, what physical parameters control this dependence?

Does the geometrical complexity of fault zones at the surface extend to great depth, or do active faults become thin, simple planes at depth?

What is the origin of pronounced slip-parallel streaks of seismicity, as seen in relocated hypocenters? Are these due to asperities (lithologic indenters) on the fault surface or to slip-parallel variations in some other physical parameter, such as pore pressure?

Does mantle deformation as inferred from seismic anisotropy consist of two additional components, e.g., a sub-Moho alignment that is parallel to fault slip and a deeper (>60–80 km) alignment that follows global plate motions?

How is crustal deformation accommodated within the lithospheric mantle? Do the crust and underlying mantle decouple in convergent orogenic settings, as suggested beneath the Tibetan Plateau? Do most crustal faults in such settings flatten with depth to become subhorizontal crustal detachments?

Is the LAB directly connected with faults at then surface via continuous zones of predomantly simple shear (Wernicke 1985), or does the LAB sole into the Moho?

ACKNOWLEDGMENTS

We are grateful to the dedicated staff of the Dahlem Conferences, who make these stimulating meetings possible. M. Handy provided encouragement and general guidance. All three authors are indebted to their respective coauthors in previous publications who have generously shared ideas. Reviews by I. Artemieva, G. S. Chulick, M. Handy, N. Okaya, and B. Wernicke are gratefully acknowledged. M. Coble (USGS) assisted materially in all aspects of assembling this paper, and N. Okaya assisted with figure preparation; their efforts are greatly appreciated. Support for this study has come from the U.S. National Earthquake Hazards Reduction Program, the U.S. National Science Foundation, and the Deutsche Forschungsgemeinschaft (German Research Foundation).

REFERENCES

Abers, G.A. 2000. Hydrated subducted crust at 100–200 km depth. *Earth Planet. Sci. Lett.* **176**:323–330.

Alinaghi, A., G. Bock, R. Kind et al. 2003. Receiver function analysis of the crust and upper mantle from the North German Basin to the Archean Baltic Shield. *Geophys. J. Intl.* **155**:641–652.

Ammon, C.J., G.E. Randall, and G. Zandt. 1990. On the nonuniqueness of receiver function inversions. *J. Geophys. Res.* **95**:15,303–15,318.

Avouac, J.P. 2003, Mountain building, erosion, and the seismic cycle in the Nepal Himalaya. *Advances in Geophysics* **46**, pp. 1–80. Amsterdam: Elsevier.

Babuska, V. 1981. Anisotropy of Vp and Vs in rock-forming minerals. *J. Geophys.* **50**:1–6.

Bai, D.H., and M.A. Meju. 2003. Deep structure of the Longling-Ruili fault underneath Ruili Basin near the eastern Himalayan syntaxis: Insight from magnetotelluric imaging. *Tectonophysics* **364**:135–146.

Beaudoin, B.C., N.J. Godfrey, S.L. Klemperer et al. 1996. Transition from slab to slabless: Results from the 1993 Mendocino triple junction seismic experiment. *Geology* **24**:195–199.

Ben-Zion, Y. 1998. Properties of seismic fault zone waves and their utility for imaging low velocity structures. *J. Geophys. Res.* **103**:12,567–12,585.

Ben-Zion, Y., Z. Peng, D. Okaya et al. 2003. A shallow fault zone structure illuminated by trapped waves in the Karadere-Duzce branch of the North Anatolian fault, western Turkey. *Geophys. J. Intl.* **152**:699–717.

Ben-Zion, Y., and C.G. Sammis. 2003. Characterization of fault zones. *Pure & Appl. Geophys.* **160**:677–715.

Bohm, M., S. Lueth, H. Echtler, et al. 2002. The Southern Andes between 36 degrees and 40 degrees S latitude: Seismicity and average seismic velocities. *Tectonophysics* **356**:275–289.

Bostock, M.G. 1998. Seismic stratigraphy and evolution of the Slave province. *J. Geophys. Res.* **103**:21,183–21,200.

Bostock, M.G. 1999. Seismic imaging of lithospheric discontinuities and continental evolution. *Lithosphere* **48**:1–16.

Bousquet, R., B. Goffe, P. Henry, X. Le Pichon, and C. Chopin. 1997. Kinematic, thermal, and petrologic model of the central Alps: Lepontine metamorphism in the upper crust and eclogitization of the lower crust. *Tectonophysics* **273**:105–127.

Braathen, A., P.T. Osmundsen, and R.H. Gabrielsen. 2004. Dynamic development of fault rocks in a crustal-scale detachment: An example from western Norway. *Tectonics* **23**:TC4010, doi: 10.1029/2003TC001558.

Brewer, J.A., and D.K. Smythe. 1986. Deep structure of the foreland to the Caledonian orogen, NW Scotland: Results of the BIRPS Winch profile. *Tectonics* **5**:171–194.

Brown, L.D., W. Zhao, K.D. Nelson et al. 1996. INDEPTH deep seismic reflection observation of a regionally extensive high-amplitude basement reflector and associated "bright spots" beneath the northern Yadong-Gulu rift, Tibet. *Science* **274**:1688–1690.

Caine, J.S., J.P. Evans, and C.B. Forster. 1996. Fault zone architecture and permeability structure. *Geology* **24**:1025–1028.

Catchings, R.D., M.R. Goldman, W.H.K. Lee, M.J. Rymer, and D.J. Ponti. 1998. Faulting apparently related to the 1994 Northridge, California, earthquake and possible coseismic origin of surface cracks in Potrero Canyon, Los Angeles County, California. *Bull. Seism. Soc. Am.* **88**:1379–1391.

Chastel, Y.B., P.R. Dawson, H.-R. Wenk, and K. Bennett. 1993. Anisotropic convection with implications for the upper mantle. *J. Geophys. Res.* **98**:17,757–17,771.

Chen, W.-P., and C.-Y. Chen. 2004. Seismogenic structures along continental convergent zones: From oblique subduction to mature collision. *Tectonophysics* **385**:105–120.

Chester, F.M., J.P. Evans, and R.L. Biegel. 1993. Internal structure and weakening mechanism of the San Andreas fault. *J. Geophys. Res.* **98**:771–786.

Collettini, C., and R.E. Holdsworth. 2004. Fault zone weakening and character of slip along low-angle normal faults: Insight from the Zuccale fault, Elba, Italy. *J. Geol. Soc. London* **161**:1039–1051.

Ellsworth, W.L. 1990. Earthquake History, 1769–1989. In: The San Andreas Fault System, California, ed. R.E. Wallace, Prof. Paper 1515, pp. 153–181. Reston, VA: U.S. Geol. Survey.

Ewing, M., and J.M. Worzel. 1948. Long Range Sound Transmission, Memoir 27, pp. 1–35. Boulder, CO: Geol. Soc. Am.

Famin, V., P. Philippot, L. Jolivet, and P. Agard. 2004. Evolution of hydrothermal regime along a crustal shear zone, Tinos Island, Greece. *Tectonics* **23**: TC5004, doi:10.1029/2003TC001509.

Faulkner, D.R., A.C. Lewis, and E.H. Rutter. 2003. On the internal structure and mechanics of large strike slip fault zones: Field observations of the Carboneras fault in southeastern Spain. *Tectonophysics* **367**:235–251.

Felzer, K.R., and G.C. Beroza. 1999. Deep structure of a fault discontinuity. *Geophys. Res. Lett.* **26**:2121–2124.

Fialko, Y., D. Sandwell, D. Agnew et al. 2002. Deformation on nearby faults induced by the 1999 Hector Mine earthquake. *Science* **297**:1858–1862.

Flack, C.A., and M.R. Warner. 1990. Three-dimensional mapping of seismic reflections from the crust and upper mantle northwest of Scotland. *Tectonophysics* **173**:469–481.

Fohrmann, M., H. Igel, G. Jahnke, and Y. Ben-Zion. 2004. Guided waves from sources outside faults: An indication for shallow fault zone structure. *Pure & Appl. Geophys.* **161**:1–13.

Fuis, G.S., R.W. Clayton, P.M. Davis et al. 2003. Fault systems of the 1971 San Fernando and 1994 Northridge earthquakes, southern California: Relocated aftershocks and seismic images from LARSE II. *Geology* **31**:171–174.

Fuis, G.S., T. Ryberg, N.J. Godfrey, D.A. Okaya, and J.M. Murphy. 2001. Crustal structure and tectonics from the Los Angeles basin to the Mojave Desert, southern California. *Geology* **29**:15–18.

Fukao, Y., S. Hori, and M. Ukawa. 1983. A seismological constraint on the depth of the basalt–eclogite transition in a subducting oceanic crust. *Nature* **303**:413–415.

Gebrande, H. 1982. Elastic wave velocities and constants of elasticity of rocks and rock-forming minerals. In: Physical Properties of Rocks, Landolt-Boernstein Numerical Data and Functional Relationships in Science and Technology, ed. G. Angenheister, subvol. B, pp. 1–99. Berlin: Springer.

Grocott, J., K.J.W. McCaffrey, G.K. Taylor, and B. Tikoff. 2004. Vertical coupling and decoupling in the lithosphere. In: Vertical Coupling and Decoupling in the Lithosphere, ed. J. Grocott, K.J.W. McCaffrey, G. Taylor, and B. Tikoff, Spec. Publ. 227, pp. 1–7. London: Geol. Soc.

Haberland, C., A. Agnon, R. El-Kelani et al. 2003. Modeling of seismic guided waves at the Dead Sea Transform. *J. Geophys. Res.* **108(B7)**:1–12.

Handy, M.R., and J.-P. Brun. 2004. Seismicity, structure, and strength of the continental lithosphere. *Earth Planet. Sci. Lett.* **223**:427–441.

Hearne, T.M. 1996. Anisotropic Pn tomography in the western United States. *J. Geophys. Res.* **101**:8403–8414.

Henstock, T.J., and A. Levander. 2000. Lithospheric evolution in the wake of the Mendocino Triple Junction: Structure of the San Andreas fault system at 2 Ma. *Geophys. J. Intl.* **140**:233–247.

Hill, D.P., J.P. Eaton, and L.M. Jones. 1990. Seismicity, 1980–1986. In: The San Andreas Fault System, California, ed. R.E. Wallace, Prof. Paper 1515, pp. 115–151. Reston, VA: U.S. Geol. Survey.

Hirn, A., J.-C. Lepine, G. Jobert et al. 1984a. Crustal structure and variability of the Himalayan border of Tibet. *Nature* **307**:23–25.

Hirn, A., A. Nercessian, M. Sapin et al. 1984b. Lhasa block and bordering sutures: A continuation of a 500 km Moho traverse through Tibet. *Nature* **307**:25–27.

Holdsworth, R.E., M. Hand, J.A. Miller, and I.S. Buick. 2001a. Continental reactivation and reworking: An introduction. In: Continental Reactivation and Reworking, ed. J. Miller, R.E. Holdsworth, I.S. Buick, and M. Hand, Spec. Publ. 184, pp. 1–12. London: Geol. Soc.

Holdsworth, R.E., M. Stewart, J. Imber, and R.A. Strachan. 2001b. The structure and rheological evolution of reactivated continental fault zones: A review and case study. In: Continental Reactivation and Reworking, ed. J. Miller, R.E. Holdsworth, I.S. Buick, and M. Hand, Spec. Publ. 184, pp. 115–138. London: Geol. Soc.

Hole, J.A., B.C. Beaudoin, and, T.J. Henstock. 1998. Wide-angle seismic constraints on the evolution of the deep San Andreas plate boundary by Mendocino Triple Junction migration. *Tectonics* **17**:802–818.

Janssen C., A. Hoffmann-Rothe, S. Tauber, and H. Wilke. 2002. Internal structure of the Precordilleran fault system (Chile): Insight from structural and geophysical observations. *J. Struct. Geol.* **24**:123–143.

Kind, R., G.L. Kosarev, L.I. Makeyeva, and L.P. Vinnik. 1985. Observations of laterally inhomogeneous anisotropy in the continental lithosphere. *Nature* **318**:358–361.

Kind, R., X. Yuan, J. Saul et al. 2002. Seismic images of crust and upper mantle beneath Tibet: Evidence for Eurasian plate subduction. *Science* **298**:1219–1221.

Klemperer, S.L., and R. Hobbs. 1991. The BIRPS Atlas: Deep seismic reflection profiles around the British Isles. Cambridge: Cambridge Univ. Press.

Klosko, E.R., F.T. Wu, H.J. Anderson et al. 1999. Upper mantle anisotropy in the New Zealand region. *Geophys. Res. Lett.* **26**:1497–1500.

Kosarev, G., R. Kind, S.V. Sobolev, X. Yuan, and W. Hanka. 2001. Seismic evidence for a detached Indian lithosheric mantle beneath Tibet. *Science* **283**:1306–1309.

Kumar, P., R. Kind, W. Hanka et al. 2004. The lithosphere of the NW Atlantic and the Iceland Plume Track. *Earth Planet. Sci. Lett.* **236**:249–257.

Kummerow, J., R. Kind, O. Oncken et al. 2004. Identification of a major active thrust fault system in the deep Adriatic crust of the Eastern Alps. *Earth Planet. Sci. Lett.* **225**:115–129.

Langston, C.A. 1979. The structure under Mount Rainier, Washington, inferred from teleseismic body waves. *J. Geophys. Res.* **84**:4749–4762.

Leary, P., Y.G. Li, and K. Aki. 1987. Observations and modeling of fault zone fracture anisotropy. I. P, SV, SH travel times. *Geophys. J. Roy. Astron. Soc.* **91**:461–484.

Lemonnier, C., G. Marquis, F. Perrier et al. 1999. Electrical structure of the Himalaya of central Nepal: High conductivity around the mid-crustal ramp along the MHT. *Geophys. Res. Lett.* **26**:3261–3264.

Li, Y.G., P. Leary, K. Aki, and P. Malin. 1990. Seismic trapped modes in the Oroville and San Andreas fault zones. *Science* **249**:763–766.

Li, Y.G., J.E. Vidale, K. Aki, and F. Xu. 2000. Depth-dependent structure of the Landers fault zone from trapped waves generated by aftershock. *J. Geophys. Res.* **105**:6237–6254.

Li, Y.G., J.E. Vidale, K. Aki, F. Xu, and T. Burdette. 1998. Evidence of shallow fault zone strengthening after the 1992 *M* 7.5 Landers, California, earthquake. *Science* **279**:217–219.

Li, Y.G., J.E. Vidale, D.D. Oglesby, S.M. Day, and E. Cochran. 2003. Multiple-fault rupture of the *M* 7.1 Hector Mine, California, earthquake from fault zone-trapped waves. *J. Geophys. Res.* **108**: doi:10.1029/2001JB001456.

Li, X., R. Kind, X. Yuan, I. Wölbern, and W. Hanka. 2004. Rejuvenation of the lithosphere by the Hawaiian plume. *Nature* **427**:827–829.

Maercklin, N., C. Haberland, T. Ryberg, M. Weber, Y. Bartov, and the DESERT Group. 2004. Imaging the Dead Sea Transform with scattered seismic waves. *Geophys. J. Intl.* **158**:179–186.

Matthews, D.H. 1986. Seismic reflections from the lower crust around Britain. In: The Nature of the Lower Crust, ed. J.B. Dawson, D.A. Carswell, J. Hall, and K.H. Wedepohl, Spec. Publ. 24, pp. 11–21. London: Geol. Soc.

Matthews, D., and the BIRPS Group. 1990. Progress in BIRPS deep seismic reflection profiling around the British Isles. *Tectonophysics* **173**:387–396.

McBride, J.H. 1995. Does the Great Glen fault really disrupt the Moho and upper mantle structure? *Tectonophysics* **14**:422–434.

McNally, K.C., and T.V. McEvilly. 1977. Velocity contrast across the San Andreas fault in central California: Small-scale variations from P-wave nodal plane distortion. *Bull. Seism. Soc. Am.* **67**:1565–1576.

Meissner, R., W.D. Mooney, and I. Artemieva. 2002. Seismic anisotropy and mantle creep in young orogens. *Geophys. J. Intl.* **149**:1–14.

Minshull, T.A., R.S. White, J.C. Mutter et al. 1991. Crustal stucture of the Blake Spur fracture zone from expanding spread profiles. *J. Geophys. Res.* **96**:9955–9984.

Mooney, W.D., and T.M. Brocher. 1987. Coincident seismic refraction and reflection measurements of the continental lithosphere: A global review. *Rev. Geophys.* **25**:723–742.

Mooney, W.D., and A. Ginzburg. 1986. Seismic measurements of the internal properties of fault zones. *Pure & Appl. Geophys.* **124**:141–157.

Nabelek, J., X.-Q. Li, S. Azevedo et al. 1993. A high-resolution image of the Cascadia subduction zone from teleseismic converted phases recorded by a broadband seismic array (abstract). *EOS Trans. AGU* **Suppl. 74**:431.

Nelson, K.D., W.J. Zhao, L.D. Brown et al. 1996. Partially molten middle crust beneath southern Tibet: A synthesis of project INDEPTH initial results. *Science* **274**:1684–1688.

Okaya, N., R. Freeman, E. Kissling, and S. Mueller. 1996. A lithospheric cross-section through the Swiss Alps. Part 1: Thermokinematic modeling of the Neoalpine orogeny. *Geophys. J. Intl.* **125**:504–518.

Oppenheimer, D.H., W.H. Bakun, and A.G. Lindh. 1990. Slip partitioning of the Calaveras fault, California, and prospects for future earthquakes. *J. Geophys. Res.* **95**:8483–8498.

Owens, T.J., S.R. Taylor, and G. Zandt. 1987. Crustal structure at regional seismic test network stations determined from inversion of broadband teleseismic P waveforms. *Bull. Seismol. Soc. Am.* **77**: 631–662.

Peng, Z., Y. Ben-Zion, A.J. Michael, and L. Zhu. 2003 Quantitative analysis of seismic trapped waves in the rupture zone of the 1992 Landers, California, earthquake: Evidence for a shallow trapping structure. *Geophys. J. Intl.* **155**:1021–1041.

Pfiffner, O.A., P. Lehner, P. Heitzmann, S. Mueller, and A. Steck, eds. 1997. Deep Structure of the Swiss Alps: Results of NRP-20. Basel: Birkhäuser.

Polet, J., and H. Kanamori. 2002. Anisotropy beneath California: Shear wave splitting measurements using a dense broadband array. *Geophys. J. Intl.* **149**:313–327.

Potter, C.J., W.C. Day, D.S. Sweetkind, and R.P. Dickerson. 2004. Structural geology of the proposed site area for a high-level radioactive waste repository, Yucca Mountain, Nevada. *Geol. Soc. Am. Bull.* **116**:858–879.

Rabbel, W., and W.D. Mooney. 1996. Seismic anisotropy of the crystalline crust: What does it tell us? *Terra Nova* **8**:16–21.

Ranalli, G., and D.C. Murphy. 1987. Rheological stratification of the lithosphere. *Tectonophysics* **132**:281–296.

Ritter, O., T. Ryberg, U. Weckmann et al. 2003. Geophysical images of the Dead Sea Transform in Jordan reveal an impermeable barrier for fluid flow. *Geophys. Res. Lett.* **30**: doi 10.1029/2003GL017541.

Rubin, A.M., D. Gillard, and, J.-L. Got. 1999. Streaks of microearthquakes along creeping faults. *Nature* **400**:635–641.

Ruempker, G., T. Ryberg, G. Bock, and the DESERT Group. 2003. Boundary layer mantle flow under the Dead Sea Transform fault from seismic anisotropy. *Nature* **425**:497–501.

Savage, M.K. 2003. Seismic anisotropy and mantle deformation in the western United States and southwestern Canada. In: The Lithosphere of Western North America and Its Geophysical Characteristics, ed. S.L. Klemperer and W.G. Ernst, Intl. Book Ser. vol. 7, pp. 421–445. Boulder, CO: Geol. Soc. Am.

Savage, M.K., K.M. Fischer, and C.E. Hall. 2004. Stain modeling, seismic anisotropy, and coupling at strike-slip boundaries: Application in New Zealand and the San Andreas fault. In: Continental Reactivation and Reworking, ed. J. Miller, R.E. Holdsworth, I.S. Buick, and M. Hand, Spec. Publ. 184, pp. 9–39. London: Geol. Soc.

Savage, M.K., and P.G. Silver. 1993. Mantle deformation and tectonics: Constraints from seismic anisotropy in western United States. *Phys. Earth Planet. Inter.* **78**:207–228.

Schaff, D.P., G.H.R. Bokelmann, G.C. Beroza, F. Waldhauser, and W.L. Ellsworth. 2002. High resolution image of Calaveras fault seismicity. *J. Geophys. Res.* **107**:2186, doi:10.1029/2001JB000633.

Schaff, D.P., G.H.R. Bokelmann, W.L. Ellsworth et al. 2004. Optimizing correlation techniques for improved earthquake location. *Bull. Seism. Soc. Am.* **94**:705–721.

Schmid, S., and E. Kissling. 2000. The arc of the western Alps in the light of geophysical data on deep crustal structure. *Tectonics* **19**:62–85.

Schmid, S.M., O.A. Pfiffner, N. Froutzhaim, G. Schoenborn, and E. Kissling. 1996. Geophysical-geologic transect and tectonic evolution of the Swiss-Italian Alps. *Tectonics* **15**:1036–1064.

Schultz, S.E., and J.P. Evans. 2000. Mesoscopic structure of the Punchbowl fault, southern California, and the geologic and geophysical structure of active strike-slip faults. *J. Struct. Geol.* **22**:913–930.

Shearer, P.M. 2002. Parallel fault strands at 9-km-depth resolved on the Imperial fault, southern California. *Geophys. Res. Lett.* **29**:doi:10.1029/2002GL015302.

Shi, D., W. Zhao, L. Brown et al. 2004. Detection of southward intracontinental subduction of Tibetan lithosphere along the Bangong-Nujiang suture by P-to-S converted waves. *Geology* **32**:209–212.

Sibson, R.H. 2003. Thickness of the seismic slip zone. *Bull. Seism. Soc. Am.* **93**:1169–1178.

Silver, P.G., and W.W. Chan. 1991. Shear wave splitting and subcontinental mantle deformation. *J. Geophys. Res.* **96**:16,429–16,454.

Snyder, D.B., and C.A. Flack. 1990. A Caledonian age for reflectors within the mantle lithosphere north and west of Scotland. *Tectonics* **9**:903–922.

Stephenson, W.J., J.K. Odum, R.A. Williams, and M.L. Anderson. 2002. Delineation of faulting and basin geometry along a seismic reflection transect in urbanized San Bernardino Valley, California. *Bull. Seism. Soc. Am.* **92**:2504–2520.

Stern, T.A., and J.H. McBride. 1998. Seismic exploration of continental strike-slip zones. *Tectonophysics* **286**:63–78.

Storti, F., R.E. Holdsworth, and F. Salvini. 2003. Interplate strike-slip deformation belts. In: Intraplate Strike-Slip Deformation Belts, ed. F. Shorti, R.E. Holdsworth, and F. Salvini, Spec. Publ. 210, pp. 1–14. London: Geol. Soc.

ten Brink, U.S., N. Schoenberg, R.L. Kovach, and Z. Ben-Avraham. 1990. Uplift and a possible Moho offset across the Dead Sea Transform. *Tectonophysics* **180**:71–85

Teyssier, C., and B. Tikoff. 1998. Strike-slip partitioned transpression of the San Andreas fault system: A lithosphere-scale approach. In: Continental Transpressional and Transtensional Tectonics, ed. R.E. Holdsworth, R.A. Strachan, and J.F. Dewey, Spec. Publ. 135, pp. 143–158. London: Geol. Soc.

Thurber, C., S. Roecker, K. Roberts et al. 2003. Earthquake locations and three-dimensional fault zone structure along the creeping section of the San Andreas fault near Parkfield, CA.: Preparing for SAFOD. *Geophys. Res. Lett.*:doi:10.1029/2002GL016004.

Thurber, C., S. Roecker, H. Zhang, S. Baher, and W. Ellsworth. 2004. Fine-scale structure of the San Andreas fault zone and location of the SAFOD target earthquakes. *Geophys. Res. Lett.* **31**:L12S02, doi:10.1029/2003GL019398.

Tikoff, B., R. Russo, C. Teyssier, and A Tommasi. 2004. Mantle-driven deformation of orogenic zones and clutch tectonics. In: Vertical Coupling and Decoupling in the Lithosphere, ed. J. Grocott, K.J.W. McCaffrey, G. Taylor, and B. Tikoff, Spec. Publ. 227, pp. 41–64. London: Geol. Soc.

Tilmann, F., J. Ni, and INDEPTH III Seismic Team. 2003. Seismic imaging of the downwelling India lithosphere beneath central Tibet. *Science* **300**:1424–1427.

Unsworth, M.J., P.E. Malin, G.D. Egbert, and J.R. Booker. 1997. Internal structure of the San Andreas fault at Parkfield, California. *Geology* **25**:359–362.

Vauchez, A., and A. Tommasi. 2003. Wrench faults down to the asthenosphere: Geological and geophysical evidence and thermo-mechanical effects. In: Intraplate Strike-Slip Deformation Belts, ed. F. Shorti, R.E. Holdsworth, and F. Salvini, Spec. Publ. 210, pp. 15–34. London: Geol. Soc.

Vauchez, A., A. Tommasi, and G. Barruol. 1998. Rheological heterogeneity, mechanical anisotropy, and deformation of the continental lithosphere. *Tectonophysics* **296**:61–86.

Victor, P., O. Oncken, and J. Glodny. 2004. Uplift of the western Altiplano plateau: Evidence from the Precordillera between 20° and 21° S (northern Chile). *Tectonics* **23**:TC4004, doi:10.1029/2003TC001519.

Vidale, J., and Y.G. Li. 2003. Damage to the shallow Landers fault from the nearby Hector Mine earthquake. *Nature* **421**:524–526.

Vinnik, L.P., L.I. Makeyeva, A. Milev, and A.Y. Usenko. 1992. Global pattern of azimuthal anisotropy and deformation in the continental mantle. *Geophys. J. Intl.* **111**:433–447.

von Blanckenburg, F., and J.H. Davis. 1995. Slab breakoff: A model for syncollisional magmatism and tectonics in the Alps. *Tectonophysics* **14**:120–131.

Waldhauser, F., and W.L. Ellsworth. 2000. A double-difference earthquake location algorithm: Method and application to the northern Hayward fault, California. *Bull. Seism. Soc. Am.* **90**:1353–1368.

Waldhauser F., W.L. Ellsworth, D.P. Schaff, and A. Cole. 2004. Streaks, multiplets, and holes: High-resolution spatio-temporal behavior of Parkfield seismicity. *Geophys. Res. Lett.* **31**:L18608, doi:10.1029/2004GL020649.

Waldhauser, F., E. Kissling, J. Ansorge, and S. Mueller. 1998. Three-dimensional interface modeling with two-dimensional seismic data: The Alpine crust–mantle boundary. *Geophys. J. Intl.* **135**:264–278.

Wang, Y.X., W.D. Mooney, X.C. Yuan, and N. Okaya. 2007. Crustal structure of the northeastern Tibetan Plateau from the southern Tarim basin to the Sichuan basin. *J. Geophys. Res.*, in press.

Warner, M.R., and S. McGeary. 1987. Seismic reflection coefficients from mantle fault zones. *Geophys. J. Roy. Astron. Soc.* **89**:223–230.

Weber, M., K. Abu-Ayyash, A. Abueledas, and other DESERT group members. 2004. The crustal structure of the Dead Sea transform. *Geophys. J. Intl.* **156**:655–681.

Wernicke, B. 1985. Uniform-sense normal simple shear of the continental lithosphere. *Canad. J. Earth Sci.* **22**:108–125.

Whitmarsh, R.B., and A.J. Calvert. 1986. Crustal structure of Atlantic fracture zones. I. The Charlie-Gibbs fracture zone. *Geophys. J. Roy. Astron. Soc.* **85**:107–138.

Wittlinger, G., V. Farra, and J. Vergne. 2004. Lithospheric and upper mantle stratification beneath Tibet: New insights from Sp conversions. *Geophys. Res. Lett.* **31**: doi:10.1029/2004GL020955.

Wittlinger, G., P. Tapponier, G. Poupinet et al. 1998. Tomographic evidence for localized lithospheric shear along the Altyn Tagh fault. *Science* **282**:74–76.

Wittlinger, G., J. Vergne, P. Tapponnier et al. 2001. Teleseismic imaging of subducting lithosphere and Moho offsets beneath western Tibet. *Earth Planet. Sci. Lett.* **221**:117–130.

Wortel, M.J.R., and W. Spakman. 2000. Subduction and slab detachment in the Mediterranean–Carpathian region. *Science* **290**:1910–1917.

Yuan, X., J. Ni, R. Kind, J. Mechie, and E. Sandvol. 1997. Lithospheric and upper mantle structure of southern Tibet from a seismological passive source experiment. *J. Geophys. Res.* **102**:27,491–27,500.

Yuan, X., S.V. Sobolev, R. Kind et al. 2000. Subduction and collision processes in the Central Andes constrained by converted seismic phases. *Nature* **408**:958–961.

Zanzerkia, E.E. 2003. Towards an Understanding of Seismic Triggering through Precise Earthquake Locations. Ph.D. thesis. Stanford. Stanford Univ.

Zhao, J.M., W.D. Mooney, X.K. Zhang et al. 2006. Crustal structure across the Altyn Tagh range at the northern margin of the Tibetan Plateau and implications for tectonic deformation processes. *Earth Planet. Sci. Lett.* **241**:804–814.

Zhao, W.J., J. Mechie, L.D. Brown et al. 2001. Crustal structure of central Tibet as derived from project INDEPTH wide-angle seismic data. *Geophys. J. Intl.* **145**:486–498.

Zhao, W.J., K.D. Nelson, and Project INDEPTH Team. 1993. Deep seismic-reflection evidence for continental underthrusting beneath southern Tibet. *Nature* **366**:557–559.

Zhu, L.P. 2000. Crustal structure across the San Andreas fault, southern California, from teleseismic converted waves. *Earth Planet. Sci. Lett.* **179**:183–190.

Zhu, L.P., and D.V. Helmberger. 1998. Moho offset across the northern margin of the Tibetan Plateau. *Science* **281**:1170–1172.

Zoback, M.L. 1992. First- and second-order patterns of stress in the lithosphere: The World Stress Map Project. *J. Geophys. Res.* **97**:11,703–11,728.

3

Strain Localization within Fault Arrays over Timescales of 10^0–10^7 Years

Observations, Explanations, and Debates

PATIENCE A. COWIE[1], GERALD P. ROBERTS[2],
and ESTELLE MORTIMER[1, 3]

[1]School of GeoSciences, Grant Institute of Earth Sciences, Edinburgh University,
West Mains Road, Edinburgh, EH9 3JW, U.K.
[2]Research School of Geological and Geophysical Sciences, Birkbeck College
and University College London, Gower Street, London, WC1E 6BT, U.K.
[3]Institute für Geowissenschaften, Universität Potsdam, Karl-Liebknecht-Strasse,
14476 Potsdam, Germany

ABSTRACT

Statistical characterization of fault networks, combined with an analysis of geodetic data and the location of historical earthquakes, is a method commonly used to quantify the degree of strain localization in a given tectonic setting. However, such analyses do not address the fundamental questions of why, how, and when (i.e., after what percent total strain) does localization occur on a lithospheric scale. Many studies suggest that the initial phase of crustal deformation is characterized by distributed strain accumulation and structural complexity and that the system evolves towards highly localized deformation on a small number of discrete fault zones. What controls the transition from one regime to the other within a rheologically layered lithosphere? Observations of the evolution of fault networks over a range of spatial and temporal scales may help us to understand the underlying controls on the localization process. Dip-slip faults, especially moderate to high-angle extensional structures, inherently provide the best conditions for preserving such temporal information over geological time because they generate adjacent sedimentary depocenters that usually remain undeformed by subsequent movement on the fault. The aim of this paper is (a) to review recent observations of extensional fault growth, (b) to summarize conclusions drawn from these observations concerning the underlying controls on strain localization in extensional settings, and (c) to discuss the relevance of these observations to other tectonic settings. In particular, several of the ideas that have been derived from studies of strain localization in extensional settings are used to reexamine existing theories concerning strike-slip fault evolution.

INTRODUCTION

Crustal deformation in the brittle field is largely accommodated by slip on discrete fault or fracture surfaces. When the majority of the total strain (>80%) is accounted for by a small number (e.g., <3) of the largest faults in the population, the deformation is usually described as being strongly localized. Ben Zion and Sammis (2003) suggest that "mature" fault networks are strongly localized, to the extent that major crustal-scale faults can be modeled as Euclidean objects embedded within a continuum. This simplification is very helpful for solving related problems, such as coseismic rupture propagation on an existing fault surface. But to what extent do we really understand the processes that control fault pattern evolution and strain localization on a crustal and/or lithospheric scale; that is, how does a "mature" fault network emerge?

Brittle fault patterns develop through one or more of the following processes: (a) nucleation of a new fault by rock fracture, (b) tip propagation of an existing fault, (c) fault coalescence (linkage). The third mechanism is the most efficient way to produce long sublinear fault zones along which large magnitude earthquakes may occur. The displacement-length scaling relationship for brittle fault populations indicates that longer (larger) faults do accommodate greater geologic offsets (e.g., Schlische et al. 1996). Thus, if fault coalescence/ linkage is the dominant growth process, strain localization may be geologically "rapid" and a mature fault network produced after a relatively small amount of total strain.

One profitable way in which to study the evolution of strain localization on a lithospheric scale is to look at areas that have undergone low to moderate total strain, otherwise the record of strain localization is lost or is too difficult to interpret. There is a wealth of information on fault array evolution being compiled for extensional settings. Fault growth can be inferred from sedimentation patterns within the adjacent fault-controlled depocenters. This approach is most informative when the sediment supply to the basin exceeds the rate of tectonic subsidence so that the true fault movement is given by the stratigraphic thickness. Depending on the available stratigraphic age constraints, a temporal evolution of the fault pattern may be reconstructed to quantify the rates of tip propagation and date fault linkage events. In this paper, we review this information and discuss whether general insights can be derived that are relevant to other tectonic settings, concerning either the stage of localization and/or the type of observations necessary to identify the processes that could be operating.

DEFINING THE ISSUE

Figure 3.1 shows an example of "rapid" strain localization in a numerical model of extensional deformation. It is a two-dimensional finite element model that simulates elastoplastic deformation of a heterogeneous medium with a strain-softening Von Mises rheology (Hardacre and Cowie 2003). Cumulative size–frequency distributions for the fault population have been calculated after different amounts of total strain (see Figure 3.1). In this case, fault size refers to the geometric moment; that is, displacement × length. The distributions of active and inactive faults are plotted as well as that for the entire fault population at several stages of the deformation. Active faults are those that record increased plastic strain in the previous 0.2% increment of extension. The graphs (Figure 3.1c) show that a large number of faults (~200) nucleate and are active initially, described approximately by a power-law size–frequency distribution. After ~1% extension, the size distributions of the active fault population and the total fault population are indistinguishable. As the total strain increases further, the number of active faults decreases. Large faults form by coalescence of smaller faults and eventually penetrate through the entire layer. By 1.75% total strain, the size–frequency distribution of the active faults is no longer power-law and there are relatively few of them (~20). In contrast, the bulk of the faults that initiated earlier in the extension history have become inactive. The inactive faults follow a power-law distribution and, because they are more numerous, the fault population overall exhibits approximately power-law size–frequency scaling. An exponential distribution best fits the largest scale-range of active faults, indicating that a characteristic fault size has emerged, although there are still some faults of all sizes that remain active. During this experiment, the mean extensional stress increases initially to a peak of 3.5 MPa at ~1% strain and then declines rapidly to ~2.5 MPa by 1.6% strain. For different realizations of the strength heterogeneity the bulk strain softening varies between 1 MPa and 3 MPa, but the overall pattern of fault population evolution shown in Figure 3.1 is similar for all the experiments (Hardacre and Cowie 2001).

Figure 3.2 shows a field example of progressive strain localization within an extensional fault population, very similar to the model results of Figure 3.1. In this example, the accumulation of bio-stratigraphically dated sediments adjacent to the faults provides a "tape recorder" of fault activity and growth. The

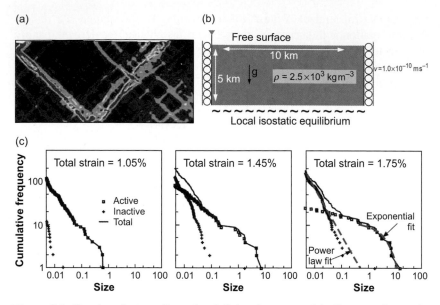

Figure 3.1 Results of a two-dimensional finite element model of progressive strain localization during extension of an elasto-plastic layer (Hardacre and Cowie 2003). The deforming medium is modeled using a strain-softening Von Mises rheology, with a Gaussian heterogeneity in yield strength distributed randomly in space. The degree of heterogeneity in this simulation is ~10% of the mean strength. (a) Strain-rate distribution within the layer ~1.5% total strain: warm colors indicate high strain-rate zones which we interpret as faults, dark blue areas indicate zero strain rate. Model setup is shown in (b). The three panels in (c) show, from left to right, the evolution of the fault population with increasing total extensional strain, as reflected by the size–frequency distributions. The faults are extracted using a clustering algorithm described in Hardacre and Cowie (2003). The size of each fault is the total plastic strain that it accommodates, which is equivalent to its geometric moment. Active faults are those that have recorded increased plastic strain in the previous 0.2% increment of regional extension. The size–frequency distributions for the total fault population (thin line), the active faults only (squares) and the inactive faults (crosses) are plotted separately at each stage to highlight the strain localization process. Theoretical fits to the active and inactive fault populations are also shown, in (c), for 1.75% total strain (see text for discussion).

interpretation shown in Figure 3.2 comes from the Inner Moray Firth Rift Basin, northern North Sea, and is typical of the evolution of faulting seen in many rift basins worldwide (see section on Evidence for Strain Localization Associated with Fault Linkage below).

The study summarized in Figure 3.1 suggests that after only 1–2% total strain the fault array changes from being "immature" to "mature." This result compares well with other studies of localization in single rheology, albeit

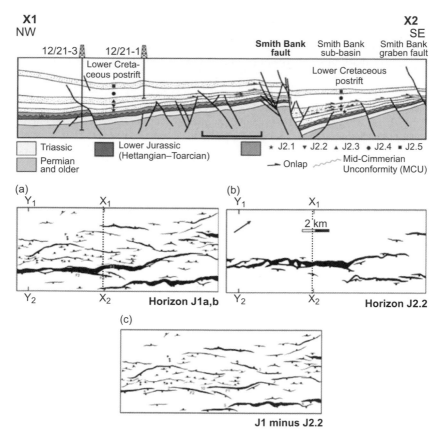

Figure 3.2 An example of strain localization as recorded by stratigraphy within the Inner Moray Firth Rift Basin, northern North Sea (modified from Gupta et al. 1998). The top panel shows an interpreted seismic line perpendicular to the strike of extensional faults in this basin. Location of line is shown by dashed lines in maps (a) and (b) below, taken from Walsh et al. (2003). The interpretation has been made by integrating three-dimensional seismic and core data derived from hydrocarbon exploration of this area. The earliest syn-rift sequences (J1, J2.1, and J2.2) are dissected by numerous small faults (throws generally < 100 m); see map (a). Later sequences (J2.3 and J2.4) are only affected by a few large faults and show pronounced stratigraphic expansion into these faults; see map (b). Map (c) shows the fault population at the onset of extension, obtained by displacement back-stripping. Clearly cessation of activity on smaller faults corresponds to the onset of strain localization on a few large, long-lived faults, such as the Smith Bank fault (throw ≈ 1.5 km). Using biostratigraphy it is possible to show that localization occurred approximately 5 Ma after the onset of extension when the total strain was < 10%.

heterogeneous materials, with fairly simple boundary conditions, for example, rock deformation experiments. However, other numerical and analogue modeling of localization within a rheologically layered lithosphere has shown that a large number of factors, apart from total strain, control the rate and degree of strain localization:

- heterogeneity in strength of crustal rocks (e.g., Hardacre and Cowie 2003; Figure 3.1);
- strength of fault zones relative to the surrounding rock mass, which is fundamentally controlled by H_2O (through effective stress and reaction softening, e.g., Wibberley 1999) and total strain (e.g., Hardacre and Cowie 2003);
- rate of fault healing (e.g., Lyakhovsky et al. 2001);
- presence of a viscous or viscoelastic lower crust (e.g., Davy et al. 1995; Heimpel and Olson 1996; Roy and Royden 2000);
- onset of plastic or brittle failure in the lower crust, which depends strongly on strain rate, temperature, and composition (e.g., Behn et al. 2002; Pérez-Gussinyé et al. 2003; Ellis and Stöckhert 2004); and
- boundary conditions driving the deformation (e.g., Heimpel and Olson 1996).

There are certainly some boundary conditions and lower crustal rheologies for which fault coalescence is suppressed relative to the other growth processes and strain never localizes (e.g., Heimpel and Olson 1996; Buck et al. 1999; Roy and Royden 2000).

GROWTH OF DIP-SLIP FAULTS

At the beginning of the 1990s the fault growth literature was focused on the idea of bilateral propagation of a single fault. This was quantified in terms of the ratio (γ) of maximum accumulated displacement, d, to the dimensions of the fault plane, typically its along-strike length, L. The majority of these data come from "high-angle" (>45°) normal faults and thrust faults within foreland propagating fold-and-thrust belts. The position of the maximum displacement is usually located towards the mid-point of a fault and the displacement dies out to zero at either tip. The point of maximum displacement was generally thought, at that time, to indicate the location of fault nucleation and that a fault should grow in a self-similar way along a growth trajectory defined by the d-L scaling relationship. We now know that, in general, this is not the case. In the last ten years studies have shown that, although fault growth by tip propagation does occur, in many settings it is of limited importance to regional strain accumulation (for a review, see Gupta and Cowie 2000)). It has been recognized and quantified in some field studies of both normal faults (e.g., Jackson and Leeder 1994; Taylor et al. 2004) and thrust faults (e.g., Benedetti et al. 2000), but it is not the process by which most faults substantially increase their length. Linking together two en échelon faults is obviously a much more efficient process of growth. When deformation initiates it is observed that new faults nucleate in large numbers due to the heterogeneous strength distribution of the crust (e.g., Figures 3.1 and 3.2c). Thus, there is a high probability of fault "interaction" and eventual linkage, at least under conditions of coaxial deformation. In addition, new smaller-scale faults nucleate preferentially beyond the tip of a developing fault, due to the tip stress concentration, and are "cannibalized" to form a larger structure (e.g., the model for normal fault growth by Wu and Bruhn 1994). Of course, growth by tip propagation and growth by linkage are not mutually exclusive processes; the evolution of a fault is likely due to a combination of the two, but the impact of linkage events on the growth trajectory dominates over the effects of tip propagation. Thus the typical fault growth trajectory consists of step-like deviations about the globally averaged *d-L* scaling relationship and ideal self-similar growth is rarely observed (e.g., Mansfield and Cartwright 2001). Consequently, there is a lot of scatter in displacement-length correlation plots: the average value of γ is 0.03 over nine orders of magnitude, but it ranges from ~0.005–0.1.

Insights from Fault Displacement Profiles

Several studies of incipiently linking or linked normal fault arrays have shown how a simple bell-shaped displacement profile, which typifies isolated faults (Dawers et al. 1993), becomes modified by prelinkage fault interaction and subsequent linkage (e.g., Peacock and Sanderson 1991; Willemse 1997; Gupta and Scholz 2000). The concept of fault interaction is used by some workers to indicate the mutual influence of faults that physically intersect; that is, there is a geometrical constraint on fault movement (e.g., Jackson and McKenzie 1983). Interaction also arises, however, because a fault growing in the shallow crust perturbs the stress field and, consequently, influences the rate of development of neighboring structures, even if they do not intersect. In most of the fault growth and fault mechanics literature, the word *interaction* refers to this latter process (e.g., Willemse 1997).

Asymmetric fault displacement profiles are usually a good indication of fault interaction (e.g., Gupta and Scholz 2000). Interaction causes steepening of displacement gradients at the interacting fault tips, and the d/L ratio of the individual fault segments also increases (Willemse 1997). Once two faults have linked, the combined structure may be "underdisplaced"; that is, it has an anomalously low d/L ratio, and further displacement accumulates with little tip propagation (e.g., Cartwright et al. 1995). Local displacement minima and irregularities in the fault plane, along an otherwise through-going extensional structure, usually mark the site where linkage has occurred. In summary, these studies have shown that dip-slip fault growth dominated by segment linkage is not self-similar (see sections on Evidence for Strain Localization Associated with Fault Linkage and Evidence for Prelinkage Interaction on Crustal-scale Normal Faults). The widely observed linear *d-L* scaling relationship for faults represents the average scaling of these quantities but does not define the growth trajectory in this case. The only study that finds evidence for self-similar growth of faults with significant displacement profile asymmetry is Manighetti et al.'s

(2001) work on extensional faults in Afar. In this case the asymmetry is interpreted to be related to the overall propagation direction of the Aden rift (i.e.,
tapering fault displacement in the direction of propagation).

Evidence for Strain Localization Associated with Fault Linkage

McLeod et al. (2000) used a combination of three-dimensional seismic and
well-data in the northern North Sea to document growth of the >60 km
Strathspey-Brent-Statfjord fault over a time period of ~25 Ma with a time resolution of 2–3 Ma. The maximum displacement on this fault is ~2.5 km. Using
syn-rift strata, McLeod et al. (2000) showed that the growth of this structure
occurred by the coalescence of fault segments, which had nucleated approximately contemporaneously (similar to that shown in Figure 3.2c). There were
four key observations that came out of this study:

1. Linkage along the entire length of the structure appeared to be a geologically "rapid" process (<2–3 Ma).
2. Subsequent to segment linkage, the accumulation of displacement was almost uniform along the length of the entire fault with very little tip propagation and no evidence for persistent postlinkage displacement deficits where
   linkage had occurred.
3. The slip rate on the linked fault increased by more than 3-fold compared to
   the prelinkage slip rate.
4. The increase in slip rate appeared to coincide with cessation of activity on
   other nearby structures.

Walsh et al. (2002) suggested that rapid segment linkage along the entire length
of a crustal-scale fault could indicate that its growth was controlled by a preexisting zone of weakness inherited from an earlier tectonic episode. Although
this could explain the first two observations made by McLeod et al. (2000), it is
difficult to see how it explains the other two. Moreover, field observations from
the Gulf of Suez, published by Sharp et al. (2000) and Gawthorpe et al. (2003),
support McLeod et al.'s observations. In particular, Sharp et al. (2000) documented the presence of small fault-controlled depocenters, typically 50–100 m
deep and a few kilometers in length, which controlled early syn-rift deposition.
The onset of more rapid basin deepening and the formation of the major Gulf of
Suez extensional faults coincided with the abandonment of many of these small
basins, as Sharp et al. (2000) convincingly demonstrated. Note that this same phenomenon is shown in Figure 3.2: the rate of slip on the Smith Bank fault, as recorded by sediment accumulation in the hanging wall, increases at approximately
the same time that adjacent faults cease to be active (Gupta et al. 1998). It is clear
from all of these examples that the increase in fault slip rate is not due to an
overall increase in basin extension rate but rather can be attributed to the cessation of activity on a large number of small-scale faults (e.g., Figures 3.1 and 3.2).

A recent study of the Rangitaiki fault in the Whakatane Graben, New Zealand, constitutes one of the most detailed and extensive records of fault array evolution in three dimensions in an extensional setting. Taylor et al. (2004) show that this ~20 km long fault has been actively growing for the last 1.3 Ma. Using a combination of regional and high-resolution seismic reflection and core data, these authors are able to (a) resolve the growth history of the un-linked fault segments during the first 1 Ma of activity, (b) pin down the timing of fault linkage to between 300 ka and 18 ka, and (c) show that at the time of fault linkage the average slip rate increased from 0.52 ± 0.18 mm yr^{-1} to 1.41 ± 0.31 mm yr^{-1}.

Evidence for Prelinkage Interaction on Crustal-scale Normal Faults

Taken together, the observations summarized in the previous section suggest that, within extensional settings at least, fault linkage is intimately associated with the way that strain is partitioned on a regional scale. However, the docu-mented increase in fault slip rate could either be a cause or a consequence of fault linkage. In other words, it is possible that the rate only increases once a through-going (smoother, lower friction?) structure has formed that can ac-commodate large displacements more easily. Alternatively, the rate may in-crease prior to linkage due to fault interaction and it is the interaction that drives the coalescence of the fault segments.

Extensional faulting in central Italy appears to be an example of acceler-ated fault growth due to prelinkage interaction. Figure 3.3(a) shows the pat-tern of extensional faulting in the Abruzzo area. It consists of a dominant set of NW trending fault segments with lengths in the range 20–40 km and throws of up to ~2 km. Faulting initiated approximately 3 Ma and the present-day extension rate across the whole Abruzzo area, determined from geodetic ob-servations, is ~3–6 mm yr^{-1} (Hunstad et al. 2003). Profiles of total fault throw along strike indicate that the fault segments are not yet fully linked because the throw diminishes to approximately zero between the fault tips. Incipient linkage is inferred from the presence of active faults, such as the Tre Monte fault (TMF, Figure 3.3a), lying between the tips of the main fault segments and orientated at a high angle to the overall strike of the array. Note that the throw rate averaged over the last 18 ka along the main fault segments mimics the profile of total throw indicating that these faults are not growing in a self-similar way (Figure 3.3b). The throw and throw rates on the linking struc-tures, such as the Tre Monte fault, are an order of magnitude lower than those measured at the centers of the adjacent main fault segments (Morewood and Roberts 2000).

In Figure 3.4 and Table 3.1 estimates of throw rate on the Fucino fault (FFS in Figure 3.3a) are summarized and used to reconstruct an internally consistent throw versus time plot for this fault. The Fucino fault is located near the center

(a)

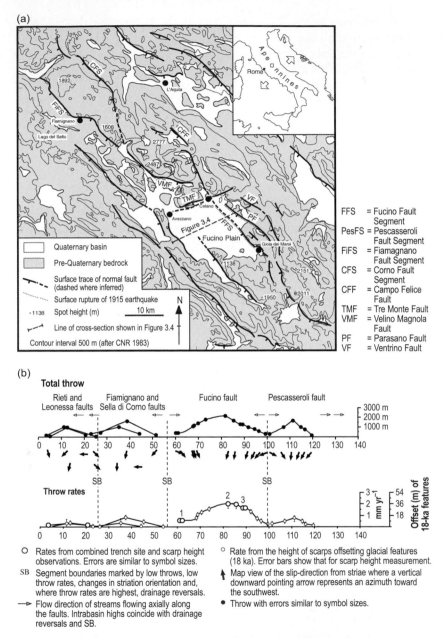

FFS	= Fucino Fault Segment
PesFS	= Pescasseroli Fault Segment
FiFS	= Fiamagnano Fault Segment
CFS	= Corno Fault Segment
CFF	= Campo Felice Fault
TMF	= Tre Monte Fault
VMF	= Velino Magnola Fault
PF	= Parasano Fault
VF	= Ventrino Fault

Figure 3.3 Extensional fault array in Lazio-Abruzzo, central Italy. (a) Map view of fault pattern from Morewood and Roberts (2000). (b) Profiles of total throw (in meters) and throw rate (in mm yr^{-1}) since 18 ka along the length of each fault segment. Note that both total throw and throw rate decrease to zero at the ends of each segment (full details of data sources given in Roberts and Michetti 2004). Numbers 1, 2, 3 refer to trench sites (Table 3.1).

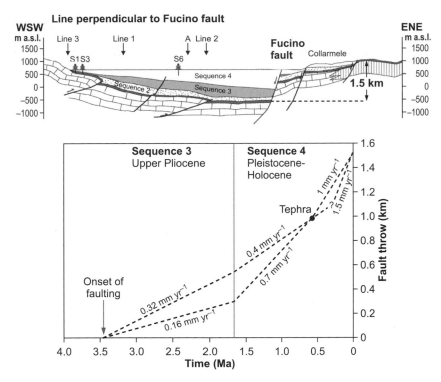

Figure 3.4 Throw versus time for the Fucino fault since its initiation in the Pliocene (~3.4 Ma). Cross section across Fucino Basin is shown above (from Cavinato et al. 2002). Note that the small thrust faults seen in the cross section predate the formation of the basin. Data sources given in Table 3.1.

Table 3.1 Data used to construct the throw versus time plot shown in Figure 3.4.

Time interval	Throw rate (mm yr^{-1})	Description	Reference
3.4–1.6 Ma	0.16–0.32	Basin stratigraphy (biostratigraphy; Seq 3)	Cavinato et al. (2002)
1.6 Ma–present	0.46–0.7	Basin stratigraphy (biostratigraphy; Seq 4)	Cavinato et al. (2002)
0.54 Ma–present	0.9	Tephra deposit	Cavinato et al. (2002)
18 Ka–present	1.38	Wavecut lake terrace + tephra	This study; Michetti et al. (1996)
18 Ka–present	>1.0	Postglacial fault scarps	Roberts and Michetti (2004)
1 375 yr–present	1.45–1.81	Trench dates	Michetti et al. (1996)

of the Abruzzo fault array. Figure 3.4 clearly shows that the average slip rate on the Fucino fault increased during its evolution and that the acceleration began between 1.6 and 0.5 Ma. Cessation of activity on faults elsewhere within the Abruzzo fault array has been shown to be consistent with the timing of acceleration on the Fucino fault (Roberts and Michetti 2004).

There are two things that are particularly significant about these data from central Italy. The first is that the increase in slip rate (Figure 3.4) clearly precedes the formation of a through-going linked fault system. This is important for understanding the mechanism causing this phenomenon. If the individual faults, shown in Figure 3.3a, were linked at depth, then the Holocene slip rate would not mimic the throw profiles but would be approximately uniform along strike (e.g., McLeod et al. 2000). Furthermore, Cowie and Roberts (2001) showed that the magnitude of the increase in slip rate varies systematically across the fault array in a way that is consistent with the predictions of elastic interaction models (e.g., Willemse 1997). Such spatial and temporal variations in slip rate across the fault array would not be expected if they were due to a regional increase in strain rate. The second point to note is that significant changes in fault slip rate have been shown to occur on a sub-million-year timescale in this area. Therefore, this study provides an example of the quantity and quality of data that is necessary to resolve such temporal changes. Sufficient spatial resolution is also required to begin to understand this system. The GPS network in this area has a typical baseline length of 30–40 km and is only able to capture the overall extension across the entire area, not the lateral variations within the array.

Postlinkage Displacement Accumulation in Segment Boundaries

There has been considerable debate concerning the possibility of *long-term displacement deficits* at segment boundaries along extensional faults. The study by McLeod et al. (2000) suggests that, although there may be displacement minima associated with relict zones of linkage, displacement accumulation postlinkage is approximately uniform along strike. The displacement minima reflect the distributed nature of strain accumulation in the segment boundary prior to linkage (i.e., 3D strain accommodated by folding as well as faulting). In the example from central Italy (see above), the displacement variations postlinkage may be significant because of the prelinkage interaction history. Recent data from the Wasatch fault in Utah, however, support the idea that even though segment linkage may result in an irregular fault trace, with splays and bifurcations in the fault surface, these zones rarely remain as sites of persistent displacement deficit subsequent to linkage (Armstrong et al. 2004). This implies that by the time linkage occurs there are no significant variations in average strength or rheology between segment boundaries and the adjacent fault segments.

Evidence for Migrating Fault Activity

Systematic Migration

The detailed studies of fault interaction and linkage discussed earlier were obtained from areas that had undergone <10% extensional strain. At increasing levels of strain, inelastic components of lithospheric rheology clearly influence extensional fault growth, and larger-scale strain localization is observed. One of the key pieces of evidence for this comes from observations of systematic migration of fault activity at the basin scale over millions of years. Goldsworthy and Jackson (2001) reviewed this phenomenon along the southern coast of the Gulf of Corinth, Greece, and Gawthorpe et al. (2003) described a similar pattern of migration of activity between faults along the eastern seaboard of the Gulf of Suez. These studies focus on migration of activity from one block-bounding fault to another block-bounding fault over several tens of kilometers.

Figures 3.5 and 3.6 show an example from the Late Jurassic phase of extension of the northern North Sea where both fault-linkage and basin-scale strain localization are observed. The extensional strain rate during this rifting event was of the order of $\sim 10^{-16}$ s^{-1}, and the observed migration of fault activity occurs over ~ 30 Ma. Note that the Brent-Statfjord fault labeled in Figure 3.5b was the focus of the McLeod et al. (2000) study described above (see section on Evidence for Strain Localization Associated with Fault Linkage). Stages 1 and 2 of Figure 3.6 summarize the observations of McLeod et al. (2000) for that fault. Cowie et al. (2005) present the three-dimensional seismic interpretations and strain-rate inversion results that reveal the migration summarized in Figure 3.6. These data show that maximum fault displacement, duration of fault activity, and maximum fault slip rate all increase towards the center of the basin (Figures 3.5a and 3.6d). Thus the basin initially developed as a broad zone of deformation with many faults active (Figure 3.6a). The dominant inward-dipping fault set, seen in Figure 3.5, only emerged as the main controlling structures approximately 10 Ma after rifting began, after $\sim 20\%$ extensional strain. These inward-dipping faults switched off in a systematic way, with those close to the basin margins switching off before those closer to the basin center. As strain rates across the whole basin were consistently very low ($<3 \times 10^{-16}$ s^{-1}), this pattern of strain localization cannot be attributed simply to thermally induced strength loss associated with lithospheric necking (Newman and White 1999). Therefore, Cowie et al. (2005) interpret the migration of fault activity as being due to a feedback between two coupled effects: (a) near-surface strain localization, driven by brittle (or plastic) failure on faults, and (b) the evolving thermal structure of the thinning lithosphere. The thermal structure strongly influences the development of the fault pattern (both fault dip direction and slip rate), and, in turn, extensional faulting focuses the advection of heat. Together, these contribute to strain localization on a lithospheric scale.

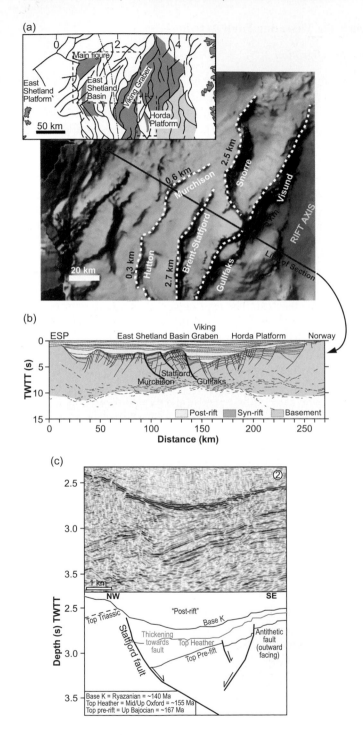

◄ **Figure 3.5** (a) Extensional faults within the East Shetland Basin on the western flank of the North Viking Graben, northern North Sea (inset shows location). Colors represent depth to top syn-rift (base-Cretaceous reflector): blue-purples are deep, and red-yellows are shallow. This surface represents the basin bathymetry at the end of Late Jurassic extension. Image is illuminated from the NW. The maximum throw of each fault is shown in red text. (b) Cross section (cf. red line in (a) for location) perpendicular to the strike of the basin, showing the position, depth extent, and geometry of the faults highlighted in (a). (c) Example of three-dimensional seismic data and line drawing of the interpretation used for understanding the fault evolution (dashed black line in (a) indicates location of seismic line). TWTT: Two-Way Travel Time. Modified from Cowie et al. (2005).

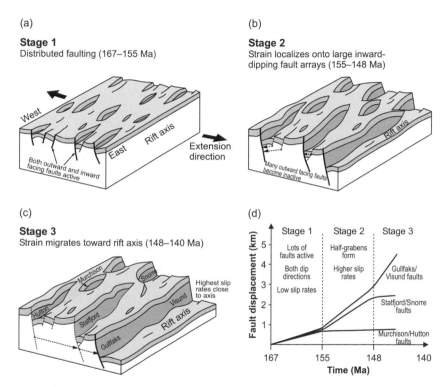

Figure 3.6 Illustration of the temporal evolution of faulting for the area shown in Figure 3.5 (modified from Cowie et al. 2005).

There is evidence for evolving fault zone properties revealed by seismic reflection data across extensional basins that helps to understand strain *and* strain-rate softening of these North Sea Basin faults. In their study of seismic reflection and refraction data across the Galicia Interior Basin (GIB), Pérez-Gussinyé et al. (2003) noted that the depth extent of fault planes increases with the total amount of extension. Faults in the GIB appear to be confined to the brittle upper crust at <50% total strain, but to extend down into the lower crust as the total strain increases. The authors suggest that this is evidence for "embrittlement" of the lower crust and attribute this process to cooling and decompression associated with finite crustal extension.

What is particularly significant about "embrittlement" of the lower crust is that fluids may enter more easily the fault zone as a result of increased fracture permeability and thus dramatically alter the strength of the faults (see Chapter 12). Pérez-Gussinyé et al. (2001, 2003) focused on the effects of serpentinization of the upper mantle by fluid infiltration, but the hydration of feldspar to produce mica within the middle and lower crust could have a similar effect on the frictional strength of faults (e.g., Wibberley 1999; Gueydan et al. 2003; Holdsworth 2004). These fluids could originate from dehydration reactions in the lower crust and enter the fault zone from below, or they could be meteoric fluids entering the fault zone in the shallow crust and percolating downwards, as discussed by Gratier and Gueydan (Chapter 12). In fact, faults may become imaged in seismic reflection profiles only when there are fluids present. The result of fluid infiltration is the development of relatively weak extensional fault zones that penetrate through the entire crust, which could remain active at dip angles as low as ~20° (Pérez-Gussinyé et al. 2001, 2003; Gueydan et al. 2003). These faults accommodate the bulk of the total strain and, as suggested by the evolution shown in Figure 3.6, are most likely to record higher slip rates and remain active for longer compared to neighboring structures.

Ellis and Stöckhert (2004) have proposed another mechanism of "brittle damage" within the lower crust that is applicable to all tectonic settings, not just extension. They show that, during coseismic slip on a fault within the brittle crust, strain rates as high as 10^{-11} s^{-1} are generated within the ductile lower crust beneath the lower tip of the ruptured fault. Such high strain rates result in a significant, although transient, increase in differential stress (up to 300 MPa), which is sufficient to cause a change in deformation mechanism within the ductile layer directly below the coseismic rupture zone. They cite examples of seismic reflections from fault planes within the lower crust, this time within compressional settings, as evidence for this phenomenon. Higher slip rate faults generally experience shorter recurrence intervals. Thus, once strain begins to localize within a fault network, greater levels of brittle damage to the lower crust could occur by this process, resulting in further weakening and greater localization.

Temporal Variations in Fault Slip Rate: Loreto Fault, Baja California Sur

There is an increasing body of literature that presents field evidence for nonuniform moment release rate on faults over geologic time (e.g., Rockwell et al. 2000; Friedrich et al. 2003; Palumbo et al. 2004; Chevalier et al. 2005). The Pliocene Loreto Basin (formed between ~3.5 Ma and ~2.0 Ma) is a commonly cited example for episodic slip on an active basin-bounding fault. Dorsey et al.

(1997) argued that temporal clustering of earthquake activity on the Loreto fault is the best explanation for the high-frequency fluctuations in sedimentation observed in the hanging wall. The Loreto Basin is an extensional half-graben located on the western margin of the Gulf of California, bounded by the east-dipping Loreto fault (Figure 3.7). Within the basin there is a ~650 m thick sequence of stacked coarse-grained delta progradational units of mid-Pliocene age that are back-tilted into the fault (Figure 3.7a, b). The deltaic units, each one 20–35 m thick, are separated by flooding surfaces. The sequence is temporally constrained by the presence of two tuff horizons. Mortimer et al. (2005) recently redated the tuffs, using single crystal Ar/Ar methods, and confirmed the basic conclusion of Dorsey et al. (1997): the average duration of the 16 progradational units (14.1 ±5 ka) is much shorter than the dominant period of eustatic fluctuations during the Pliocene (41 ka). In fact, they have shown that the majority of these units have durations significantly less than this average (Mortimer et al. 2005).

In addition, Mortimer et al. have documented in detail the stratigraphic architecture of the deltas themselves and have identified a repeated pattern of evolving delta geometry that can only be explained by a tectonic driving mechanism. It is well understood that the height of a delta foreset records the water depth into which the delta is prograding (Postma 1990). Mortimer et al. have used this constraint to reconstruct the water depth at the delta front through time. What they have observed is that the height of the foresets in the Loreto Basin increase systematically (from ~5 m to ~25 m over a distance of 500–750 m) in the direction of delta progradation. Furthermore, the deltas increase in thickness and become sigmoidal in geometry; that is, the deltas are aggrading as well as prograding (Figure 3.7c, d). This pattern of foreset development results in a concave up shoreline trajectory (the position of the shoreline in space and time) associated with each delta progradational unit. Through an analysis of fluvial paleoslopes within the alluvial section of the basin, Mortimer et al. have also demonstrated that back-tilting was actively occurring during deposition of the delta units. They argue that the shoreline trajectory, coupled with the observations from the alluvial deposits, can only be explained by varying the rate of back-tilting and tectonic subsidence due to the fault.

In the original interpretation of Dorsey et al. (1995), it was argued that the deltas prograded when the fault was not active and that the delta tops were transgressed during periods of fault activity. In contrast, Mortimer et al. have calculated that progradation of coarse-grained deltas initiated during periods of relatively low tectonic subsidence rate along the basin-bounding Loreto fault (~1 mm yr^{-1}). However, the continually increasing foreset height (Figure 3.7a, b) requires a gradual acceleration in the rate of tectonic subsidence to ~3 mm yr^{-1}. The delta tops may be transgressed when the tectonic subsidence rate is significantly higher (>3 mm yr^{-1}). The average tectonic subsidence rate along the fault during the entire interval of delta deposition was ~2.5 mm yr^{-1}.

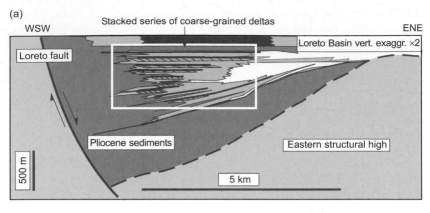

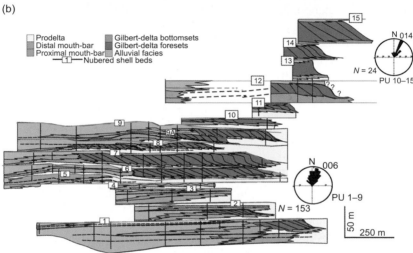

Figure 3.7 (a) Geometry and location of the Pliocene Loreto Basin within the Gulf Extensional Province, Gulf of California (after Dorsey et al. 1995). Note the high-angle Loreto fault dipping to the northeast, with a total throw of 1.5 km, and the wedge of sediments that thicken in to the fault. It appears that the "fulcrum" for the Pliocene fill was the Eastern Structural High, which is bounded by another extensional fault to the east. (b) Delta stacking pattern within Sequence 2 of the Loreto Basin fill, showing 16 cycles of delta progradation (location is noted in (a) by dashed lines). Shelly conglomerates layers mark the flooding surfaces between each cycle. Measurements of delta foreset height (c) and thickness (d) for individual progradation cycles, plotted as a function of distance from the point of delta nucleation (in c) and distance from the Loreto fault (in d). Modified from Mortimer et al. (2005).

A plausible explanation for the variable rate of fault slip recorded by the Loreto Basin stratigraphy is a temporal variation in strain partitioning between two or more potentially active structures (see also Palumbo et al. 2004). For

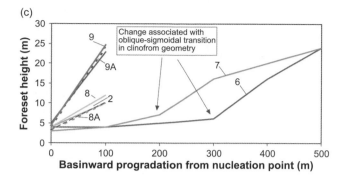

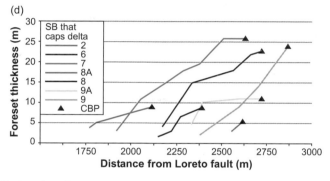

Figure 3.7 (continued)

example, bounding the eastern margin of the Loreto Basin is another extensional fault, the Eastern Structural High fault (Figure 3.7a). The High became subaerially exposed and a source of material towards the end of delta deposition within the Loreto Basin (Dorsey et al. 1995). Thus the episodic slip history inferred for the Loreto fault could be due to the interplay between these two structures with alternating periods of activity on each.

These observations are very intriguing because of what they may reveal about the controls on strain localization in extensional settings. Fault interaction, as defined earlier (see section on Insights from Fault Displacement Profiles), provides one mechanism for variable rates of slip on faults, as shown by Cowie (1998). According to this model, perturbations to the regional stress field caused by rupture events on actively growing structures lead to complex spatial and temporal variations in the loading history of any one particular fault as it interacts with its neighbors in the active fault population. As a result, the slip history on an individual fault within the model is very variable due to the history of stress interaction as it evolved. Another alternative mechanism for episodic fault slip comes from the need to maintain kinematic compatibility during finite strain. A hierarchy of faults is required to accommodate three-dimensional strain, but it is unlikely that all the faults in the population will be optimally orientated to accommodate strain at a uniform rate over geologic time. Thus variations in strain partitioning in space and time may be a general feature of faulted terrains undergoing three-dimensional strain. The transtensional nature of the Gulf Coast Extensional Province during the Pliocene, when the Loreto Basin was forming, may have promoted spatial and temporal changes in strain partitioning, compared to an area undergoing coaxial deformation. In regions that accommodate large finite strains, kinematic compatibility constraints may even lead to a situation where there are phases of strain delocalization if new fault systems develop in response to the imposed deformation.

STRIKE-SLIP FAULT ARRAY EVOLUTION

The evolution of crustal-scale strike-slip fault systems over geologic time is typically much more difficult to characterize than extensional deformation because reliable data on fault offsets in these settings are sparse in both space and time. Also, continued movement on a strike-slip fault usually destroys information of its earlier evolution. The classic paper by Wesnousky (1988) still underpins a lot of the thinking about the evolution of strike-slip fault systems. Wesnousky (1988) outlined the idea that the segmentation structure of strike-slip faults is fundamentally controlled by the total offset across the fault. The typical segment boundary width on a strike-slip fault is 1–5 km (measured perpendicular to fault strike) and they are usually marked by a localized topographic high (push-up) or low (pull-apart) along the fault trace. Wesnousky showed that strike-slip faults with displacements of a kilometer or so are in most cases highly segmented, whereas faults like the San Andreas fault, which has >150 km of offset, are virtually unsegmented.

Underlying Wesnousky's model is the assumption that the mapped trace of the fault is the surface expression of a through-going localized shear zone at depth. Accumulation of shear strain at depth leads to smoothing of the fault zone observed at the surface. In this sense strike-slip and normal fault arrays do appear to be different: It is not uncommon to observe strike-slip fault zones that can be traced quite easily for several tens to >100 km along strike in spite of the fault trace itself being clearly segmented. The ease of lateral correlation, in spite of prominent discontinuities, derives from the fact that the width of the segment boundaries along strike-slip fault zones is typically much smaller than (e.g., <1/5) the adjacent segment lengths. In contrast, it is usually very difficult to correlate along strike within extensional fault arrays prior to segment linkage (e.g., Figure 3.3). Subsequent to linkage of two normal fault segments, most of the deformation is localized along a well-defined throughgoing fault surface even though that surface may remain as an irregular three-dimensional plane.

Controls on Strain Localization in Strike-Slip Settings

In the light of the data presented by Wesnousky (1988), and subsequently by Stirling et al. (1996), a number of questions arise: if this "deep-seated" strain localization emerges early in the whole evolution of strike-slip deformation, over what depth range does it occur and how is it produced? If strain localization occurs first within the ductile lower crust, this presents a conceptual problem. All the mechanisms that have been proposed for producing strain localization within the ductile regime invoke *locally* high (coseismic?) strain rates or finite shear strain (= 0.2), which suggests that localization within the ductile regime must lag behind localization within the brittle regime. For example, shear heating, grain-size reduction by recrystallization, brittle-damage (plus reaction softening), and shear-induced mineral segregation are all processes that may result in weakening within ductile materials. However, for these processes to operate they require either the deformation to be already localized or for strain rates to be generated that are much larger than interseismic rates (for a summary, see Montesi and Hirth 2003). The presence of melt may be one possible mechanism for strain localization in the ductile regime that does not require these conditions (Rosenberg et al., this volume), but it is unclear whether this is a general mechanism.

Insights from a Displacement-length Scaling Argument

An alternative explanation to that discussed above is that strain localization within strike-slip fault arrays is initiated by brittle processes occurring near the base of the brittle upper crust prior to strain localization nearer to the Earth's surface and also prior to the formation of a ductile shear zone. Evidence that might support this explanation comes from the displacement-length scaling relationship derived for strike-slip faults documented by Stirling et al. (1996). Within their data set of ~30 strike-slip faults, some are clearly plate boundary faults and thus their lengths are controlled by plate geometry rather than displacement and must be excluded. After further excluding those faults for which there is no constraint on length (i.e., they extend offshore), sixteen faults remain that range in length from 30 km (Tanna fault, Japan) to 1600 km (Altun fault, central China). Note that the length in this case does not refer to segment length but the entire length of each fault zone identified by Stirling et al. (1996). The sixteen remaining faults have an average displacement length ratio of 0.04, which, given the scatter inherent in these data, is basically indistinguishable to the global average (0.03) obtained for other fault data sets worldwide (Schlische et al. 1996). This suggests, therefore, that over several orders of magnitude strike-slip faults scale in exactly the same way as dip-slip faults. The physical interpretation for this scaling relationship is based on fracture mechanics principles, that is, brittle deformation processes (Cowie and Scholz 1992). Fault populations that scale in this way are therefore interpreted to have been governed largely by the mechanics of brittle deformation during their growth.

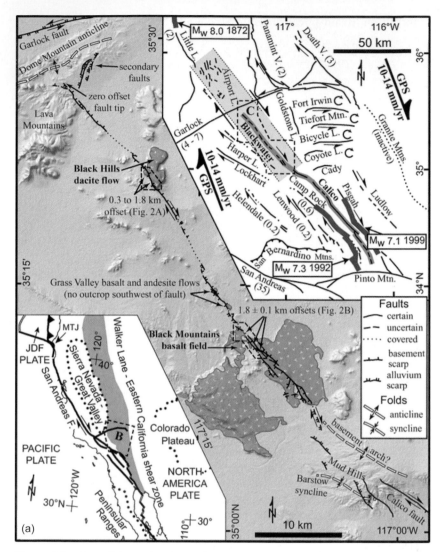

Figure 3.8 Map of the 55 km long Blackwater strike-slip fault in the Mojave desert (from Oskin and Iriondo 2004). Note the offset on the fault (1.8 ± 0.1 km) since 3.77 Ma, when this fault probably formed. The offset decreases to zero at the northern fault tip. At the southern end of the Blackwater fault there is a 15 km wide compressional step across to the Calico fault, which has an offset of 8.2 km and a length of ~ 125 km (Stirling et al. 1996). (b) Line-of-sight displacement rate across the Blackwater fault derived from InSAR combined with GPS data (Peltzer et al. 2001). In (b), black dots and error bars (2σ) are horizontal velocities observed by GPS; solid line is the running mean of InSAR measurements; short dashed line is the estimated interferometric baseline; long dashed line is velocity model of Peltzer et al. (2001), which includes modeled slip on the Blackwater fault.

The suggestion that strike-slip faults might scale in the same way as dip-slip faults has further implications. At the most basic level it obviously implies that the displacement on a strike-slip fault decreases to zero at the ends of the fault! Many studies of strike-slip systems assume that displacement is approximately constant along fault strike and that the end of the fault is controlled by the intersection with another strike-slip fault (e.g., block rotation models). The scaling of displacement with length implies, moreover, that there are two different types of strike-slip segment boundary:

- Type 1: Segmentation along individual faults; these segments root into the same fault plane at depth (cf. Wesnousky 1988).
- Type 2: Segment boundaries between the tips of individual faults that mark sites of incipient linkage between previously separate structures.

Because there are only sparse data about displacement variations along strike-slip faults, discriminating between these two types is difficult, although it seems reasonable to suggest that small steps (<5 km) may be examples of the first type, whereas offsets of ~10 km or greater (i.e., the seismogenic thickness) may be examples of the second type. If this twofold classification is correct, the fault displacement profile should diminish (possible to ~ zero displacement) towards a Type 2 segment boundary, whereas it is likely to be fairly constant across a Type 1 boundary. The faults on either side of a Type 2 boundary may also have distinctly different maximum displacements, consistent with their respective lengths.

Figure 3.8a shows a *possible* example of a Type 2 segment boundary between the Blackwater and Calico faults in the Mojave Desert. The Mud Hills mark the site of a 15 km wide compressional step between these two faults.

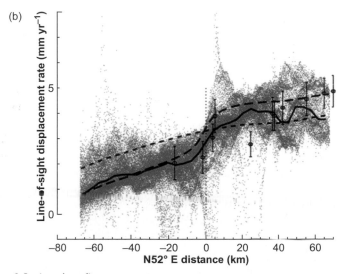

Figure 3.8 (continued)

The Calico fault has a maximum displacement of 8.2 km and is ~125 km long (d/L ~ 0.06; Stirling et al. 1996), whereas the Blackwater fault has a maximum displacement of 1.8 km and is 50 km long (d/L = 0.036; Oskin and Iriondo 2004). Oskin and Iriondo (2004) argue that the displacement on the Blackwater fault dies out to zero at its northern end. If we assume that the fault displacement profile along this fault has a symmetrical bell-shape similar to that observed for many normal faults (e.g., Figure 3.3b), then the displacement on this fault is expected to diminish significantly towards the Mud Hills in the south, over the same distance as that observed at the northern fault tip. Therefore, it seems reasonable to suggest that the Blackwater and Calico faults may be distinct structures that developed independently. Is it possible that tectonic uplift within the Mud Hills is evidence for incipient linkage between these two structures? Evidence for compressional deformation in the step-over comes from uplifted and exposed Quaternary fan gravels in the Mud Hills (Oskin and Iriondo 2004). The faults are known to be much older than Quaternary in age. The Blackwater fault is estimated to be ~3.7 Ma and 6–10 Ma is the current best estimate for the age of the Calico fault. Thus the age of the step-over appears to be significantly younger than the age of the faults on either side, consistent with the idea that linkage of the two faults is not yet complete.

Implications of Displacement-length Scaling for Strike-slip Fault Array Evolution

If the Blackwater and Calico faults are beginning to link, then this linkage event could change the way that strain is partitioned on a regional scale, just as has been described for normal fault arrays (see sections Evidence for Strain Localization Associated with Fault Linkage and Evidence for Prelinkage Interaction on Crustal-scale Normal Faults). There is a simple way of estimating the magnitude of this effect by considering the geometric moment of the faults prior to and subsequent to linkage. The strain represented by a fault of length, L, and down-dip width, W, depends on the geometric moment, $M_g = d.A$, where d is displacement. Fault plane area $A = LW$, and W is the thickness of the faulted layer (Kostrov 1974). Let us assume for simplicity that the dimension W is a constant for both faults. Prior to linkage, the geometric moment of the Calico fault is 1025 W, whereas the Blackwater fault has a geometric moment of 90 W, giving a total geometric moment prior to linkage of 1025 W + 90 W = 1115 W. The linked structure will have a length of ~180 km (125 + 55). According to the d-L scaling relationship, the linked fault should accrue further displacement commensurate with this new length. Taking the existing d/L ratios for the two faults (0.036 and 0.06), the predicted average displacement of the linked fault is 6.5–10.8 km, so that the geometric moment postlinkage will be in the range 1170 W to 1944 W. Thus the strain accommodated by the linked fault, once it has reestablished its prelinkage d/L ratio, is 5–74% greater than prior to linkage. For the regional strain rate to remain constant, other faults must become inactive. If we take the upper limit of this percentage change, we would predict that linkage of the Blackwater and Calico faults could, for example, result in cessation of activity on another fault that is 117–150 km in length with a displacement of 5.5–7 km (i.e., a fault with a geometric moment of 820 W). There are several strike-slip faults lying to the west of the Blackwater–Calico fault system that could be potential candidates for deactivation (e.g., Harper, Lockhart, Lenwood, and Helendale faults), and indeed they appear to have lower slip rates.

There is much speculation in the literature about the possibility that a new through-going fault system may indeed be gradually emerging within the Mojave Desert (for a review of the debate, see Rockwell et al. 2000). Much of this discussion arose after the occurrence of a sequence of large earthquakes in the area, most notably the M_w 7.3 Landers event in 1992 and the M_w 7.1 Hector Mine event in 1999. These two earthquakes occurred on strike-slip faults, subparallel and adjacent to the Calico fault, as shown in Figure 3.8a. The postulated zone of incipient strain localization is shown by the shaded areas in insets to Figure 3.8a and is thought to represent the southern extension of the Eastern California Shear Zone. It is shown extending from Owen's Valley into the Mojave Desert across the left lateral Garlock fault, approximately orthogonal to that structure. However, paleoseismological studies do not support the idea that strain is becoming localized within this zone. Instead, the paleoseismological data indicate that the deformation within the Mojave Desert is distributed across a large number of low slip rate faults with the total offset on any one fault being <10–15 km (Rockwell et al. 2000); that is, there is no obvious right lateral structure that has a higher slip rate and/or higher levels of seismic activity. Moreover, the Garlock fault, which cuts right across the center of the proposed zone of strain localization, has a paleoseismologically determined slip rate of ~7 mm yr^{-1} (McGill and Sieh 1993). This area of the central Mojave Desert, and its postulated northward extension into Owen's Valley, would therefore constitute a prime area for studying strike-slip fault array evolution, using a wide variety of geological, geophysical, and geodetic techniques, because it may be a good example of a fault array in a state of incipient localization.

Constraints from Geodetic Observations

Geodetic surveys across the Mojave Desert are able to resolve the overall shear strain accommodated across the entire area (~150 km wide) but are unable to resolve lateral variations within the fault array. Analyses of InSAR data have thus far provided the most detailed information regarding recent fault movements within this area (e.g., Peltzer et al. 2001; Fialko et al. 2002). In this context there is a particularly interesting feature associated with the Blackwater fault, shown in Figure 3.8b. Crustal deformation measurements over an eight-year time window (1992–2000), obtained by averaging InSAR data, indicates that there is a significant contemporary strain rate (~1.5 mm yr^{-1}) in a narrow zone aligned along the Blackwater fault (Peltzer et al. 2001). This phenomenon cannot be explained by postseismic relaxation because there is no geologic evidence for a latest Holocene rupture on this fault (Oskin and Iriondo 2004). It could, however, be a postseismic response to historic earthquakes on nearby faults. Even if it is a transient phenomenon it will still affect the loading history and subsequent rupture of the overlying locked portion of the fault as illustrated by the model of Huc et al. (1998).

Peltzer et al. (2001) interpret the strain accumulation pattern along the Blackwater fault as evidence that this fault is creeping (at a rate of $\sim 7 \pm 3$ mm yr^{-1}) below 5–7 km depth (Peltzer et al. 2001). An important issue is the time evolution of the strain-rate signal shown in Figure 3.8b to determine whether it has decayed over time consistent with postseismic relaxation. It would be particularly important to investigate whether the time evolution of the signal can discriminate between deformation of a ductile shear zone versus aseismic creep within the brittle regime, and thus provide constraints on the rheology of this zone. If the signal is originating within a ductile shear zone at depth, then given the small offset of the Blackwater fault (1.8 km), it indicates that strain localization within the ductile regime develops after a relatively small amount of shear offset. Based on experimental studies, Herwegh and Handy (1996, 1998) suggest that strain localization on a granular scale in the viscous lithosphere can occur after shear strains ~ 0.2. Using this value, and the known offset the width of the ductile shear zone beneath the Blackwater fault would be ≈ 10 km.

SUMMARY AND CONCLUSIONS

In this chapter we have reviewed recent observations of extensional fault growth, summarized conclusions drawn from these observations concerning the underlying controls on strain localization in extensional settings, and discussed the relevance of these observations to fault growth in strike-slip settings. Our primary aim has been to argue that observations of fault growth in extensional settings may provide some interesting insights into understanding strike-slip deformation, in spite of the fact that geologic data on strike-slip fault evolution are sparse.

We have shown that significant advances in understanding strain localization in extensional settings have come from detailed analysis of sedimentation patterns within fault-controlled depocenters. Using stratigraphic marker horizons, a temporal evolution of the fault pattern may be reconstructed to determine relative rates of fault movement and fault tip propagation as well as timing of fault segment linkage. It is the knowledge of the strain rates in space and time that allow rheological controls to be inferred; geometry alone is insufficient. Strike-slip fault systems that have developed with contemporaneous sediment accumulation (e.g., transtensional systems) may lend themselves to similar analysis, particularly those that currently lie below sea/lake level and can therefore be imaged using high-resolution three-dimensional seismic reflection surveys (cf. Taylor et al. 2004). Surface exposure age dating using cosmogenic nuclides is also proving to be a very valuable tool for constraining slip rates on faults over the last ~100 ka and is likely to lead to significant new advances in understanding temporal variations in strain partitioning within fault arrays (e.g., Palumbo et al. 2004; Chevalier et al. 2005).

Recognizing the role of segment linkage has been a major step forward in our understanding of the evolution of extensional fault growth. Thus far, research on this topic has largely focused on fault displacement distributions and changes in strain partitioning associated with linkage. Future work could usefully focus more on understanding the changes in fluid flow and thus fault rock rheology that might accompany fault linkage events because of the changes in crustal permeability that are likely to be produced.

We identified active strike-slip deformation occurring within the central Mojave Desert as a prime study area where some key questions concerning strike-slip fault evolution may be addressed. In the section on STRIKE-SLIP FAULT ARRAY EVOLUTION, we reviewed evidence suggesting that this is an area of incipient strain localization, possibly representing the southward extension of the Eastern California Shear Zone. A great deal of information could be obtained from more detailed, continuous geodetic measurements across this region. Based on what we have learned from extensional deformation in central Italy, we suggest that greater spatial resolution in geodetic coverage is necessary to understand the evolution of a relatively immature fault array, such as the Mojave fault array, than to understand the seismic cycle of major plate boundary faults. In addition to increasing spatial resolution, continuous monitoring is vital if we are to place better constraints on lithospheric structure and rheology. In particular, we suggest that identifying and analyzing the postseismic response of fault zones that are perturbed by large earthquakes occurring nearby, could yield new insights into fault zone rheology and the state of stress in the crust (see also Chapter 4).

It is also clear that numerical modeling is a useful approach to investigate what might be the underlying cause of localization in strike-slip settings. Roy and Royden (2000) and Lyakhovsky et al. (2001) have developed models that simulate the evolution of strike-slip fault zones. However, in neither model does the rheology of the viscous lower crust evolve as a function of total strain or strain rate except where the threshold for brittle failure is reached locally (Roy and Royden 2000). An important next step is to include in these, or similar, models strain weakening within the viscous layer due to the accumulated strain or strain rate fluctuations generated by brittle failure processes at shallow depths. For example, the effects of grain-size reduction by recrystallization, reaction softening, and shear-induced mineral segregation should be explicitly included in such models to observe how localization within the ductile regime may feed back into localization within the brittle layer above. This type of approach has been applied in extensional settings to model the development of low-angle normal faults in the middle crust (Gueydan et al. 2003), but it has not yet been used to model strain localization in strike-slip settings. With such a modeling approach it would be possible to investigate whether strain localization phenomena, such as may be occurring currently within the Mojave Desert, are driven by changes in boundary conditions, or whether they arise as spontaneous self-organization of the fault system.

ACKNOWLEDGMENTS

This paper benefited from the stimulating discussion at the Dahlem Workshop. In particular, comments from Mark Handy, Chris Wibberley, Greg Hirth, Frederic Gueydan, and Mark Behn significantly improved aspects of the discussion.

REFERENCES

Armstrong, P.A., A.R. Taylor, and T.A. Ehlers. 2004. Is the Wasatch fault footwall (Utah, U.S.A.) segmented over million year time scales? *Geology* **32**:385–388.

Behn, M.D., J. Lin, and M.T. Zuber. 2002. A continuum mechanics model for normal faulting using a strain-rate softening rheology: Implications for thermal and rheological controls on continental and oceanic rifting. *Earth Planet. Sci. Lett.* **202**:725–740.

Benedetti, L., P. Tapponier, G.P. King, B. Meyer, and I. Manighetti. 2000. Growth folding and active thrusting in the Montecello region, Veneto, northern Italy. *J. Geophys. Res.* **105**:739–766.

Ben-Zion, Y., and C.G. Sammis. 2003. Characterization of fault zones. *Pure & Appl. Geophys.* **160**:77–715.

Buck, W.R., L. Lavier, and A. Poliakov. 1999. How to make a rift wide. *Phil. Trans. R. Soc. Lond.* **357**:671–693.

Cartwright, J.A., B.D. Trudgill, and C.S. Mansfield. 1995. Fault growth by segment linkage: An explanation for scatter in maximum displacement and trace length data from the Canyonlands Grabens of SE Utah. *J. Struct. Geol.* **17**:1319–1326.

Cavinato, G.P., C. Carusi, M. Dall'Asta, E. Miccadei, and T. Piacentini. 2002. Sedimentary and tectonic evolution of Plio-Pleistocene alluvial and lacustrine deposits of the Fucino Basin (central Italy). *Sediment. Geol.* **148**:29–59.

Chevalier, M.-L., F.J. Ryerson, P. Tapponnier et al. 2005. Slip-rate measurements on the Karakorum fault may imply secular variations in fault motion. *Science* **307**:411–414.

Cowie, P.A. 1998. A healing–reloading feedback control on the growth rate of seismogenic faults. *J. Struct. Geol.* **20**:1075–1087.

Cowie, P.A., and G.P. Roberts. 2001. Constraining slip rates and spacings for active normal faults. *J. Struct. Geol.* **23**:1901–1915.

Cowie, P.A., and C.H. Scholz. 1992. Physical explanation for the displacement-length scaling relationship for faults using a post-yield fracture mechanics model. *J. Struct. Geol.* **14**:1133–1148.

Cowie, P.A., J.R. Underhill, M.D. Behn, J. Lin, and C. Gill. 2005. Spatio-temporal evolution of strain accumulation derived from multi-scale observations of Late Jurassic rifting in the northern North Sea: A critical evaluation of models for lithospheric extension. *Earth Planet. Sci. Lett.* **234**:401–419.

Davy, P., A. Hansen, E. Bonnet, and S.-Z. Zhang. 1995. Localisation and fault growth in layered brittle-ductile systems: Implications for deformation of the continental lithosphere. *J. Geophys. Res.* **100**:6281–6289.

Dawers, N.H., M.H. Anders, and C.H. Scholz. 1993. Growth of normal faults: Displacement-length scaling. *Geology* **21**:1107–1110.

Dorsey, R.J., P.J. Umhoefer, and P.D. Falk. 1997. Earthquake clustering inferred from Pliocene Gilbert-type fan deltas in the Loreto basin, Baja California Sur, Mexico. *Geology* **25**:679–682.

Dorsey, R.J., P.J. Umhoefer, and P.R. Renne. 1995. Rapid subsidence and stacked Gilbert-type fan deltas, Pliocene Loreto Basin, Baja California Sur, Mexico. *Sediment. Geol.* **98**:181–204.

Ellis, S., and B. Stöckhert. 2004. Elevated stresses and creep rates beneath the brittle-ductile transition caused by seismic faulting in the upper crust. *J. Geophys. Res.* **109**:B05407 doi:10.1029/2003JB002744.

Fialko, Y., D. Sandwell, D. Agnew et al. 2002. Deformation on nearby faults induced by the 1999 Hector Mine earthquake. *Science* **297**:1858–1862.

Friedrich, A.M., B.P. Wernicke, N.A. Niemi, R.A. Bennett, and J.L. Davis. 2003. Comparison of geologic and geodetic data from the Wasatch region, Utah, and implications for the spectral character of Earth deformation at periods of 10 to 10 million years. *J. Geophys. Res.* **108**:2109 doi:10.1029/2001JB000682.

Gawthorpe, R.L., C.A.L. Jackson, M.J. Young et al. 2003. Normal fault growth, displacement localisation, and the evolution of normal fault populations: The Hammam Faraun fault block, Suez Rift, Egypt. *J. Struct. Geol.* **25**:1347–1348.

Goldsworthy, M., and J. Jackson. 2001. Migration of fault activity within normal fault systems: Examples from the Quaternary of mainland Greece. *J. Struct. Geol.* **23**:489–506.

Gueydan, F., Y.M. Leroy, L. Jolivet, and P. Agard. 2003. Analysis of continental mid-crustal strain localization induced by reaction-softening and microfracturing. *J. Geophys. Res.* **108**:2064 doi:10.1029/2001JB000611.

Gupta, A., and C.H. Scholz. 2000. A model of normal fault interaction based on observations and theory. *J. Struct. Geol.* **22**:865–880.

Gupta, S., and P.A. Cowie. 2000. Processes and controls in the stratigraphic development of extensional basins. *Basin Res.* **12**:185–194.

Gupta, S., P.A. Cowie, N.H. Dawers, and J.R. Underhill. 1998. A mechanism to explain rift-basin subsidence and stratigraphic patterns through fault array evolution. *Geology* **26**:595–598.

Hardacre, K.H., and P.A. Cowie. 2001. Variability in fault size scaling due to rock strength heterogeneity: A finite element investigation. *J. Struct. Geol.* **25**:1735–1750.

Hardacre, K.H., and P.A. Cowie. 2003. Controls on strain localization in a two-dimensional elastoplastic layer: Insights into size–frequency scaling of extensional fault populations. *J. Geophys. Res.* **108**:2529 doi:10.1029/2001JB001712.

Heimpel, M., and P. Olson. 1996. A seismodynamical model of lithospheric deformation: Development of continental and oceanic rift networks. *J. Geophys. Res.* **101**:16,155–16,176.

Herwegh, M., and M.R. Handy. 1996. The evolution of high-temperature mylonitic microfabrics: Evidence from simple shearing of a quartz analogue (norcamphor). *J. Struct. Geol.* **18**:689–710.

Herwegh, M., and M.R. Handy. 1998. The origin of shape preferred orientations in mylonite: Inferences from *in situ* experiments on polycrystalline norcamphor. *J. Struct. Geol.* **20**:681–694.

Holdsworth, R.E. 2004. Weak faults: Rotten cores. *Science* **303**:181–182.

Huc, M., R. Hassani, and J. Chery. 1998. Large earthquake nucleation associated with stress exchange between middle and upper crust. *Geophys. Res. Lett.* **25**:551–554.

Hunstad, I., G. Selvaggi, N. D'Agostino et al. 2003. Geodetic strain in peninsular Italy between 1875 and 2001. *Geophys. Res. Lett.* **30**:1181 doi:10.1029/2002GL016447.

Jackson, J., and M. Leeder. 1994. Drainage systems and the development of normal faults: An example from Pleasant Valley, Nevada. *J. Struct. Geol.* **16**:1041–1059.

Jackson, J., and D. McKenzie. 1983. The geometrical evolution of normal fault systems. *J. Struct. Geol.* **5**:471–482.

Kostrov, V. 1974. Seismic moment and energy of earthquakes, and seismic flow of rock. *Izv. Acad. Sci. USSR Phys. Solid Earth* **1**:23–44.

Lyakhovsky, V., Y. Ben-Zion, and A. Agnon. 2001. Earthquake cycle faults and seismicity patterns in rheologically layered lithosphere. *J. Geophys. Res.* **106**:4103–4120.

Manighetti, I., G.C.P. King, Y. Gaudemer, C.H. Scholz, and C. Doubre. 2001. Slip accumulation and lateral propagation of active normal faults in Afar. *J. Geophys. Res.* **106**:13,667–13,696.

Mansfield, C.S., and J.A. Cartwright. 2001. Fault growth by linkage: Observations and implications from analogue models. *J. Struct. Geol.* **23**:745–763.

McGill, S.F., and K. Sieh. 1993. Holocene slip rate of the central Garlock fault in southeastern Searles Valley. *J. Geophys. Res.* **98**:14,217–14,231.

McLeod, A.E., N.H. Dawers, and J.R. Underhill. 2000. The propagation and linkage of normal faults: Insights from the Strathspey-Brent-Statfjord fault array, northern North Sea. *Basin Res.* **12**:263–284.

Michetti, A.M., F. Brunamonte, L. Serva, and E. Vittori. 1996. Trench investigations of the 1915 Fucino earthquake fault scarps (Abruzzo, central Italy): Geological evidence of large historical events. *J. Geophys. Res.* **101**:5921–5936.

Montesi, L.G.J., and G. Hirth. 2003. Grain size evolution and the rheology of ductile shear zones: From laboratory experiments to post-seismic creep. *Earth Planet. Sci. Lett.* **211**:97–110.

Morewood, N.C., and G.P. Roberts. 2000. The geometry, kinematics, and rates of deformation within an en échelon normal fault segment boundary, central Italy. *J. Struct. Geol.* **22**:1027–1047.

Mortimer, E.J., S. Gupta, and P.A. Cowie. 2005. Nucleation and growth of clinoforms in coarse-grained, syn-rift deltas: A response to episodic accelerations in fault displacement. *Basin Res.* **17**:337–359.

Newman, R., and N. White. 1999. The dynamics of extensional sedimentary basins: Constraints from subsidence inversion. *Phil. Trans. R. Soc. Lond.* **357**:805–830.

Oskin, M., and A. Iriondo. 2004. Large-magnitude transient strain accumulation on the Blackwater fault, Eastern California shear zone. *Geology* **32**:313–316.

Palumbo, L., L. Benedetti, D. Bourlès, A. Cinque, and R. Finkel. 2004. Slip history of the Magnola fault (Apennines, central Italy) from ^{36}Cl surface exposure dating: Evidence for strong earthquakes over the Holocene. *Earth Planet. Sci. Lett.* **225**:163–176.

Peacock, D.C.P., and D.J. Sanderson. 1991. Displacements, segment linkage and relay ramps in normal fault zones. *J. Struct. Geol.* **13**:721–733.

Peltzer, G., F. Crampé, S. Hensley, and P. Rosen. 2001. Transient strain accumulation and fault interaction in the Eastern California shear zone. *Geology* **29**:975–978.

Pérez-Gussinyé, M., C.R. Ranero, T.J. Reston, and D. Sawyer. 2003. Mechanisms of extension at non-volcanic margins: Evidence from the Galicia interior basin, west of Iberia. *J. Geophys. Res.* **108**:B5, 2245.

Pérez-Gussinyé, M., and T.J. Reston. 2001. Rheological evolution during extension at passive non-volcanic margins: Onset of serpentinization and the development of detachments to continental breakup. *J. Geophys. Res.* **106**:3961–3975.

Postma, G. 1990. Depositional architectures and facies of river and fan deltas: A synthesis. In: Coarse Grained Deltas, ed. A. Colella and D.B. Prior, pp. 13–27. Intl. Assoc. Sedimentologists Special Publ. 10.

Roberts, G.P., and A.M. Michetti. 2004. Spatial and temporal variations in growth rates along active normal fault systems: An example from the Lazio-Abruzzo Apennines, central Italy. *J. Struct. Geol.* **26**:339–376.

Rockwell, T.K., S. Lindvall, M. Herzberg et al. 2000. Paleoseismology of the Johnson Valley, Kickapoo, and Homestead Valley faults: Clustering of earthquakes in the Eastern California shear zone. *Bull. Seismol. Soc. Am.* **90**:1200–1236.

Roy, M., and L.H. Royden. 2000. Crustal rheology and faulting at strike-slip plate boundaries: 2. Effects of lower crustal flow. *J. Geophys. Res.* **105**:5599–5613.

Schlische, R.W., S.S. Young, R.V. Ackermann, and A. Gupta. 1996. Geometry and scaling relations of a population of very small rift-related faults. *Geology* **24**:683–686.

Sharp, I.R., R.L. Gawthorpe, B. Armstrong, and J.R. Underhill. 2000. Propagation history and passive rotation of mesoscale normal faults: Implications for syn-rift stratigraphic development. *Basin Res.* **12**:285–306.

Stirling, M.W., S.G. Wesnousky, and K. Shimazaki. 1996. Fault trace complexity, cumulative slip, and the shape of the magnitude frequency distribution for strike-slip faults: A global survey. *Geophys. J. Intl.* **89**:5849–5857.

Taylor, S.K., J.M. Bull, G. Lamarche, and P.M. Barnes. 2004. Normal fault growth and linkage in the Whakatane Graben, New Zealand, during the last 1.3 Myr. *J. Geophys. Res.* **109**:B02408.

Walsh, J., C. Childs, J. Imber et al. 2003. Strain localisation and population changes during fault system growth within the Inner Moray Firth, northern North Sea. *J. Struct. Geol.* **25**:307–315.

Walsh, J.J., A. Nicol, and C. Childs. 2002. An alternative model for the growth of faults. *J. Struct. Geol.* **24**:1669–1675.

Wesnousky, S. 1988. Seismological and structural evolution of strike-slip faults. *Nature* **335**:340–342.

Wibberley, C.A.J. 1999. Are feldspar-to-mica reactions necessarily reaction-softening processes in fault zones? *J. Struct. Geol.* **21**:1219–1227.

Willemse, E.J.M. 1997. Segmented normal faults: Corrrespondence between three-dimensional mechanical models and field data. *J. Geophys. Res.* **102**:675–692.

Wu, D., and R.L. Bruhn. 1994. Geometry and kinematics of active normal faults, South Oquirrh Mountains, Utah. *J. Struct. Geol.* **16**:1061–1075.

Front, left to right: Mark Handy, Tuncay Taymaz, Jean-Pierre Brun, Walter Mooney. Back, left to right: Patience Cowie, Christian Teyssier, Alain Vauchez, Kevin Furlong, Gregory Beroza, Brian Wernicke

4

Group Report: Nucleation and Growth of Fault Systems

KEVIN FURLONG, Rapporteur

GREGORY C. BEROZA, JEAN-PIERRE BRUN, PATIENCE A. COWIE,
MARK R. HANDY, WALTER D. MOONEY, TUNCAY TAYMAZ,
CHRISTIAN TEYSSIER, ALAIN VAUCHEZ, and BRIAN WERNICKE

INTRODUCTION

Faults are key components of the Earth system. They are sites of great earth-quakes, which have an obvious, direct impact on the biosphere, including human societies. Faults have repeatedly affected the evolution of the Earth's lithosphere and surface. Mountain belts, rift valleys, mid-ocean ridges, and other features at plate boundaries are dominated by faults. In addition, faulting triggers surface motion on a broad range of scales (Figure 4.1). Thus, it is critical that we understand how faults form.

The focus of our discussion group was on finding ways to improve our understanding of how the structure of major fault systems evolves in space and time. Key questions relevant to understanding the characteristics of fault systems include:

1. What is the behavior of fault systems on different timescales?
2. What is the pattern(s) of fault system localization?
3. What drives localization?
4. Do different types of faults (strike-slip, thrust, normal) interact with the lithosphere in fundamentally different ways?
5. What is the style of faulting at different lithospheric levels, and how do transitions occur?

Identifying these basic questions is straightforward, but answering them is problematic. There have been substantial advances in our understanding of fault

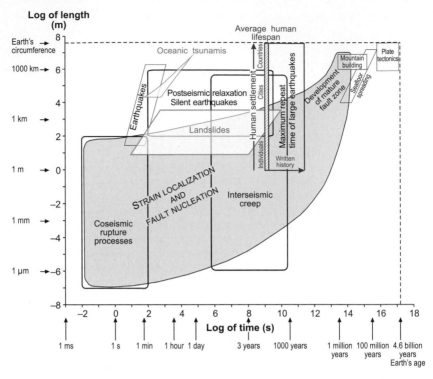

Figure 4.1 Length (m) versus time (s) on a log-log plot showing scales of fault processes (black, gray) and surface processes triggered by faulting (green) compared to the human scale (red). Diagram includes information from the following sources: Coseismic, postseismic and interseismic processes (modified from Figure 7.2), mountain building (Pfiffner and Ramsay 1982), landslides from rock falls to earth flow (www.landslides.usgs, www.planat.ch) and tsunamis (wwww.walrus.wr.usgs.gov/tsunami). Gray area in background shows the large range of conditions for faulting, from nucleation and localization to maturation on the scale of plate boundaries. Diagram contributed by M. R. Handy.

systems within the lithosphere, but we are faced with some fundamental limitations. For one, we cannot directly observe deformation along faults except near the Earth's surface. Despite the fact that faulting has direct impacts on human civilization (see Chapter 10), most processes related to faulting occur on time- and/or length scales other than the human scale, as shown in Figure 4.1. For example, faults nucleate on very short timescales, but reach structural maturity only after millions of years. The repeat time of large earthquakes on some seismogenic faults exceeds the human life span. We therefore rely on a combination of tools to study faults—from geophysical imaging of the fingerprints of active processes, to integrated field studies of exhumed, inactive (fossil) fault zones—and to see the deformation itself, even after faulting has ended. Unfortunately, many key observations are still missing; techniques to improve the resolving power of our deep structural images need improvement, as does our understanding of the physical processes that drive faulting.

The questions raised above are interlinked. We cannot isolate temporal from spatial behavior; deformation in fault zones at shallow levels can have profound impacts on the style and amount of deformation at depth, and vice versa. Fault systems develop under widely varied kinematic and stress regimes, producing fundamentally different styles of deformation in convergent, divergent, and transform fault systems. However, there are apparent similarities among these fault systems. Possibly, we can exploit these similarities to discover the fundamental processes underlying the development and evolution of fault systems. In so doing, we must recall that these processes may only be apparently similar; we may be misled to assume that all faults have common factors controlling their development.

The suite of questions above falls under three major themes, which directed our deliberations. The timescales of fault development (question 1) was our first theme. Localization of fault systems, both from an observational and process-oriented perspective (questions 2 and 3), provided our second. Our third theme focused on how fault systems interact within the lithosphere, both in terms of how fault types (strike-slip, thrust, normal faults) interact with compositional, thermal, and rheologic layering of the lithosphere (question 4) and how fault behavior varies with depth, not just within the lithosphere but also at the lithosphere–asthenosphere boundary (question 5).

In addressing these themes, we reviewed the current state of knowledge, identified critical open questions, and attempted to formulate strategies to address these questions. In this report, we focus primarily on the latter two components of our discussion. Much of the current state of knowledge is provided in the collection of background papers in this volume. However, to put our discussions of open questions and strategies into context, we interject brief discussions of some background material and provide short descriptions of key examples and processes.

TIMESCALES OF FAULT DEVELOPMENT

Faulting occurs over a wide range of timescales (Figure 4.1): Earthquakes accomplish significant displacements in seconds to at most a few minutes. At the other end of the time spectrum, shear zones deforming by aseismic (i.e., ductile or viscous) creep mechanisms develop over thousands to millions of years. Faulting on the scale of orogens, continental rifts and seafloor-spreading systems lasts even longer. Although it is clear that the earthquakes are episodic by nature, determining the degree to which other fault-related processes are episodic is an important research frontier in the study of active and ancient natural systems, as well as in laboratory and numerical experiments on fault system behavior.

Two new developments provide insight into this issue and summon a re-evaluation of our concepts of the timescales of fault zone behavior: the deployment of continuous GPS (Global Position System) networks and the application of cosmogenic nuclide dating to fault-related surfaces.

Monitoring of active faults with continuous and semi-continuous GPS instruments shows a variety of transient deformational events. The recognition of creep events on large areas of several major subduction zone interfaces shows that, for at least some parts of some subduction zone systems, much interplate deformation occurs neither continuously for years nor suddenly during earthquakes, but rather over days to months. Sometimes these events appear to be quasi-periodic, strain release that makes up a large proportion of the total strain release along the plate boundary. Although the location and geometry of these so-called silent or slow earthquakes are becoming understood (Dragert et al. 2001), the details of what goes on at depth are still unresolved.

Other evidence for transient deformation from continuous GPS networks includes intraplate extension in the northern Basin and Range Province (western U.S.A.) on a characteristic timescale of years to decades (Figure 4.2). Although the tectonic setting of this transience is well understood, there is still no satisfactory explanation for the spatial extent or timing of such events.

We are beginning to recognize transient behavior on longer timescales associated with the interaction of individual faults within fault systems. For example, the San Andreas and San Jacinto faults in southern California together accommodate a significant portion of the oblique convergence between the North American and Pacific Plates. However, the partitioning of this plate motion on the two faults has not been constant through time, but has alternated between them on a hundred to thousand to million year timescale (Bennett et al. 2004).

The use of cosmogenic radionucleides to date erosional surfaces near faults together with paleoseismic studies indicates that individual fault segments can have complex histories. Clustering of earthquakes on timescales of a thousand years and longer is an increasingly common observation in many fault zones. How this clustering affects the loci of deformation through time on large fault systems, such as in the Basin and Range Province, is an important, unanswered question.

Although it is evident that some fault zones show transient behavior on timescales ranging from seconds to perhaps millions of years, we do not fully understand the mechanics that govern such transients. Are there previously unrecognized mechanisms of lithospheric deformation?

If we can distinguish periods of fault movement with different strain rates, we may find that there are transients on longer timescales. For example, is it possible that periods of contrasting strain rates reflect changes in the relative activity of competing deformation mechanisms at depth? An example of this is

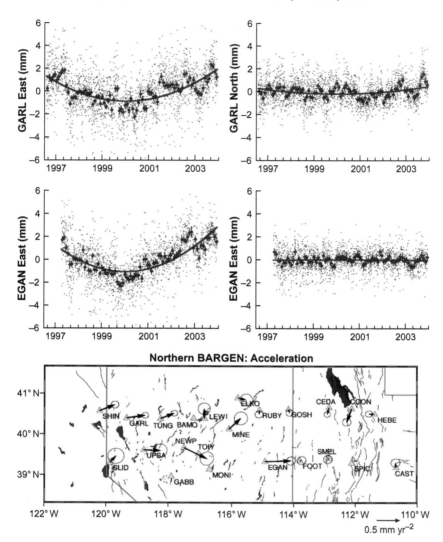

Figure 4.2 Top: Time series for continuous GPS sites GARL and EGAN in the intraplate rift setting of the northern Basin and Range Province from 1997–2004, with linear trend subtracted to show deviations from steady state. Accelerations of the sites from regression of second-order polynomial are 0.2 and 0.4 mm yr^{-1}. Bottom: Vectoral accelerations of continuous GPS sites showing regionally coherent acceleration of western sites relative to eastern sites of ~0.2 mm yr^{-2}. Accelerations are not correlated with earthquakes or patterns of regional seismicity. Source: J. L. Davis and B. Wernicke, unpublished data.

the possibility that the development of S–C fabrics in mylonite triggers accelerated fault motion at depth, loading the fault system in the upper crust and potentially leading to large earthquakes (see Chapter 6).

There are some basic strategies that could prove useful in addressing the problem of deformational transients. Clearly, we need to improve the spatial resolution of our observational methods. Increasing the density of observational networks (e.g., GPS networks) can provide both higher resolution locations and much lower detection levels in seismic magnitude. Combining seismic and geodetic observations has proven to be a powerful tool in providing observational continuity across broadly differing timescales of deformation. Extending monitoring beyond the plate-boundary zones to adjacent areas may provide different insight into the temporal scales of deformation. We must move beyond seismology and geodesy to incorporate detailed geologic studies, providing us with information on even longer timescales. Such studies may help us understand whether transient behavior results from time-dependent rheology or from changes in the kinematic boundary conditions.

As mentioned above, deformational fabrics may provide constraints on strain rate, not only strain. For example, microstructures in dynamically recrystallized quartz aggregates have been calibrated with laboratory experiments to yield estimates of temperature and strain rates during dislocation creep in natural shear zones (Stipp et al. 2002). Similarly, the calculation of heat production and dissipation rates in pseudotachylites has placed limits on paleostress and slip rate during earthquakes (e.g., Sibson 1975; Otsuki et al. 2003). Calibrations of yet other deformation mechanism(s) using laboratory data and theoretical considerations may yield constraints on stress and strain rates over a broader range of pressures, temperatures, and strain rates in nature. In particular, combining laboratory experiments with field studies of natural fault systems near the brittle-to-viscous transition may give us information on rate dependence of combined frictional and viscous deformation.

Field areas with a well-constrained thermo-barometric history are best suited as natural laboratories for this kind of work. Advances in geological dating techniques are allowing us to place closer constraints on the episodicity of slip (cf. Chapter 10). Also required is a better understanding of the deformation mechanisms during transient behavior. Do ductile fault rocks deform episodically—we know that brittle deformation can be episodic; is ductile deformation also episodic? If so, are there microstructures that can diagnose episodicity? This calls for a closer experimental investigation of transients—how they develop and what signature they leave in the rock record (see Chapters 5 and 7).

We should continue to develop tools for sampling Earth's response to sudden loading and unloading events, such as large earthquakes. These may provide important information about the time dependence of fault processes in the crust and upper mantle that are not accessible to direct observation.

Related to the timescales of localized faulting is a broader issue of the how large-scale deformation of the crust and mantle behaves with time. For example, large areas of distributed deformation in the Aegean Sea, the Basin and Range Province (western U.S.A.), and Tibet reflect homogeneous flow of the lithosphere. The mechanism by which such flow initiated or accelerated is largely unknown. If we knew this, we would have additional quantitative constraints on the interaction of faults and shear zones with the broadly deforming area.

LOCALIZATION OF FAULTS
THROUGHOUT THE LITHOSPHERE

We have very few constraints or examples of the actual structural evolution of fault zones from inception to maturity. How do fault systems localize? Are spatial or temporal patterns of localization characteristic of the processes involved in their nucleation and growth, or do these patterns also reflect the fault type and tectonic setting?

The most complete set of observations of fault evolution and localization in the brittle upper crust come from extensional fault systems (Chapter 3). However, a basic open question is how to determine the pattern of fault localization for strike-slip and thrust fault systems. Progress has been made by studying large, deeply eroded Pre-Cambrian and Early Paleozoic oblique-slip fault systems in shield areas (e.g., southern Madagascar; Martelat et al. 2000). In many cases, however, the late stages of faulting tend to overprint the record of nucleation and early growth, making it difficult to discern the mechanisms active at the onset of faulting.

For example, although we have very good constraints on the rate at which the San Andreas fault system has lengthened during northwestward migration of the Mendocino triple junction, we have only a rudimentary understanding of how the primary fault strands formed (e.g., Burgmann and Freed 2004). One strategy that can be applied along actively propagating fault systems is to exploit the idea that their lateral propagation in map view preserves progressively older stages of their evolution along their strike. In a study of the Red Sea Rift system, Favre and Stampfli (1992) showed that progressively younger stages of rifting are exposed in the direction of present-day rift propagation. This space-for-time substitution has also been applied to the San Andreas fault system, whose age decreases toward the Mendocino triple junction. Therefore, we can study the early stages of fault system development near the triple junction and the more mature stages further to the south (e.g., in the vicinity of San Francisco). This approach mitigates somewhat the problem that arises when faulting erases traces of the early deformational phases.

A second strategy is to study finite strain gradients within fossil fault zones that have undergone differential exhumation and therefore expose different paleodepths of the fault at the surface. This allows the study of fault mechanisms and their propagation both along their strike and down dip. A good example of this is the Periadriatic fault system, an array of late orogenic transpressional faults in the Tertiary European Alps (Figure 4.3). The structures

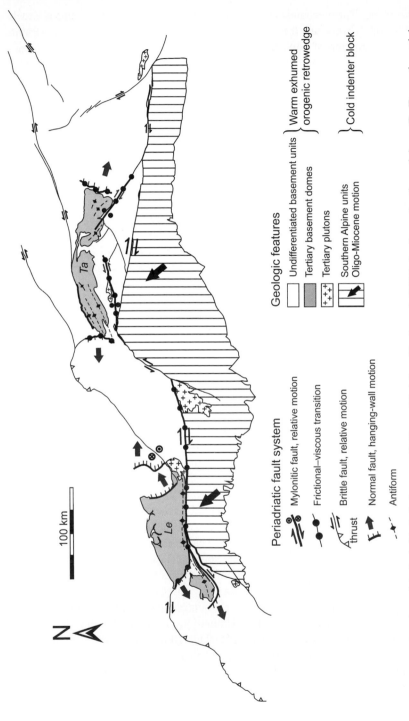

Figure 4.3 Map view of the Periadriatic fault system (PFS) modified from Handy et al. (2005). Le: Lepontine thermal dome; Ta: Tauern thermal dome.

preserved in this fossil fault zone indicate that localization in the upper, brittle crust involved fracturing and cataclasis, whereas in the lower crust, it involved buckling and viscous mylonitic shearing of existing mechanical anisotropies like compositional banding, schistosities, and even older mylonitic fault systems (see Box 4.1). Below the brittle-to-viscous transition, the transpressional shear zone terminates at its western end in a series of large constrictional folds and splayed, oblique extensional shear zones that deform the boundaries of previously formed basement nappes. This example underscores the important role of existing anisotropies on fault propagation and strain partitioning on the crustal scale.

Substituting space for time in the analysis of large fault systems to determine their evolution is not without problems. First, in the case of fossil fault systems, the interpretation of a finite strain gradient in terms of progressive localization is only valid if the strain gradient coincides with an isotopic record of progressive crystallization or cooling of synkinematic minerals (e.g., micas). The temporal resolution of the isotopic systems may not be sufficient to capture rapid localization events. Moreover, movement on fault systems with no, or only minor, dip-slip component is not recorded by differential cooling and closure of isotopic systems on either side of the fault. New *in situ* dating methods, preferably using highly retentive isotopic systems, are necessary to determine synkinematic formational ages over the complete spectrum of metamorphic temperatures. Second, the problem of selective microstructural memory is more acute in some tectonometmorphic settings than in others. Whereas fossil

---

Box 4.1 The Periadriatic fault system (PFS), European Alps.

The PFS delimits the retro-wedge of the Tertiary European Alps along a length of several hundred kilometers. During oblique convergence of the European and Adriatic Plates during the Oligo-Miocene, this fault system accommodated about 100–150 km of dextral strike-slip and several tens of km of shortening in a direction normal to the fault trace. In the vicinity of the Lepontine and Tauern thermal domes, it also accommodated up to 25 km of N-block-up exhumation. Due to differential exhumation and erosion, the PFS currently exposes fault segments that were active from near the surface to about 25 km depth. The fossil frictional-to-viscous transition during the Oligo-Miocene is marked with circles in Figure 4.3. Below this fossil frictional-viscous transition, strain localization involved buckling and shearing of preexisting anisotropies, assisted by the ingress of metamorphic fluids and melt into the mylonitic rock. Buckling was transitional, in both time and space, to passive folding, viscous mylonitic shearing, and strain localization in the fold limbs. The strain ultimately localized in a network of oblique, high-angle thrust faults and low-angle normal faults within the warm, exhuming orogenic block (see Figure 6.14). The PFS may be a good analogue for segments of large strike-slip faults at depth, like the San Andreas or North Anatolian fault systems.

extensional fault systems preserve the best record of fault evolution, thrust systems lose most structural and isotopic traces of fault nucleation and early growth due to thermal and deformational overprinting during prograde metamorphism. Third, structures in fossil fault zones are time-integrated products of the deformation history. Deformation at any given time during the activity of the shear zone may have been accommodated within only a part of the entire exposed shear zone before jumping to other parts of the same shear zone at a later time. Fourth, the structure of a fault reflects not only the finite strain, but also changes in kinematic boundary conditions. For example, the San Andreas fault system developed as the result of a change from subduction to lateral translation between the Pacific and North American Plates. Likewise, the Periadriatic fault system probably originated as a rift-related transform fault prior to becoming intra-orogenic strike-slip fault during the oblique convergence of Europe and Adria (Schmid et al. 1989). The transtensional setting at the western end of North Anatolian fault system in the northern Aegean Sea (Figure 4.4) is quite different from its transpressional environment in Anatolia during most of its formation and propagation. Determining whether there was a fundamental change in fault development across and along its trace is difficult in areas with a complex overprinting history (see Box 4.2).

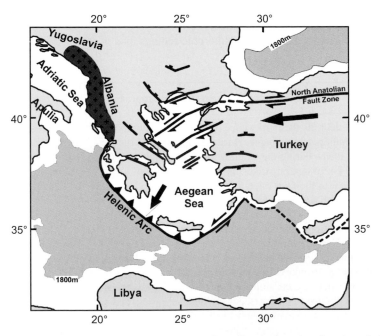

Figure 4.4 Tectonic overview of the Aegean–Anatolian region showing the relationship of Hellenic trench-arc system, Aegean extension, and motion of the North Anatolian fault (Taymaz et al. 1991).

Box 4.2 The North Anatolian fault zone (NAFZ).

NAFZ is an approximately 1500 km long, broadly arcuate, dextral strike-slip fault system that extends from Greece to eastern Turkey. It is predominantly a single zone of a few hundred meters to 40 km width. Along much of its length, this fault zone consists of a few shorter, subparallel fault strands that sometimes anastomose. The age and cause of dextral motion along NAFZ is controversial, and there are basically four different views:

1. The right-lateral motion commenced by the middle Miocene and resulted from westward lateral extrusion of Anatolia away from the collisional zone between the Arabian indenter and Eurasian Plate.
2. The NAFZ did not initiate until the latest Miocene or Early Pliocene.
3. The NAFZ initiated in eastern Anatolia during the Late Miocene and propagated westwards, reaching the Sea of Marmara area during the Pliocene.
4. The NAFZ initiated at ~16 Ma or more in the east, but at less than 3 Ma in the west.

Geologic mapping of offset markers along the NAFZ yields total displacements ranging from a maximum of 85 ±5 km to as little as 20–25 km. Geological data supports the view that the NAFZ slips at rates ranging from 5–10 to 17 ±2 mm yr^{-1}, whereas plate motions and seismological data suggest rates of 30–40 mm yr^{-1}. This discrepancy arises from the exaggerated slip rate obtained by treating the intense seismicity on the NAFZ during 1939–1967 as typical of all time. Recent GPS data indicate present-day rates of about 15–25 mm yr^{-1}. Extrapolating these rates back to the Early Pliocene yields a total displacement of 75–125 km, which is in close agreement with the maximum 85 ±5 km estimate based on offset markers.

Over the past 60 years, the earthquakes along different segments of the NAFZ are atypical of long faults. Beginning with the 1939 Erzincan earthquake ($M = 7.9$ to 8.0), which produced about 350 km of ground rupture, the NAFZ ruptured eleven times during moderate to large earthquakes ($M > 6.7$), forming a surface rupture trace more than 1000 km long. Most of the earthquakes occurred sequentially, from east to west.

Therefore, space-for-time substitution only works in field laboratories where kinematics have not varied significantly during the time span of fault evolution. To distinguish whether all faults follow similar movement paths during their evolution, we need to obtain similar data sets from various fault types and settings. In addition, we need to analyze fault systems of different sizes, because it is not clear whether fault systems behave and evolve in a self-similar fashion. For example, should we expect that lithosphere-scale extensional fault systems, such as the Basin and Range Province, have similar deformation histories to the much smaller extensional faults in regions such as the volcanic tablelands of eastern California?

Implicit in many models of fault zone evolution and localization is the assumption that similar mechanisms drive localization, irrespective of scale. Is localization at shallow levels within the brittle (seismogenic) crust simply a response to localized shearing at depth? Or does localization in the brittle crust drive localization in the deeper ductile shear zones of the lower crust and lithospheric mantle? Numerical modeling has shown that both processes are viable. However, significant feedbacks are likely between the shallow, dominantly brittle faults and the deeper, dominantly ductile shear zones of the lower crust and lithospheric mantle. One system may lead or lag behind the other on short a timescale; however, on longer timescales they likely develop together as strain compatibility is maintained.

The Blackwater fault in the Mojave Desert of southeastern California is part of a broad fault network (Chapter 3). The displacement along individual faults in the array varies, even though these faults are crustal-scale structures some tens of kilometers long. Geodetic data show that strain has accumulated across the entire Mojave Desert fault array, but the data lack the spatial and temporal resolution to show how individual faults are being loaded. They could be loaded by localized shear at depth, although this has not been documented for most faults in the array (cf. discussion of the Blackwater fault in Chapter 3). Alternatively they could be loaded horizontally by stress transfer between adjacent faults. Yet another possibility is that the loading of these faults is strongly influenced by active deformation to the north in Owen's Valley, such that deformation is propagating from north to south (G. King, pers. comm.). Recent earthquake activity in this area occupies a narrow zone within the array (e.g., Landers and Hector Mine events). The small spacing of these events has led some authors to suggest that the fault array reflects a stage of incipient localization. Thus far, however, there is no clear evidence to suggest that the recent seismic activity is due to localized strain accumulation at depth. How can we determine the degree of strain localization reached within this fault array? If it is in a state of incipient localization, then what observations or measurements should be made to constrain the driving mechanism of localization?

The mechanism of fault-loading obviously has a direct bearing on seismicity. For example, the theory of rate- and state-dependent friction indicates that the loading rate is as important as the load magnitude for producing failure (Tse and Rice 1986). Still, many seismologists argue that strike-slip faults are loaded from below by aseismic creep within a deep-seated shear zone (see Chapter 7). Indeed, Handy et al. (Chapter 6) have argued that accelerated creep associated with shearing instabilities (shear bands) within ductile shear zones potentially trigger earthquakes in the brittle, upper crust. The loading history of any given fault is expected to vary due to its interaction with nearby faults (Stein 1999) as well as to time-dependent gradients in the pore fluid pressure (e.g., Miller 2002).

We should not expect that the processes driving localization are the same on all length scales. Considered on the lithospheric scale, fault development is driven largely by mechanical instabilities on the order of kilometers to hundreds of kilometers in length. However, as one moves down to the scale of grains and grain aggregates, the determinants of strain localization are metamorphic fluids, cracks opened by local decreases in the effective pressure, or preexisting structures like grain boundaries. Localized domains on the granular scale can then coalesce to form large-scale fault zones ("upscaling" shown in Figure 6.10b), such that the change in length scale is associated with a change in the factors and mechanism of localization.

Speculation on how these various factors may influence localization in different depths is as follows: If the lower crust and/or upper mantle are nominally dry and/or melt-free, then strain localization below the seismogenic zone primarily reflects the effects of the temperature field and the spatial distribution of lithologies on rheology. The length scale of this localization will be long compared to that of hydrous and/or partially melted crust, where fluids enhance weakening due to embrittlement and the formation of intrinsically weak, hydrous minerals (e.g., micas) (Chapter 11). Brittle faults will propagate ephemerally (coseismically) downward from the seismogenic zone (Chapter 6), thereby introducing fluids to the top of the viscous crust and shortening the length scale of localization there. The localization pattern may then be controlled by stresses and fluid feedback systems in the upper, brittle crust; the rate control on localization is therefore "from above."

If the lower crust and/or upper mantle are hydrous or contain a melt (either intruded from below or derived by partial melting), then this can induce hydrofracturing (e.g., Davidson et al. 1994) and will likely shorten the length scale of localization in the viscous lithosphere with respect to the dry condition (above). Episodic hydrofracturing in the presence of a metamorphic fluid or melt is very fast, occurring at rates of meters to kilometers per second (Handy et al. 2001, and references therein). Localization will therefore propagate upward, probably sporadically, at the rate of deformationally induced, upward fluid advection. The frequency of these upwardly penetrating localization events is controlled by the overall fluid production rate at depth; the rate control on localization is therefore "from below."

In both of the above scenarios, surface processes (erosion, transport, sedimentation) only affect faulting on short length scales to superficial depths (say, 1–2 km: Chapter 8), i.e., to levels at which topographic stresses dominate the composite stress field (tectonic stress + lithostatic stress + topographic stress). The faulting geometries at these shallow depths are important inasmuch as they affect the mode of faulting and hence, the effective surface area exposed to erosion. They will govern erosion rates (Chapter 9). On longer time- and length scales, high erosion rates can influence the dynamic stability of the lithosphere and thus enhance exhumation rates.

In this conceptual model, there are three main controlling factors on fault zone evolution: (a) the rate of fluid production and location of fluid reservoirs; (b) the kinematics of faulting (thrusting, extension, or strike-slip); and (c) the length scales and geometry of preexisting anisotropies (faults, lithologies, foliations) with respect to size of the elastic stress field.

The question whether strain localization progresses top-down or bottom-up probably does not have a single, simple answer. Different fault systems behave differently and change through time as feedbacks modify the system. The style of localization also potentially affects the behavior of brittle faults. One can imagine that the manner in which a fault is loaded—from below where a shear zone has developed beneath the brittle fault, or from the side if the fault is significantly more localized than its substratum—will affect its seismogenic character.

To distinguish the relative roles of these factors during localization, it would be useful to compare continental faults with faults in oceanic environments (oceanic transforms, fracture zones), where localization evidently occurs even in the absence of fluids. The same is true of continental strike-slip fault systems, where rocks are neither significantly buried nor exhumed and therefore do not dehydrate or hydrate. In contrast, continental thrust faults and normal faults experience a continuous flux of fluids as rocks are subjected to changing $P–T$ conditions during burial and exhumation. It is likely that strain localization is achieved in different ways in hydrous and anhydrous environments.

INTERACTION OF FAULT ZONES WITH LITHOSPHERIC AND ASTHENOSPHERIC LAYERING

Do different types of faults (strike-slip, thrust, normal) interact with thermal and lithological layering in the lithosphere in fundamentally different ways? A common assumption in geodynamic modeling is that lithological layering is horizontal. Likewise, isotherms are considered to be subhorizontal in a steady-state thermal regime. Together, these thermal and lithological boundaries form a predominantly subhorizontal mechanical layering, commonly termed rheological stratification. Faults with a dip-slip component of motion obviously interact with such layering differently than subvertical strike-slip faults. Thrust and normal faults reflect the rheologic stratification, at least in the upper crust, by adopting a ramp-flat geometry. Still, we do not fully understand the mechanics of the interaction between fault geometry, kinematics, and mechanical layering, particularly at depth, within the ductile regime.

During thrusting and extension, changes in crustal thickness are accompanied by temperature changes, leading to changes in the rheology of the lithosphere. The thermo-mechanical feedbacks that characterize thrusting and normal faulting at the lithospheric scale are far less important in strike-slip fault zones, where fault displacement is parallel to horizontal isotherms and overall lithological layering.

Of course, lithological contacts are not always horizontal and, especially in orogens, they have been multiply folded during earlier deformation. Likewise, deformation and exhumation rates are usually high compared to the thermal equilibration rate of the orogenic crust, as manifest by clockwise $P–T$ loops derived from petrological studies of metamorphic mineral parageneses. In such cases, faults are expected to evolve in thickness, length, and style.

Temperature in the lower crust and lithospheric mantle almost certainly has a major influence on the width and orientation of highly deformed zones in the lithosphere. Fossil prograde amphibolite- to granulite-facies shear zones in deeply eroded Precambrian crust show that localization occurs at the scale of tens to hundreds of kilometers (Hoffman 1987; Martelat et al. 2000; Vauchez and Tommasi 2003). At Moho temperatures less than 800°C, the strength of the upper mantle exceeds that of the middle and/or lower crust. The ductile lower crust becomes a large-scale décollement horizon between the upper crust and the upper mantle (Figure 4.5). Analogue and numerical modeling have shown that the amount of coupling between the upper brittle crust and the ductile décollement layer strongly controls the dynamics of upper crust faulting, with considerable variations in the duration, vergence, sequence, and spacing of the primary thrust and normal faults. Deformation below the Moho remains poorly understood, in part due to the lack of high-resolution geophysical imaging and the paucity of large tracts (= 100 km^2) of well-exposed, exhumed mantle rocks at Earth's surface.

Petrological data and numerical modeling indicate that the formation of metamorphic core complexes in large extensional domains, such as the Basin and Range Province or the Aegean Sea, occurs at Moho temperatures of 1000°C or more. There, the lithospheric mantle was probably replaced by the asthenosphere. However, the modes of faulting and ductile shearing leading to such a replacement remain unclear.

Geologic data shows that the Death Valley region within the Basin and Range Province has accommodated the greatest amount of highly localized, upper crustal extension (150 km) in this province. Seismic imaging captures the trace

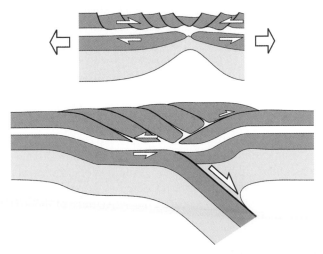

Figure 4.5 Models of rheological stratification in systems undergoing lateral extension (above) and thickening (below), showing weak a lower crustal layer that decouples deformation in the upper mantle from the faulted upper crust (J.-P. Brun, unpublished data).

of a "mylonite front" exposed at the surface in the extended area to a depth of 20 km beneath the Sierra Nevada and demonstrates that the Moho is subhorizontal across the lateral gradient in upper crustal extensional strain between the Death Valley and the Sierra Nevada. The mylonite front is interpreted to reflect the counterflow of "subcrustal asthenosphere" from beneath the Sierra Nevada toward the center of the extended terrain. The compensating layer must have the properties of a weak, possibly molten or partially molten, Newtonian fluid that is unable to maintain horizontal pressure gradients for significant lengths of time. Seismic imaging and sampling of mantle xenoliths from the Sierra Nevada shows that while rifting was occurring in the Death Valley region (between 12 and 3 Ma), a large part of the dense, Sierran mantle lithosphere detached and began to sink into the asthenosphere. At present, very little mantle lithosphere is present, with a localized region of pronounced velocity contrast (>4%) in a cylindrical area that persists to a depth of >200 km beneath the central Sierra Nevada/Great Valley block. These observations suggest that, under at least some circumstances, fault localization occurs in an environment where the lateral flow of extremely weak crustal layers and buoyancy-driven removal of the mantle lithosphere are likely important mechanical controls.

Throughout our discussions of localization, the question arose as to whether we have the tools to measure the degree of strain localization at depth. Are there limits to our ability to resolve the length scale of deforming zones within the lithosphere and asthenosphere?

Shear zones are commonly perceived to broaden with depth. Yet, the resolution of geophysical imaging techniques decreases with depth such that the length scale (e.g., width) of resolvable features also grows with depth (Figure 4.6). This raises the disturbing possibility that conceptual models of deformation broadening with depth may be partly or largely an artifact of limitations in developing high-resolution images of the deformation field.

Studies of fossil fault zones that expose shallow to lower crustal levels have provided such high-resolution views of how deformation varies with depth, as noted above. Drilling of active fault zones (e.g., the SAFOD project on the San Andreas Fault) provide *in situ* information complementary to that obtained from exhumed fault zones. We must, however, continue to refine our tools to provide high-precision geophysical imaging of fault systems in the crust and mantle. The application of seismic techniques such as "Double-Difference Earthquake locations" reveal that microseismicity along several active strands of the San Andreas fault system are restricted to extremely narrow domains (e.g., Schaff et al. 2002). In some cases, the active trace of these fault strands is only 5–10 m wide, and certainly less than 50–100 m wide at depths of 5–10 km in the crust.

In analyzing the interaction of fault systems, it is useful to consider whether deformation occurs in an open or closed system at the lithospheric scale: Is mass conserved or removed from the system? Clearly, the prevalence of tectonic

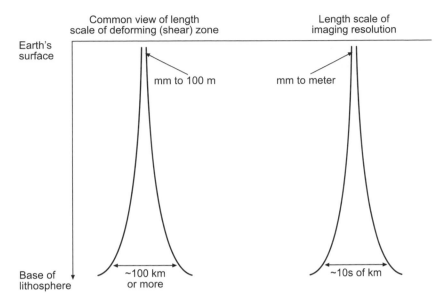

Figure 4.6 Length scale of fault zone versus depth in comparison to the depth dependence of spatial resolution in geophysical imaging (from G. Beroza).

erosion or accretion at convergent margins indicates that subduction zones are open systems from a deformational perspective. But are orogenic systems closed? The Death Valley–Sierra Nevada coupling may reflect a situation in which faulting and its localization has driven a previously closed deformational system into an open mode in which substantial mass (the entire lithospheric mantle?) is removed from the system.

Most major fault zones accommodate motion that is oblique to their traces at the surface. In the shallow parts of large transpressional fault systems, this oblique motion is often partitioned into strike-slip faults and separate convergent structures (thrust faults, folds). For example, along the Alpine fault system of South Island, New Zealand, oblique plate motion, with respect to the orientation of the primary upper crustal fault (the Alpine fault), leads to the juxtaposition of strike-slip motion on the fault with substantial crustal shortening and thickening of crust of the adjacent Pacific Plate (Stern et al. 2000; Sutherland et al. 2000; Walcott 1998). The ~50–100 km wide upper crustal deformational zone that produced the Southern Alps of New Zealand contrasts with the ~1 km wide mylonite zone in the middle crust that is exposed along the exhumed, eastern side of the Alpine fault. Deformation in the middle crust appears to have been much more localized than in the brittle upper crust, where strain partitioning is very pronounced. The component of displacement normal to the plate boundary has increased over the past 5–12 Ma as the motion between the Pacific and Australian Plates has become increasingly oblique with respect to their mutual boundary. What is not clear, however, is how the lithosphere, particularly the subcrustal lithosphere, responds to this increasingly transpressional regime. Does the ductile part of the fault also deform in a transpressional manner, producing a broad deformation zone that mimics the width of the upper crustal convergence zone? Or does the ductile shear zone change its orientation in response to changing plate motions, allowing it to accommodate simple shear while linking the subduction boundaries at both of its ends?

Similarly, the nature of deformation in the mantle lithosphere within large-scale orogenic and intracontinental extension/rift systems is not well understood. Space problems can develop in such settings unless the lithospheric mantle deforms substantially or is removed from the system. Few constraints currently exist to provide definitive answers to this question. What observations are needed, what tools need to be developed, and what disciplines linked in order to address the question of how fault zones behave in different levels of the lithosphere?

There is a definite need for improved geophysical images of mantle deformation. Currently the primary imaging tool is passive, teleseismic information, especially the direction of split shear waves, which provide information on the bulk fabric orientation of upper mantle rocks beneath the observational site (e.g., Tommasi et al. 1999). Less commonly employed, but also effective, are approaches that exploit the pronounced P-wave anisotropy of upper mantle rocks. Combining electrical conductivity anisotropy measurements (e.g., magnetotelluric, MT) with seismic anisotropy measurements is a powerful approach (e.g., Maercklin et al., in press). The frequency-depth kernels for MT measurements have been used to evaluate the depth dependence of any observed anisotropy. These observations are critical to defining patterns and amounts of deformation but are not without limitations. Interpreting seismic wave velocity anisotropy in terms of deformation involves making assumptions about deformation mechanisms and the conditions of deformation. As mentioned above, the volume of rock sampled by the seismic waves broadens with depth (Figure 4.6), and as an integral over the ray path, it is often difficult to isolate the depth interval at which the imaged fabric exists. To complicate matters, the superposition of deformational events at depth may either obscure earlier events or produce a composite structure that reflects none of the individual deformational events.

Other seismological tools include mantle tomography, receiver function analyses, and active source imaging (see Chapter 2, and references therein). All of these can provide constraints on both the three-dimensional structure and the nature of thermal and/or compositional character of the crust and mantle. Improvements in spatial resolution and coverage of all of these tools are needed. Innovative combinations of measured properties (e.g., the ratio of P- and S-wave velocities, Vp/Vs) are already used to map the temperature distribution in the lithosphere. Probably the most useful "advances" will be the integration of results from all of these geophysical techniques with detailed field and laboratory observations of structures in active and fossil fault zones.

SUMMARY

Our deliberations on how large fault systems nucleate and grow led to a consensus that progress will entail identifying the determinants of transient motion during faulting and, more generally, regarding structure and rheology in a broad kinematic context, from the Earth's surface down into its asthenosphere. Episodicity associated both with seismic and aseismic faulting ("slow" or "silent" earthquakes) is not restricted to the upper brittle crust and, indeed, may originate in the viscous lower crust and upper mantle. The origins of episodic fault activity are still not known, and several mechanisms were considered, from viscous instabilities nucleating on existing anisotropies to volume changes and fluid flux associated with phase transformations. The consideration of how stress fields interact with existing structures and thermal regimes at depth, and with erosional and depositional regimes at the surface, is crucial to understanding how faults link to form a throughgoing network that weakens the entire lithosphere. Progress will require a multifaceted approach, in which high-resolution imaging of active faults (involving satellite-based and deep geophysical sounding methods) is combined with detailed study of "frozen-in" structures in exhumed, fossil fault zones. Although these disparate methods are individually well established, using them in concert is a challenge that will require adaptation from each specialized part of the Earth Science community. Therein lies the future of research on fault dynamics.

REFERENCES

Bennett, R.A., A.M. Friedrich, and K.P. Furlong. 2004. Co-dependent histories of the San Andreas and San Jacinto fault zones from inversion of fault displacement rates. *Geology* **32**:961–964, doi:10.1130/G20806.1

Davidson, C., S.M. Schmid, and L.S. Hollister. 1994. Role of melt during deformation in the deep crust. *Terra Nova* **6**:133–142.

Dragert, H., K. Wang, and T.S. James. 2001. A silent slip event on the deeper Cascadia subduction interface. *Science* **292**:1525–1528.

Favre, P., and G.M. Stampfli. 1992. From rifting to passive margin: Examples of the Red Sea, Central Atlantic, and Alpine Tethys. *Tectonophysics* **215**:69–97.

Freed, A.M., and R. Bürgmann. 2004. Evidence of power law flow in the Mojave desert mantle. *Nature* **430**:548–551, doi:10.1038/nature02784.

Handy, M.R., J. Babist, R. Wagner, C. Rosenberg, and M. Konrad. 2005. Decoupling and its relation to strain partitioning in continental lithosphere: Insight from the Periadriatic fault system (European Alps). In: Deformation Mechanisms, Rheology, and Tectonics: From Minerals to the Lithosphere, ed. D. Gapais, J.-P. Brun, and P.R. Cobbold, Spec. Publ. 243, pp. 249–276. London: Geol. Soc.

Handy, M.R., A. Mulch, M. Rosenau, and C.L. Rosenberg. 2001. The role of fault zones and melts as agents of weakening, hardening, and differentiation of the continental crust: A synthesis. In: The Nature and Tectonic Significance of Fault Zone Weakening, ed. R.E. Holdsworth, R.A. Strachan, J.F. Magloughlin, and R.J. Knipe, Spec. Publ. 186, pp. 303–330. London: Geol. Soc.

Hoffman, P.F. 1987. Continental transform tectonics, Great Slave Lake shear zone (ca. 1.9 Ga), northwest Canada. *Geology* **15**:785–788.

Maercklin, N., P.A. Bedrosian, C. Haberland, et al. 2005. Characterizing a large shear zone with seismic and magnetotelluric methods: The case of the Dead Sea Transform. *Geophys. Res. Lett.* **32**:L15303, doi:10.1029/2005GL022724

Martelat, J.-E., J.-M. Lardeaux, C. Nicollet, and R. Rakotondrazafy. 2000. Strain pattern and late Precambrian deformation history in southern Madagascar. *Precambrian Res.* **102**:1–20.

Miller, S.A. 2002. Properties of large ruptures and the dynamical influence of fluids on earthquakes and faulting. *J. Geophys. Res.* **107**:B9, doi: 10.1029/2000JB000032.

Otsuki, K., N. Monzawa, and T. Nagase. 2003. Fluidization and melting of fault gouge during seismic slip: Identification in the Nojima fault zone and implications for focal earthquake mechanisms. *J. Geophys. Res.* **108**:B4, 2192, doi: 10.1029/2001JBB001711.

Pfiffner, O.A., and J.G. Ramsay. 1982. Constraints on geological strain rates: Arguments from finite strain rates of naturally deformed rocks. *J. Geophys. Res.* **87**:B1, 311–321.

Schaff, D.P., H. Götz, R. Bokelmann, and G.C. Beroza. 2002. High-resolution image of Calveras fault seismicity. *J. Geophys. Res.* **107**:B9, 2186, doi:10.1029/2001JB000633.

Schmid, S.M., H.R. Aebli, F. Heller, and A. Zingg. 1989. The role of the Periadriatic Line in the tectonic evolution of the Alps. In: Alpine Tectonics, ed. M.P. Coward, D. Dietrich, and R. Park, Spec. Publ. 45, pp. 153–171. London: Geol. Soc.

Sibson, R.H. 1975. Generation of pseudotachylite by ancient seismic faulting. *Geophys. J. Roy. Astron. Soc.* **43**:775–784.

Stein, R.S. 1999. The role of stress transfer in earthquake occurrence. *Nature* **402**: 605–609.

Stern, T., P. Molnar, D. Okaya, and D. Eberhart-Phillips. 2000. Teleseismic P wave delays and modes of shortening the mantle lithosphere beneath South Island, New Zealand. *J. Geophys. Res.* **105**:21,615–21,631.

Stipp, M., H. Stünitz, R. Heilbronner, and S.M. Schmid. 2002. Dynamic recrystallization of quartz: Correlation between natural and experimental conditions. In: Deformation Mechanisms, Rheology, and Tectonics: Current Status and Future Perspectives, ed. S. De Meer, M.R. Drury, J.H.P. De Bresser, and G.M. Pennock, Spec. Publ. 200, pp. 171–190. London: Geol. Soc.

Sutherland, R., F. Davey, and J. Beavan. 2000. Plate boundary deformation in South Island, New Zealand, is related to inherited lithospheric structure. *Earth Planet. Sci. Lett.* **177**:141–151.

Taymaz, T., J.A. Jackson, and D. McKenzie. 1991. Active tectonics of the north and central Aegean Sea. *Geophys. J. Intl.* **106**:433–490.

Tommasi, A., B. Tikoff, and A. Vauchez. 1999. Upper mantle tectonics: Three-dimensional deformation, olivine crystallographic fabrics, and seismic properties. *Earth Planet. Sci. Lett.* **168**:173–186.

Tse, S.T., and J.R. Rice. 1986. Crustal earthquake instability in relation to the depth variation of frictional slip properties. *J. Geophys. Res.* **91**:9452–9472.

Vauchez, A., and A. Tommasi. 2003. Wrench faults down to the asthenosphere: Geological and geophysical evidence and thermo-mechanical effects. In: Intraplate Strike-Slip Deformation Belts, ed. F. Storti, R.E. Holdsworth, and F. Salvini, Spec. Publ. 210, pp. 15–34. London: Geol. Soc.

Walcott, R.I., 1998. Modes of oblique compression: Late Cenozoic tectonics of the South Island of New Zealand. *Rev. Geophys.* **36**:1–26.

5

Seismic Fault Rheology and Earthquake Dynamics

JAMES R. RICE[1] and MASSIMO COCCO[2]

[1]Department of Earth and Planetary Sciences and Division of Engineering and
Applied Sciences, Harvard University, 224 Pierce Hall,
Cambridge, MA 02138, U.S.A.
[2]Istituto Nazionale di Geofisica e Vulcanologia, Seismology and Tectonophysics
Department, Via di Vigna Murata 605, 00143 Rome, Italy

ABSTRACT

As preparation for this Dahlem Workshop on *The Dynamics of Fault Zones*, specifically on the subtopic "Rheology of Fault Rocks and Their Surroundings," we addressed critical research issues for understanding the seismic response of fault zones in terms of the constitutive response of fault materials. This requires new concepts and a host of new observations and experiments to document material response, to understand the shear localization process and the inception of earthquake instability, and especially to understand the mechanisms of fault weakening and dynamics of rupture tip propagation and arrest during rapid, possibly large, slip in natural events. We examine in turn the geological structure of fault zones and its relation to earthquake dynamics, the description of rate and state friction at slow rates appropriate to the interseismic period and earthquake nucleation, and the dynamics of fault weakening during rapid slip. The last topic gets special attention in view of the important recent advances in theoretical concepts and experiments to probe the range of slip rates prevailing during earthquakes. We then address the assembly of the constitutive framework into viable, but necessarily simplified, conceptual and computational models for description of the dynamics of crustal earthquake rupture. This is done principally in the slip-weakening framework, and we examine some of the uncertainties in doing so, and issues of how new understanding of the rapid large slip range will be integrated to model the traction evolution and the weakening process during large slip episodes.

INTRODUCTION

The mechanics of earthquakes and faulting is currently investigated by collecting observations and performing theoretical analyses and laboratory experiments on three main physical processes: tectonic loading, fault interaction through stress transfer, and rheological response of fault zones. Earthquakes are certainly one of the most important manifestations of faulting, and the understanding of dynamic fault weakening during the nucleation and propagation of a seismic rupture is a major task for seismologists and other Earth scientists. In the literature, the study of the initiation, propagation, and arrest of an earthquake rupture is associated with the understanding of coseismic processes. However, the three main physical processes cited above also control fault behavior during the interseismic period (i.e., the time interval between two subsequent earthquakes) and the postseismic period (i.e., the time interval immediately following a seismic event). In this chapter we focus primarily on the most recent advances and progresses of research in the understanding of rheological and constitutive properties of active fault zones. Fault rheology and constitutive properties help to understand the depth extent of the seismogenic zone. They are associated with the characterization of thickness of the seismogenic zone (see Sibson 2003, and references therein), the material properties of fault gouge, and the properties of contact surfaces within the slipping zone (Dieterich 1979; Ohnaka 2003). The latter are our primary focus here.

We present geological observations of fault zones to constrain a model of a seismogenic structure capable of localizing strain and generating earthquakes. We review results of laboratory experiments aimed at understanding fault-constitutive properties. In the context of this study, earthquakes are considered as instabilities of a complex dynamic system governed by assigned frictional laws and other constitutive laws (e.g., for the damage regions bordering the zone of concentrated slip). Therefore, we briefly present the physical origin and analytical expressions of these friction laws and discuss the different competing physical mechanisms that contribute to dynamic fault weakening during earthquakes. In particular, there is now the awareness that, although the properties of the contact surface play a relevant role in controlling dynamic slip episodes, frictional heat, thermal pressurization of pore fluids, and mechanical lubrication can contribute to explain dynamic fault weakening and to control fault friction at high slip rates (Sibson 1973; Lachenbruch 1980; Mase and Smith 1987; Tsutsumi and Shimamoto 1997; Kanamori and Brodsky 2001; Andrews 2002; Goldsby and Tullis 2002, 2003; Tullis and Goldsby 2003; Di Toro et al. 2004; Fialko 2004; Hirose and Shimamoto 2005; Rice 2004, 2006). The definition of a fault zone model (i.e., to determine the size of the slipping zone with respect to the surroundings) and its characterization in terms of the dominant physical processes are extremely important tasks which will be a focus for future scientific research.

To describe a fault zone, we must address two main problems: (a) the definition of the geometrical and mechanical properties of a fault zone (see reviews by Ben-Zion and Sammis 2003; Biegel and Sammis 2004) and (b) the understanding of the spatial and temporal scale dependence of relevant physical processes. Only by clarifying these issues can we integrate the sometimes conflicting evidence on fault zone properties. The effect of fault geometry is important because it concerns the three-dimensional structure of a fault zone (i.e., thickness of the slipping region, of the fault core, or the damage zone), the fragmentation of the rupture surface (i.e., bending, branching, step over, etc.), as well as the complexity of the fracture network (i.e., fractal distribution of fractures). Scale dependence is extremely important because it allows us to establish a hierarchy to characterize where (i.e., the spatial extension) and when (i.e., the temporal evolution) the different physical processes govern crustal faulting. The contribution of each physical process controlling dynamic fault weakening and fault evolution depends on the size of the slipping zone as well as on the values of the main physical parameters (such as thermal and hydraulic diffusivity, permeability, and porosity) that appear within the mathematical representation of fault zones. In this chapter, we discuss these topics in an effort to describe a fault zone model, to discuss the nature and the dominant physical processes of fault zones, and to determine their scale of relevance. Our primary task is to stimulate a discussion and an exchange of views and ideas to clarify perspectives and new horizons in fault mechanics.

GEOLOGICAL OBSERVATIONS OF FAULT ZONES AND THEIR RELATION TO RUPTURE DYNAMICS

Many recent investigations have focused on the internal structure of fault zones to improve knowledge of microscale processes, fault zone rheology, and dynamic weakening processes. These studies pointed out that coseismic slips on mature, highly slipped fault zones often occur within ultracataclastic, possibly clayey, zones of order tens to hundreds of millimeters thick, but that the zone of principal seismic shearing may be localized to a thickness less than 1–5 mm width within that ultracataclasite core. A broad damage zone, on the order of one to hundreds of meters thick, surrounds the fault core and it is characterized by highly fractured and possibly granulated materials which, because of their porosity, must usually be assumed to be fluid-saturated. Figure 5.1a displays a sketch of the fault zone model discussed here. Evidence for such a model has been collected from the Punchbowl fault (southern California; see Chester and Chester 1998, and Figure 5.1b), the nearby North Branch San Gabriel fault (Chester et al. 1993), the Median Tectonic Line (Japan; see Wibberley and Shimamoto 2003), the Nojima fault (which ruptured in the 1995 Kobe earthquake in Japan and has been penetrated by drill holes; Lockner et al. 2000), as

(a)

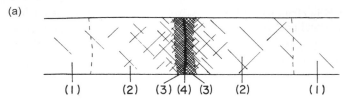

(1) Undamaged host rock
(2) Damage zone, highly cracked; 10s m to 100 m wide, minor faults may reach 1 km
(3) Gouge or foliated gouge; 1 m to 10s m wide
(4) Central ultracataclasite shear zone, may be clay rich; 10s mm to 100s mm wide
(5) [within (4), not marked above] Prominent slip surface; may be < 1 to 5 mm wide

(b)

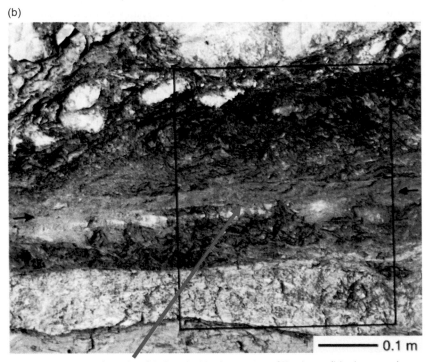

Prominent slip surface (pss) is located in the center of the layer (black arrows)

Figure 5.1 Internal structure of a major fault zone: (a) schematic representation of the inner structure of a fault zone within the brittle upper crust based on Chester et al. (1993); (b) picture of the ultracataclastic layer (with prominent slip surface indicated by the arrows) at the Punchbowl fault, taken from Chester and Chester (1998); see also Sibson (2003).

well as from other observations summarized in Sibson (2003), Ben-Zion and Sammis (2003), and Biegel and Sammis (2004). According to these observations, slip is generally accommodated along a single, nearly planar surface.

The damage zone is characterized by a higher fracture density than the surroundings. At a much larger scale of a plate boundary, the fault zone model described above can be part of a broader shear zone, whose thickness can be on the order of several hundreds of meters, which is composed by many different fracture zones and is responsible for stress reorientation with respect to the direction of remote tectonic loading. However, this "large-scale" fault zone model may describe only mature faults such as San Andreas or plate margin transform faults. Although we point out that the spatial scale at which we attempt to characterize the fault zone structure does matter, here we focus our attention on the fault zone at a local scale such as that depicted in Figure 5.1a. It is important to point out that the width and the complexity of fault zones inferred from the analysis of surface ruptures depend on the faulting mechanism (i.e., whether reverse, normal, strike-slip, or oblique) and can be affected by the presence or absence of sedimentary cover as well as by other free surface effects.

The geological observations of fault zones presented above raise several important issues that must be addressed to understand the mechanical properties of faults as well as the dynamic weakening processes occurring during earthquakes. The first concerns *frictional heating* caused by *large slip* (≥ 1 m) during an earthquake within such an *extremely thin slipping zone*. The temperature increase caused by a meter of slip within a few millimeters-thick slipping zone would be larger than 1000°C over most of a seismogenic zone under the assumption of uniform adiabatic shearing, at least if it occurred (as recent evidence suggests it may not; see below) in the absence of any mechanism to reduce strength rapidly once slip begins. Such a sudden frictional-driven temperature change should lead to melting and formation of pseudotachylites. The understanding of this issue may contribute to solving the well-known heat flow paradox along the San Andreas fault; that is, to explain the absence of measurably enhanced heat outflow along the fault. Moreover, if melts produced during frictional heating have a low viscosity, they may lubricate faults and thus reduce dynamic friction (Sibson 1975; Spray 1993; Brodsky and Kanamori 2001; Kanamori and Brodsky 2001). Such rapid changes in friction at high slip rates, not all due to melting, have also been observed in laboratory experiments (Tsutsumi and Shimamoto 1997; Goldsby and Tullis 2003; Tullis and Goldsby 2003; Di Toro et al. 2004). However, according to Sibson (2003) the evidence of localized slip and the apparent scarcity of pseudotachylites suggest that pseudotachylite is rarely preserved within mature fault zones, or that melting is rare because other phenomena play a dominant role in controlling dynamic fault weakening. We discuss some of these phenomena and, in particular, we address frictional heating and presently known thermal weakening mechanisms (flash heating at micro-contacts, thermal pressurization of pore fluid) in more detail in the next section.

A second issue concerns the relatively *simple structure of the slipping zone*. If the slipping zone of real faults is *extremely thin*, comparable to the thickness of gouge layers in some laboratory experiments, we may ask what is the contribution of rupture surface topography (and contact properties) in controlling fault friction during sliding? The resolution of this issue is related to the interpretation of results of numerous laboratory experiments on fault friction (distributed vs. localized strain within the gouge layer; grain rolling/sliding vs. grain fracture), and it is also related to the previous issue concerning flash heating on a localized zone versus broadly distributed shear of gouge layers. In general, the principal slipping zone contains wear materials or gouge, which can be cohesive or incohesive. These products of faulting have formed by macroscopic fracturing, frictional wear, and cataclastic comminution, in combination with alterations by reactions with fluids and mineral deposition (Chester et al. 1993; Sibson 2003). Thus, it is important to understand the gouge layer evolution both during single earthquakes (coseismically) and during the interseismic periods. It is likely that repeated slip episodes continuously modify the grain size and the properties of gouge materials. Power and Tullis (1991) found that roughness of natural fault surfaces in the slip direction has considerably smaller amplitude than the roughness in the direction normal to the slip. Therefore, we may expect that cumulative slip tends to smooth the fault surface, an expectation that is well documented at larger scales (Stirling et al. 1996). This view of the gouge layer evolution is commonly associated with other precepts that gouge texture is fractal (Steacy and Sammis 1991) and that gouge surface energy yields a negligible contribution to the earthquake energy balance. However, Wilson et al. (2005) have recently investigated the gouge texture of two seismic faults having very different cumulative slip and extension and report that both faults display similar gouge characteristics and the grain-size distribution is not fractal. These authors also proposed that fracture surface energy is a nonnegligible part of the energy budget, a view questioned by Chester et al. (2005), and suggested that gouge evolution is not related to quasi-static cumulative slip, but rather formed by dynamic rock pulverization during the propagation of a single earthquake. Thus, we have two competing interpretations of the gouge and damage zone evolution: the former can be defined as a large-scale geological view in which both coseismic and interseismic processes are involved; the latter solely invokes coseismic processes during individual events.

A third issue related to the previous two concerns the *presence of fluids* within a fault zone and the way in which they control fault strength evolution during and after a dynamic slip episode. One of the most important parameters to model the evolution of shear zone fluid pressure is the *hydraulic diffusivity*, which depends on the permeability, fluid viscosity, and fluid-pore compressibility. Several investigations have attempted to constrain the values of hydraulic diffusivity within the fault zone as well as to estimate the value of permeability (see Lockner et al. 2000; Wibberley and Shimamoto 2003; Sulem et al. 2004, and references therein). These studies have shown that the gouge zones forming the fault core have a much lower permeability than that measured in the surrounding damage zone, which can be highly variable. Permeability within the fault core ($\sim 10^{-19}$ m^2) can be three orders of magnitude smaller that that in the damage zone ($\sim 10^{-16}$ m^2). In both regions, permeability is reduced as the effective normal stress is increased. For example, in the case of the ultracataclastic gouge core material containing the slip zone in the Median Tectonic Line, permeability is 10^{-19} m^2 at 10 MPa effective confining stress, 10^{-20} m^2 at 70 MPa, 4×10^{-21} m^2 at 120 MPa, and 3×10^{-21} m^2 at 180 MPa (Wibberley and Shimamoto 2003); assuming hydrostatic pore pressure, 126 MPa corresponds to the effective overburden stress at 7 km, a representative centroidal depth for the slip zone of crustal earthquakes. It is important to assess the hydrodynamic behavior of fault zone fluids during dynamic slip episodes, which requires a detailed examination of permeability and poroelastic properties of fault core as well as their variations with effective pressure. In fact, hydraulic diffusivity can change during a dynamic slip episode because of effective normal stress and temperature changes: an increase in temperature can decrease fluid viscosity, therefore increasing hydraulic diffusivity during sliding. Moreover, as discussed above, porosity can evolve not just during an earthquake but also during the interseismic period. The latter may allow a creep-compaction mechanism which isolates and pressurizes fluids, as proposed to explain weakness of mature faults (Sleep and Blanpied 1992).

A fourth issue with mature fault zones is that they present *variations in elastic and seismic properties* in the *direction perpendicular to the slip surface*. At large scale, a major fault may have brought distinct lithologies into contact with one another, for example, seafloor crust and accreted sediments, along a subduction fault. At the scale of the damage zone, the highly variable materials, as just discussed, will cause property variations. That brings a new ingredient into rupture dynamics, because spatially inhomogeneous slip (like during earthquake rupture) along a fault, which is not a plane of mirror symmetry, alters not only the shear stress but also the normal stress along the fault. The effects have been shown to allow extremely unstable behavior even along faults that have been idealized to have a constant friction coefficient f (Weertman 1980; Andrews and Ben-Zion 1997; Cochard and Rice 2000) and hence would be stable if in a configuration with mirror symmetry. This means that it is critical to understand fault zone rheological response when there is *rapid change in normal stress*. In fact, it has been established (Cochard and Rice 2000; Ranjith and Rice 2001) that deviations from the classical formulation of Coulomb friction, of a type seen in shock wave experiments which deliver an abrupt change in normal stress to a slipping surface (Prakash and Clifton 1992; Prakash 1998), allow models of rupture along dissimilar material interfaces to be well posed mathematically. The classical formulation is not well posed in that circumstance. Nevertheless, the critical feature seen in the shock wave experiments, namely, that an abrupt change in normal stress does not cause a

corresponding abrupt change in shear strength, has obviously not been dupli-
cated in much lower speed friction experiments (Linker and Dieterich 1992;
Boettcher and Marone 2004) with less abrupt changes of normal stress. In real-
ity, rupture of a fault separating dissimilar materials elicits both the effect of
coupling slip to alteration of normal stress discussed above and the effect of
weakening of friction with slip or slip rate that would exist for a fault between
identical materials. Both effects seem important to a complete computational
and experimental description of rupture along dissimilar material interfaces
(Harris and Day 1997; Xia et al. 2005).

Fifth, calculations (Poliakov et al. 2002; Andrews 2005; Rice et al. 2005)
suggest that even if primary shear is confined to a thin zone, the adjoining
damage zone is likely to *deform inelastically* as the rupture tip passes by because of the localized high stressing near the tip. The effect becomes particularly marked as the rupture propagation speed approaches the Rayleigh speed.
The resulting inelastic straining is likely to interact with stressing and energy
flow to the slipping process, in a way analogous to what has been studied over
many years for tensile crack growth in elastic–plastic solids (typically, structural metals). Thus it is crucial to understand the high strain-rate constitutive
response of the highly cracked and granulated material of the damage zone
and then to address the mechanics of interaction between inelastic response
there and on the main fault. A subtle interaction, yet to be quantified, is that
nonlinear constitutive response off the fault plane (which would, in any event,
inevitably be asymmetric relative to that plane because of the dependence of
shear strength on normal stresses) has the generic effect of altering the normal
stress on the fault plane itself. It is not yet known if this has a negligible or
perhaps major effect on the shear rupture dynamics.

In this section we have presented a model of a fault zone based on geological observations that are sometimes corroborated by results of laboratory
experiments. Such a fault zone model is also consistent with seismological
observations based on the analysis of fault zone trapped waves (see Li et al.
1994). These studies have a resolution of meters and show that the damage
zone is characterized by lower body wave velocities than the surrounding host
rocks, a difference that can reach 50%. The thickness of the damage zone
at depth inferred from these studies is consistent with geological observations
(10–100 m wide).

The observations presented in this section allow the proposition of a fault
zone model characterized by the presence of localized slip in a thin zone, the
presence of frictional wear or gouge, a fault core composed of cataclasite and
ultracataclasite, and a broader damage zone (highly fractured, anisotropic, and
poroelastic). Other observational evidence comes from laboratory experiments
on fault friction, which is discussed in the next section. Thereafter we review
observational and experimental evidence to shed light on the conflicting and
supporting interpretations of data and theoretical modeling.

FAULT MATERIAL RESPONSE TO INTERSEISMIC AND EARTHQUAKE STRESSING

Aseismic to Seismic Transition, Interseismic Stressing, and Earthquake Nucleation

Two concepts concerning what underlies the *transition from ductile to brittle fault response* and the *depth extent of seismogenesis* are to be found in the current literature. The concepts are not obviously identical, but are surely interrelated: an issue is to understand how they are interrelated, and to incorporate both into a proper understanding of seismic phenomena.

The older concept, of a transition from *localized friction* to *broadly distributed creep*, is rooted in ideas of Brace, Evans, Goetz, Kohlstedt, Meissner, Strehlau, and Sibson; for recent assessments see Handy and Brun (2004) and Chapter 6. It leads to the famous pine tree-like plots of crustal strength versus depth. One asks what stress distribution would allow the shallow lithosphere to adjust to remotely imposed plate motions by temperature-dependent creep processes (dislocation and/or grain boundary creep, or fluid-assisted dissolution and transport). Where that stress is less than the friction strength, which is generally assumed to increase linearly with depth, the material is declared ductile; where creep strength is greater than friction strength, it is assumed that the latter dominates and that the response of that part of the lithosphere is brittle, occurring in earthquakes. A problem with this interpretation is that widths over which plate motions would be accommodated by pure creep processes, and hence the strain rates, are not readily estimated; the creep laws are somewhat forgiving on that since stress depends on strain rate raised to a low exponent. Greater problems are that not all localized frictional sliding is unstable and that the distribution of creep strains cannot really be analyzed independently of the episodic stressing pulses (and hence rapid transient creep) delivered by earthquakes to the lithosphere below.

The somewhat more recent concept is that of a transition from *potentially unstable* to *inherently stable* but still localized friction (Tse and Rice 1986; see also Scholz 1990, 1998). This built on earlier, unpublished, modeling concepts by Mavko, on the Brace and Byerlee (1970) results of an absence of stick slip at higher temperatures, and on high-temperature friction data from Stesky et al. (1974) and Stesky (1978) (see Tse and Rice 1986). This approach assumes that over some depth range extending below that where earthquakes can nucleate, the deformation is strongly localized, so that we can discuss response in terms of a constitutive relation between slip rate, temperature, appropriate state variables, and stress. Laboratory studies then show that the constitutive law may exhibit either steady-state rate-weakening ($a - b < 0$, in the formulation given below), or rate-strengthening ($a - b > 0$). If the latter is the case then, at least for the simple one-state-variable class of constitutive laws, earthquakes cannot nucleate. If $a - b < 0$, then they can nucleate provided that a large enough patch of fault—greater than the "nucleation size" (see below)—is made to slip. Thus the depth of seismogenesis is, from this view, not limited by a transition from *localized friction to distributed creep*, but rather by a transition from *potentially unstable* $(a - b < 0)$ to *inherently stable* $(a - b > 0)$ localized friction. Based on data from Blanpied et al. (1991, 1995) that transition is expected to take place, for a wet granitic gouge composition, at temperatures around 350°C. The difficulty here is that the model assumes localized deformation at all depths considered. We must therefore ask if there is a region slightly downdip of the seismogenic zone in which deformation, while possibly being indeed localized when large earthquakes rupture downward into it, does nevertheless exhibit broadly distributed deformation throughout the interseismic period (see related discussion in Chapter 6).

Both approaches result in a temperature limit for depth of seismogenesis, which can be made to agree within the uncertainty in choosing constitutive parameters and might reflect the same micromechanisms (e.g., onset plastic flow in wet quartz). Also, both mechanisms contribute to limiting seismic ruptures to shallow depths, because they allow for continuing interseismic creep deformation below the locked seismogenic zone. That means the stress prevailing beneath the seismogenic zone at the time of an earthquake is smaller than it must be to allow rapid frictional slip to begin. Thus as the tip of a propagating rupture begins its downward penetration into the hotter material, a negative stress drop (stress becomes higher during rupture) develops that weakens the stress concentration at the tip and soon stops the downward penetration.

Rate and State Frictional Constitutive Laws

These laws begin with the empirical observations and formulations by Dieterich (1978, 1979) and Ruina (1983) and focused on the slow slip range appropriate for nucleation of slip instability under slowly increasing load. They were soon applied broadly to interpretation of laboratory experiments (Tullis 1986), crustal earthquake sequences (Tse and Rice 1986), and descriptions of aftershocks and induced seismicity rate changes (Dieterich 1994). The basic form for these laws, as understood more recently (see Rice et al. 2001, and references therein) is that there is a thermally activated slip process at the stressed microscopic asperity contacts and that contact properties evolve with the maturity of a contact, in a way that strengthens the contact with increase in its age θ. Frictional strength is regarded as being due to atomic bonding (e.g., like at a defect-rich, high-angle grain boundary) at those contacts. The coherence of the contact increases with its age θ, whether due to creep within the contact region which drives the contact towards a less misfitting, lower energy, boundary configuration at the atomic scale, or to desorption of impurities (e.g., water molecules) which were trapped at the contact at levels beyond their equilibrium concentration when it was first formed, and whose desorption thus also lowers energy of the boundary. Those processes would cause the activation energy E for shear processes within the contact to increase with age θ. Also, local creep flow in the vicinity of the contact, at the scale of the contact diameter, can allow the contact area to grow in time (e.g., Dieterich and Kilgore 1996) so that, for a given macroscopic normal stress σ, the average normal stress σ_c at the contacts decreases with age θ.

These contacts may be between roughness asperities on nominally bare surfaces or may be the places where particles of a gouge contact one another. Shear strength τ_c is very high at the contacts and is estimated to be of order $0.1 \times$ shear modulus μ in minerals in which dislocation motion is difficult. (Such estimates can be obtained by using $\tau / \tau_c = \sigma / \sigma_c$, where τ and σ are the macroscopic shear and normal stresses and where both sides of the equation correspond to the ratio of contact area to nominal area, and by measurement of friction coefficient $f = \tau / \sigma = \tau_c / \sigma_c$ and estimate of σ_c from microhardness measurements or, consistently in transparent materials, from optical inference of true contact area; Dieterich and Kilgore 1994, 1996). For the constitutive law, we consider atomic scale thermally activated jumps over energy barriers within the contact zone, with activation energy written as $E - \tau_c \Omega$. Here E is the barrier in absence of bias by the contact shear stress, and Ω is an activation volume.

Thus, if V_1 is the pre-factor in an Arrhenius description (V_1 is estimated to be of order shear wave speed times the fraction of the contact area that slipped by a lattice spacing in an elementary activated event; Rice et al. 2001), the slip rate is

$$V = V_1 \left(\exp[-(E - \tau_c \Omega) / k_B T] - \exp[-(E + \tau_c \Omega) / k_B T] \right) \quad (5.1)$$

The second term, representing backward jumps (in direction opposite to the driving shear traction) is generally negligible except at small positive, or at negative, applied stress. We can usually neglect it. Using $f = \tau / \sigma = \tau_c / \sigma_c$ as above, the Arrhenius law, with backward jumps neglected, is thus equivalent to writing $f = a \ln(V / V_1) + E / \sigma_c \Omega$, which is of a familiar structure in the rate and state formulation. The direct-response parameter a, characterizing the response to a sudden change in V at fixed state of the contacts, is thereby identified as $a = k_B T / \sigma_c \Omega$. This relation together with experimental constraints on a, f and σ_c for quartzite and granite at room temperature led to estimates of Ω equal to a few atomic volumes and $E \sim 1.7–1.8$ e.v. in Rice et al. (2001). Those results had to rely on an estimate of the pre-factor V_1 and could be improved by appropriate experiments to determine response to a stress jump as a function of temperature.

The intrinsic resistance of the contact to shear, represented by E, will increase with the maturity of the contact, and those maturing processes will depend on temperature of the fault zone, on the fluid environment outside the contact, and, of course, on the lifetime θ of the contact. Likewise, growth of the contact area due to creep processes taking place on the scale of the contact diameter corresponds to a reduction in σ_c with increase of θ. The more precise elucidation of those variations with θ is an important goal for tribological research. Assuming that the temperature and external fluid environment remain unchanged during the lifetime of a given contact, we may then assume that $E/\sigma_c\Omega$ is some increasing function of contact lifetime θ; such a function must presently be represented empirically. That naturally introduces a state variable θ into the formulation and, to simplify, it is usual to associate θ with the average lifetime of the contact population. (It may be more fundamental to have the parameters in $E/\sigma_c\Omega$ depend not directly on θ but rather on a measure of lifetime that is stretched or contracted according to temperature, e.g., like $\theta\exp(-Q/k_BT)$ (see Blanpied et al. 1995, 1998, for a related concept). Also, it is well known (Ruina 1983; Tullis 1986) that more than one state variable does better than one at fitting experiments; that may reflect either diversity in the contact populations or presence of more than one maturing process, or both.) Making the assumption that $E/\sigma_c\Omega$ grows logarithmically with contact age θ then leads to a law which can be put into the form of the classic Dieterich-Ruina "ageing," or "slowness," law $f = f_0 + a\ln(V/V_0) + b\ln(\theta/\theta_0)$, where b is a new dimensionless parameter (actually, a function of T). Here V_0 is an arbitrarily chosen reference value, θ_0 is the contact lifetime during sustained steady sliding at rate V_0, and f_0 is the associated steady-state friction coefficient. The evolution law for θ is now postulated as a simple law which meets the requirements that (a) $d\theta/dt = 1$ when $V = 0$, and (b) θ scales inversely with V in sustained steady sliding (so that in steady state $V\theta = L$, a constant—often denoted by d_c—equal to the sliding distance to renew the contact population; then $f = f_0 + (a - b)\ln(V/V_0)$ in steady state). A simple law which accomplishes these features is the one normally used in the ageing formulation (Ruina 1983), namely, $d\theta/dt = 1 - V\theta/L$. The above logarithmic form in the expression for f is not sensible for V near zero or negative, but we can handle all such cases by not then neglecting the backward jumps in the original Arrhenius law, leading to the more general expression $f = a\,\text{arcsin}\,h((V/2V_1)\exp[E/a\sigma_c\Omega])$; that is the version, with logarithmic dependence of $E/\sigma_c\Omega$ on θ like assumed above, sometimes used in simulations of earthquake sequences (e.g., Rice and Ben-Zion 1996; Lapusta et al. 2000).

It will be important to determine how to generalize those concepts to the regime of rapid slips, at rates much greater than those examined in developing the rate and state formulation discussed above, a subject on which there has been some attention already (Prakash 1998). Some rapid-slip models, below, have so far advanced to giving the steady state f as a function of slip rate V, but a description of friction in the form $f = f(V)$ with rate-weakening ($df(V)/dV < 0$) has been shown to either provide an ill-posed model, for which no mathematical solutions generally exist to problems of sliding between elastically deformable continua, or in a limited parameter range when solutions do exist, to predict nonobserved phenomena like rupture fronts that propagate faster than the fastest elastic wave speed (Rice et al. 2001). Thus it is critical to bring any viable constitutive description with steady state rate-weakening to the form $f = f(V, \text{state})$ where "state" represents some set of evolving state parameters like θ and perhaps also contact temperature, and where $a > 0$ (where we recall that $a = V\partial f(V, \text{state})/\partial V$).

Existing treatments of fault response under variable normal stress, following Linker and Dietrich (1992), attempt to map changes in normal stress into changes in θ, because those changes alter the contact population, but more fundamental work is needed too on that. Further, in existing formulations of rate and state friction, it is considered that the deformation is always localized to some thin zone of dimension set by smaller-scale physics, so that overall slip, and not the strain distribution through the thickness of the fault zone, is the only needed descriptor of deformation. However, an issue still to be resolved arises when considering the shear of granular layers that show steady-state rate-weakening. Then because $a > 0$, it is possible that response to a rapid increase in overall slip rate would create broad shear strain throughout the granular layer, whereas the steady-state rate-weakening would take over in sustained slip at that faster rate, promoting highly localized shear. Such effects, to the extent that they matter, have not yet been incorporated into existing models.

Dynamic Weakening Processes (Thermal, Fluid) during Seismic Slip

The materials physics of dynamic weakening had been largely ignored in theoretical modeling of earthquake processes up to relatively recent times. It is now an area of vigorous research. Given that earthquake slips are often accommodated within thin zones, but that evidence of melting is not pervasive, especially at the shallow depths of activity represented by surface exposures, it is reasonable to suspect that strong weakening mechanisms must exist during rapid, large slip. A combination of observation and theory has now identified some important candidates, summarized here. There may be others. The first of these requires rapid but not necessarily large slip. All are discussed in primarily theoretical terms here, citing experimental evidence, although specific challenges for experimental resolution are outlined in the next section. Some of these mechanisms imply a rate-weakening of (steady-state) friction that is much stronger than can be inferred by extrapolating the above rate and state laws to the seismic regime. Such a stronger rate dependence is likely to have a significant role (Cochard and Madariaga 1996; Beeler and Tullis 1996; Zheng and Rice 1998; Nielsen and Carlson 2000) in inducing the *self-healing* rupture mode, which has been advocated on the basis of seismic slip inversions for large events by Heaton (1990). In that mode, slip at a point effectively ceases at a time after passage of the rupture front which is much smaller than the overall event duration.

Flash Heating and Weakening of Micro-asperity Contacts

As noted, shear strength τ_c is very high (estimated to be of order $0.1 \times$ shear modulus μ) at contacts in typical rock systems, and thus when forced to shear, they generate intense but highly localized heating during their limited lifetime, which is of order L/V (here again L is the slip needed to renew the asperity contact population, and V is slip rate). If slip is fast enough, the significantly heated zone is just a thin (relative to contact diameter) region adjoining the contact. The contact's shear strength is diminished by temperature increase, but because the affected zone is thin, the capacity of the contact to support normal stress, and also the net area of contact, are not much affected. Thus the friction coefficient reduces with slip rate V. An elementary first model (Rice 1999, 2004) considers contacts of uniform size L, hence lifetime L/V, and assumes that their shear strength remains at the low-temperature value τ_c until temperature has reached a weakening value T_w, above which shear strength is taken to be zero. (Their temperature rise is estimated from a simple one-dimensional heat conduction analysis, with heating rate $\tau_c V$ per unit area at the sliding contact interface.) The modeling thus identifies a critical slip rate V_w, such that there is no weakening if $V < V_w$, but strong weakening if $V > V_w$. That is, the friction coefficient f (precisely, a *steady state* friction coefficient at slip rate V), which has the value f_0 at low slip rates, is given in this simple model by

$$f = f_0 \;\; \text{if} \;\; V < V_w \;, \;\; f = f_0 \frac{V_w}{V} \;\; \text{if} \;\; V > V_w \;;$$

$$\text{here} \;\; V_w = \left(\frac{\pi \alpha}{L}\right)\left[\frac{\rho c (T_w - T_f)}{\tau_c}\right]^2 \tag{5.2}$$

where α is thermal diffusivity, ρc is heat capacity per unit volume and T_f is the average temperature of the fault surface; T_f increases gradually due to the heat streaming in at the sliding contacts. [Beeler and Tullis (2003) and Beeler et al. (2006) assume that some low contact strength is retained at $T > T_w$ and modify the latter to $f = f_w + f_0 V_w / V$ where $f_w < f_0$.] Goldsby and Tullis (2003) and Rice (1999, 2006) have estimated V_w to range from 0.1 to 0.5 m s^{-1}, when it is recognized that $\tau_c \sim 0.1\mu$. Since the average slip rate in an earthquake is thought from seismic slip inversions (Heaton 1990) to be of order 1 m s^{-1}, the theoretical expectation is that f is reduced significantly from its low speed value f_0 during seismic slip. Laboratory experiments imposing rapid slip (Tsutsumi and Shimamoto 1997; Hirose and Shimamoto 2005; Goldsby and Tullis 2003; Tullis and Goldsby 2003; Prakash 2004) are indeed consistent which such an anticipated friction reduction at higher V. The results suggest the possibility that f of order 0.2 to 0.3 may prevail at average seismic slip rates for rocks whose low-speed f is of order 0.6 to 0.7. The weakened f implies a slower heating rate (i.e., that more slip is needed to achieve a given temperature change). When combined with the next thermal weakening mechanism, this leads to effective values of the slip-weakening parameter D_c (see later) which may reach a size ~0.2 m (Rice 2004, 2006).

Thermal Pressurization of Pore Fluid

This mechanism (Sibson 1973; Lachenbruch 1980; Mase and Smith 1985, 1987) assumes that fluids (water, typically) are present within the fault gouge which shears, and that the shear strength τ during seismic slip can still be represented by the classical effective stress law $\tau = f(\sigma - p)$, where σ is normal stress and p is pore pressure (see Cocco and Rice 2002). Frictional heating then would cause the fluid, if it was unconstrained, rather than caged by the densely packed solid particles, to expand in volume much more than would the solid cage. Thus, unless shear-induced dilatancy of the gouge cage overwhelms the thermal expansion effect, or unless the gouge is highly permeable, a pressure increase must be induced in the pore fluid. Since σ typically remains constant during slip, strength τ is reduced, ultimately towards zero, as shear heating continues to raise temperature so that p approaches σ. Calculations by Rice (2006) evaluated this mechanism using permeability and poroelastic properties and shear zone thicknesses based on properties of the Median Tectonic Line fault (Wibberley 2002; Wibberley and Shimamoto 2003), Nojima fault (Lockner et al. 2000), and Punchbowl fault (Chester and Chester 1998; Chester and Goldsby 2003; Chester et al. 2003). Rice (2004) as well as Cocco and Bizzarri (2004) found that predictions based on thermal pressurization enabled plausible estimates of the fracture energy of earthquakes, as have been established independently in seismological studies (for recent summaries, see Abercrombie and Rice 2005, Rice et al. 2005; Tinti et al. 2005; Rice 2006; Bizzarri and Cocco 2006a, b), and could explain why strength loss over all but deeper portions of crustal seismogenic zones is too rapid for melting to take place. That seems consistent with general conclusions that fault zone pseudotachylites of tectonic earthquake origin have generally formed deep in the seismogenic zone.

An illustration is provided by the analysis (following Rice 2004, 2006) for slip at speed V on the fault plane $y = 0$ (zero thickness shear zone) in a poroelastic solid under constant normal stress σ_n. With standard simplifications, the pore pressure p and temperature T then satisfy:

$$\text{In } |y| > 0 \ , \quad \frac{\partial T}{\partial t} = \alpha_{th} \frac{\partial^2 T}{\partial y^2} \quad \text{and} \quad \frac{\partial p}{\partial t} - \Lambda \frac{\partial T}{\partial t} = \alpha_{hy} \frac{\partial^2 p}{\partial y^2} \ , \quad (5.3)$$

with conditions on

$$y = 0^\pm \ , \quad -\rho c \alpha_{th} \frac{\partial T}{\partial y} = \pm \frac{1}{2} f(\sigma_n - p)V \quad \text{and} \quad \frac{\partial p}{\partial y} = 0 \ .$$

Here α_{th} and α_{hy} are the respective thermal and hydraulic diffusivities, Λ is the value of dp/dT under undrained conditions, and ρc is the specific heat per unit mass. The two partial differential equations express energy conservation, assuming conductive heat transfer but neglecting advective transfer, and conservation of fluid mass during the increase of T and p and Darcy transport. The boundary conditions express that the frictional work rate provides the heat input at the fault, and that there is no fluid outflow from a vanishingly thin zone. These can be solved for the case of constant V and f, and when the solution is written in terms of slip δ ($= Vt$), the p and T on the fault plane $y = 0$ are

$$p - p_0 = (\sigma_n - p_0)\left[1 - \exp\left(\frac{\delta}{L*}\right)\mathrm{erfc}\left(\sqrt{\frac{\delta}{L*}}\right)\right]$$

$$T - T_0 = \left(1 + \sqrt{\frac{\alpha_{hy}}{\alpha_{th}}}\right)\frac{p - p_0}{\Lambda}$$

(5.4)

where $$L* = \frac{4}{f^2}\left(\frac{\rho c}{\Lambda}\right)^2 \frac{\left(\sqrt{\alpha_{hy}} + \sqrt{\alpha_{th}}\right)^2}{V},$$

and where p_0 and T_0 are the initial values; T_0 would be the ambient temperature but, if we sensibly assume that the onset of shear should be associated with some inelastic dilatancy of material near the fault zone, inducing a sudden suction in it (Segall and Rice 1995), then p_0 should be assumed to be reduced from the ambient value by that suction. The shear stress transmitted across the fault plane is thus predicted to be

$$\tau = f(\sigma_n - p) = f(\sigma_n - p_0)\exp\left(\frac{\delta}{L*}\right)\mathrm{erfc}\left(\sqrt{\frac{\delta}{L*}}\right).$$

(5.5)

This shows continued weakening at an ever-decreasing rate over a very broad range of size scales, Figure 5.2. How large is $L*$, the single length scale which enters into the description of the weakening process under the conditions considered? Lachenbruch (1980) gives $\rho c \approx 2.7$ MPa °C^{-1} and $\alpha_{th} \approx 1$ mm^2 s^{-1}. Supplementing his data set by results from Wibberley (2002) and Wibberley and Shimamoto (2003) for gouge from the central slip zone of the Median Tectonic Line at 130 MPa effective confining stress ($\approx$ effective overburden at 7 km depth, for hydrostatic pore pressure), with porosity 0.04 (Wibberley, pers. comm.), gives $\Lambda \approx 0.8$ MPa °C^{-1} and $\alpha_{hy} \approx 1.8$ mm^2 s^{-1}, using a permeability of 10^{-20} m^2. That gives, with $V = 1$ m s^{-1} and $f = 0.25$ to allow for flash heating, $L* = 4$ mm. That estimate directly uses lab data on undisturbed gouge samples, whereas in the natural situation there may be initial dilatant deformation and

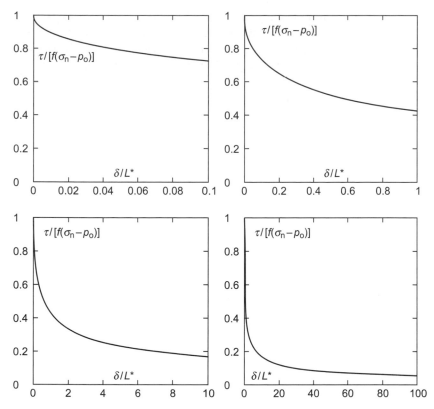

Figure 5.2 Prediction of shear strength τ versus slip δ, due to *thermal pressurization of pore fluid* during slip on a plane, at constant rate V and with constant friction coefficient f, in a fluid-saturated solid. The four panels show the solution in Eq. 5.5 for different δ/L^* ranges, extending from 0 to, respectively, 0.1 (upper left), 1.0 (upper right), 10 (lower left), and 100 (lower right). After Rice (2004, 2006). Note the multiscale nature of the weakening. Parameter L^* (see text) is estimated to be in range 4–30 mm at representative centroidal depths, ~ 7 km, of the slipping region during crustal earthquakes. Here σ_n is fault-normal stress and p_o is the pore pressure just after its reduction from ambient pressure by any dilatancy at onset of shear.

damage near the rupture front (e.g., Poliakov et al. 2002; Rice et al. 2005) at the start of shear. As guesses as to how such effects might change parameters, a tenfold greater permeability, 10^{-19} m^2, and doubled pressure-expansivity of the pore space (making $\Lambda \approx 0.5$ MPa °C^{-1} and $\alpha_{hy} \sim 12$ mm^2 o^{-1}), would instead give $L^* = 33$ mm. (These estimates were adopted from a preliminary version, Rice (2004), of those published in Rice [2006]. The final 2006 version has a much fuller analysis of the underlying data, including its T and p dependence. The resulting poro-thermo-elastic parameter values in the final version are different from those just quoted but, nevertheless, $L* = 4$ mm and 30 mm remain plausible "low end" and "high end" values, as representative of properties of intact gouge and of gouge with the guessed effects of damage as above, respectively.) Thus the upper right panel in Figure 5.2, showing slip up to $\delta = L*$, could correspond to a maximum slip between 4 and perhaps 30 mm slip, and the lower right panel, showing slip up to $\delta = 100L*$, to a maximum slip between 0.4 and 3 m.

Bizzarri and Cocco (2006a,b) have performed three-dimensional simulations of the spontaneous nucleation and propagation of a dynamic rupture governed either by slip-weakening or rate- and state-dependent laws (defined in more detail below), thus including a cohesive zone where dynamic weakening occurs and accounting for frictional heating (by solving the heat flow equation; see Fialko 2004) and thermal pressurization of pore fluids (by solving the above fluid diffusion equation, like in Andrews 2002). These authors analyzed the traction evolution as a function of time or slip for different values of fault zone thickness and hydraulic diffusivity. In these simulations, slip velocity evolves spontaneously and is not constant. Bizzarri and Cocco (2006a,b) have shown that fault zone thickness and hydraulic diffusivity modify the shape of the traction versus slip curves and affect the stress drop and the critical slip-weakening distance (D_c, defined below; see Figure 5.4). For particular configurations they found that traction evolution shows a gradual and continuous weakening such as those shown in Figure 5.2 and predicted by Eq. 5.5.

The fracture energy G associated with an event with slip δ may be calculated from a slip-weakening function $\tau = \tau(\delta)$ by

$$G = G(\delta) = \int_0^\delta [\tau(\delta') - \tau(\delta)] d\delta' \, ,$$

that definition (Abercrombie and Rice 2005) generalizes Ida (1972) and Palmer and Rice (1973) to forms of $\tau(\delta)$ for which weakening continues at an ever-decreasing rate to slips greater than that in the event. Then, for the above $\tau(\delta)$, there results (Rice 2006)

$$G(\delta) = f(\sigma_n - p_o)L* \left[\exp\left(\frac{\delta}{L*}\right) \text{erfc}\left(\sqrt{\frac{\delta}{L*}}\right)\left(1 - \frac{\delta}{L*}\right) - 1 + 2\sqrt{\frac{\delta}{\pi L*}} \right] . \quad (5.6)$$

This is plotted as the solid lines in Figure 5.3, for the low and high poromechanical estimates of $L*$ above, using $V = 1$ m s^{-1}, $f = 0.25$ as an approximate representation of flash heating (Hirose and Shimamoto 2005; Prakash 2004; Goldsby and Tullis 2003), and $\sigma_n - p_o = 126$ MPa (initial effective overburden at 7 km depth, a representative centroidal depth for slip during crustal events). Seismic data assembled recently by Abercrombie and Rice (2005) and, for the seven large events of Heaton (1990), by Rice et al. (2005) is also shown there. The seismic data and theoretical modeling have many uncertainties.

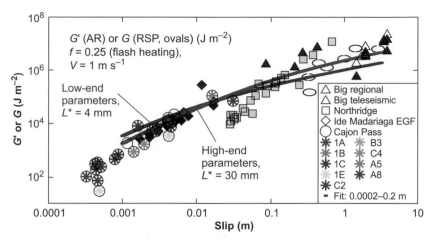

Figure 5.3 Lines show theoretical predictions of earthquake fracture energy G versus slip δ in the event, based on combined effects of *thermal pressurization* of pore fluid and *flash heating*, with simplified representation at constant friction coefficient f and slip rate V. Symbols represent estimates of G from seismic data. The basic plot, to which the curves and oval symbols have been added, is from Abercrombie and Rice (2005); it shows their parameter G', thought to be of the same order as G ($G' = G$ when final dynamic sliding strength and final static stress coincide). Oval symbols at large slip from Rice et al. (2005) for the seven large earthquakes with slip inversions reported by Heaton (1990). (The figure is a preliminary version, taken from Rice (2004), of a corresponding figure in Rice (2006); the 2006 version compares theoretical predictions to an enlarged seismic data set for large earthquakes, including results from Tinti et al. (2005) and other sources.)

Nevertheless, the rough coincidence does lend credibility to the concept that flash heating together with thermal pressurization of pore fluid may be dominant processes in fault weakening during earthquakes. (Figure 5.3 is from a preliminary version, taken from Rice (2004), of the corresponding figure in Rice (2006); the final 2006 version is similar but compares to a larger seismic data set including results from Tinti et al. (2005).)

There is a critical need for laboratory tests of this mechanism. Also, it is important to understand if and how it might operate in conditions that might exist at midcrustal depths, possibly involving an extensively mineralized pore space with isolated, unconnected pockets of liquid water, for which the classical dependence of strength on $\sigma - p$ might then not hold. For example, if shear of such a zone has been initiated by a propagating rupture tip, so that a possibility of fluid connectivity is reestablished, can it then be assumed that the effective stress characterization of strength, $\tau = f(\sigma - p)$, holds again? Further, we need to understand how to represent the shear strength of a dense, rapidly shearing, gouge with $\sigma = p$. Is it sufficiently small to be negligible compared to $f(\sigma - p_{initial})$? Or does it represent a substantial fraction of that value? Can the condition of p

approaching σ actually be achieved on a real fault? Or will any such highly pressurized pore fluids hydraulically crack their way into the already damaged walls of the fault, (since p will then be at least as great, and generally greater, than the least principal compressive stress)? Will such hydraulic cracking effectively increase permeability and bring the fault weakening to a halt (Sibson 1973), perhaps abruptly halting slip?

Finally, it is essential to constrain better the dilatancy of a gouge under shear (Marone et al. 1990; Segall and Rice 1995), and also the effect of dilatancy and shear on the instantaneous permeability and poroelastic moduli. Those values are essential inputs for quantitative estimates of fault weakening by thermal pressurization, but at present they are known only for nondeforming gouge. It is also essential to understand how the evolution of porosity is related to that of surface contact properties, if we aim to constrain a constitutive behavior for the principal slipping zone, as well as to determine where porosity evolution is of relevance (within the damage zone, the fault core, or the slip zone).

Silica Gel Formation

This mechanism, reported by DiToro et al. (2004) (see also Goldsby and Tullis 2002; Roig Silva et al. 2004), and supported by observations of fracture surface morphologies, applies most directly for large slip (>0.5–1.0 m) and moderately rapid slip (>1 mm s^{-1}), in the presence of water. It assumes that there is a silica component (quartz) within the shearing zone. Present understanding is linked to the observation in friction experiments on a quartzite, Arkansas novaculite, of "now solidified ... flow-like textures that make it ... evident that at the time the deformation was going on, a thin layer coating the sliding surface was able to flow with a relatively low viscosity" (Tullis, pers. comm.). The concept is that granulation within the shear zone produces fine silica particles which adsorb water to their surfaces and form a gel. It is weak but would gradually consolidate into a strong, amorphous solid if shear was stopped. However, the presence of shear continuously disrupts particle bonding (thixotropic response) so that the fluidized gel mass deforms at low strength. Fuller limits to their range in which the mechanism is active have not yet been identified, and it is not known if it could contribute during seismic slip. Nevertheless, for a given shear and within a given velocity range for which the mechanism was plausibly established for pure quartzite rocks (Arkansas novaculite), weakening for other rock types seems to be ordered by their silica content (Roig Silva et al. 2004):

quartzite (novaculite) > granite > gabbro.

Granite and a pure albite-feldspar rock with nearly identical silica content show essentially identical weakening, although the "solidified flow structures have so far only been seen for novaculite" (Tullis, pers. comm.). A gel can be thought of for some purposes as a water-infiltrated porous medium. That raises the question of whether there could be any connection between this weakening mechanism and the Sibson (1973) mechanism of weakening by thermal elevation of pore fluid pressure. Nevertheless, even if so, a new ingredient here is the strong thixotropic response, causing a fractionally much greater regain of strength after cessation of shear than what might be expected that a granular gouge would show on a comparable timescale.

Melting

This is the ultimate mechanism of thermal weakening, but it is not a simple mechanism. Comparison of strength during rapid shear in the range prior to, versus shortly after, the transition to macroscale melting (i.e., when a coherent melt layer has formed along the whole sliding surface) shows that at or near the transition, there is an abrupt increase in frictional strength (Tsutsumi and Shimamoto 1997; Hirose and Shimamoto 2005). At least, that increase is observed for the small normal stresses σ in experiments reported thus far; Tsutsumi and Shimamoto (1997) find that, at $\sigma = 1.5$ MPa, the friction coefficient $f = \tau/\sigma$ shoots up to ~0.9. Similar behavior is known in other parts of the tribological literature. There is complex response on the way to the transition: Sizeable blobs of melt form near the larger frictional contact asperities, as an extreme form of the flash heating process. These get smeared out along the sliding surfaces and rapidly solidify at least when the average temperature of those surfaces is low enough, so that the surfaces are spot-welded together. During macroscopic melting, the fault strength is $\tau = \eta V/h$ where h is the thickness of the shearing layer and viscosity $\eta = \eta(T$, melt composition). In the early phases of macroscale melting, h is small and T is low, compared to what it will become with continued shear, and thus τ is relatively high. It is only with increasing h, so that shear rate decreases, or with increasing T, so that η decreases, that the melted fault zone weakens to a friction level comparable to what it showed (due to flash heating) at a comparable slip rate shortly before the transition to macroscale melting.

Important problems for the future are to understand and quantify shear resistance in the transition range (Hirose and Shimamoto 2005; Di Toro et al. 2006) and the macroscale melting range. With the present understanding, as outlined here, if water or other fluids are completely absent, we would have only frictional heating (at a reduced f due to flash heating) and then a transition to melting. Melting itself is complex and takes place under strongly non-equilibrium conditions. For example, whereas a hydrous granitic composition could be expected to equilibrium melt over an ~50°C interval, pseudotachylite studies (Otsuki et al. 2003; Di Toro et al. 2005) suggest a 500–600°C melting interval (starting at 750°C) for the melting of a granitic composition on the earthquake timescale (~1 s for 1 m of slip, since the average V is inferred to be ~1 m s^{-1}; that is based on seismic slip inversions for large events, Heaton 1990, and is obtained by dividing his inferred slip by inferred slip duration at a point). For the macroscale melting range, a primary issue is to describe the evolution of h, T, and melt composition (including reactions with initial pore water of the fault zone). Presumably h is set by the balance between melt generation and either freezing or loss by melt injection into cracks in the fault wall. Concerning injection, we must recognize that the melt layer is under pressure $p = \sigma$, the fault-normal stress, and that in general the propensity to inject should increase with the increasing difference between σ and the least principal compressive stress at the fault wall.

New Advances Needed in Laboratory Experiments on Fault Friction

Rapid Slip

New advances have been reported quite recently in achieving slip in rocks at high rates, approaching or exceeding what is thought to be the average slip rate of order 1 m s^{-1} during seismic instability. The work achieves rates up to 0.3 m s^{-1} (Goldsby and Tullis 2003) using a hydraulically-driven mechanical testing machines capable of torsional loading, and rates of 3 to 30 m s^{-1} (Prakash 2004) based on sudden unloading of a pre-torqued Kolsky bar or, for the higher rates, adaptation of the oblique shock impact arrangement of Prakash and Clifton (1992) and Prakash (1998). These new approaches have the promise in opening a new chapter in making laboratory rock friction studies relevant not just to nucleation of unstable slip, but to characterizing the fault constitutive response during the dynamic rupture process itself. They, or such other new experimental techniques as may be devised, need to be more widely adopted to develop this important phase of the subject. They will be needed to confidently characterize such mechanisms of weakening as flash heating at microscopic asperity contacts and thermal pressurization of pore fluids discussed above.

Fault Strengthening

A severe limitation of current frictional constitutive modeling in the rate and state framework is the almost completely empirical manner by which time dependent fault strengthening is brought into the description. For example, in the commonly adopted ageing, or slowness, version of the Dieterich-Ruina friction law discussed earlier, f is assumed to increase in proportion to $\log(\theta)$ where θ is interpreted as contact lifetime (or at least as its average value). The physical basis of actual healing, and its dependence on θ or other appropriate variables to be identified, needs to be better quantified by appropriate experiments. The physical chemistry of water is relevant here; water-assisted processes have been shown, in variable humidity studies, to play a major role in time-dependent strengthening at least at room temperature (Dieterich and Conrad 1984; Frye and Marone 2002). Exclusion of water seems to remove most of that time-dependent strengthening. In the

Earth, very long times of (nearly) stationary contact are of interest for the interseismic period, and their effect on strengthening needs fuller experimental clarification, and basic physical understanding, at temperatures up to those slightly higher than for the seismogenic depth range. Significant contact strengthening by mineral deposition, alterations by reactions with fluids (the kinetics of hydration is critical for that; see Chapter 12), and creep flow may be important; those processes are also of interest for overall energy balances in faulting. Creep compaction of fine fault gouge and depositional processes may also seal off fluids into noncommunicating pore spaces and drive pore pressure above hydrostatic values, allowing low-strength shear (Sleep and Blanpied 1992). Constitutive response after such long hold times at high temperature is directly relevant to nucleation of large events (e.g., Tse and Rice 1986; Lapusta and Rice 2003), which usually occurs towards the lower depth range of the effectively locked seismogenic zone. It is also relevant to evaluating aftershock production (Dieterich 1994).

Localized versus Distributed Shear

In many attempts to represent fault gouge effects, by shear of granular layers in the lab (e.g., Marone 1998; Scruggs and Tullis 1998), there is a variation in deformation response between stable shearing that may span the full width of the granular layer versus localized shearing that may take place on a family of imperfectly aligned shear structures, or may take place on a through-going surface. This remains an important process to characterize and understand, especially at higher shear rates typical of seismic stressing, for which thermal weakening processes may contribute to localization in an otherwise stable granular shear flow. Beeler et al. (1996) proposed that considering the combined dependence of strength on slip and slip rate might be of relevance and developed a relationship between the change of slip zone thickness and the strength change in such a formulation.

A more fundamental starting point might be to begin with a formulation in terms of shear rate and evolving state parameters, including porosity, of the fault gouge. Nevertheless, in any such study, we run up against a pervasive difficulty, recognized for many years in the macroscopic mechanics of ductile and granular materials, which is that of representing the full transition from distributed deformation to highly localized deformation like in a shear band or fault. This must generally be addressed by appeal to length parameters of the smaller-scale physics that limits localization zone thickness, for example, by adding positive-definite strain gradient terms to the strength expression (by dimensional considerations, their coefficients will necessarily introduce a length scale), or using a spatially nonlocal relation between strength and deformation, or by setting the cell size of a finite element or finite difference computational grid to the length parameter, or in cases for which it is appropriate like adiabatic shear localization in temperature-weakening materials, by including transport phenomena (e.g., α_{th}/V provides a length scale). In all but the latter approach, the present level of understanding is such that these must generally carried out as ad hoc procedures, often without clear identification of the relevant microscopic physics and length scales, or convincing demonstration that it leads to the localization-limiter procedure adopted. It is generally accepted that localized shear zones span ~5–30 $d_{50\%}$ in aggregates of relatively equiaxed granular particles like for sands (Desrues and Viggiani 2004), and possibly as much as 200 $d_{50\%}$ in fine-grained clays with platelet particles (Morgenstern and Tschalenko 1967; Vardoulakis 2003), where $d_{50\%}$ is the particle diameter dividing the aggregate into two equal masses. However, we do not presently know how such results generalize to real fault gouges, with a wide particle size distribution (Chester et al. 1993; Ben-Zion and Sammis 2003) from, say, 0.01–100 μm (the diameter to be used is unknown), with particles that are often angular and are susceptible to cracking.

Variable Normal Stress

This arises in slip propagation along dissimilar material interfaces (Weertman 1980; Andrews and Ben-Zion 1997; Harris and Day 1997; Cochard and Rice 2000; Xia et al. 2005), because gradients in slip then induce changes in normal stress, and also when the propagating front of a rupture encounters a bend or branch in the fault path (Poliakov et al. 2002; Kame et al. 2003), rapidly increasing both normal and shear stress there. It is also relevant to rapid alterations of pore pressure due to shear heating and thermal pressurization of pore fluid. There remains conflicting evidence on its effect of frictional strength: Precise laser diagnosis of oblique shock wave experiments at slip rates of order 10 m s^{-1} involving abrupt normal stress changes by a reflected sharp shock front (Prakash and Clifton 1992; Prakash 1998) suggest that there is no correspondingly abrupt change in shear strength, but only an evolution with continuing slip (over a few μm scale). In contrast, conventional slow friction studies with much less abrupt change of normal stress (Linker and Dieterich 1992; Boettcher and Marone 2004) suggest that shear strength changes on the same timescale as the normal stress change, but changes only partly towards what it will ultimately evolve to after a few μm of further slip.

Scaling from Lab to Natural Faults

It is commonly asserted that results of laboratory experiments must be "scaled," in some manner yet to be determined, to the geometrically much larger natural fault scale. That would surely be a valid point of view if the lab experiments were done on other than natural fault materials (or hopefully similar lithologies), and were done to produce laboratory analogs of crustal earthquakes. However, the laboratory studies that we have discussed are aimed at determining local constitutive response of fault materials. A viable proposition to be discussed is that they require no scaling whatsoever and are directly applicable to the Earth, as descriptors of local response on a fault. All that may be called "scaling" is then just a description of those predictions which result when appropriate boundary and initial value problems are set for mathematical models, incorporating those constitutive relations, to predict earthquake behavior. The tacit assumption, then, is that the experimental fault rock or gouge zone responds to any given history of slip rate and normal stress with the same history of shear traction as that which would occur along the actual fault zone. That is a reasonable assumption when there is a clear separation between the scales of microphysical processes determining the local response and the (presumably) more macroscopic scales over which the locally averaged stress and slip vary. An area of some uncertainty here, however, involves the fractal-like roughness of faults (Power and Tullis 1991). Friction response as we understand it seems to be controlled at the multi-micron size scale by properties of contacts. Is there then any effect, on what we use for constitutive laws, of the nonplanarity of faults at larger size scales? Or is it properly accommodated (not that such is yet done in practice) by regarding the normal stress supported by the fault as a variable which has an effectively random component at larger size scales?

That concept—that there may be no scaling to be imposed on laboratory (constitutive equation) results—should not be confused with a scaling-like compromise that fault modelers must generally make to fit their problems on current-day computer systems. Computational tractability of crustal scale earthquake models requires use of far larger constitutive length scales L in description of frictional weakening than can be justified experimentally. That is because the required size of numerical discretization cells, for a numerical solution to represent the solution of the underlying continuous system of equations on a rate-weakening fault, must be small compared to the size of a slipping patch at the transition from aseismic to seismic slip (which is called the nucleation size). That nucleation size scales roughly as (Lapusta et al. 2000) L times a large number, of order $5\mu/[(b-a)(\sigma-p)]$ where μ is the elastic shear modulus and other parameters are as above. At 10 km depth, identifying σ as overburden and taking p as hydrostatic, and taking $b-a=0.004$ (Blanpied et al. 1991), ends up meaning that the grid size should be small compared to $2 \times 10^5 L$, which would mean small compared to a 1–5 m nucleation size if $L = 5$–50 μm, representative lab values. Such resolution is unattainable in simulations which try to resolve the entire crustal scale. The conventional approach is then to choose an artificially large L so as to make the required discretization size large enough, and then to try to get some sense of what happens as L is reduced as close as one can get to the laboratory range. However, it has recently been realized that this is a nontrivial extrapolation, in that new populations of smaller events emerge near the downdip end of the seismogenic zone as L is reduced in size (Lapusta and Rice 2003), a complication which does nevertheless have compatibility with natural observations of earthquake locations.

REPRESENTATION OF FAULT CONSTITUTIVE BEHAVIOR FOR PREDICTIVE EARTHQUAKE MODELS

Slip Weakening

To use experimental and theoretical results on constitutive response, as in the last section, for predicting large scale fault rupture behavior, it is necessary to contend with the art of simplification, that is, of identifying simple, tractable modeling procedures, but not so simplified as to lose essential features. To that aim, a main result that is relevant to characterizing dynamic fault weakening during an earthquake is the traction evolution. Weakening during rupture propagation is now conventionally represented by the traction drop associated with slip increase (see Figure 5.4), resulting in the well known slip-weakening model. Different physical processes can yield a traction evolution consistent with that behavior (Cocco and Bizzarri 2002). This shear stress degradation during dynamic propagation occurs in a finite extended zone at the crack front called the cohesive zone (Barenblatt 1959; Ida 1972; Palmer and Rice 1973; see Figure 5.5). Although the shape of the slip-weakening curve can differ among different constitutive formulations, such a traction variation with slip must be common to any constitutive relation proposed to model rupture propagation. Because of its simplicity and in order to prescribe the traction evolution within the cohesive zone (see Figure 5.5), the slip-weakening model has been widely used as a

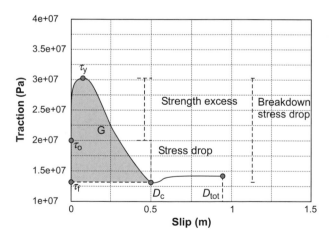

Figure 5.4 Traction evolution as a function of slip obtained by a numerical experiment of spontaneous dynamic propagation on a fault. This kind of evolution is common to different constitutive formulations (see Bizzarri and Cocco 2003, and references therein). τ_0 is the initial stress, τ_y the upper yield stress, and τ_f the kinetic friction level; the characteristic slip-weakening distance is D_c, and D_{tot} is the final slip value. The difference $(\tau_y - \tau_0)$ is usually named the strength excess, while $(\tau_0 - \tau_f)$ is the stress drop. The shaded area indicated by G yields an estimate of the fracture energy.

constitutive relation to model dynamic rupture with theoretical and numerical approaches (Andrews 1976a, b). The main parameters of this model are the initial stress (τ_o), the upper yield stress (τ_y) or peak strength, the kinetic friction level (τ_f) or residual strength, and the characteristic slip-weakening distance D_c (see Figures 5.4 and 5.5). Dependence of traction on slip has been observed in dynamic laboratory experiments (e.g., Ohnaka and Yamashita 1989).

A set of important questions arise as we try to unite this established rupture dynamics methodology with new physical understanding and laboratory documentation of friction behavior. This has been partly addressed, as we explain later, for interpreting dynamic predictions based on the now classical, logarithmic, rate and state friction laws in terms of slip-weakening concepts. However, the recent theories and experiments for the range of rapid, large slip involve, in some cases, much stronger steady-state rate-weakening at seismic slip rates than predicted by extrapolation of the logarithmic laws. Could that bring on new types of dynamic response, like inducement of self-healing of the rupture (Cochard and Madariaga 1996; Beeler and Tullis 1996; Zheng and Rice 1998; Nielsen and Carlson 2000), which might be obscured by mapping laws for the range of rapid, large slips into the slip-weakening framework? Also, some of the theoretical constitutive modeling of weakening in large slips by thermal

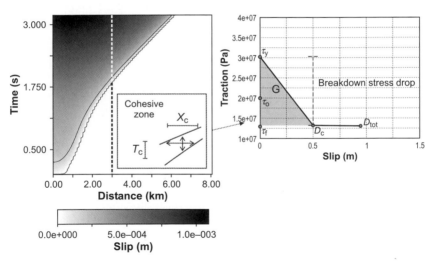

Figure 5.5 Spatiotemporal evolution of slip obtained by a numerical experiment of spontaneous dynamic propagation on a two-dimensional fault obeying to a slip-weakening law stated in Eq. 5.8 and shown in the right panel. This sketch allows the identification of the cohesive or breakdown zone, which is defined as the region shear stress degradation from the upper yield stress to the kinetic friction level. The spatial dimension of the cohesive or breakdown zone (X_c) is different from the critical slip-weakening distance D_c. The difference ($\tau_y - \tau_f$) is the breakdown strength drop in the terminology of Ohnaka and Yamashita (1989).

pressurization of fluid (Figures 5.2 and 5.4), as well as seismic attempts to look at the scaling of fracture energy with slip in an earthquake (Abercrombie and Rice 2005; Rice et al. 2005; Tinti et al. 2005), suggest that the effective slip-weakening law might have a multiscale character not envisioned in the classical formulations of slip weakening. That multiscale character means that discernible weakening continues, at an ever-diminishing rate with slip, out to large slips, say, of order of a meter. Such response, when fitted to classical linear slip-weakening models (Figure 5.5), has led instead to the interpretation (Ohnaka 2003) that D_c depends on the slip in an earthquake.

Different Constitutive Formulations

The choice of a fault constitutive law is necessary to solve numerically the elastodynamic equations and to model spontaneous nucleation and propagation of an earthquake rupture. A constitutive law relates the total dynamic traction to fault friction and allows the absorption of a finite fracture energy (named G in Figures 5.3 and 5.5) at the crack tip. Different constitutive relations have been proposed in the literature. They can be grouped in two main classes: slip-dependent (Andrews 1976a, b; Ohnaka and Yamashita 1989) and rate- and state-dependent (R&S) laws (Dieterich 1979; Ruina 1983). The former assumes that friction is a function of the fault slip only, whereas the latter implies that the friction is a function of slip velocity and state variables. (The former can in fact be considered as the time- and rate-insensitive limit of a law formulated in a general R&S framework.) The analytical expression of the classical slip-weakening law (Andrews 1976a, b) is:

$$\tau = \begin{cases} \tau_y - \left(\tau_y - \tau_f\right)\dfrac{\Delta u}{D_c} & \Delta u < D_c \\[2mm] \tau_f & \Delta u \geq D_c \end{cases} \tag{5.7}$$

where Δu is the slip. The traction evolution associated to this law is shown in Figure 5.5 and it is characterized by a constant and linear traction decay and by a constant kinetic friction level (τ_f). The analytical expression of R&S friction laws is, as explained earlier, composed by two equations (written here generally, and then in their most commonly used forms): the strength and the evolution laws. These are, respectively,

$$\tau = \Im(V,\Phi,\sigma_n^{\text{eff}}) = f(V,\Phi)\cdot\sigma_n^{\text{eff}}$$

$$= \left[f_* + a\ln\left(\frac{V}{V_*}\right) + b\ln\left(\frac{V_*\Phi}{L}\right)\right]\cdot\sigma_n^{\text{eff}} \tag{5.8}$$

$$\frac{d\Phi}{dt} = g(V,\Phi) = 1 - \frac{\Phi V}{L}$$

where f is the coefficient of friction, σ_n^{eff} is the effective normal stress, V is the slip velocity, a and b are constitutive parameters, together with length L, and the reference parameters f_* and V_* (of which one can be chosen arbitrarily). Φ is the state variable (like denoted by θ earlier), which provides a memory of previous slip episodes and its evolution equation guarantees a time dependence of friction. In particular, the analytical expression stated in Eq. 5.8 is the ageing (slowness) law proposed by Ruina (1983) and Dieterich (1986), and motivated by Dieterich (1979). It represents one possible formulation among different ones in the literature (see Beeler et al. 1994).

The characteristic length scale parameters of these two constitutive formulations are the slip-weakening distance D_c and the parameter L: the former represents the slip required for traction to drop; the latter is the characteristic length for the renewal of a population of contacts along the sliding surface and controls the evolution of the state variable. These two length scale parameters are different (Cocco and Bizzarri 2002): Bizzarri and Cocco (2003) proposed analytical relations to associate slip weakening and R&S constitutive parameters. Such an association is possible because the commonly used versions of R&S laws lead to predictions which mimic those of the slip-weakening description when applied to a situation of rapid increase in slip rate, like at a propagating rupture front. In fact, the most general form of R&S law is broad enough that by choice of $f(V,\Phi)$ and $g(V,\Phi)$, we can duplicate an arbitrarily chosen slip-weakening law $f = f(\text{slip})$; e.g., set $g(V,\Phi) = V$ so that Φ is the slip, and choose $f(V,\Phi) = f(\Phi)$.

R&S Friction Laws Applied to Earthquake Modeling

The R&S constitutive formulation allows the modeling of rupture nucleation (Lapusta et al. 2000, and references therein), dynamic rupture propagation (Bizzarri et al. 2001, and references therein) as well as fault restrengthening during the interseismic period; therefore, it has been used to simulate repeated seismic cycles. R&S constitutive laws allow the definition of different frictional regimes: an earthquake is associated with an instability occurring in a velocity weakening field, meaning that there is negative change of fault strength with slip rate, $d\tau_{ss}/d\ln(V) < 0$ (where τ_{ss} is the steady-state traction), which corresponds to $(a - b) < 0$. Theoretical and observational evidence shows that, while spontaneous nucleation and dynamic propagation of an earthquake rupture are governed by the constitutive behavior (and the analytical form of Eq. 5.8 does matter), rupture arrest is more likely associated with geometrical complexities as well as frictional or rheological heterogeneities, which can stop the propagating dynamic rupture front. In the framework of the R&S laws, a velocity-strengthening regime (defined by $d\tau_{ss}/d\ln(V) > 0$, or $(a - b) > 0$, as occurs at higher temperatures [Blanpied et al. 1991, 1995]) can arrest the rupture, although it can also participate in a dynamic rupture if loaded enough and $(a - b)$ is relatively small. Rice (1993) and Boatwright and Cocco (1996) discussed the frictional control of crustal faulting and presented a classification of fault heterogeneity in terms of nonuniform distribution of the R&S constitutive parameters on the fault plane. This represents understanding of rate-weakening effects that it is important to extend to stronger weakening processes, for example, based on the flash heating mechanism discussed earlier which seems likely to be important at seismic slip rates.

Short Slip Duration

Another feature of the constitutive relation and overall fault model concerns their prediction of the duration of dynamic slip. Seismological observations demonstrate that slip duration is relatively short compared to the duration of the whole rupture propagation, and a mechanism has to be identified to explain the healing of slip. Two different interpretations have been recently proposed: one associates the healing of slip with strong heterogeneity of stress or strength on the fault plane (Beroza and Mikumo 1996; Day et al. 1998), whereas the other associates healing with strong rate-weakening in the constitutive relation (Cochard and Madariaga 1996; Beeler and Tullis 1996; Zheng and Rice 1998; Nielsen and Carlson 2000). These two mechanisms yield different traction evolutions, as shown in Figure 5.6. If the healing of slip is caused by the constitutive relation a fast restrengthening occurs immediately after the dynamic weakening. On the contrary, if strength or stress heterogeneity controls slip duration, the stress

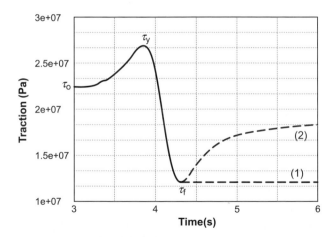

Figure 5.6 Sketch showing the traction evolution as a function of time in a generic point on the fault: dynamic traction increases from its initial value (τ_o) to the upper yield stress (τ_y) and therefore drops to the kinetic friction level (τ_f). The two subsequent evolutions refer to case (1), when slip occurs at a constant kinetic friction level, or to case (2) which is characterized by a fast restrengthening causing the healing of slip. In these two configurations, the final traction when slip is healed is different.

seems to remain near to or somewhat below (dynamic overshoot) the kinetic friction level. Effects of different elastic properties on the two sides of the fault plane can contribute to healing (Weertman 1980; Andrews and Ben-Zion 1997). Yet another possibility is that abrupt cessation of a weakening mechanism, for example, by hydraulic cracking of thermally pressurized fluid into the fault walls (Sibson 1973) would induce healing. There is evidence in lab studies (Tsutsumi and Shimamoto 1997; Hirose and Shimamoto 2005) that frictional resistance might increase abruptly in association with the earliest phases of melting and formation of a continuous melt layer; that too is a possible basis for healing, provided that it is effective before a broader, hotter and weaker melt layer can develop (Fialko 2004). Understanding the mechanisms controlling slip duration and estimating its value during real earthquakes is relevant to estimating frictional heating and radiated seismic energy, and therefore to the earthquake energy balance.

Slip Inversions and Rupture Parameters

Seismological observations and the modeling of ground motion waveforms recorded during large magnitude earthquakes allow the imaging of the slip-time history and its distribution on the fault plane. Numerous papers show that slip and rupture time distribution on the fault plane are heterogeneous, thus supporting the complexity of the processes controlling the mechanics of faulting and earthquakes. The availability of kinematic slip models (see Mai 2004) obtained by inverting geophysical data and by fitting observations makes feasible the estimate of the dynamic parameters strength excess ($\tau_y - \tau_0$), dynamic stress drop ($\tau_0 - \tau_f$) or strength breakdown ($\tau_y - \tau_f$), and critical slip-weakening distance D_c (Ide and Takeo 1997; Guatteri and Spudich 2000; Piatanesi et al. 2004). Seismological estimates of D_c yield large values of this parameter (of the order of 0.1–1 m) and suggest that it is a large fraction (up to 80%) of the total slip during the earthquake (Dalguer et al. 2002, among others). That might reflect a multiscale weakening process in which weakening continues at an ever decreasing rate with respect to slip, out to large slip (Abercrombie and Rice 2005), a feature also found in some thermal pressurization models (Rice 2004, 2006; Figures 5.2 and 5.4). These findings raise important issues which affect the estimate of fracture energy and the earthquake energy balance. The first concerns the problem of bridging laboratory and seismological estimates of length scale parameters. This is equivalent to the problem of interrelating microscale and macroscale processes controlling dynamic fault weakening. The second issue concerns the definition of fracture energy or the understanding of the way in which the work done during sliding is spent, including the much neglected possibility of dissipation in inelastic deformation of highly stressed material in the damage zone. Slip inversions for large earthquakes often lead to the inference of supershear rupture propagation (e.g., 1999 Izmit, Turkey; 2001 Kunlun, Tibet; 2002 Denali, Alaska), a phenomenon also seen in laboratory studies

(Rosakis 2002; Xia et al. 2004), and according to existing theoretical under-standing of the conditions for transition to that regime (Andrews 1976b; Dun-ham et al. 2003), those observations should place some at least loose constraints on fault stressing and strength at the times of the events.

DISCUSSION

To face the problems of comprehensively describing earthquake phenomena, we have emphasized here the critical need for new types of laboratory and natural fault observations, together with theory, for moving the conceptual back-ground beyond what is now available. This is needed to address, for example, at short timescales, the weakening during rapid, large slips of significant seis-mic events as well as, at long timescales, the interseismic restrengthening of fault zone materials, including the effects of temperature and fluid reactions, and consequences for subsequent earthquake nucleation and (to go back to short timescales) reinitiation of failure at a propagating rupture tip. The present com-mon formulations of R&S laws have represented a milestone advance for the field, but they have been proposed to interpret the results of laboratory experi-ments at low slip velocity (e.g., velocity stepping experiments usually with $V < 1$ mm s^{-1}), and usually under constant normal stress. Moreover, different interpretations exist in the literature concerning the state variable and its evolu-tion with slip and time, all of which have only the barest basic physical, versus empirical, foundations. The two most common interpretations rely on consid-ering the state variable as representative of the evolution of properties of the micro-asperity contacts during sliding (Dieterich 1986), or of the granular pack-ing density within the shear zone (Sleep and Blanpied 1992; Sleep 1997; Sleep et al. 2000). Both of these are expected to change during dynamic slip.

An important implication for future research is the understanding of the different temporal evolutions of those and other processes (like flash heating, thermal pressurization, gouge gelation, and local and macroscale melting), and of where, in the complex fault zone model here discussed (Figure 5.1), and under what conditions, those processes occur. Along with the necessity for a new generation of laboratory experiments—some performed at high slip rates and under widely variable normal stress conditions, some under long hold times at temperature—it will be most convincing to use, in the laboratory, rock samples taken from fault cores from depth, through drilling programs, as well as from exhumed faults. The properties of natural fault zones, while in need of further elucidation, seem from recent studies (Chester et al. 1993; Chester and Chester 1998; Wibberley 2002; Ben-Zion and Sammis 2003; Chester and Goldsby 2003; Chester et al. 2003; Otsuki et al. 2003; Sibson 2003; Wibberley and Shimamoto 2003; Sulem et al. 2004) to be more complex than usually believed. We should account for that not just in choice of laboratory materials, but certainly also in our theoretical and numerical interpretations.

ACKNOWLEDGMENT

The authors thank Rachel Abercrombie, Joe Andrews, Michael Ashby, Nick Beeler, Andrea Bizzarri, Judy Chester, Giulio Di Toro, Eiichi Fukuyama, David Goldsby, Greg Hirth, Laurent Jacques, Nadia Lapusta, Stefan Nielsen, Vikas Prakash, Alan Rempel, John Rudnicki, Charles Sammis, Paul Spudich, Toshi Shimamoto, Elisa Tinti, Terry Tullis, and Chris Wibberley for discussions, and thank Mark Handy and Paul Segall for discussions as well as review comments on the first version of the manuscript from them and, on a later version, from an anonymous reviewer. Support is gratefully acknowledged by JRR to NSF-EAR grants 0125709 and 0510193, and to the Southern California Earthquake Center (SCEC), funded by NSF Cooperative Agreement EAR-0106924 and USGS Cooperative Agreement 02HQAG0008; this is SCEC contribution number 890.

REFERENCES

Abercrombie, R.E., and J.R. Rice. 2005. Can observations of earthquake scaling constrain slip weakening? *Geophys. J. Intl.* **162**:406–424, doi: 10.1111/j.1365-246X.2005.02579.x.

Andrews, D.J. 1976a. Rupture propagation with finite stress in antiplane strain. *J. Geophys. Res.* **81**:3575–3582.

Andrews, D.J. 1976b. Rupture velocity of plane strain shear cracks. *J. Geophys. Res.* **81**:5679–5687.

Andrews, D.J. 2002. A fault constitutive relation accounting for thermal pressurization of pore fluid. *J. Geophys. Res.* **107**:B12, 2363, doi: 10.1029/2002JB001942.

Andrews, D.J. 2005. Rupture dynamics with energy loss outside the slip zone. *J. Geophys. Res.* **110**:B1, B01307, doi: 10.1029/2004JB003191.

Andrews, D.J., and Y. Ben–Zion. 1997. Wrinkle-like slip pulse on a fault between different materials. *J. Geophys. Res.* **102**:B1, 553–571.

Barenblatt, G.I. 1959. The formation of brittle cracks during brittle fracture. General ideas and hypotheses. Axially-symmetric cracks. *Appl. Math. Mech.* **23**:1273–1282.

Beeler, N.M., and T.E. Tullis. 1996. Self-healing slip pulses in dynamic rupture models due to velocity-dependent strength. *Bull. Seismol. Soc. Am.* **86**:1130–1148.

Beeler, N.M., and T.E. Tullis. 2003. Constitutive relationships for fault strength due to flash-heating. In: 2003 SCEC Annual Meeting Proceedings and Abstracts, vol. 13, p. 66. Los Angeles: Southern California Earthquake Center, Univ. of Southern California.

Beeler, N.M., T.E. Tullis, M.L. Blanpied, and J.D. Weeks. 1996. Frictional behavior of large displacement experimental faults. *J. Geophys. Res.* **101**:B4, 8697–8715.

Beeler, N.M., T.E. Tullis, and D.L. Goldsby. 2006. Constitutive relationships and physical basis of fault strength due to flash-heating. *J. Geophys. Res.*, in press.

Beeler, N.M., T.E. Tullis, and J.D. Weeks. 1994. The roles of time and displacement in the evolution effect in rock friction. *Geophys. Res. Lett.* **21**:1987–1990.

Ben–Zion, Y., and C.G. Sammis. 2003. Characterization of fault zones, *Pure & Appl. Geophys.* **160**:677–715.

Beroza, G., and T. Mikumo. 1996. Short slip duration in dynamic rupture in the presence of heterogeneous fault properties. *J. Geophys. Res.* **101**:22,449–22,460.

Biegel, R.L., and C.G. Sammis. 2004. Relating fault mechanics to fault zone structure. *Adv. Geophys.* **47**:65–111.

Bizzarri, A., and M. Cocco. 2003. Slip–weakening behavior during the propagation of dynamic ruptures obeying to rate- and state-dependent friction laws. *J. Geophys. Res.* **108**:B8, 2373, doi: 10.1029/2002JB002198.

Bizzarri, A., and M. Cocco. 2006a. A thermal pressurization model for the spontaneous dynamic rupture propagation on a 3D fault. Part I: Methodological approach. *J. Geophys. Res. 111:* B05303, doi:10.1029/2005JB003862.

Bizzarri, A., and M. Cocco. 2006b. A thermal pressurization model for the spontaneous dynamic rupture propagation on a 3D fault. Part II: Traction evolution and dynamic parameters. *J. Geophys. Res.* **111**: B05304, doi:10.1029/2005JB003864.

Bizzarri, A., M. Cocco, D.J. Andrews, and E. Boschi. 2001. Solving the dynamic rupture problem with different numerical approaches and constitutive laws. *Geophys. J. Intl.* **144**:656–678.

Blanpied, M.L., D.A. Lockner, and J.D. Byerlee. 1991. Fault stability inferred from granite sliding experiments at hydrothermal conditions. *Geophys. Res. Lett.* **18**:609–612.

Blanpied, M.L., D.A. Lockner, and J.D. Byerlee. 1995. Frictional slip of granite at hydrothermal conditions. *J. Geophys. Res.* **100**:B7, 13,045–13,064.

Blanpied, M.L., C.J. Marone, D.A. Lockner, J.D. Byerlee, and D.P. King. 1998. Quantitative measure of the variation in fault rheology due to fluid-rock interactions. *J. Geophys. Res.* **103**:B5, 9691–9712.

Boatwright, J., and M. Cocco. 1996. Frictional constraints on crustal faulting. *J. Geophys. Res.* **101**:B6, 13,895–13,909

Boettcher, M.S., and C. Marone. 2004. Effects of normal stress variation on the strength and stability of creeping faults. *J. Geophys. Res.* **109**:B3, B03406.

Brace, W.F., and J.D. Byerlee. 1970. California earthquakes: Why only shallow focus? *Science* **168**:1573–1575.

Brodsky, E.E., and H. Kanamori. 2001. Elastohydrodynamic lubrication of faults. *J. Geophys. Res.* **106**:B8, 16,357–16,374.

Chester, F.M., and J.S. Chester. 1998. Ultracataclasite structure and friction processes of the Punchbowl fault, San Andreas system, California. *Tectonophysics* **295**:199–221.

Chester, F.M., J.P. Evans, and R.L. Biegel. 1993. Internal structure and weakening mechanisms of the San Andreas fault. *J. Geophys. Res.* **98**:771–786.

Chester, J.S., F.M. Chester, and A.K. Kronenberg. 2005. Fracture surface energy of the Punchbowl fault, San Andreas system. *Nature* **437**:133–136, doi: 10.1038/nature03942.

Chester, J.S., and D.L. Goldsby. 2003. Microscale characterization of natural and experimental slip surfaces relevant to earthquake mechanics. In: 2003 Annual Progress Report to the Southern California Earthquake Center. College Station: Texas A & M University.

Chester, J.S., A.K. Kronenberg, F.M. Chester, and R.N. Guillemette. 2003. Characterization of natural slip surfaces relevant to earthquake mechanics. *EOS Trans. AGU* **84**: Fall Meeting Suppl., abstract S42C-0185.

Cochard, A., and R. Madariaga. 1996. Complexity of seismicity due to highly rate-dependent friction. *J. Geophys. Res.* **101**:B11, 25,321–25,336.

Cochard, A., and J.R. Rice. 2000. Fault rupture between dissimilar materials: Ill-posedness, regularization, and slip pulse response. *J. Geophys. Res.* **105**:891–907.

Cocco, M., and A. Bizzarri. 2002. On the slip-weakening behavior of rate- and state-dependent constitutive laws. *Geophys. Res. Lett.* **29**:1–4.

Cocco, M. and A. Bizzarri. 2004. Dynamic fault weakening caused by thermal pressurization in an earthquake model governed by rate- and state-dependent friction. *EOS Trans. AGU* **85**: Fall Meeting Suppl., Abstract T22A-06.

Cocco, M., and J.R. Rice. 2002. Pore pressure and poroelasticity effects in Coulomb stress analysis of earthquake interactions. *J. Geophys. Res.* **107**:B2, 2030, doi:10.1029/2000JB000138.

Dalguer, L.A., K. Irikura, W. Zhang, and J.D. Riera. 2002. Distribution of dynamic and static stress changes during 2000 Tottori (Japan) earthquake: Brief interpretation of the earthquake sequences; foreshocks, mainshock and aftershocks. *Geophys. Res. Lett.* **29**: doi: 10.1029/2001GL014333.

Day, S.M., G. Yu, and D.J. Wald. 1998. Dynamic stress changes during earthquake rupture. *Bull. Seismol. Soc. Am.* **88**:512–522.

Desrues, J., and G. Viggiani. 2004. Strain localization in sand: An overview of the experimental results obtained in Grenoble using stereophotogrammetry. *Intl. J. Numer. Anal. Meth. Geomech.* **28**:279–321.

Dieterich, J.H. 1978. Time-dependent friction and the mechanics of stick slip. *Pure & Appl. Geophys.* **11**:790–806.

Dieterich, J.H. 1979. Modeling of rock friction. 1. Experimental results and constitutive equations. *J. Geophys. Res.* **84**:2161–2168.

Dieterich, J.H. 1986. A model for the nucleation of earthquake slip. In: Earthquake Source Mechanics, ed. S. Das, J. Boatwright, and C.H. Scholz, Geophys. Monograph vol. 37, Maurice Ewing Ser. no. 6, pp. 37–47. Washington D.C.: Am. Geophys. Union.

Dieterich, J.H. 1994. A constitutive law for rate of earthquake production and its application to earthquake clustering. *J. Geophys. Res.* **99**:2601–2618.

Dieterich, J.H., and G. Conrad. 1984. Effect of humidity on time-dependent and velocity-dependent friction in rocks. *J. Geophys. Res.* **89**:B6, 4196–4202.

Dieterich, J.H., and B.D. Kilgore. 1994. Direct observation of frictional contacts: New insights for state-dependent properties. *Pure & Appl. Geophys.* **143**:283–302.

Dieterich, J.H., and B.D. Kilgore. 1996. Imaging surface contacts: Power law contact distributions and contact stresses in quartz, calcite, glass and acrylic plastic. *Tectonophysics* **256**:219–239.

Di Toro, G., D.L. Golbsby, and T.E. Tullis. 2004. Friction falls toward zero in quartz rock as slip velocity approaches seismic rates. *Nature* **427**:436–439.

Di Toro, G., T. Hirose, S. Nielsen, G. Pennacchioni, and T. Shimamoto. 2006. Natural and experimental evidence of melt lubrication of faults during earthquakes. *Science* **311**:647–649.

Di Toro, G., S. Nielsen, and G. Pennacchioni. 2005. Earthquake rupture dynamics frozen in exhumed ancient faults. *Nature* **436**:1009–1012.

Dunham, E.M., P. Favreau, and J.M. Carlson. 2003. A supershear transition mechanism for cracks. *Science* **299**:1557–1559.

Fialko, Y.A. 2004. Temperature fields generated by the elastodynamic propagation of shear cracks in the Earth. *J. Geophys. Res.* **109**:1–14, art. no. B01303, doi: 10.1029/2003JB002497.

Frye, K.M., and C. Marone. 2002. Effect of humidity on granular friction at room temperature. *J. Geophys. Res.* **107**:B11, 2309, doi: 10.1029/2001JB000654.

Goldsby, D.L., and T.E. Tullis. 2002. Low frictional strength of quartz rocks at subseismic slip rates. *Geophys. Res. Lett.* **29**:1844.

Goldsby, D.L., and T.E. Tullis. 2003. Flash heating/melting phenomena for crustal rocks at (nearly) seismic slip rates. In: 2003 SCEC Annual Meeting Proceedings and Abstracts, vol. 13, pp. 88–90. Los Angeles: Southern California Earthquake Center, Univ. of Southern California.

Guatteri, M., and P. Spudich. 2000. What can strong-motion data tell us about slip-weakening fault-friction laws? *Bull. Seismol. Soc. Am.* **90**:98–116.

Handy, M.R., and J.-P. Brun. 2004. Seismicity, structure, and strength of the continental lithosphere. *Earth Planet. Sci. Lett.* **223**:427–441.

Harris, R., and S. Day. 1997. Effects of a low velocity zone on a dynamic rupture. *Bull. Seismol. Soc. Am.* **87**:1267–1280.

Heaton, T.H. 1990. Evidence for and implications of self-healing pulses of slip in earthquake rupture. *Phys. Earth & Planet. Inter.* **64**:1–20.

Hirose, T., and T. Shimamoto. 2005. Growth of a molten zone as a mechanism of slip weakening of simulated faults in gabbro during frictional melting. *J. Geophys. Res.* **110**:B05202, doi:10.1029/2004JB003207.

Ida, Y. 1972. Cohesive force across the tip of a longitudinal-shear crack and Griffith's specific surface energy. *J. Geophys. Res.* **77**:3796–3805.

Ide, S., and M. Takeo. 1997. Determination of constitutive relations of fault slip based on seismic wave analysis. *J. Geophys. Res.* **102**:B12, 27,379–27,391.

Kame, N., J.R. Rice, and R. Dmowska. 2003. Effects of pre-stress state and rupture velocity on dynamic fault branching. *J. Geophys. Res.* **108**:B5, 2265, doi:10.1029/2002JB002189.

Kanamori, H., and E.E. Brodsky. 2001. The physics of earthquakes. *Phys. Today* **54**:34–39.

Lachenbruch, A.H. 1980. Frictional heating, fluid pressure, and the resistance to fault motion. *J. Geophys. Res.* **85**:B11, 6097–6122.

Lapusta, N., and J.R. Rice. 2003. Nucleation and early seismic propagation of small and large events in a crustal earthquake model. *J. Geophys. Res.* **108**:B4, 2205, doi: 10.1029/2001JB000793.

Lapusta, N., J.R. Rice, Y. Ben-Zion, and G. Zheng. 2000. Elastodynamic analysis for slow tectonic loading with spontaneous rupture episodes on faults with rate- and state-dependent friction. *J. Geophys. Res.* **105**:23,765–23,789.

Li, Y.G., J. Vidale, K. Aki, C. Marone, and W.K. Lee. 1994. Fine structure of the Landers fault zone: Segmentation and the rupture process. *Science* **265**:367–380.

Linker, M.F., and J.H. Dieterich. 1992. Effects of variable normal stress on rock friction: Observations and constitutive equations. *J. Geophys. Res.* **97**:B4, 4923–4940.

Lockner, D., H. Naka, H. Tanaka, H. Ito, and R. Ikeda. 2000. Permeability and strength of core samples from the Nojima fault of the 1995 Kobe earthquake. In: Proc. Intl. Workshop on the Nojima Fault Core and Borehole Data Analysis, Tsukuba, Japan, Nov. 22–23, 1999, ed. H. Ito, H. Fujimoto, H. Tanaka, and D. Lockner, Open File Report 00-129, pp. 147–152. Menlo Park, CA: U.S. Geol. Survey.

Mai, M. 2004. Database of finite-source rupture models. http://www.seismo.ethz.ch/srcmod/, ETH, Zurich.

Marone, C. 1998. Laboratory-derived friction laws and their application to seismic faulting. *Ann. Rev. Earth & Planet. Sci.* **26**:643–696.

Marone, C., C.B. Raleigh, and C.H. Scholz. 1990. Frictional behavior and constitutive modeling of simulated fault gouge. *J. Geophys. Res.* **95**:B5, 7007–7025.

Mase, C.W., and L. Smith. 1985. Pore-fluid pressures and frictional heating on a fault surface. *Pure & Appl. Geophys.* **92**:6249–6272.

Mase, C.W., and L. Smith. 1987. Effects of frictional heating on the thermal, hydrologic, and mechanical response of a fault. *J. Geophys. Res.* **92**:B7, 6249–6272.

Morgenstern, N.R., and J.S. Tschalenko. 1967. Microscopic structures in kaolin subjected to direct shear. *Géotechnique* **17**:309–328.

Nielsen, S.B., and J.M. Carlson. 2000. Rupture pulse characterization; self-healing, self-similar, expanding solutions in a continuum model of fault dynamics. *Bull. Seismol. Soc. Am.* **90**:1480–1497.

Ohnaka, M., and T. Yamashita. 1989. A cohesive zone model for dynamic shear faulting based on experimentally inferred constitutive relation and strong motion source parameters. *J. Geophys. Res.* **94**:4089–4104.

Ohnaka, M.A. 2003. A constitutive scaling law and a unified comprehension for frictional slip failure, shear fracture of intact rock, and earthquake rupture. *J. Geophys. Res.* **108**:B2, 2080, doi: 10.1029/2000JB000123.

Otsuki, K., N. Monzawa, and T. Nagase. 2003. Fluidization and melting of fault gouge during seismic slip: Identification in the Nojima fault zone and implications for focal earthquake mechanisms. *J. Geophys. Res.* **108**:B4, 2192, doi: 10.1029/2001JB001711.

Palmer, A.C., and J.R. Rice. 1973. The growth of slip surfaces in the progressive failure of over-consolidated clay. *Proc. R. Soc. Lond. A* **332**:527–548.

Piatanesi, A., E. Tinti, M. Cocco, and E. Fukuyama. 2004. The dependence of traction evolution on the earthquake source time function adopted in kinematic rupture models. *Geophys. Res. Lett.* 31:L04609, doi: 10.1029/2003GL019225.

Poliakov, A.N.B., R. Dmowska, and J.R. Rice. 2002. Dynamic shear rupture interactions with fault bends and off-axis secondary faulting. *J. Geophys. Res.* **107**:B11, 2295, doi: 10.1029/2001JB000572.

Power, W.L., and T.E. Tullis. 1991. Euclidean and fractal models for the description of rock surface roughness. *J. Geophys. Res.* **96**:415–424.

Prakash, V. 1998. Frictional response of sliding interfaces subjected to time varying normal pressure. *J. Tribology* **120**:97–102.

Prakash, V. 2004. Pilot studies to determine the feasibility of using new experimental techniques to measure sliding resistance at seismic slip rates. In: 2004 Annual Progress Report to the Southern California Earthquake Center. Cleveland, OH: Case-Western Reserve Univ.

Prakash, V., and R.J. Clifton. 1992. Pressure-shear plate impact measurement of dynamic friction for high speed machining applications. In: Proc. VII Intl. Congress on Experimental Mechanics, pp. 556–564. Bethel, CT: Soc. Experimental Mechanics.

Ranjith, K., and J.R. Rice. 2001. Slip dynamics at an interface between dissimilar materials. *J. Mech. Phys. Solids* **49**:341–361.

Rice, J.R. 1993. Spatio-temporal complexity of slip on a fault. *J. Geophys. Res.* **98**: 9885–9907.

Rice, J.R. 1999. Flash heating at asperity contacts and rate-dependent friction. *EOS Trans. Amer. Geophys. Union* **80**: Fall Meeting Suppl., F6811.

Rice, J.R. 2004. Thermal weakening in large seismic slips and effects on rupture dynamics. In: 2004 Annual Progress Report to the Southern California Earthquake Center. Cambridge, MA: Harvard University.

Rice, J.R. 2006. Heating and weakening of faults during earthquake slip. *J. Geophys. Res.* **111**: B05311, doi:10.1029/2005JB004006.

Rice, J.R., and Y. Ben-Zion. 1996. Slip complexity in earthquake fault models. *PNAS* **93**:3811–3818.

Rice, J.R., N. Lapusta, and K. Ranjith. 2001. Rate and state dependent friction and the stability of sliding between elastically deformable solids. *J. Mech. Phys. Solids* **49**:1865–1898.

Rice, J.R., C.G. Sammis, and R. Parsons. 2005. Off-fault secondary failure induced by a dynamic slip-pulse. *Bull. Seismol. Soc. Am.* **95**:109–134, doi: 10.1785/0120030166.

Roig Silva, C., D.L. Goldsby, G. Di Toro and T.E. Tullis 2004.The role of silica content in dynamic fault weakening due to gel lubrication. *EOS Trans. AGU* **85**: Fall Meeting Suppl., Abstract T21D-06.

Rosakis, A.J. 2002. Intersonic shear cracks and fault ruptures. *Adv. Phys.* **51**:1189–1257.

Ruina, A.L. 1983. Slip instability and state variable friction laws. *J. Geophys. Res.* **88**:B12, 10,359–10,370.

Scholz, C.H. 1990. The Mechanics of Earthquake and Faulting. New York: Cambridge Univ. Press.

Scholz, C.H. 1998. Earthquakes and friction laws. *Nature* **391**:37–41.

Scruggs, V.J., and T.E. Tullis. 1998. Correlation between velocity dependence of friction and strain localization in large displacement experiments on feldspar, muscovite, and biotite gouge. *Tectonophysics* **295**:15–40.

Segall, P., and J.R. Rice. 1995. Dilatancy, compaction, and slip instability of a fluid-infiltrated fault. *J. Geophys. Res.* **100**:22,155–22,171.

Sibson, R.H. 1973. Interaction between temperature and pore-fluid pressure during earthquake faulting: A mechanism for partial or total stress relief. *Nature* **243**:66–68.

Sibson, R.H. 1975. Generation of pseudotachylite by ancient seismic faulting. *Geophys. J. Roy. Astr. Soc.* **43**:775–794.

Sibson, R.H. 2003. Thickness of the seismic slip zone. *Bull. Seismol. Soc. Am.* **93**: 1169–1178.

Sleep, N.H. 1997. Application of a unified rate and state friction theory to the mechanics of fault zones with strain localization. *J. Geophys. Res.* **102**:B2, 2875–2895.

Sleep, N.H., and M.L. Blanpied. 1992. Creep, compaction, and the weak rheology of major faults. *Nature* **359**:687–692.

Sleep, N.H., E. Richardson, and C. Marone. 2000. Physics of friction and strain rate localization in synthetic gouge. *J. Geophys. Res.* **105**:25,875–25,890.

Spray, J.G. 1993. Viscosity determinations of some frictionally generated silicate melts: Implications for fault zone rheology at high strain rates. *J. Geophys. Res.* **98**: 8053–8068.

Steacy, S.J. and C.G. Sammis. 1991. An automaton for fractal patterns of fragmentation. *Nature* **353**:250–252.

Stesky, R.M. 1978. Rock friction effect of confining pressure, temperature, and pore pressure. *Pure & Appl. Geophys.* **116**:690–704.

Stesky R.M., W.F. Brace, D.K. Riley, and P.Y.F. Robin. 1974. Friction in faulted rock at high-temperature and pressure. *Tectonophysics* **23**:177–203.

Stirling, M.W., S.G. Wesnousky, and K. Shimazaki. 1996. Fault trace complexity, cumulative slip, and the shape of the magnitude-frequency distribution for strike-slip faults: A global survey. *Geophys. J. Intl.* **124**:833–868.

Sulem, J., I. Vardoulakis, H. Ouffroukh, M. Boulon, and J. Hans. 2004. Experimental characterization of the thermo-poro-mechanical properties of the Aegion fault gouge. *C. R. Geoscience* **336**:455–466.

Tinti, E., P. Spudich, and M. Cocco. 2005. Earthquake fracture energy inferred from kinematic rupture models on extended faults. *J. Geophys. Res.* **110**:B12, B12303, doi: 10.1029/2005JB003644.

Tse, S.T., and J.R. Rice. 1986. Crustal earthquake instability in relation to the depth variation of frictional slip properties. *J. Geophys. Res.* **91**:9452–9472.

Tsutsumi, A., and T. Shimamoto. 1997. High velocity frictional properties of gabbro. *Geophys. Res. Lett.* **24**:699–702.

Tullis, T.E., ed. 1986. Friction and faulting, topical issue. *Pure & Appl. Geophys.* **124**.

Tullis, T.E., and D.L. Goldsby. 2003. Laboratory experiments on fault shear resistance relevant to coseismic earthquake slip. In: 2003 Annual Progress Report to the Southern California Earthquake Center. Providence, RI: Brown Univ.

Vardoulakis, I. 2003. Dynamic thermo-poro-mechanical analysis of catastrophic landslides. *Géotechnique* **53**:523–524.

Weertman, J. 1980. Unstable slippage across a fault that separates elastic media of different elastic constants. *J. Geophys. Res.* **85**:1455–1461.

Wibberley, C.A.J. 2002. Hydraulic diffusivity of fault gouge zones and implications for thermal pressurization during seismic slip. *Earth Planets Space* **54**:1153–1171.

Wibberley, C.A.J., and T. Shimamoto. 2003. Internal structure and permeability of major strike-slip fault zones: The Median Tectonic Line in Mid Prefecture, Southwest Japan. *J. Struct. Geol.* **25**:59–78.

Wilson, B., T. Dewers, Z. Reches, and J. Brune. 2005. Particle size and energetics of gouge from earthquake rupture zones. *Nature* **434**:749–752.

Xia, K., A.J. Rosakis, and H. Kanamori. 2004. Laboratory earthquakes: The sub-Rayleigh-to-supershear transition. *Science* **303**:1859–1861.

Xia, K., A.J. Rosakis, H. Kanamori, and J.R. Rice. 2005. Inhomogeneous faults hosting earthquakes in the laboratory: Directionality and supershear. *Science* **308**:681–684.

Zheng, G., and J.R. Rice. 1998. Conditions under which velocity-weakening friction allows a self-healing versus a cracklike mode of rupture. *Bull. Seismol. Soc. Am.* **88**:1466–1483.

6

Continental Fault Structure and Rheology from the Frictional-to-Viscous Transition Downward

MARK R. HANDY[1], GREG HIRTH[2], and ROLAND BÜRGMANN[3]

[1]Department of Earth Sciences, Freie Universität Berlin, Malteserstr. 74–100, 12249 Berlin, Germany
[2]Department of Geology and Geophysics, Woods Hole Oceanographic Institution, MS#8, WH01, Woods Hole, MA 02543, U.S.A.
[3]Department of Earth and Planetary Science, University of California, Berkeley, 389 McCone Hall, Berkeley, CA 94720, U.S.A.

ABSTRACT

Faulting is an expression of the interaction between rock rheology, kinematic boundary conditions, and associated stress fields. The structure and rheology of faults vary with depth, such that pressure-dependent frictional behavior predominating in the upper, brittle part of the crust is transitional to strongly temperature- and rate-dependent behavior in the lower part of the crust and mantle. This frictional-to-viscous transition (FVT) is characterized by changes in rock structure, rheology, and fluid activity that are closely tied to the earthquake cycle. As such, the FVT is a first-order decoupling zone, whose depth and lateral extent vary in time. Brittle, sometimes seismic, instabilities perturb the ambient stress field within the lithosphere on timescales ranging from seconds to years. These instabilities are measurable as transient motions of the Earth's surface and are manifest both at, and below, the FVT by the development of structural anisotropies (fractures, foliations). Surface motion studies of plate-boundary strike-slip faults indicate that shearing below the FVT is more localized in the lower crust than in the upper mantle. Structural investigations of exhumed shear zones reveal that this localization involves the nucleation of fractures at the FVT, as well as the buckling and rotation of existing foliations below the FVT. In some cases, rotation of these surfaces can initiate transient deformation, transferring stress upward and potentially triggering earthquakes. The networking of shear zones on several length scales allows them to function as decoupling horizons that partition three-dimensional strain within the lithosphere. The simplification of fault geometry with progressive strain lends justification to the use of laboratory-derived flow laws to estimate the bulk rock rheology on length scales at which strain is homogeneous. In general, the longer the time- and length scales of faulting considered, the greater the potential influence of the kinematic and thermal history on the rheology of the fault system. Taken together, studies suggest that future fault modeling must include parameters that quantify the thermal and structural aspects of rock history, as well as the fluid activity in and around faults.

INTRODUCTION

This chapter addresses processes controlling the structure and rheology of fault zones in the lithosphere, from the frictional-to-viscous transition (FVT) down- wards. We define the FVT as the change from fracture and frictional sliding on one or more discrete surfaces to thermally activated creep within zones of vis- cous, solid-state flow. Such zones of continuous, distributed flow are variously referred to as shear zones, ductile shear zones, or ductile faults. Cataclastic fault rocks predominate above the FVT, whereas mylonitic rocks form at the FVT and below (e.g., Schmid and Handy 1991). In accordance with conven- tion, we use "fault" for any brittle shear surface and "shear zone" for a zone of viscous, predominantly mylonitic creep, irrespective of the ductility (i.e., distributedness) of the deformation on a given length scale.

In the familiar lithospheric strength model, the FVT occupies a depth interval in the lithosphere between brittle deformation at the surface and predominantly viscous flow at greater depths in the crust and mantle. The FVT is usually depicted to coincide with a strength maximum between the brittle upper crust and the vis- cously deforming layers of the lower crust and mantle (Brace and Kohlstedt 1980). Such rheological stratification provides a first-order explanation of the architec- ture of orogens and passive continental margins on timescales of 10–100 Ma (e.g., Ranalli and Murphy 1987). However, as illustrated for an idealized strike-slip fault in Figure 6.1, this model needs to be modified to explain a variety of fault- related processes active on a much broader range of time- and length scales. The cyclical nature of seismicity provides a convenient measure of time for our discus- sion of fault evolution: Long, interseismic periods of near steady-state motion alternate with short periods of transient motion (coseismic, postseismic periods in Figure 6.1). We emphasize, however, that the FVT does not necessarily coincide with the transition from seismic to aseismic slip in the crust, which can represent a strain- and time-dependent change from unstable, stick-slip motion to stable frictional sliding on fault surfaces above the FVT (see Chapter 5).

This paper is structured around a series of questions addressing some basic issues in fault dynamics. In raising these questions, we highlight some of the discrepancies between nature, experiment, and theory regarding the behavior of Earth's lithosphere and upper mantle. Our goal in this paper is to show how current thinking and state-of-the-art approaches may be applied over a range of scales to yield better insight into fault dynamics at, and below, the FVT.

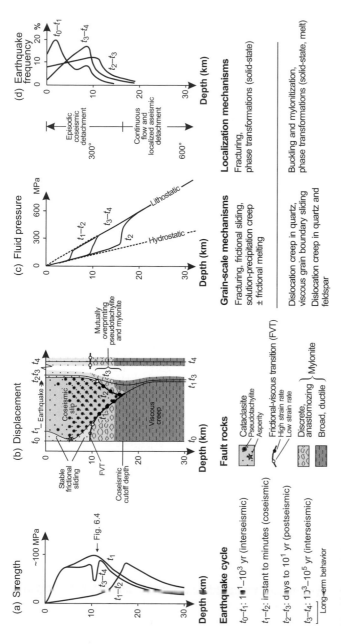

Figure 6.1 Characteristics and processes of an idealized strike-slip fault in continental crust during one earthquake cycle. (a) Strength vs. depth: arrow indicates depth for deformation mechanism map in Figure 6.4. (b) Displacement vs. depth: each line represents the locus of points along the fault surface at a given time (one line each for times t_0 to t_4), but the time interval between any two neighboring lines is not identical (modified from Tse and Rice 1986). Star marks fault nucleation on an asperity. (c) Pore-fluid pressure vs. depth. (d) Earthquake frequency (in % of earthquakes) vs. depth. Figure modified from Handy and Brun (2004). Maximum strength estimate in (a) taken from Handy et al. (1999). Earthquake frequency curves in (d) adapted from Rolandone et al. (2004).

RESEARCH TRENDS AND PROBLEMS

What Goes On at the Frictional-to-Viscous Transition?

Most knowledge of the FVT derives from experiments and microstructural studies of naturally deformed rocks. The view of the FVT as the site of a lithospheric strength maximum is based on the extrapolation of laboratory-derived constitutive equations for steady-state frictional sliding and power-law creep to natural strain rates and temperatures. The result is a strength-depth envelope similar in form to curve t_1 in Figure 6.1a. Taken at face value, some observations are consistent with this simple model. For example, stress measurements near the surface show an increase in stress with depth (Zoback and Townend 2001), and earthquake frequency and coseismic stress drop often peak within the same depth interval as the purported strength maximum (Meissner and Strehlau 1982; Scholz 1998). Also, stress-dependent microstructures in exhumed mylonitic fault rocks from near the FVT indicate higher differential stresses (~ 100 MPa, Stipp et al. 2002) than in originally warmer, hydrous crustal rocks from deeper crustal levels (e.g., 1 to tens of MPa, Etheridge and Wilkie 1981, see below).

When assessing this simple model, it is important to be aware of the assumptions upon which it rests, as well as of some frequently held misconceptions. Implicit in the construction of strength-depth envelopes is the view that rheology is solely a function of temperature, effective pressure and strain rate, and in some cases, grain size. Other simplifying assumptions include a steady-state geotherm, a uniform strain rate (or stress), and rock strength that is independent of length scale. In most tectonic settings none of these assumptions are likely to be strictly correct, as discussed in Handy and Brun (2004). Furthermore, strength-depth envelopes are commonly calculated for a single deformation mechanism in the viscous deformational regime (most often dislocation creep). Microstructural studies of fault rocks, however, indicate that deformation mechanisms at the FVT evolve with strain, and that several grain-scale deformation mechanisms with different temperature, pressure, and strain-rate dependencies accommodate the bulk strain and strain rate, as shown in Figures 6.2 and 6.3, and discussed below. Competing deformation mechanisms at the FVT are also observed in experiments, for example, on granite (Dell Angelo and Tullis 1996), quartzite (Hirth and Tullis 1994), marble (Fredrich et al. 1989), halite (Chester 1988), and mica-quartz aggregates (Bos et al. 2000; Bos and Spiers 2002). Finally, there has been a tendency in the literature to relate seismicity to rock strength (e.g., Jackson 2002) despite the fact that seismicity is at best an ambiguous indicator of strength (Handy and Brun 2004). The realization that frictional properties and permeability within fault zones are time dependent has led to the notion of the FVT as a depth interval marked by significant fluctuations in strength (Scholz 1998) and fluid pressure (Sibson and Rowland 2003), especially on short time- and length scales. These fluctuations are closely tied to the earthquake cycle, as depicted schematically in Figure 6.1c.

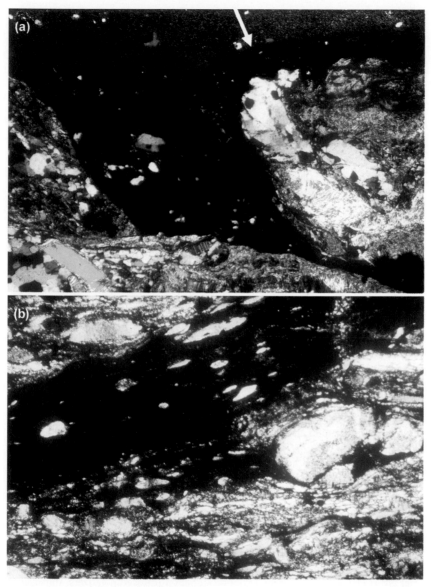

Figure 6.2 Microstructures of the coseismic frictional-to-viscous transition (FVT) in which a pseudotachylite vein cuts a mylonite in (a) and is itself overprinted in (b) by mylonite during postseismic to interseismic mylonitic creep. The arrow in (a) points to the boundary between glassy chilled margin and aphanitic, devitrified vein interior. The protolith of the black ultramylonite in (b) is the pseudotachylite vein in (a). Note the rounded clasts of mica in (b) indicative of mylonitic overprinting. Crossed polarizers, dimensions 2.0 × 1.3 mm). Samples taken from the Pogallo ductile fault zone, southern Alps, northern Italy (Handy 1998).

Interseismic strain leads to the buildup of stress, with differential stress at the FVT approaching a maximum (t_0 to t_1; Figures 6.1a, 6.1b). This stress is released suddenly along a rupture surface that nucleates at an irregularity in the fault surface (marked with a star in Figure 6.1b). This irregularity can be an asperity, that is, a rough or uneven part of the fault, a pocket of anomalous pore-fluid pressure, or a change of mineralogy along the fault surface. The effect of the irregularity is to allow local stress levels to attain the fracture (or frictional) strength of the rock or to reduce fracture strength to the local stress level. Either case favors failure and rapid localization of strain.

During coseismic faulting, fractures propagate down into crust (curves t_1 to t_2; Figure 6.1b), which deforms viscously during interseismic periods. Figure 6.2 shows a natural example of this: mylonite is cut by pseudotachylite (Figure 6.2a), which is itself subsequently deformed by mylonite (Figure 6.2b). The depth and duration of this downward penetration depend on the area and amount of coseismic slip, reflected in the magnitude of the main earthquake event. For example, the $M_W = 7.3$ 1992 Landers earthquake along the San Andreas fault system in the southwestern United States led to a ~3 km depression of the base of the seismogenic zone (Figure 6.1d), which gradually decayed over a time of four to five years (Rolandone et al. 2004; see also Figure 2.11 in Chapter 2). The much smaller, $M_W = 6.2$ Morgan Hill earthquake resulted in only a ~500 m depression that recovered within two years (Schaff et al. 2002). Coseismic stress drops as great as 100 MPa (Bouchon 1997) have been recorded locally, but earthquake seismology indicates that the stress drop is usually one to two orders of magnitude less than this value and is unrelated to the magnitude of the main earthquake event (e.g., Kanamori and Heaton 2000). Indeed, the possibility that complete stress drops occur during earthquakes remains a controversial interpretation of the static stress drop observed during earthquakes (e.g., Scholz 1992). Geologic evidence for a coseismic transition

---

Figure 6.3 Microstructures of the aseismic frictional-to-viscous transition (FVT): (a) Veins ▶ with fibrous quartz-cut fine-grained mylonitic matrix comprising syntectonically recrystallized chlorite, white mica, and quartz. Note that the older, less-deformed vein at top right is at a higher angle to foliation than more deformed vein at bottom (crossed polarizers, dimensions 22.6 × 15 mm). (b) Rotated quartz veins with various degrees of mylonitic overprint are stretched parallel to the mylonitic foliation (crossed polarizers, dimensions 5.6 × 3.8 mm). The inference of coeval fracturing, fluid flow, and mylonitic creep in this sample comes from the mutual overprinting of veins and mylonitic foliation: Veins cut the foliation but are themselves increasingly deformed with decreasing angle to the mylonitic foliation The inferred deformation mechanisms are dislocation creep and viscous granular flow including pressure solution in the fine-grained matrix, fracturing, veining, and precipitation of quartz in the presence of a fluid phase in the fibrous quartz veins, and dislocation creep (subgrain rotation and grain-boundary migration recrystallization of quartz fibers) in the veins. Sample was taken from the same shear zone as in Figure 6.2.

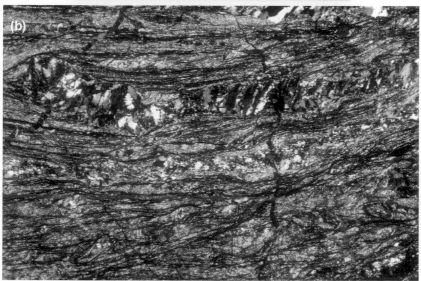

from viscous to brittle deformation comes from mutually overprinting pseudot-achylite and mylonite (e.g., Hobbs et al. 1986 and Figure 6.2) and from micro-structures in quartz-rich crustal rocks indicative of transient differential stresses much higher than average flow stresses in the viscous crust (= 300–500 MPa, Küster and Stöckhert 1999). Küster and Stöckhert (1999) and Trepmann and Stöckhert (2003) interpret these microstructures to have formed during coseismic elastic loading followed by rapid postseismic relaxation in the vicinity of a rupture surface, although no traces of such a surface leading down from a fossil FVT have been found so far in their example.

Postseismic deformation is highly transient, with the magnitude of change in the strain rate of the viscous crust determined by the amount of coseismic slip in the brittle, upper crust (t_2 to t_3 in Figure 6.1b). During reinitiation of long interseismic periods (t_3 to t_4 in Figure 6.1) the FVT is inferred to be a site of localized shearing and therefore to have reduced strength relative to the ad-jacent over- and underlying rocks. This hypothesis is supported by the preva-lence at the FVT of very fine grained (1–10 µm), mylonitic fault rocks with microstructures diagnostic of viscous granular flow as well as by the presence of recrystallized hydrous minerals that are intrinsically weak (e.g., biotite, white mica). Figure 6.3 shows an example of the aseismic FVT, where dislocation creep, viscous granular flow, fluid-assisted veining, and precipitation of quartz are inferred to have occurred simultaneously. Abundant geologic evidence in-dicates that anastomozing shear zones at the FVT act as detachment surfaces within the crust (Handy and Brun 2004).

To illustrate the possible effects of stress- and strain-rate fluctuations during faulting, we show a deformation mechanism map in Figure 6.4 for quartz-rich rocks at conditions for the FVT in Figure 6.1. Three of the mechanisms identi-fied in the example above are represented in Figure 6.4 (cataclasis, dislocation creep, viscous granular flow), although explicit constitutive formulations are not possible for all of them due to the limited rheological data (sources in the caption). In particular, the field for pressure-solution creep must be regarded as broadly representative of several different types of viscous granular flow in-volving fluid-assisted, intergranular diffusion (e.g., Paterson 1995). Neverthe-less, experiments allow us to place the following general constraints on rock strength (e.g., Kohlstedt et al. 1995), as depicted in Figure 6.4: First, the transi-tion from localized (faulting) to distributed (ductile) deformation ("semi-brittle flow" in experimentalist's parlance) occurs when rock strength becomes less than that required for stable frictional sliding. A reasonable upper bound for this stress is obtained from Byerlee's (1978) empirical relation, often referred to as "Byerlee's law," modified for effective stress at hydrostatic pore-fluid pressure, as shown in Figure 6.4. This frictional stress is an upper bound to rock strength because pore-fluid pressure can exceed hydrostatic values and because some fault zone materials have very low frictional coefficients that do not obey Byerlee's law (e.g., some clays [Byerlee 1978], serpentine [Moore

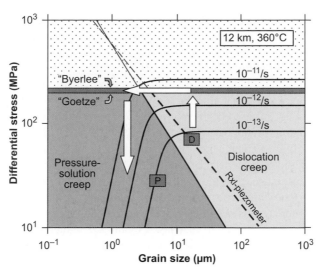

Figure 6.4 Deformation mechanism map for quartz-rich rock at the FVT at 12 km and 360°C, corresponding to the depth indicated with the arrow in Figure 6.1a. Strain-rate contours in the dislocation creep field are from Hirth et al. (2001), in the viscous granular flow field from Kenis et al. (2005) for pressure-solution creep. The thick line separates these mechanism fields, whereas the dashed line labeled "rxl-piezometer" is the quartz recrystallized grain-size piezometer of Stipp and Tullis (2003). Thin continuation of all these lines in the stippled area is only valid for creep at pore-fluid pressures less than hydrostatic. Boxes labeled P and D indicate deformational conditions, respectively, for fine-grained siliciclastic rocks undergoing pressure-solution creep (Kenis et al. 2005; Schwarz and Stöckhert 1996) and for coarser-grained quartzite undergoing dislocation creep (Dunlap et al. 1997). The line labeled "Byerlee" shows frictional stress for a strike-slip fault along a hydrostatic pore-fluid pressure gradient. Stresses in the stippled field above this line exceed the fracture strength and hence are not realizable at hydrostatic pore-fluid pressure. The line labeled "Goetze" represents conditions at which differential stress equals the least principal stress. White arrows illustrate a possible stress–grain-size path during and just after an earthquake, corresponding to intervals t_1–t_2 and t_2–t_3 in Figure 6.1a. See text for explanation. Fluid-assisted veining is not shown explicitly in Figure 6.4 but occupies the stippled field above the horizontal line representing Goetze's criterion. An increase of pore-fluid pressure above hydrostatic values would shift this line downward, consistent with embrittlement at lower differential stresses.

et al. 1997]). Both conditions would shift the horizontal line in Figure 6.4 marked "Byerlee" to lower values of differential stress. Second, viscous creep is expected when rock strength ($\Delta\sigma = \sigma_1 - \sigma_3$) becomes less than the least principal stress (σ_3), an empirical relation referred to as "Goetze's criterion." For the stress state on strike-slip faults outlined by Zoback and Townend (2001), Goetze's criterion is satisfied when $\Delta\sigma = (2\rho gh)/3$, where ρ is rock density, g is the acceleration of gravity, and h is depth or height of the rock column. As illustrated in

Figure 6.4, this differential stress is similar to that predicted for frictional sliding by Byerlee's law at hydrostatic pore-fluid pressure, lending credence to a strength maximum at the FVT, at least for some intervals of the interseismic period (near t_1 in Figure 6.1a).

During the long interseismic periods, rock strength from the FVT downward is limited by viscous granular flow in fine-grained micaceous rocks as well as by dislocation creep in relatively coarse-grained, quartz-rich rocks (boxes P and D in Figure 6.4). This is consistent with experimental data indicating that diffusion creep is not rate competitive in pure quartz aggregates at geologic conditions (Rutter and Brodie 2004). However, major changes in grain-scale mechanisms are expected to occur during co- and postseismic intervals, as shown by the white arrow in Figure 6.4. The increase of strain rate during coseismic rupture results in rapidly increased stress and therefore effects a transition from dislocation creep to cataclasis. Grain-size reduction and gouge formation due to cataclasis and fluid-enhanced mineral reactions promote a transition to viscous granular flow (Figure 6.4, e.g., Handy 1989) and an associated increase in strain rate. Depending on the extent of grain-size reduction, postseismic creep rates can be high while occurring at relatively low differential stresses. Creep rates may be further enhanced by increased porosity owing to dilation during rupture (Sleep 1995). The subsequent strength evolution depends on postseismic microstructural changes. Fault-sealing involving porosity reduction and cementation, or syntectonic grain growth (Gratier and Gueydan, see Chapter 12), can induce a return to dislocation creep with a related increase in strength. Alternatively, rock strength at the FVT may remain low, as shown in Figures 6.1a and 6.4, if newly crystallized minerals (e.g., phyllosilicates) pin grain boundaries; the average grain size remains small, promoting continued dominance of viscous granular flow. At greater depths below the FVT, the stress increase associated with an earthquake rupture above may not be sufficient to induce embrittlement. There, dislocation creep will be favored, even where long-term creep involves viscous granular flow.

In summary, the FVT is a zone marked by spatiotemporal variations in rock strength, mineralogy, and fluid activity that are related to the earthquake cycle, as illustrated in Figure 6.1. These fluctuations make the FVT a first-order mechanical discontinuity that often serves as a major decoupling horizon within the lithosphere.

The model shown in Figure 6.1 pertains to a strike-slip fault with no significant dip-slip motion. Large thrust faults in subduction zones and extensional faults in rifted margins may deviate from this behavior, especially due to changes in fluid availability in the vicinity of metamorphic reactions. For example, prograde reactions in rocks undergoing burial are associated with a decrease in volume and tend to release volatile phases, potentially leading to local embrittlement, and possibly even triggering seismicity, as proposed for intermediate-depth earthquakes along megathrusts in some subducting slabs (Hacker et al. 2003). In such depth intervals, the transition from viscous creep to frictional behavior (and back) is ephemeral, strain dependent, and controlled by reaction kinetics and permeability evolution (Miller 2002; Miller et al. 2003). Conceivably, such faults have several short-lived FVTs beneath the primary, long-term FVT described above, with strength and pore-fluid pressure curves that evolve differently from those shown in Figure 6.1.

How Can the Fossil Frictional-to-Viscous Transition Be Identified in Nature?

Criteria for identifying the fossil FVT in naturally deformed rocks include the structural continuity of fractures, cataclasite and mylonite formed at identical metamorphic conditions, as well as mutually overprinting relationships between fractures, cataclasite and mylonite on a given length scale of observation (Figure 6.5, Handy 1998). These criteria can be used on the outcrop and regional scales, as well as on the grain scale (Figures 6.2, 6.3). Unfortunately, tracing exhumed tracts of a fossil FVT over areas of 10–100 km^2 is hampered by later fault reactivation and by the difficulty of distinguishing coexisting cataclastic and mylonitic fault rocks at the FVT from cataclastic overprinting that all mylonitic fault rocks experience during exhumation. Recognizing the FVT on such large length scales requires excellent exposure and considerable topographic relief.

Exhumed FVTs are best preserved along the tops of the footwalls of low-angle normal faults (the so-called metamorphic carapace) within metamorphic core complexes, as depicted in Figure 6.6. There, the preservation of FVT structures is

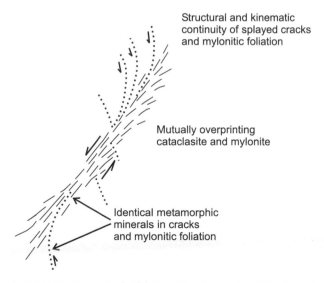

Figure 6.5 Diagnostic feature of the frictional-to-viscous transition (FVT) on the outcrop and map scales.

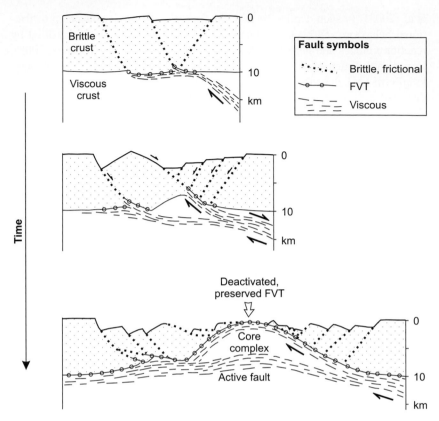

Figure 6.6 Progressive exhumation of the frictional-to-viscous transition (FVT) during crustal extension. Figure modified from Brun et al. (1994).

facilitated by strain localization during extensional denudation and cooling of the core, as shown in detail in Figure 6.7a (see also Handy 1986). The spatial distribution of microstructures within the fault reflects the fault orientation and rate of shearing with respect to the transient isothermal surfaces in the footwall of the fault. The extent of microstructural preservation depends on the rock's trajectory and, hence, on its thermomechanical history during exhumation (Figure 6.7b). Generally, preservation of the FVT in quartz-feldspar-mica rocks is best in a narrow domain of discrete, anastomozing mylonite (marked B in Figure 6.7) that experienced episodic embrittlement during exhumation through the 300–500°C range. The domains on either side of B either never experienced episodic embrittlement (C), or were affected by various degrees of brittle overprinting (A) or annealing (D). Progressive cooling and hardening of the top of the exhuming core complex can lead to a downward jump of deformation into weaker, partially melted layers (bottom of Figure 6.6). This deactivates faults along the crest of the arched core complex and so protects the FVT from brittle overprinting.

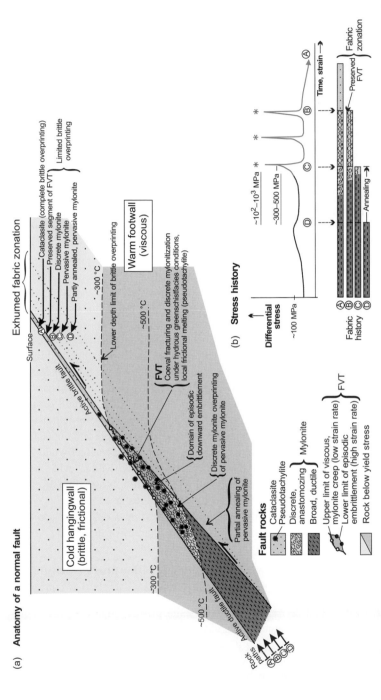

Figure 6.7 Generic cross section through a normal fault, based on an example in the Southern European Alps (Handy 1998): (a) zonation of fault rock fabrics, including the frictional-to-viscous transition (FVT), with respect to isotherms and the surface; (b) hypothetical stress and fabric histories for rock exhumation paths A–D in different parts of the fault in (a). Differential stress estimates in (b) taken from Etheridge and Wilkie (1981) and Otsuki et al. (2003).

Ancient thrust systems are less likely to preserve relics of the FVT due to overprinting during footwall collapse and/or subsequent folding, whereas faults with a large strike-slip component are steep, rendering exposure of the subhorizontal FVT fortuitous. Only where blocks of crust containing an older strike-slip fault have been rotated and/or differentially eroded, for example, in younger orogenic belts or on rift shoulders, can one expect to find continuous outcrops of the tilted and exhumed FVT.

Searches for fossil FVTs can be mounted using a combination of structural and thermochronological methods. The 270–350°C temperature range of the FVT in quartz-rich rocks (Dunlap et al. 1997; Handy et al. 1999; Stöckhert et al. 1999) coincides generally with the diffusional closure temperatures of the Rb-Sr biotite, Ar-Ar white mica, and biotite isotopic systems (Villa 1998) for a range of common grain sizes. Thus, mica cooling-age isochrons constructed from many analyses across metamorphic domes can delimit potential, temporal traces of the FVT in exhumed pelitic and granitic rocks. Once a fossil FVT is identified, dating it can be a challenge fraught with analytical and interpretational difficulties. For example, conventional Ar-Ar dating and *in situ* laser mapping of pre- and syntectonic white micas from a fossil FVT along an exhumed extensional fault yielded a range of post-tectonic Ar-Ar ages within compositionally homogeneous grains (Mulch et al. 2002); the variable ages turned out to reflect defect-enhanced diffusive argon loss controlled by intragranular microstructures formed during deformation. These microstructures reduced the effective grain size to less than the grain size observed in thin section (Mulch et al. 2002). In a different example, however, a close correlation of intragranular age and composition in different white mica microstructures formed during sequential crystallization faithfully documented progressive deformation (Mulch and Cosca 2004). Obtaining a reliable deformational age requires identifying the deformation mechanism(s) in the sample and understanding the effect of this mechanism on composition and grain size (Reddy and Potts 1999). Age versus grain-size relationships, microsampling (Müller et al. 2000), and *in situ* laser mapping of single grains all yield potentially valuable information on the relationship between measured isotopic ages and the time at which microstructures were formed and then effectively frozen. Constraining this relationship is crucial to determining the thermomechanics of the FVT.

How Do Mechanical Instabilities in the Viscous Lithosphere Nucleate and Grow?

Strain localization can be treated as a two-stage process involving the nucleation and growth of a mechanical instability, such as a shear zone. The inherent weakness of a shear zone relative to its surrounding, less-deformed, or undeformed wall rock drives localization. The deformational system is defined as the localizing shear zone plus its stable (or conditionally stable) surroundings.

Localizing instabilities can nucleate along material heterogeneities at scales ranging from individual grains to the lithosphere. In experiments, localization instabilities nucleate at material heterogeneities; these heterogeneities are either inherent to the experiment from the outset (e.g., the piston–sample interface) or must be introduced during the experiment (soft inclusions in an otherwise homogeneous medium). In fact, even in experiments designed for homogeneous deformation, it is sometimes difficult to avoid the nucleation of instabilities at the boundaries of the sample.

Continuum mechanical approaches have been developed to evaluate how strain localizes in the absence of an existing heterogeneity. Although such approaches may seem of little practical applicability to heterogeneous materials like rocks, they nevertheless bear some interesting implications for natural rock deformation. Continuum approaches to localization involve analysis of three separate criteria for the nucleation of a shearing instability (Hobbs et al. 1990): (a) stability, (b) bifurcation potential, and (c) bifurcation mode. A system becomes unstable if a small perturbation in its deformation amplifies, such that the strain energy or work performed decreases in the next increment of strain. A system can bifurcate if the deformational response to an applied stress is non-unique, that is, if deviation from a homogeneous, incremental strain field becomes possible. The bifurcation mode specifies the path of this deviation, that is, the manner and degree of strain localization. Important from the standpoint of rock deformation is that any deformation which involves a non-unique relationship between stress state and strain rate may become heterogeneous, usually while weakening, and in the case of dilatant materials, even while hardening (e.g., Rudnicki and Rice 1975). In viscous materials, the nucleation of an instability coincides with weakening of the deforming rock. For composite materials with a tendency to dilate (e.g., mylonitic rock with fractures or brittle inclusions), weakening may only initiate after the instability has grown to the point where it forms a throughgoing network subparallel to the shearing plane (Handy 1994a).

At conditions near the FVT, structural evolution associated with brittle and viscous mechanisms may lead to localization and concomitant weakening. For example, as described above, cataclasis may result in a reduction of grain size that is sufficient to promote viscous granular flow. Grain-scale weakening mechanisms that could promote localization without significant dilation include dynamic recrystallization by grain-boundary migration, texture development (crystallographic preferred orientation), and shear heating. Of these, only shear heating has yet to be unequivocally identified in shear zones. Perhaps this is because the adiabatic condition required for runaway thermal weakening only applies at anomalously high strain rates just below the FVT (Hobbs et al. 1986) where its traces are masked by postseismic and interseismic creep.

The role of syntectonic metamorphic reactions in localizing strain has been documented in nature and experiments for some time (e.g., Brodie and Rutter 1985, 1987), it but has been widely overlooked in modeling studies of lithosphere-scale faulting and orogenesis. The nucleation of fine-grained products can weaken a rock drastically, especially if the reactants are deformed outside of their *P-T* stability field (thereby enhancing reaction rates) and if the reaction is exothermic (preventing it from freezing; Green and Burnley 1990; Kirby et al. 1991). The stress drop in the shear zone is larger for reactions with a large change in molar volume, which together with the exothermic condition, limits this type of seismic instability to prograde reactions at great depth in the upper mantle, for example, in subducting lithospheric slabs (e.g., Karato et al. 2001). Endothermic reactions, especially those involving a fluid phase, can also induce rapid weakening during initial fracturing and phase nucleation (e.g., eclogite-facies pseudotachylite and shear zones in granulites; Klaper 1990; Austrheim and Boundy 1994), but beyond this, grain growth tends to stabilize creep.

At a larger scale, buckling, or active folding of existing anisotropies, is a ubiquitous strain localization mechanism in the viscous lithosphere. Metamorphic rocks generally have one or more mechanical anisotropies (compositional banding, sedimentary layering, schistosity) that develop buckling instabilities, even if they are oriented obliquely to the greatest incremental shortening axis. In strain fields near simple shear, progressive rotation of the dominant anisotropy towards the macroscopic shearing plane—whether achieved by active or passive folding—is accompanied by a reduction in bulk rock strength as stress within the layers becomes increasingly uniform. Termed rotational weakening (Cobbold 1977), this process eventually leads to pervasive mylonitic shearing as the folds become isoclinal and the dominant anisotropy is transposed into the shearing plane. As discussed below, mechanical anisotropies on all scales tend to acquire stable orientations that are related to the kinematic framework.

Networking of shear zones is another growth mechanism that weakens rock over a wide range of scales. Individual shear zone strands rotate and coalesce to form interconnected, anastomozing weak layers that envelop and boudinage lozenges of less-deformed or undeformed wall rock. The proportion of weak shear zone to stronger lozenge rock increases until the entire deformational system approaches steady-state creep at a lower bulk strength than in its initial, unnetworked state (Handy 1994b).

Can Strain Localization below the Frictional-to-Viscous Transition Induce Transience? Can It Trigger Earthquakes?

Whereas progressive folding can lead to strain homogenization and weakening as outlined above, mylonitic shearing can remain heterogeneous to high shear strains. Evidence for this persistent heterogeneity comes from the prevalence of S, C, and C' shear surfaces at very high shear strains ($\gamma > 10$) in mylonitic fault rocks (Berthé et al. 1979; Platt and Vissers 1980). The example shown in Figure 6.8 comes from a retrograde, greenschist-facies, crustal-scale shear zone formed just below the FVT transition (Palm 1999). The shear bands (C' surfaces)

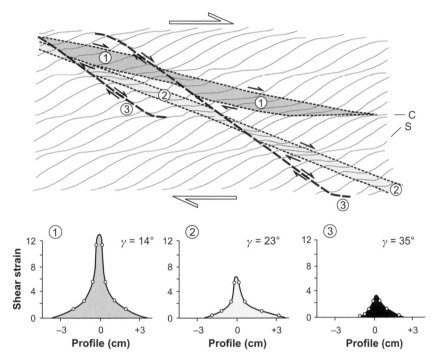

Figure 6.8 Coeval strain localization in newly nucleated shear bands, and broadening and lengthening of older, rotating shear bands in a retrograde, greenschist-facies mylonite (modified from Palm 1999). (a) Sketch of three sets of C' surfaces (shear bands), labeled 1–3 in order of decreasing age based on cross-cutting relationships. The shear bands are inferred to rotate anticlockwise, that is, antithetically, with respect to the overall dextral shear vorticity. (b) Shear strain calculated as a function of distance along profiles normal to the shear bands numbered in (a) using the foliation-deflection method of Ramsay and Graham (1970).

nucleated either as shear fractures or, more often, as hybrid, extensional shear fractures at moderate angles (20–35°) to the shearing plane (Figure 12 in Bauer et al. 2000). They then broadened, lengthened, and rotated antithetically (shear band 1, Figure 6.8) before being cut and displaced by successively younger C' surfaces (shear bands 2 and 3, Figure 6.8). The antithetical rotation of the shear bands reflects shortening normal to the shearing plane (C surfaces) during general (dextral) noncoaxial deformation, as predicted by the kinematic model of Platt (1984). Shear strain on the shear bands increased as they rotated away from their initial orientation and toward concordance with the macroscopic shearing plane, represented by the C surfaces (Figure 6.8).

The coeval nucleation and rotation of broadening shear bands are interpreted to have caused transient mechanical behavior on the length scale of these heterogeneities. The orientations of these shear surfaces, depicted in Figure 6.9a,

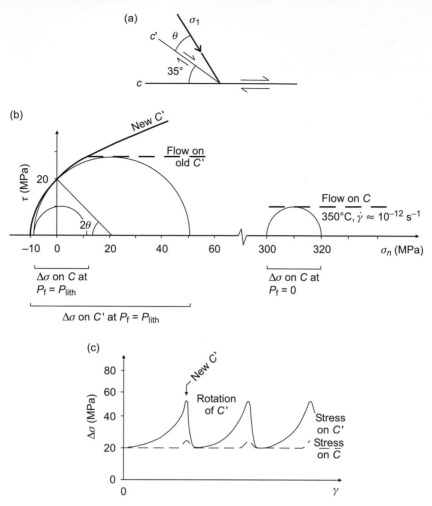

Figure 6.9 Geometry and stress states on shear surfaces for the rock in Figure 6.8. (a) Schematic representation of C and C' shear surfaces at the moment of formation of C' fractures. (b) Mohr diagram showing stress states used to constrain the stress drop at the moment of fracturing along the C' direction, as discussed in the text. (c) Episodic fluctuations of stress resulting from episodic shear band formation during continuous shearing.

form the basis for a first-order estimate of the local stress drop associated with the formation of a shear band. This estimate is obtained by subtracting the ambient, mylonitic flow stress on the S and C surfaces (6–34 MPa derived from paleopiezometry in dynamically recrystallized quartz grains in the same example above, Handy 1986) from the differential stress during hybrid fracturing across the S and C foliations ($4\sigma_T < \Delta\sigma < 2\sigma_T\sqrt{8}$, where σ_T is the tensile strength of the unfractured rock, Secor 1965). Taking average values of 20 MPa for the mylonitic flow stress, 10 MPa for σ_T (Etheridge 1983) and an average fracture angle, Θ, of 22.5° for hybrid and shear fractures in the vicinity of the natural example above (Figure 6.9a), we obtain a stress drop of ~36 MPa at the moment of fracturing on C' (Figure 6.9). Stress drops of up to ~100 MPa are possible for higher values of σ_T (20 MPa, Etheridge 1983), lower values of ambient flow stress and/or for shear bands that initiate as shear fractures ($30° > \Theta > 22.5°$) at the confining pressure of fault activity (~300 MPa; Handy 1986). Regardless of the magnitude of stress drop, antithetic rotation of the shear bands away from their initial, favorable orientation leads to progressive hardening until the differential stress locally exceeds the fracture strength of the rock and the next generation of shear bands develops (Figure 6.9c).

The example above serves to illustrate two points. First, new anisotropies nucleate in strongly anisotropic rocks when the existing anisotropies (e.g., S and C' surfaces in Figure 6.8) either become unfavorably oriented for creep and harden, or are insufficient in number to accommodate strain compatibly. Von Mise's criterion of ductility predicts that at least three independent slip systems are required for compatible plane strain deformation, and this minimum number of slip systems increases to five for general, three-dimensional deformation. Second, stress drops associated with the cyclical formation and rotation of new anisotropies are episodic on the length scale of the existing anisotropy (Figure 6.9c). However, are strength drops on the order of tens to a hundred MPa sufficiently large to trigger large-magnitude earthquakes?

The answer to this question is, conceivably, yes—provided that the shear surfaces develop quickly and on a sufficiently large length scale. As pointed out in the following section, C' surfaces in exhumed mountain belts exist on scales of one to tens of kilometers, possibly even hundreds of kilometers. A mechanism by which instabilities below the FVT might trigger seismic slip in the upper crustal part of a strike-slip fault is shown qualitatively in Figure 6.10. This diagram is modified from Figure 6.1b to illustrate the effect of weakening during a single episode of shear band formation on the crustal scale. From times t_{1a} to t_{1b}, weakening enhances the displacement rate below the FVT, thereby transferring horizontal displacement and stress upward into the brittle crust until the rock strength at the asperity (star in Figure 6.10) is exceeded at time t_{1b}. Vertical stress transfer, therefore, can trigger seismic instabilities in the upper crust earlier than if no shear bands had developed and displacement and stress had accrued continuously during homogeneous flow. Obviously, this mechanism is most effective when near-lithostatic pore fluid at the top of the viscous crust reduces the effective stress on potential fault surfaces. The feasibility of this mechanism awaits testing with appropriate mechanical models.

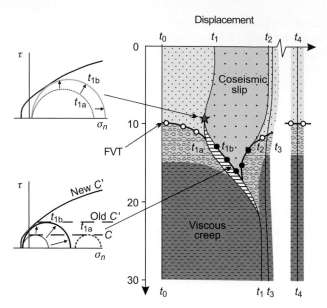

Figure 6.10 Displacement versus depth diagram as in Figure 6.1b, but illustrating accelerated displacement during a single episode of shear band development below the FVT from times t_{1a} to t_{1b} (vertically hatched domain). Vertical stress transfer during this time potentially triggers coseismic slip at an asperity (star) in the upper, brittle crust at time t_{1b} (see text for explanation). Mohr diagrams contain semicircles representing stress states at times t_{1a} (dashed semicircles) and t_{1b} (solid semicircles).

What Factors Affect the Geometrical Scaling Properties of Shear Zones?

Natural deformation is heterogeneous on some scales but homogeneous on others. Our goal is to understand under which conditions the assumption of homogeneous deformation is reasonable and how bulk rheology changes when strain becomes heterogeneous.

The scales at which strain becomes localized are often controlled by existing mechanical heterogeneities in the crust and/or mantle. For example, centimeter-scale mylonitic shear bands scale with harmonics of the characteristic spacing between feldspar clasts in the protolith (Dutruge et al. 1995). In contrast, the spacing and aspect ratio of S and C shear surfaces are fractal on scales ranging from 10^3 μm to 1 km (Hippert 1999). Some strike-slip faults several hundreds of kilometers long transect entire mountain chains at low angles to the orogenic trend (e.g., North American Cordillera; see Figure 1 in Eisbacher 1985) and may represent C' surfaces (i.e., shear bands) at the largest scale. If so, the heterogeneities that promote strain localization on such long scales may reside in the upper mantle. Alternatively, they may reflect the distribution and amounts of metamorphic fluid and/or melt in the lower crust (Chapter 13).

Yardley and Baumgartner (Chapter 11) mention several mechanisms by which fluids can weaken rock, including hydrofracturing due to increased pore-fluid pressure, hydrolytic weakening of silicate minerals, or transformation of hard reactant minerals to intrinsically weak, hydrous mineral products (e.g., micas). Fluids released during prograde metamorphic reactions at depth may advect to sites of ongoing shearing, further weakening the rock and shortening the length scale of localization (see Chapter 4). The growth of mechanical instabilities may also be influenced by strain-dependent transitions in the active deformation mechanism(s), changes in the kinematics of deformation (i.e., shape of the strain ellipsoid, degree of noncoaxiality), or some combination of the above.

Two end-member scenarios describing the evolution of strain localization in a strike-slip fault system are illustrated in Figure 6.11. In strike-slip fault systems, most displacement is accommodated along a single master fault or a limited number of fault strands oriented parallel to the macroscopic shearing plane; only minor displacements are taken up by the surrounding complex of shear surfaces. The evolution of this geometry involves either increasing strain localization as the surrounding shear surfaces are deactivated and strain is concentrated on a narrow master fault (Figure 6.11a, homogenization of strain on a smaller scale than that of the initial fault array), or strain delocalization as shear surfaces making up an initially smaller fault array rotate, broaden, and coalesce to form the master fault (Figure 6.11b, homogenization of strain on the scale of the initial fault array). The former evolution, that is, "down-scaling" of strain homogeneity, has been observed in analogue studies of fault systems in brittle and elastoplastic materials (Tchalenko 1970) as well as in multi-layered, brittle-viscous composites (Schreurs 1994). "Up-scaling" of strain homogeneity has

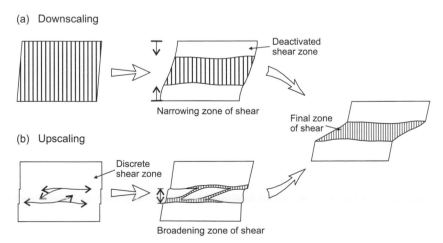

(a) Downscaling

Deactivated shear zone

Narrowing zone of shear

Final zone of shear

(b) Upscaling

Discrete shear zone

Broadening zone of shear

Figure 6.11 Two end-member evolutions of strain localization: (a) down-scaling; (b) up-scaling.

been inferred from structural studies on natural shear zones (Fusseis et al. 2006) and observed on the granular scale in *in situ* deformation experiments (White et al. 1985). The implication of both types of strain evolution is that the relatively simple kinematic and dynamic boundary conditions of structurally mature fault systems justifies the extrapolation of laboratory-derived flow laws to the largest length scale at which shearing is homogeneous.

These structural observations provide a context from which to develop rheological models of large-scale fault systems. For continuum mechanical modeling of heterogeneous deformation, the challenge of modeling scale-dependent deformation is to define a representative elementary volume (REV), the smallest length scale or volume of rock that is statistically homogeneous. The REV comprises all constituent features of a rock that may be averaged and therefore treated as a structural and mechanical continuum on larger scales. Because rock is heterogeneous at scales smaller than that of the REV, the material parameters at larger scales must be determined from models that account for the properties of the components of the REV. Thus, the properties of the REV might be constrained by averaging the rheological parameters of the constituent phases. This would involve volume-weighting of the constituents' parameters, similar to the approach employed in modeling of polymineralic viscous aggregates (Tullis et al. 1991; Handy et al. 1999). In the case of heterogeneous shearing, however, the mechanical phases correspond not to different minerals but to sheared and less-sheared rock volumes with correspondingly different structures and rheologies (Handy 1994b).

What Is the Stress State in Shear Zones?

In situ stress measurements in deep boreholes indicate that the continental crust down to the FVT is at a critical or near-critical stress state at hydrostatic pore-fluid pressure (KTB borehole: Dresen et al. 1997; Zoback and Harjes 1997); that is, even small variations in the stress applied can lead to failure and a marked drop in strength. At the San Andreas Fault Observatory Drillhole (SAFOD), the measured stresses are consistent with those obtained from Byerlee's Law (Byerlee 1978, regime 3) for the thrust fault orientation in that area (Hickman and Zoback 2004). These observations suggest that brittle strength is invariant with length scale, at least over several orders of magnitude. However, it is unlikely that a measured stress state, critical or not, pertains to all length scales. For example, the anastomozing nature of all faults, above and below the FVT, indicates that yielding is attained only at some length scales and not at others. Therefore, criticality with regard to stress state appears to vary with time- and length scales of strain localization.

From the FVT downwards, differential stress estimates on the order of ~10–100 MPa are obtained from grain-size paleopiezometers for dynamically recrystallized quartz, and by extrapolating experimentally derived constitutive relations for dislocation creep of quartz to natural strain rates (Etheridge and Wilkie 1981; Hirth et al. 2001; Stipp et al. 2002). Similar values have been obtained with identical methods for olivine-rich rocks of the upper mantle. These stress values are based on the assumption of mechanical and microstructural steady state for the length scale of observation (usually a thin section). In some cases, this assumption appears reasonable based on the preservation of relatively homogeneous microstructures at scales of 10–100 m (e.g., Dunlap et al. 1997). In other cases, however, microstructures preserve evidence of very large differential stress gradients on the millimeter scale (Handy 1994a).

The 10–100 MPa range cited above for rocks undergoing dislocation creep probably represents maximum, near-steady-state values, because any contribution from viscous granular flow mechanisms would reduce the creep strength, as inferred from microstructural studies in many shear zones. As mentioned above, such flow mechanisms are enhanced relative to dislocation creep when grain boundaries are pinned by phases produced during syntectonic reactions (Stünitz and Fitzgerald 1993). Most of the crustal faults sampled for paleo-piezometric studies are retrograde, amphibolite- and green-schist-facies shear zones. Their measured differential stress levels are probably greater than in higher grade shear zones, where high temperatures favor lower stresses. The extrapolation of experimental flow laws for power-law creep to temperatures at the base of the continental crust suggests that differential stresses there may be as low as 1–10 MPa, although nominally dry rocks (e.g., granulites) deforming at the same conditions may support flow stresses of an order of magnitude greater (e.g., Kohlstedt et al. 1995). This is corroborated by the observation of very fine (≤ 10 μm), dynamically recrystallized quartz and feldspar grains in granulites of the Ivrea-Verbano Zone, northern Italy (Zingg et al. 1990).

Prograde shear zones are rarely preserved in nature and usually cannot be used for paleopiezometry because post-tectonic grain growth has modified their stress-sensitive microstructures. Prograde shear zones probably accommodated crustal thickening at low temperature and/or high differential stress during initial burial. Only where stress and temperature drop very rapidly are prograde, stress-sensitive microstructures expected to be preserved (Knipe 1989). In general, it is a rock's thermal history and the mechanical response to that history which determine whether or not its dynamic microstructures are "frozen in."

Deviations from steady-state creep are expected over short timescales (<1000 yr) in the vicinity of active faults. As cited above, peak values of differential stress during coseismic faulting may be significantly higher than steady-state values. High transient stress during creep is consistent with evidence of rapid frictional melting and quenching in the form of pseudotachylite (rock glass) in some shear zones; the formation of pseudotachylite requires very fast (~ 1 m s^{-1}), short ($\leq$ s) bursts of movement at high shear strain rate ($\gamma = 1$–1000 s^{-1}) and shear stress ($\sim 10^2$ to 10^3 MPa; Otsuki et al. 2003).

The rate of stress change following rapid loading-unloading events (time interval t_2–t_3 in Figure 6.1) depends on the initial (peak) stress as well as on the strain, the creep parameters, and the ambient temperature of the host rock. Minimum rates of stress drop for the preservation of dynamically recrystallized quartz grain size in equilibrium with stress are about 10^{-9}–10^{-11} MPa s^{-1} (Prior et al. 1990), but much higher stress rates (and correspondingly large changes in strain rate) certainly occur, as manifested by stress-dependent microstructures in mylonite that have not equilibrated with differential stress (Küster and Stöckhert 1999; Trepmann and Stöckhert 2003). A key question from the standpoint of rheology is: How much strain and/or time are necessary for a state variable (e.g., grain size or dislocation density) to remain in equilibrium with stress after a change in deformational conditions? For example, when deformation is accommodated by subequal contributions of diffusion creep and dislocation creep, the grain-size evolution after a sudden change in deformational conditions controls the subsequent rheological evolution (Montesi and Hirth 2003).

Treatment of the lower continental crust as a heterogeneous viscous body emulates the stress memory of horizontally layered crust; a postseismic stress decay on the order of 10^{-7} MPa s^{-1} is obtained during the first 20 days after an assumed shock of M_w 7.3 (Ivins and Sammis 1996). Rapid viscous unloading of the lithosphere may occur in response to one or a combination of events: (a) the networking of large shear zones leading to the formation of an interconnected layer of weak rock; (b) the reaction of strong reactant phases, particularly phases deformed at physical conditions outside of their stability field, resulting in the production of intrinsically weak phases; (c) fluid influx reducing the effective stress, favoring brittle failure and stress release. In the case of (a), experimental and theoretical work on the rheology of polycrystalline aggregates indicates that the magnitude of weakening owing to networking depends on the interconnectivity of the weak phase and on its strength contrast with the surrounding, stronger phase (Handy et al. 1999). For two phases with a strength contrast of 5:1, for example, the interconnection of only 10 vol.-% of the weak phase reduces the strength of the entire rock by some 80%. However, shear zones do not always anastomose completely and the entire lithosphere may not be cut by a single shear zone (e.g., Lister et al. 1991; Brun et al. 1994). Therefore, stress can remain very heterogeneous within the deforming lithosphere. In that case, bulk strength does not drop as far towards the uniform-stress lower bound and the strength of the system remains within 15–20% of the bulk strength of homogeneously deforming rock (Handy 1994b).

The mechanical response of the viscous crust to perturbations in stress and strain determines how stress is transferred through the lithosphere, both among shear zones below the FVT and back upwards to seismogenic fault zones in the brittle crust. To a first approximation, creep related to the downward propagation of a rupture surface into the viscous crust occurs just ahead of the surface and on either side of the surface at distances comparable to the depth of the seismogenic zone (Mavko 1981; Montesi 2004). The time- and length scales of stress transfer in the viscous crust vary with viscosity, which in turn decreases with depth as a function of the geothermal gradient at a given background strain rate. The shorter the response time and the longer the length scale of stress transfer below the FVT, the greater the potential for shear zones to (re)load faults in the overlying brittle crust and therefore to trigger earthquakes. Stress transfer among brittle faults (Stein 1999) may be controlled in part by the viscous response of their ductile continuations below the FVT (e.g., Freed and Lin 2003).

What Do Measurements of Surface Deformation Tell Us about Rheology below the Frictional-to-Viscous Transition?

Large earthquakes or other deformation sources (e.g., volcanic intrusions, water reservoir fluctuations, or glacier surges) initiate a rock mechanics experiment of lithospheric dimensions in which a sudden stress change leads to a measurable relaxation of the lower crust and upper mantle. The surface motions related to faulting can be tracked with millimeter precision using space-based geodetic techniques—global positioning systems (GPS), very long baseline interferometry (VLBI), and synthetic aperture radar interferometry (InSAR). Thus, surface measurements provide a basis for testing structural and rheological models of the lithosphere below the FVT. Of course, motions at depth are filtered through the upper, brittle crust so that using them to resolve structure and constrain rheological parameters below the FVT is complicated by the fact that several deformational processes contribute to the observed motions.

Deformation rates during the interseismic period late in the earthquake cycle do not vary significantly with time and can be used to estimate long-term geologic slip rates (Savage and Burford 1970). First-order estimates of rheological parameters have been obtained by integrating geologic and geophysical information: paleoseismic constraints on the time since the last major earthquake, average earthquake repeat intervals, and geologic fault slip rate estimates, combined with estimates of the depth to the FVT from seismic or heat flow data can all be incorporated in a Bayesian statistical approach (Segall 2002; Johnson and Segall 2004; Hilley et al. 2005). In applying this approach to the Kunlun fault in northern Tibet, Hilley et al. (2005) showed that the viscosity of the lower crust may be at least an order of magnitude greater ($> 10^{19}$ Pa s) than that estimated in channel flow models (10^{16}–10^{18} Pa s; e.g., Clark and Royden 2000). In another study using VLBI and GPS data from the southwestern U.S., Flesch et al. (2000) determined that effective, depth-averaged viscosity decreases by three orders of magnitude going from the Basin and Range area to the San Andreas fault system and the eastern California shear zone. Their estimates probably overestimate the real viscosity below the FVT, because they effectively equate strain rate with viscosity and therefore provide average viscosity estimates for the whole lithosphere.

Whether or not deformation is localized below the active FVT can be determined by comparing interseismic slip rates on aseismically moving fault segments with the overall, long-term displacement rate on those segments (Thatcher 1983). If deformation below the seismogenic zone is broadly distributed, then interseismic displacement rates should only be a fraction of the long-term rate. Accelerated motion following earthquakes on adjoining seismogenic fault segments would account for a significant portion of the long-term slip budget (Ben-Zion et al. 1993). However, observed rapid displacement rates along the central San Andreas and Hayward faults (California) approach their long-term slip rates, so deformation below the FVT is inferred to be quite localized (Bürgmann et al. 2000; Malservisi et al. 2002). A limitation of this approach is that the evolution of fault displacement rate on any given fault segment is affected by the activity of nearby seismogenic faults. Fault displacement therefore depends on the relaxation of time constants and distribution of transient viscous flow at depth (Schmidt 2002; Lynch et al. 2003).

Geodetic measurements of postseismic relaxation are more effective than interseismic displacement measurements for probing structure and rheology below the FVT. Interpretation of postseismic measurements in terms of rheology is hampered by the fact that several processes from different depths of a fault system contribute to the observed transient deformation. These include afterslip on narrow faults above (Bilham 1989; Bürgmann et al. 1997) and below (Tse and Rice 1986) the base of the seismogenic zone, viscous flow in the lower crust and upper mantle (Pollitz et al. 2000), and poroelastic rebound in the upper crust due to fluid flow in response to coseismic pressure changes (Peltzer et al. 1996; Jónsson et al. 2003). These effects are difficult to separate (Thatcher 1983; Savage 1990), especially for moderate to small earthquakes ($M_W < \sim 7$), where transient signals from below the FVT are difficult to distinguish from the multitude of shallow events (Bilham 1989; Pollitz et al. 1998; Jónsson et al. 2003). However, the signals are not as ambiguous when considering postseismic deformation near large strike-slip ruptures with a well-defined, finite length (Pollitz et al. 2000) or faults with a significant dip-slip component (Pollitz et al. 1998; Hsü et al. 2002).

Comparison of three-dimensional ground motions associated with large strike-slip earthquakes indicates that the style of deformation below the seismic–aseismic transition varies with lithology, lithospheric structure, and thermal regime. For example, the pattern of deformation after the 1992 $M_W = 7.3$ Landers, 1999 $M_W = 7.1$ Hector Mine earthquake sequence in the Mojave Desert of the southwestern U.S. (Figure 6.12) suggests that poroelastic rebound predominated in the upper crust (Peltzer et al. 1996), whereas viscous flow below the coseismic rupture was probably localized in the lower crust (Fialko 2004) and/or upper mantle (Pollitz et al. 2000, 2001; Freed and Bürgmann 2004). Freed and Bürgmann (2004) fit the spatial and temporal patterns in the GPS data following both earthquakes with a model that relied on experimentally

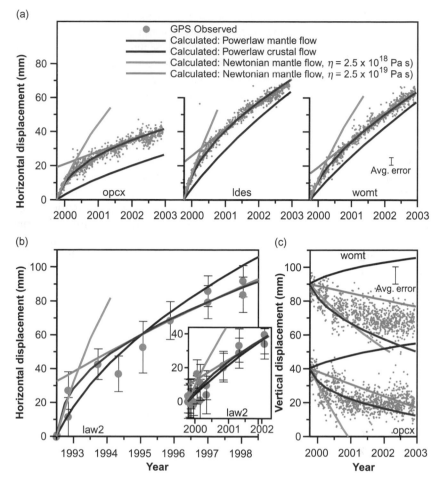

Figure 6.12 Observed and calculated postseismic displacement time series following the 1992 Landers and 1999 Hector Mine earthquakes (from Freed and Bürgmann 2004): (a) horizontal displacements at three continuously monitored GPS stations following the 1999 Hector Mine earthquake; (b) horizontal displacements at campaign GPS station law2 following the 1992 Landers earthquake. Inset shows campaign data at law2 following the Hector Mine earthquake; (c) vertical motions at two continuously monitored stations following 1999 Hector Mine earthquake. The power-law mantle flow model (red curves) is valid for an aplitic lower crust, a mantle with the rheology of nominally wet olivine (Hirth and Kohlstedt 2003), and a geotherm near the upper bound permitted ($T_{40\,km} = 1225°C$) by heat flow data. Neither the power-law crustal model (blue curves), which assumes wet quartzite in the lower crust and a cooler dry olivine mantle ($T_{40\_km} = 1100°C$), nor any models that assume dominant creep in the lower crust, fit the vertical motion data. The Newtonian models incorporate mantle flow with a low viscosity (2.5×10^{18} Pa s, cyan curves) and an order-of-magnitude greater viscosity (green curves) that match early and late time series slopes, respectively.

determined nonlinear viscous creep parameters of nominally wet olivine (Hirth and Kohlstedt 2003) at temperatures consistent with surface heat-flow measurements (Figure 6.12). In light of microstructural evidence for viscous granular flow in naturally deformed lower crustal and upper mantle rocks (e.g., Jaroslow et al. 1996; Stünitz and Fitzgerald 1993; Handy and Stünitz 2002), it is interesting to consider whether the same GPS data could be modeled with a composite dislocation creep-diffusion creep rheology in which the average grain size evolves with time (e.g., Montesi and Hirth 2003). Grain growth at high homologous temperatures is likely in fine-grained aggregates produced by high coseismic stresses.

Irrespective of the creep mechanisms, the lower crust beneath the earthquakes appears to contribute little to the postseismic transients of the first few years, suggesting that Mojave Desert crust comprises relatively mafic and/or dry lithologies. Thus, the viscous contribution to relaxation following the Landers–Hector Mine earthquake sequence appears to originate from flow of a hot, possibly wet asthenospheric mantle—a conclusion also reached for the 2002 $M_w = 7.8$ Denali earthquake in southern Alaska (Bürgmann et al. 2003). Both of these earthquakes occurred in broad, active plate-boundary settings that lack the deep lithospheric root found beneath cratonic, continental interiors (Dixon et al. 2004; Hyndman et al. 2005).

In contrast to the cases above, very localized shearing extending below the coseismic rupture to depths of 35 km to the base of the crust is inferred from surface deformation following the 1999 $M_w \approx 7.4$ Izmit earthquake along the North Anatolian fault in Turkey (Bürgmann et al. 2002). A velocity-strengthening frictional rheology or a tabular zone of very low effective viscosity in the lower crust provides good fits to the observed surface motions (Hearn et al. 2000). A similar conclusion was reached by Kenner and Segall (2003) based on modeling of several decades-old accelerated postseismic surface deformation following the 1906 $M_w \approx 7.7$ San Andreas fault earthquake. Likewise, Johnson and Segall (2004) found that deformation data in northern and southern California are best fit by a single model parameterization, combining a lower crustal, aseismic shear zone beneath the San Andreas fault with an underlying viscous mantle. This is qualitatively consistent with evidence for distributed shear in the upper mantle from shear wave splitting (Hartog and Schwartz 2001) and magnetotelluric studies (Maerklin et al. 2005).

With the exception of studies of subduction zone events, there are few well-studied postseismic deformation studies for thrust faults and normal faults. Two earthquake studies thus far illustrate differences in the surface response to thrusting in differently aged, convergent settings: a young active fold-and-thrust belt in Taiwan versus intraplate deformation of a cold, thick cratonic lithosphere on the Indian peninsula. In the former case, rapid and extensive surface deformation following the $M_w = 7.8$ Chi-Chi thrust earthquake in Taiwan indicates the strong contribution of afterslip extending down-dip from the base of the coseismic rupture surface (Hsü et al. 2002), possibly into the viscous crust. In contrast, the equally large 2001 Bhuj intraplate thrust earthquake, a deep event located in a reactivated Proterozoic rift structure in northwestern India, was followed by only minor transient deformation in the first six months (Jade et al. 2002).

Normal faulting during the 1959 Hebken Peak earthquake in the northern Basin and Range province (southwestern U.S.) led to broad postseismic uplift that is consistent with viscous relaxation in the upper mantle, but not with lower crustal relaxation or localized afterslip in the viscous crust down-dip of the rupture (Nishimura and Thatcher 2003). These results are consistent with other analyses of lithospheric rheology in the Basin and Range province based on deformation associated with reservoir impoundment (Kaufmann and Amelung 2000) and Holocene rebound of shorelines along Lake Bonneville (Bills et al. 1994).

Mineralogy, geothermal gradient, and fluid availability govern the rheology of the lithosphere, and thus the nature of deformation during the earthquake cycle. Where the bulk viscous rheology below the FVT is nonlinear, as suggested by surface deformation studies and microstructural studies of exhumed mylonitic rocks, viscosities can be highly stress and, hence, time dependent. A strong time dependence of bulk viscosity is also expected for lithosphere containing networks of fine-grained mylonitic rocks undergoing viscous granular creep, especially if syntectonic grain size increases with time during the post- and interseismic intervals (Montesi and Hirth 2003). Thus, effective viscosities determined by measurements of surface deformation at various stages of postseismic deformation cannot be directly compared, either with each other or to estimates derived from long-term geologic features. Better and more data are needed to constrain fully the evolution of postseismic deformation in time and space.

Owing to the relatively small strains during postseismic deformation, it is unclear how much of the deformational response measured at the Earth's surface is accommodated by transient creep processes. For example, experimental work indicates that although the temperature and stress dependence of high-temperature transient creep in olivine is similar to that of steady-state creep, the effective viscosity may be a factor of 2–5 less (Chopra 1997). A promising approach for understanding the physics of transient creep involves formulating equations of state that bridge the gap between steady-state and transient creep processes (Stone et al. 2004). This may significantly improve our ability to relate creep rates inferred from geodetic measurements to deformational conditions recorded by microstructures in exhumed fault rocks.

How Is Decoupling across Mechanical Interfaces Related to Strain Partitioning in the Lithosphere?

"Coupling" and "decoupling" are widely used and misleading terms because they mean different things to different specialists. In geological parlance, decoupling refers to discontinuities in a velocity field, manifested as a displacement of markers across a surface (i.e., a fault or shear zone). As noted in the discussion of Figure 6.1, faulting can involve episodic decoupling. The degree of decoupling can be defined as the ratio of displacement on a fault to the total displacement accommodated by the system (fault plus adjacent rock) over a specified time interval. This definition obviously depends on the time- and length scales of observation. By contrast, in seismological parlance, "seismic coupling" refers to the proportion of total displacement on a surface accommodated by coseismic movement. Here, confusion arises because decoupling zones (in a geological context) during an earthquake are referred to by seismologists as seismic coupling zones.

Similarly, "partitioning" is a kinematic term describing the division of a far-field strain or displacement field into different vectorial components on a system of variably oriented faults (discussion in Dewey et al. 1998). Partitioning necessarily involves decoupling, but not all decoupling surfaces partition displacement. Displacements on faults that are coincident with the overall displacement field are said to be unpartitioned. These definitions are important when attempting to draw mechanical inferences from displacement fields.

First-order decoupling surfaces within the continental lithosphere are the sediment–basement contact, the FVT, the crust–mantle boundary, and, where present, the boundary between the base of the hydrous quartz-rich crust and the mafic lower crust (Handy and Brun 2004). These are inferred to be weak horizons within the lithosphere. However, either coupling or decoupling, in the absence of partitioning between the continental crust and upper mantle, is required to explain nearly coincident displacement fields for the surface and the upper mantle inferred from geodetic and seismic anisotropy data in eastern Tibet (Holt 2000). In subduction zone systems, the relationship between plate kinematics and fault displacements is more complex: oblique convergence between lithospheric plates is partitioned into thrusting at high angles to the plate boundary on low-angle megathrusts and strike-slip motion on trench-parallel faults within the magmatic arc (Fitch 1972; Jarrard 1986).

Rheological models incorporating time-dependent brittle behavior of the upper crust predict that, while viscous creep prevails at depths below about 15 km, fault motion above this level is punctuated by episodic, coseismic displacement (Figure 6.1). In the case of strike-slip faulting, such models adopt two, end-member, dynamic-kinematic boundary conditions: lateral drive, in which forces are exerted by the opposite motion of blocks on either side of the fault (e.g., Tse and Rice 1986), and basal drive involving drag of the blocks from below by creep of a viscous substratum (e.g., Li and Rice 1987). Molnar (1992) argued that viscous creep of the lower crust transmits displacements from the strong, ductile upper mantle to the upper, brittle crust, while smoothing displacement discontinuities between these two layers. He inferred the lower depth limit of strain partitioning in the crust to be the interface between brittle and viscous crustal domains (the FVT) because flow of the weak, viscous crust causes the principle stress axes to be nearly orthogonal to the horizontal base of the over-lying crustal blocks. Still, most rocks in the lithosphere are strongly aniso-tropic and folding of anisotropies below the FVT can cause the principal stress axes to be oblique to both compositional layering on the lithospheric scale and to shear zones aligned parallel to the FVT. Strain continuum models (Bourne et al. 1998; Teyssier et al. 2002) predict that intracrustal strain is accommodated within zones of distributed, three-dimensional strain at the top of the viscous crust. The thickness and geometry of these accommodation zones depends presum-ably on the degree of coupling at the FVT, which in the continuum models is simply assumed to be complete. These end-member models shed valuable light on some, but not all, aspects of strain accommodation at mechanical interfaces in the lithosphere. The dilemma remains whether strain compatibility in the lithosphere is maintained by coseismic decoupling, by continuous coupling and three-dimensional flow, or by some combination of these scenarios.

A possible solution to this dilemma is that mature shear zone systems com-prise zones of decoupling which develop, usually in previously foliated rocks, in response to strain incompatibilities between adjacent lithospheric blocks. Structural studies reveal that the geometry of networked, shear zone systems in foliated rocks reflects both the flow vorticity and the shape of the finite strain ellipsoid (Figure 6.13a; Gapais et al. 1987). The orientations of mylo-nitic foliations and stretching lineations in natural fault systems support the prediction of slip-line theory for anisotropic materials (Cobbold and Gapais 1986) that active shear zones track (but do not exactly coincide with) surfaces of no finite extension within the bulk strain ellipsoid, whereas their lineations track directions of maximum shear within this ellipsoid (Figure 6.13b). These fabric orientations correspond to the shearing plane and principal finite strain axes only for simple strain configurations (e.g., simple shearing, coaxial flat-tening, Figure 6.13b) at high strains. For more general, three-dimensional strain states, they are oriented obliquely to the shearing plane. The attainment of these stable orientations for foliations and stretching lineations involves sev-eral possible mechanisms: from the growth and/or rigid body rotation of platy minerals (e.g., mica) to the buckling and shearing of layers, as discussed in the previous section.

Strain partitioning and fabric asymmetry related to decoupling at and below the FVT are shown in Figure 6.14, which summarizes the deep structure of the Periadriatic fault system in the Central Alps as exposed today at Earth's sur-face. As described by Furlong et al. (Chapter 4), this fossil fault system is an excellent natural laboratory for studying deep-seated processes. Oblique shorten-ing is accommodated by a network of kilometer-scale shear zones and folds that nucleated at existing nappe contacts, and especially along the surface of the rigid, cold orogenic block. When analyzed with the slip-line method, the orienta-tion of this network with respect to the cold indenter block (equal area projections in Figure 6.14) is consistent with dextral transpression and exhumation of the

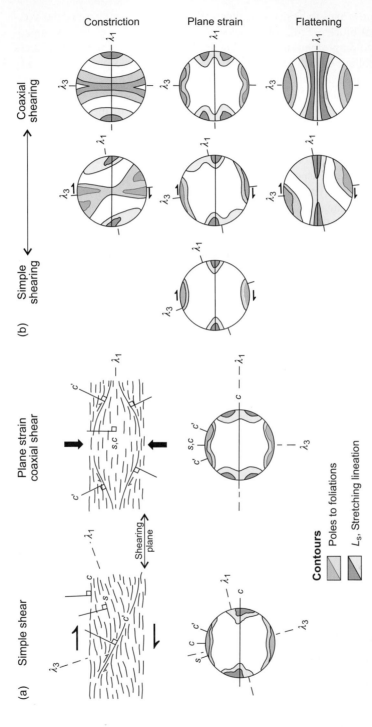

Figure 6.13 Structures and orientation distribution patterns for shear zones predicted by slip-line theory: (a) bulk simple shearing (left) and plane-strain coaxial shearing (right) with diagnostic pole patterns: black, gray = poles to mylonitic foliations (S, C, C'); hatched, dark gray = stretching lineations; (b) equal-area projections of predicted preferred fabric orientations for various finite strains and deformational histories. λ_1, λ_3 are the greatest and least principal quadratic stretching directions. Figure modified from Figure 14 of Gapais and Cobbold (1987).

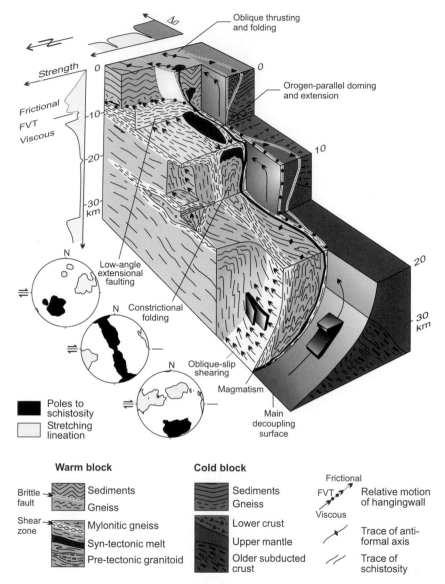

Figure 6.14 Generic model of decoupling zones related to strain partitioning along the transpressive Periadriatic fault system (modified from Handy et al. 2005). The equal area projections (bottom left) contain planar and linear fabrics in the warm block.

warm orogenic block (Handy et al. 2005). These kincmatically linked structures thus serve as vertical and horizontal decoupling horizons, partitioning strain and transferring stress from the upper mantle to the upper crust.

We note that the symmetry of the foliation and stretching lineation orientations predicted with the Cobbold–Gapais slip-line model is never lower than monoclinic, even in general three-dimensional strain fields. This contrasts with the triclinic fabric symmetries predicted for some transpressional and transtensional configurations by strain continuum models in which the foliation and lineation are assumed to be passive markers (Fossen and Tikoff 1998; Lin et al. 1998). In these models, fabrics tend to rotate toward so-called fabric attractors, that is, directions for which the rates of particle motion are lowest. Yet, despite the simple assumptions of the slip-line model (simple shearing parallel to inextensible anisotropies, no strain hardening, no mechanical control on shear zone initiation and propagation), the fact that it incorporates anisotropies which can rotate and/or grow at each other's expense renders it arguably more realistic than existing, more complex three-dimensional strain models for homogeneous materials. Excepting some high-grade shear zones in which strength contrasts are presumably negligible, most natural rocks have one or more anisotropies (schistosity, compositional layering) comprising minerals with contrasting rheologies.

The Cobbold–Gapais model has several implications for coupling and strain partitioning in the lithosphere. First, strain partitioning is not restricted to the brittle crust; it potentially occurs in all foliated rocks of the lithosphere. Second, decoupling and strain partitioning are complementary phenomena on the time- and length scales of deformation in the viscous crust; shear strength is reduced, but never totally lost during the networking of shear zones. Finally, the orientation of structures in shear zones yields information about the strain and stress history during localization (Figure 6.13b).

These ideas pertain to the crust and upper mantle, but little is known about decoupling and strain partitioning at the interface of the lithosphere and asthenosphere. As imaged by current geophysical methods (see Chapter 2), shear zones in the upper mantle appear to be homogeneous shear zones some 100–200 km wide (Vauchez and Tommasi 2003). However, shear zones active in the upper mantle that have been mapped in exhumed mantle rocks are much narrower (widths from millimeters to hundreds of meters, e.g., Drury et al. 1991; Vissers et al. 1995; Handy and Stünitz 2002), indicating that current imaging methods are unable to resolve all length scales of localization in the mantle (see Chapter 4). Large strike-slip faults at the surface can be traced down to the upper mantle, where they are discernable as offsets in P-wave velocity anomalies at depths of 170 km (Wittlinger et al. 1996). Whereas fault geometry and kinematics in the lithosphere reflect the motion of tectonic plates, faults at the base of the lithosphere may develop in response to motion of the asthenosphere. Numerical models of convective overturn in a closed and coupled lithosphere–asthenosphere system indicate that surface displacement fields and faults may be kinematically linked to the boundaries of rapidly convecting mantle masses (Trompert and Hansen 1998).

CONCLUSIONS

Faults evolve on different time- and length scales. As such, faults must be considered as multicomponent systems that comprise one or more mechanical instabilities (weakened rock) and their surroundings (adjacent rock, the erosional surface). Faults nucleate on existing mechanical heterogeneities in the lithosphere, but it is still not clear why mechanical instabilities develop faster and grow longer on some length scales and not on others. Several factors influence the growth of instabilities, primarily the nature of stress interaction between existing heterogeneities. This interaction is itself a function of the geometry and spacing of these heterogeneities, the time- and rate-dependent rheology of the intervening rock, as well as the ambient, far-field stress state.

The rheology of fault rocks is more than just the relationship between the tensors of stress and incremental strain or strain rate; it includes aspects of rock history, such as the compositional, thermal, and structural characteristics of the rock. In general, the longer the time- and length scales of faulting considered, the greater the potential influence of the kinematic and thermal history on the rheology of the fault system.

Fracturing and folding-plus-mylonitization are the main modes of strain localization, respectively, above and below the FVT. These modes operate over the same large range of length scales, but on very different timescales, reflecting the different grain-scale mechanisms (cataclasis, creep) controlling these processes. At the FVT, localization processes are strain dependent and rate competitive; brittle and viscous mechanisms contribute subequally to the bulk strain and strain rate if regarded on sufficiently long time- and length scales. Still, fault systems never achieve mechanical steady state in their entirety. This pertains especially to parts of faults at and near the FVT, where brittle, sometimes seismic instabilities perturb the ambient stress field on timescales ranging from seconds to years. As a consequence, the notion that the crust is at a critical state of failure may only apply to length scales related to those of existing mechanical heterogeneities, and on timescales related to the mechanical response time of the crust.

Fault geometry simplifies with progressive strain, lending justification to the use of laboratory-derived flow laws to estimate the bulk rock rheology on length scales at which strain is homogeneous. The rheology of larger or smaller rock bodies for which strain is heterogeneous can be modeled either as a two-phase aggregate or a representative size or volume within which the mechanical properties of the system are statistically homogeneous. In general, the kinematics of faulting and fault geometry are as important as grain-scale deformation mechanisms in determining the rheology of fault systems on the lithospheric scale. This is consistent with observations and theoretical predictions that shear zone structures (foliations, stretching lineations) attain orientations which reflect the kinematic vorticity and the shape of the finite strain ellipsoid. Shear zones are therefore zones of strain accommodation that transmit shear stress within the lithosphere.

ACKNOWLEDGMENTS

We would like to thank our friends and colleagues of the Dahlem Workshop for animated discussion on this subject, as well as Martyn Drury, Christian Teyssier, and Torgeir Andersen for their thoughtful reviews. Martina Grundmann is thanked for patient technical support with the figures. MH acknowledges the support of the German Science Foundation (DFG) in the form of grant Ha 2403/6.

REFERENCES

Austrheim, H., and T.M. Boundy. 1994. Pseudotachylites generated during seismic faulting and eclogization of the deep crust. *Science* **265**:82–83.

Bauer, P., S. Palm, and M.R. Handy. 2000. Strain localization and fluid pathways in mylonite: Inferences from in-situ deformation of a water-bearing quartz analogue (norcamphor). *Tectonophysics* **320**:141–165.

Ben-Zion, Y., J.R. Rice, and R. Dmowska. 1993. Interaction of the San Andreas fault creeping segment with adjacent great rupture zones and earthquake recurrence at Parkfield. *J. Geophys. Res.* **98**:2135–2144.

Berthé, D., P. Choukroune, and D. Gapais. 1979. Orientations préférentielles du quartz et orthogneissification progressive en régime cisaillant: L'exemple du cisaillement Sud-Armoricain. *Bull. Minéral.* **102**:265–272.

Bilham, R. 1989. Surface slip subsequent to the 24 November 1987 Superstition Hills, California, earthquake monitored by digital creepmeters. *Bull. Seismol. Soc. Am.* **79**:424–450.

Bills, B.G., D.R. Currey, and G.A. Marshall. 1994. Viscosity estimates for the crust and upper mantle from patterns of lacustrine shoreline deformation in the Eastern Great Basin. *J. Geophys. Res.* **99**:22,059–22,086.

Bos, B., C.J. Peach, and C.J. Spiers 2000. Frictional–viscous flow of simulated fault gouge caused by the combined effects of phyllosilicates and pressure solution. *Tectonophysics* **327**:173–194.

Bos, B., and C.J. Spiers. 2002. Frictional–viscous flow of phyllosilicate-bearing fault rock: Microphysical model and implications for crustal strength profiles. *J. Geophys. Res.* **107**:B2, 1–13.

Bouchon, M. 1997. The state of stress on some faults of the San Andreas system as inferred from near-field strong motion data. *J. Geophys. Res.* **102**:11,731–11,744.

Bourne, S.J., P.C. England, and B. Parsons. 1998. The motion of crustal blocks driven by flow of the lower lithosphere and implications for slip rates of continental strike-slip faults. *Nature* **391**:655–659.

Brace, W.F., and D.L. Kohlstedt. 1980. Limits on lithospheric stress imposed by laboratory experiments. *J. Geophys. Res.* **85**:B11, 6248–6252.

Brodie, K.H., and E.H. Rutter. 1985. On the relationship between deformation and metamorphism with special reference to the behavior of basic rocks. In: Kinematics, Textures, and Deformation: Advances in Physical Geochemistry, ed. A.B. Thompson and D. Rubie, pp. 138–179. Berlin: Springer Verlag.

Brodie, K.H., and E.H. Rutter. 1987. The role of transiently fine-grained reaction products in syntectonic metamorphism. *Canad. J. Earth Sci.* **24**:556–564.

Brun, J.P., D. Sokoutis, and J. Driessche. 1994. Analogue modeling of detachment fault systems and core complexes. *Geology* **22**:319–322.

Bürgmann, R., E. Calais, A.M. Freed, J. Freymueller, and S. Hreinsdóttir. 2003. Mechanics of postseismic deformation following the 2002, $M_W = 7.9$, Denali fault earthquake. *EOS Trans. AGU* **84**.

Bürgmann, R., S. Ergintav, P. Segall et al. 2002. Time-dependent distributed afterslip on and deep below the Izmit earthquake rupture. *Bull. Seismol. Soc. Am.* **92**:126–137.

Bürgmann, R., D. Schmidt, R. Nadeau et al. 2000. Earthquake potential along the northern Hayward fault. *Science* **289**:1178–1182.

Bürgmann, R., P. Segall, M. Lisowski, and J. Svarc. 1997. Postseismic strain following the 1989 Loma Prieta earthquake from GPS and leveling measurements. *J. Geophys. Res.* **102**:4933–4955.

Byerlee, J.D. 1978. Friction of rocks. *Pure & Appl. Geophys.* **116**:615–626.

Chester, F.M. 1988. The brittle–ductile transition in a deformation mechanism map for halite. *Tectonophysics* **154**:125–136.

Chopra, P.N. 1997. High-temperature transient creep in olivine rocks. *Tectonophysics* **279**:93–111.

Clark, M.K., and L.H. Royden. 2000. Topographic ooze: Building the eastern margin of Tibet by lower crustal flow. *Geology* **28**:703–706.

Cobbold, P.R. 1977. Description and origin of banded deformation structures. II. Rheology and the growth of banded perturbations. *Canad. J. Earth Sci.* **14**:2510–2523.

Cobbold, P.R., and D. Gapais. 1986. Slip-system domains. I. Plane-strain kinematics of arrays of coherent bands with twinned fibre orientations. *Tectonophysics* **131**:113–132.

Dell Angelo, L.N., and J. Tullis. 1996. Textural and mechanical evolution with progressive strain in experimentally deformed aplite. *Tectonophysics* **256**:57–82.

Dewey, J.F., R.E. Holdsworth, and R.A. Strachan. 1998. Transpression and transtension zones. In: Continental Transpressional and Transtensional Tectonics, Spec. Publ. 135, pp. 1–14. London: Geol. Soc.

Dixon, J.E., T.H. Dixon, D.R. Bell, and R. Malservisi. 2004. Lateral variation in upper mantle viscosity: Role of water. *Earth Planet. Sci. Lett.* **222**:451–467.

Dresen, G., J. Duyster, B. Stöckhert, R. Wirth, and G. Zulauf. 1997. Quartz dislocation microstructure between 7000 m and 9100 m depth from the Continental Deep Drilling Program KTB. *J. Geophys. Res.* **102**:B8, 18,443–18,452.

Drury, M.R., R.L.M. Vissers, D. Van der Wal, and E.H. Hoogerduijn Strating. 1991. Shear localisation in upper mantle peridotites. *Pure & Appl. Geophys.* **137**:439–460.

Dunlap, W.J., G. Hirth, and C. Teyssier. 1997. Thermomechanical evolution of a midcrustal duplex and implications for midcrustal rheology. *Tectonics* **16**:983–1000.

Dutruge, G., J.-P. Burg, and J. Lapierre. 1995. Shear strain analysis and periodicity within shear gradients of metagranite shear zones. *J. Struct. Geol.* **17**:819–830.

Eisbacher, G.H. 1985. Pericollisional strike-slip faults and synorogenic basins, Canadian Cordillera. In: Strike-Slip Deformation, Basin Formation, and Sedimentation, ed. K.T. Biddle and N. Christie-Blick, SEPM Spec. Publ. 37, pp. 265–282. Tulsa, OK: Soc. Economic Paleontologists and Mineralogists.

Etheridge, M.A. 1983. Differential stress magnitudes during regional deformation and metamorphism: Upper bound imposed by tensile fracturing. *Geology* **11**:232–234.

Etheridge, M.A., and J.C. Wilkie. 1981. An assessment of dynamically recrystallized grain size as a paleopiezometer in quartz-bearing mylonite zones. *Tectonophysics* **78**:475–508.

Fialko, Y. 2004. Evidence of fluid-filled upper crust from observations of postseismic deformation due to the 1992 M_W 7.3 Landers earthquake. *J. Geophys. Res.* **109**:B08401, doi: 10.10129/2004JB002985.

Fitch, T.J. 1972. Plate convergence, transcurrent faults, and internal deformation adjacent to southeast Asia and the western Pacific. *J. Geophys. Res.* **77**:4432–4460.

Flesch, L.M., W.E. Holt, A.J. Haines, and B. Shen-Tu. 2000. Dynamics of the Pacific–North American plate boundary in the United States. *Science* **287**:834–836.

Fossen, H., and B. Tikoff. 1998. Extended models of transpression and transtension, and application to tectonic settings. In: Continental Transpressional and Transtensional Tectonics, ed. J.F. Dewey, R.E. Holdsworth, and R.A. Strachan. Spec. Publ. 135, pp. 15–33. London: Geol. Soc.

Fredrich, J.T., B. Evans, and T.-F. Wong. 1989. Micromechanics of the brittle to plastic transition in Carrara marble. *J. Geophys. Res.* **94**:4129–4145.

Freed, A.M., and R. Bürgmann. 2004. Evidence of powerlaw flow in the Mojave Desert mantle. *Nature* **430**:548–551.

Freed, A.M., and J. Lin. 2002. Accelerated stress buildup on the southern San Andreas fault and surrounding regions caused by Mojave Desert earthquakes. *Geology* **30**:6, 571–574.

Fusseis, F., M.R. Handy, and C. Schrank. 2006. Networking of shear zones at the brittle-to-viscous transition (Cap de Creus, NE Spain). *J. Struct. Geol.* **28**:228–1243.

Gapais, D., P. Bale, P. Choukroune et al. 1987. Bulk kinematics from shear zone patterns: Some field examples. *J. Struct. Geol.* **9**:635–646.

Gapais, D., and P.R. Cobbold. 1987. Slip system domains. 2. Kinematic aspects of fabric development in polycrystalline aggregates. *Tectonophysics* **138**:289–309.

Green, H.W., II, and P.C. Burnley. 1990. The failure mechanism for deep-focus earthquakes. In: Deformation Mechanisms, Rheology and Tectonics, ed. R.J. Knipe and E.H. Rutter. Geological Soc. Spec. Publ. London: 54. Geological Soc. pp. 133–141.

Hacker, B.R., S.M. Peacock, G.A. Abers, and S.D. Holloway. 2003. Subduction factory 2. Are intermediate-depth earthquakes in subducting slabs linked to metamorphic dehydration reactions? *J. Geophys. Res.* **108**:B1, doi: 10.1029/2001JB001129.

Handy, M.R. 1989. Deformation regimes and the rheological evolution of fault zones in the lithosphere: The effects of pressure, temperature, grain size, and time. *Tectonophysics* **163**:119–152.

Handy, M.R. 1994a. Flow laws for rocks containing two nonlinear viscous phases: A phenomenological approach. *J. Struct. Geol.* **16**:287–301.

Handy, M.R. 1994b. The energetics of steady state heterogeneous shear in mylonitic rock. *Mater. Sci. & Engin.* **A174**:261–272.

Handy, M.R. 1998. Fault rocks from an exhumed intracrustal extensional detachment: The Pogallo ductile fault zone. In: Fault-related Rocks: A Photographic Atlas, ed. A.W. Snoke, J.A. Tullis, and V.R. Todd, pp. 164–169. Princeton: Princeton Univ. Press.

Handy, M.R., J. Babist, R. Wagner, C. Rosenberg, and M. Konrad. 2005. Decoupling and its relation to strain partitioning in continental lithosphere: Insight from the Periadriatic fault system (European Alps). In: Deformation Mechanisms, Rheology and Tectonics: From Minerals to the Lithosphere, ed., D. Gapais, J.-P. Brun, and P.R. Cobbold, Spec. Publ. 243, pp. 249–276. London: Geol. Soc.

Handy, M.R., and J.-P. Brun. 2004. Seismicity, structure, and strength of the lithosphere. *Earth Planet. Sci. Lett.* **223**:427–441.

Handy, M.R., and H. Stünitz. 2002. Strain localization by fracturing and reaction weakening: A mechanism for initiating exhumation of subcontinental mantle beneath rifted margins. In: Deformation Mechanisms, Rheology and Tectonics: Current Status and Future Perspectives, ed. S. de Meer, M.R. Drury, J.H.P. de Bresser, and G.M. Pennock. Geological Society, Spec. Publ. 200, pp. 387–407. London: Geol. Soc.

Handy, M.R., S.B. Wissing, and J.E. Streit. 1999. Frictional-viscous flow in mylonite with varied bimineralic composition and its effect on lithospheric strength. *Tectonophysics* **303**:175–191.

Hartog, R., and S.Y. Schwartz. 2001. Depth-dependent mantle anisotropy below the San Andreas fault system: Apparent splitting parameters and waveforms. *Geophys. Res. Lett.* **106**:4155–4167.

Hearn, E.H., R. Bürgmann, and P. Reilinger. 2002. Dynamics of Izmit earthquake postseismic deformation and loading of the Duzce earthquake hypocenter. *Bull. Seismol. Soc. Am.* **92**:172–193.

Hickman, S., and M. Zoback. 2004. Stress orientations and magnitudes in the SAFOD pilot hole. *Geophys. Res. Lett.* **31**:L15S12, doi:10.1029/2003GL020043.

Hilley, G.E., R. Bürgmann, P. Molnar, and P. Zhang. 2005. Bayesian inference of plastosphere viscosities near the Kunlun Fault, northern Tibet. *Geophys. Res. Lett.* **32**:L01302, doi:10.1029/2004GL021658.

Hippert, J. 1999. Are S-C structures, duplexes, and conjugate shear zones different manifestations of the same scale-invariant phenomenon? *J. Struct. Geol.* **21**: 975–984.

Hirth, G., and D.L. Kohlstedt. 2003. Rheology of the upper mantle and the mantle wedge: A view from the experimentalists, in: Inside the Subduction Factory, ed. J. Eiler, Geophys. Monogr. Ser. 138, pp. 83–105. Washington, D.C.: Am. Geophys. Union.

Hirth, G., C. Teyssier, and W.J. Dunlap. 2001. An evaluation of quartzite flow laws based on com-parisons between experimentally and naturally deformed rocks. *Intl. J. Earth Sci.* **90**:77–87.

Hirth, G., and J. Tullis. 1994. The brittle–plastic transition in experimentally deformed quartz aggregates. *J. Geophys. Res.* **99**:11,731–11,747.

Hobbs, B.E., H.-B. Mühlhaus, and A. Ord. 1990. Instability, softening, and localization of deformation. In: Deformation Mechanisms, Rheology, and Tectonics, ed. R.J. Knipe and E.H. Rutter, Spec. Publ. 54, pp. 143–165. London: Geol. Soc.

Hobbs, B.E., A. Ord, and C. Teyssier. 1986. Earthquakes in the ductile regime? *Pure & Appl. Geophys.* **124**:309–336.

Holt, W.E. 2000. Correlated crust and mantle strain fields in Tibet. *Geology* **28**:67–70.

Hsu, Y.J., N. Bechor, P. Segall et al. 2002. Rapid afterslip following the 1999 Chi-Chi, Taiwan earthquake. *Geophys. Res. Lett.* **29**:16, doi:10.1029/2002GL014967.

Hyndman, R.D., C.A. Currie, and S.P. Mazzotti. 2005. Subduction zonne backarcs, mobile belts, and orogenic heat. *GSA Today* **15**:4–10.

Ivins, E.R., and C.G. Sammis. 1996. Transient creep of a composite lower crust, 1. Constitutive theory. *J. Geophys. Res.* **101**:B12, 27,981–28,004.

Jackson, J. 2002. Strength of the continental lithosphere: Time to abandon the jelly sandwich? *GSA Today* **12**:4–10.

Jade, S., M. Mukul, I.A. Parvez et al. 2002. Estimates of coseismic displacement and post-seismic deformation using Global Positioning System geodesy for the Bhuj earthquake of 26 January 2001. *Curr. Sci.* **82**:748–752.

Jaroslow, G.E., G. Hirth, and H.J.B. Dick. 1996. Abyssal peridotite mylonites: Implications for grain-size sensitive flow and strain localization in the oceanic lithosphere. *Tectonophysics* **256**:17–37.

Jarrard, R.D. 1986. Relations among subduction parameters. *Rev. Geophys.* **24**: 217–284.

Johnson, K.M., and P. Segall. 2004. Viscoelastic earthquake cycle models with deep stress-driven creep along the San Andreas fault system. *J. Geophys. Res.* **109**:B10403, doi:10.1029/2004JB003096.

Jónsson, S., P. Segall, R. Pedersen, and G. Björnsson. 2003. Post-earthquake ground movements correlated to pore-pressure transients. *Nature* **42**:179–183.

Kanamori, H., and T.H. Heaton. 2000. Microscopic physics of earthquakes. In: GeoComplexity and the Physics of Earthquakes, Geophys. Monogr. Ser. 120, pp. 147–163. Washington, D.C.: Am. Geophys. Union.

Karato, S., M.R. Riedel, and D.A. Yuen. 2001. Rheological structure and deformation of subducted slabs in the mantle transition zone: Implications for mantle circulation and deep earthquakes. *Phys. Earth & Planet. Inter.* **127**:83–108.

Kaufmann, G., and F. Amelung. 2000. Reservoir-induced deformation and continental rheology in the vicinity of Lake Mead, Nevada. *J. Geophys. Res.* **105**: 16,341–16,358.

Kenis, I., J.L. Urai, W. van der Zee, C. Hilgers, and M. Sintubin. 2005. Rheology of fine-grained siliciclastic rocks in the middle crust: Evidence from a combined structural and numerical analysis. *Earth Planet. Sci. Lett.* **233**:351–360.

Kenner, S., and P. Segall. 2003. Lower crustal structure in northern California: Implications from strain-rate variations following the 1906 San Francisco earthquake. *J. Geophys. Res.* **108**:doi: 10.1029/2001JB000189.

Kirby, S.H., W.B. Durham, and L.A. Stern. 1991. Mantle phase changes and deep-earthquake faulting in subducting lithosphere. *Science* **252**:216–225.

Klaper, E.M. 1990. Reaction-enhanced formation of eclogite-facies shear zones in granulite-facies anorthosites. In: Deformation Mechanisms, Rheology, and Tectonics, ed. R.J. Knipe and E.H. Rutter, Spec. Publ. 54, pp. 167–173. London: Geol. Soc.

Kohlstedt, D.L., B. Evans, and S.J. Mackwell. 1995. Strength of the lithosphere: Constraints imposed by laboratory experiments. *J. Geophys. Res.* **100**:17,587–17,602.

Knipe, R.J. 1989. Deformation mechanisms: Recognition from natural tectonites. *J. Struct. Geol.* **11**:127–146.

Küster, M., and B. Stöckhert. 1999. High differential stress and sublithostatic pore fluid pressure in the ductile regime: Microstructural evidence for short-term post-seismic creep in the Sesia Zone, Western Alps. *Tectonophysics* **303**:263–277.

Li, V.C., and J.R. Rice. 1987. Crustal deformation in Great California earthquake cycles. *J. Geophys. Res.* **92**:B11, 11,533–11,551.

Lin, S., D. Jiang, and P.F. Williams. 1998. Transpression (or transtension) zones of triclinic symmetry: Natural example and theoretical modelling. In: Continental Transpressional and Transtensional Tectonics, ed. J.F. Dewey, R.E. Holdsworth, and R.A. Strachan, Spec. Publ. 135, pp. 41–57. London: Geol. Soc.

Lister, G.S., M.A. Etheridge, and P.A. Symonds. 1991. Detachment models for the formation of passive continental margins. *Tectonics* **10**:1038–1064.

Lynch, J.C., R. Bürgmann, M.A. Richards, and R.M. Ferencz. 2003. When faults communicate: Viscoelastic coupling and earthquake clustering in a simple two-fault strike-slip system. *Geophys. Res. Lett.* **30**:doi:10.1029/2002-GL016765.

Maerklin, N., P.A. Bedrosian, C. Haberland et al. 2005. Characterizing a large shear zone with seismic and magnetotelluric methods: The case of the Dead Sea Transform. *Geophys. Res. Lett.* **32**:L15303, doi: 10.1029/2005GL022724.

Malservisi, R., C. Gans, and K.P. Furlong. 2002. Numerical modeling of creeping faults and implications for the Hayward fault, California. *Tectonophysics* **361**:121–137.

Mavko, G.M. 1981. Mechanics of motion on major faults. *Ann. Rev. Earth & Planet. Sci.* **9**:81–111.

Meissner, R., and J. Strehlau. 1982. Limits of stresses in continental crusts and their relation to the depth frequency distribution of shallow earthquakes. *Tectonics* **1**:73–89.

Miller, S.A. 2002. Properties of large ruptures and the dynamical influence of fluids on earthquakes and faulting. *J. Geophys. Res.* **107**:B9, doi: 10.1029/2000JB000032.

Miller, S.A., W. van der Zee, D.L. Olgaard, and J.A.D. Connolly. 2003. A fluid-pressure feedback model of dehydration reactions: Experiments, modelling, and application to subduction zones. *Tectonophysics* **370**:241–251.

Molnar, P. 1992. Brace-Goetze strength profiles, the partitioning of strike-slip and thrust faulting at zones of oblique convergence, and the stress–heat flow paradox of the San Andreas fault. In: Fault Mechanics and Transport Properties of Rocks: A Festschrift in Honor of W.F. Brace, ed. B. Evans, and T.F. Wong, Intl. Geophys. Ser. 51, pp. 435–460. London: Academic Press.

Montesi, L.G.J. 2004. Postseismic deformation and the strength of ductile shear zones. *Earth Planet & Space* **56**:1135–1142.

Montesi, L.G.J., and G. Hirth. 2003. Grain size evolution and the rheology of ductile shear zones: From laboratory experiments to postseismic creep. *Earth Planet. Sci. Lett.* **211**:97–110.

Moore, D., D.A. Lockner, M. Shengli, R. Summers, and J. Byerlee. 1997. Strengths of serpentinite gouges at elevated temperatures. *J. Geophys. Res.* **102**:14,787–14,801.

Mulch, A., and M.A. Cosca. 2004. Recrystallization or cooling ages: *In situ* UV laser ^{40}Ar/^{39}Ar geochronology of muscovite in mylonitic rocks. *J. Geol. Soc. Lond.* **161**:573–582.

Mulch, A., M.A. Cosca, and M.R. Handy. 2002. *In situ* UV laser ^{40}Ar/^{39}Ar geochronology of a micaceous mylonite: An example of defect-enhanced argon loss. *Contrib. Mineral. & Petrol.* **142**:738–752.

Müller, W., D. Aerden, and A.N. Halliday. 2000. Isotopic dating of strain fringe increments: Duration and rates of deformation in shear zones. *Science* **288**:2195–2198.

Nishimura, T., and W. Thatcher. 2003. Rheology of the lithosphere inferred from postseismic uplift following the 1959 Hebgen Lake earthquake. *J. Geophys. Res.* **108**:B8, doi:10.1029/2002JB002191.

Otsuki, K., N. Monzawa, and T. Nagase. 2003. Fluidization and melting of fault gouge during seismic slip: Identification in the Nojima fault zone and implications for focal earthquake mechanisms. *J. Geophys. Res.* **108**:B4, 2192 doi:10.1029/2001JBB001711.

Palm, S. 1999. Verformungsstrukturen und gesamtgesteinsgeochemische Alteration nahe dem spröd-viskosen Übergang an der Pogallo-Störungszone (Norditalien), Ph.D. Diss., Justus-Liebig-Univ. Giessen, Germany.

Paterson, M.S. 1995. A granular flow approach to superplasticity. In: Plastic Deformation of Ceramics, ed. R.C. Bradt, pp. 279–283. New York: Plenum Press.

Paterson, M.S. 2001. Relating experimental and geological rheology. *Intl. J. Earth Sci.* **90**:149–156.

Peltzer, G., P. Rosen, F. Rogez, and K. Hudnut. 1996. Postseismic rebound in fault step-overs caused by pore fluid flow. *Science* **273**:1202–1204.

Platt, J.P. 1984. Secondary cleavages in ductile shear zones. *J. Struct. Geol.* **6**:439–442.

Platt, J.P., and R.L.M. Vissers. 1980. Extensional structures in anisotropic rocks. *J. Struct. Geol.* **2**:397–410.

Pollitz, F.F., R. Bürgmann, and P. Segall. 1998. Joint estimation of afterslip rate and postseismic relaxation following the 1989 Loma Prieta earthquake. *J. Geophys. Res.* **103**:26,975–26,992.

Pollitz, F.F., G. Peltzer, and R. Bürgmann. 2000. Mobility of continental mantle: Evidence from postseismic geodetic observations following the 1992 Landers earthquake. *J. Geophys. Res.* **105**:8035–8054.

Pollitz, F.F., C. Wicks, and W. Thatcher. 2001. Mantle flow beneath a continental strike-slip fault: Postseismic deformation after the 1999 Hector Mine earthquake. *Science* **293**:1814–1818.

Prior, D.J., R.J. Knipe, and M.R. Handy. 1990. Estimates of the Rates of Microstructural Change in Mylonites, Spec. Publ. 54, pp. 309–320. London: Geol. Soc.

Ramsay, J.G., and R.H. Graham. 1970. Strain variation in shear belts. *Canad. J. Earth Sci.* **7**:786–813.

Ranalli, G., and D.C. Murphy. 1987. Rheological stratification of the lithosphere. *Tectonophysics* **132**:281–296.

Reddy, S.M., and G.J. Potts. 1999. Constraining absolute deformation ages: The relationship between deformation mechanisms and isotope systematics. *J. Struct. Geol.* **21**:1255–1265.

Rolandone, F., R. Bürgmann, and R.M. Nadeau. 2004. The evolution of the seismic–aseismic transition during the earthquake cycle: Constraints from the time-dependent depth distribution of aftershocks. *Geophys. Res. Lett.* **31**:L23610, doi:10.1029/2004GL21379.

Rudnicki, J.W., and J.R. Rice. 1975. Conditions for the localization of deformation in pressure-sensitive dilatant materials, *J. Mech. Phys. Solids* **23**:371–394.

Rutter, E.H., and K.H. Brodie. 2004. Experimental grain size-sensitive flow of hot-pressed Brazilian quartz aggregates. *J. Struct. Geol.* **26**:2011–2023.

Savage, J.C. 1990. Equivalent strike-slip earthquake cycles in half-space and lithosphere–asthenosphere earth models. *J. Geophys. Res.* **95**:4873–4879.

Savage, J.C., and R.O. Burford. 1970. Accumulation of tectonic strain in California. *Bull. Seismol. Soc. Am.* **60**:1877–1896.

Schaff, D.P., G.H.R. Bokelmann, G.C. Beroza, F. Waldhauser, and W.L. Ellsworth. 2002. High resolution image of Calaveras fault seismicity. *J. Geophys. Res.* **107**:doi:10.1029/2001JB000633.

Schmid, S.M., and M.R. Handy. 1991. Towards a genetic classification of fault rocks: Geological usage and tectonophysical implications. In: Controversies in Modern Geology, ed. K.J. Hsü, J. Mackenzie, and D. Müller, pp. 339–361. London: Academic Press.

Schmidt, D.A. 2002. The kinematics of faults in the San Francisco Bay area inferred from geodetic and seismic data. Ph.D. thesis. Berkeley: Univ. California, Berkeley.

Scholz, C. 1992. Paradigms or small change in earthquake mechanics. In: Fault Mechanics and Transport Properties of Rocks: A Festschrift in Honor of W.F. Brace, ed. B. Evans, and T.F. Wong, Intl. Geophys. Ser. 51, pp. 435–460. London: Academic Press.

Scholz, C.H. 1998. Earthquakes and friction laws. *Nature* **391**:37–42.

Schreurs, G. 1994. Experiments on strike-slip faulting and block rotation. *Geology* **22**:567–570.

Schwarz, S., and B. Stöckhert. 1996. Pressure solution in siliciclastic HP-LT metamorphic rocks: Constraints on the state of stress in deep levels of accretionary complexes. *Tectonophysics* **255**:203–209.

Secor, Jr., D.T. 1965. Role of fluid pressure in jointing. *Am. J. Sci.* **263**:633–646.

Segall, P. 2002. Integrating geologic and geodetic estimates of slip rate on the San Andreas fault system. *Intl. Geol. Rev.* **44**:62–82.

Sibson, R.H., and J.V. Rowland. 2003. Stress, fluid pressure, and structural per-meability in seismogenic crust, North Island, New Zealand. *Geophys. J. Intl.* **154**:584–594.

Sleep, N.H. 1995. Ductile creep, compaction, and rate- and state-dependent friction within major fault zones. *J. Geophys. Res.* **100**:13,065–13,080.

Stein, R.S. 1999. The role of stress transfer in earthquake occurrence. *Nature* **402**:605–609.

Stipp, M., H. Stünitz, R. Heilbronner, and S.M. Schmid. 2002. Dynamic recrystallization of quartz: Correlation between natural and experimental conditions. In: Deformation Mechanisms, Rheology, and Tectonics: Current Status and Future Perspectives, ed. S. De Meer, M.R. Drury, J.H.P. De Bresser, and G.M. Pennock, Spec. Publ. 200, pp. 171–190. Lond.: Geol. Soc.

Stipp, M., and J. Tullis. 2003. The recrystallized grain size piezometer for quartz. *Geophys. Res. Lett.* **30**:2008, doi:10.1029/2003GL018444.

Stone, D.S., T. Plookphol, and R.F. Cooper. 2004. Similarity and scaling in creep and load relaxation of single-crystal halite (NaCl). *J. Geophys. Res.* **109**:B12201, doi:10.1029/2004JB003064.

Stöckhert, B., M.R. Brix, R. Kleinschrodt, A.J. Hurford, and R. Wirth. 1999. Thermochronometry and quartz microstructures – A comparison with experimental flow laws and predictions on the brittle-plastic transition. *J. Struct. Geol.* **21**:351–369.

Stünitz, H., and J.D. FitzGerald. 1993. Deformation of granitoids at low metamorphic grade. II: Granular flow in albite-rich mylonites. *Tectonophysics* **221**:299–324.

Tchalenko, J.S. 1970. Similarities between shear zones of different magnitudes. *Geol. Soc. Am. Bull.* **81**:1625–1640.

Teyssier, C., B. Tikoff, and J. Weber. 2002. Attachment between brittle and ductile crust at wrenching plate boundaries. Stephan Mueller Spec. Publ. Ser. 1, pp. 119–144. European Geosciences Union (EGU).

Thatcher, W. 1983. Nonlinear strain buildup and the earthquake cycle on the San Andreas fault. *J. Geophys. Res.* **88**:5893–5902.

Trepmann, C., and B. Stöckhert. 2003. Quartz microstructures developed during non-steady state plastic flow at rapidly decaying stress and strain rate. *J. Struct. Geol.* **25**:2035–2051.

Trompert, R., and U. Hansen. 1998. Mantle convection simulations with rheologies that generate plate-like behaviour. *Nature* **395**:686–689.

Tse, S.T., and J.R. Rice. 1986. Crustal earthquake instability in relation to the depth variation of frictional slip properties. *J. Geophys. Res.* **91**:B9, 9452–9472.

Tullis, T.E., F.G. Horowitz, and J. Tullis. 1991. Flow laws of polyphase aggregates from end-member flow laws. *J. Geophys. Res.* **96**:B5, 8081–8096.

Vauchez, A., and A. Tommasi. 2003. Wrench faults down to the asthenosphere: geological and geophysical evidence and thermo-mechanical effects. In: Intraplate Strike-Slip Deformation Belts, ed., F. Storti, R.E. Holdsworth, and F. Salvani, Spec. Publ. 210, pp. 15–34. London: Geol. Soc.

Villa, I.M. 1998. Isotopic closure. *Terra Nova* **10**:42–47.

Vissers, R.L.M., M.R. Drury, E.H. Hoogerduijn Strating, and D. Van der Wal. 1995. Mantle shear zones and their effect on lithosphere strength during continental breakup. *Tectonophysics* **249**:155–171.

White, S.H., M.R. Drury, S.E. Ion, and F.J. Humphreys. 1985. Large strain deformation studies using polycrystalline magnesium as a rock analogue. Part I: Grain-size paleopiezometry in mylonite zones. *Phys. Earth & Planet. Inter.* **40**:201–207.

Wittlinger, G., P. Tapponnier, G. Poupinet et al. 1996. Tomographic evidence for localized lithospheric shearing along the Altyn Tagh fault. *Science* **282**:74–76.

Zingg, A., M.R. Handy, J.C. Hunziker, and S.M. Schmid. 1990. Tectonometamorphic history of the Ivrea Zone and its relation to the crustal evolution of the Southern Alps. *Tectonophysics* **182**:169–192.

Zoback, M.D., and H.-P. Harjes. 1997. Injection-induced earthquakes and crustal stress at 9 km depth at the KTB deep drilling site. Germany. *J. Geophys. Res.* **102**:B8, 18,477–18,491.

Zoback, M.D., and J. Townend. 2001. Implication of hydrostatic pore pressures and high crustal strength for the deformation of intraplate lithosphere. *Tectonophysics* **336**:19–30.

From left to right: Terry Tullis, Roland Bürgmann, Onno Oncken, Allan Rubin, Paul Segall, Sergei Shapiro, Geoff King, Greg Hirth, Chris Wibberley, Massimo Cocco, Jim Rice, Kenshiro Otsuki

7

Group Report:
Rheology of Fault Rocks
and Their Surroundings

TERRY E. TULLIS, Rapporteur

ROLAND BÜRGMANN, MASSIMO COCCO, GREG HIRTH,
GEOFFREY C. P. KING, ONNO ONCKEN, KENSHIRO OTSUKI,
JAMES R. RICE, ALLAN RUBIN, PAUL SEGALL, SERGEI A. SHAPIRO,
and CHRISTOPHER A. J. WIBBERLEY

OVERVIEW

At depth, all major faults undergo aseismic motion, and some faults slip
aseismically at shallower levels. However, most slip on shallow faults occurs
seismically, generating earthquakes. Due to the importance of earthquake-related hazards to society and because more remains to be understood about the
mechanics of earthquakes than about stable motion, the focus of this report is
on understanding the processes involved in seismic motions. In addition, we
consider the process by which stress accrues on faults, as well as the interactions between seismic events and the aseismically deforming crust at depth.

We organize the discussion around the seismic cycle, the repeated occurrence of earthquakes, and its stages: prior to an earthquake, slip increases quasi-
statically during a nucleation phase. The coseismic stage, i.e, when the earthquake occurs and nearly all slip accumulates, coincides with dynamic rupture
and is relatively short (seconds to minutes). It is followed by post- and
interseismic phases, characterized by gradually decreasing slip and healing of
the slip surface, until the next earthquake occurs. Coseismic slip increases stress
below the maximum depth of rupture, accelerating aseismic deformation during the postseismic period and reloading seismogenic portions of the fault
(Figures 6.1a, b, this volume). During much of the earthquake cycle, strain in

the vicinity of active faults accumulates linearly in time, reflecting steady loading by tectonic plate motion. Typically, successive earthquakes differ from one another in their location, nucleation, rupture, and postseismic creep, so that although the concept of an earthquake cycle is useful, it should not be interpreted as a regular series of identical events.

Although fault motion during the vast majority of earthquakes is ultimately caused by plate motion, the details of how faults are loaded differ from one tectonic setting to another. We begin our discussion with considerations of this loading and the distribution of deformation on the largest scale before focusing on the several stages of the earthquake cycle itself. We close with a discussion of the significance of off-fault damage resulting from fault growth and complex geometry, and a section on the implications of our earlier discussions for the magnitude of stresses on faults. The principles and processes discussed below apply to the three main types of faults: strike-slip, thrust, and normal. However, due to the importance of subduction thrusts as the site of the world's largest earthquakes, and because their geometry and kinematics are more complicated than the strike-slip example fault illustrated in Figure 7.1, we include a special section devoted to subduction thrusts. Within each section we focus discussion on topics of importance for future study and offer ideas as to which approaches might be taken during this study.

A schematic representation of a continental plate boundary strike-slip fault, including a more detailed view of the seismogenic portion of the fault, is illustrated in Figure 7.1. This diagram summarizes much of what we know about faults and earthquakes and also serves as a basis for discussion of what still needs to be learned. In addition, time and length scales of the processes underlying faulting and earthquakes are shown in Figure 7.2.

Figure 7.1 Schematic diagrams of a generic strike-slip fault showing details of seismogenic ▶ portion (a) and overall configuration (b). The state of current understanding and outstanding questions are presented. Although the example shown is a strike-slip fault, it highlights issues of fault zone structure, rupture initiation, and propagation that pertain to all fault types. A–B is the product of effective normal stress and $d\mu_{ss}/dt$, where μ_{ss} is the coefficient of steady-state friction; individually, A and B are the products of the effective normal stress and a and b, respectively, in the rate and state friction equation (see text). The lower depth limit of seismicity occurs where A–B becomes positive; peak strength occurs at greater depth. Between these depths, the frictional coefficient increases with sliding velocity, behavior termed velocity-strengthening. At shallow depths, it is still unclear if behavior is velocity-weakening or velocity-strengthening (A–B negative or positive). Earthquakes only nucleate where A–B is negative, but can rupture some distance into A–B positive regions. Nucleation may occur at special sites where gradually increasing static stress locally reaches static friction. During rupture propagation, rock strength τ_y at static friction may be overcome by stress that is dynamically concentrated ahead of the rupture tip in the presence of much lower tectonic stress τ_0. Stress τ is probably low behind the tip due to low coseismic friction τ_f but may increase slightly after slip ceases due to possible healing.

LARGE-SCALE VIEW OF THE EARTHQUAKE MACHINE

Figure 7.1 illustrates that slip on the seismogenic portion of the generic strike-slip fault is usually driven by plate motion. Although this general picture of faults has held for at least thirty years, several unresolved problems regarding

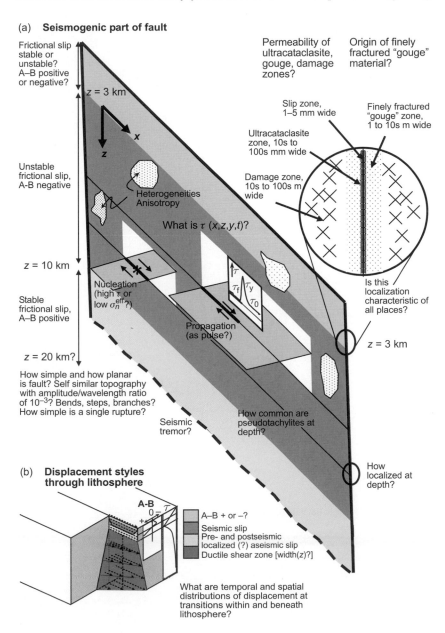

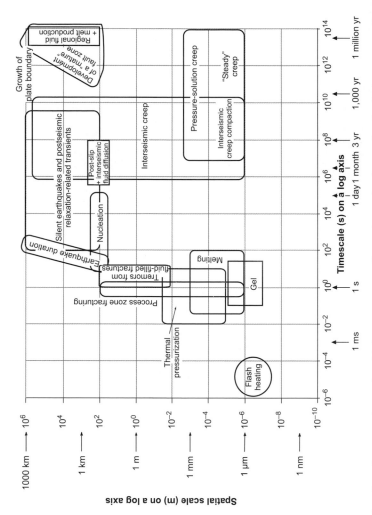

Figure 7.2 Plot of length scale versus timescale with boxes corresponding to intervals of operation for fault processes discussed in this report. The broad range of space–time intervals for these processes is part of the complexity of understanding faulting and earthquakes. Note that because fault zone dimensions parallel to the fault surface are much greater than those that are perpendicular to it, we plot the dimensions believed to be most relevant to the particular process in order to avoid covering the entire spatial scale.

the interactions between the seismogenic zone and the aseismic portions of the plate boundary remain (see Chapters 2 and 6):

- How is displacement transferred through the lithosphere to the seismogenic zone?
- To what degree is deformation beneath the seismogenic zone localized?
- What constitutive relationships apply for deformation beneath the seismogenic zone?
- How does the fault zone beneath the seismogenic zone respond to stress perturbations associated with earthquakes?

Many of these issues are currently investigated by modeling inter- and postseismic deformation measured with geodetic techniques. In particular, following a number of large earthquake ruptures, stresses transferred to the lower crust and upper mantle can be thought of as rock mechanics experiments in a natural laboratory of lithospheric proportions. Models of postseismic deformation are sometimes nonunique; that is, more than one relaxation process (e.g., shallow and deep afterslip, poroelastic rebound, and viscous flow in lower crust and upper mantle) may contribute to the observed deformation. However, together with continued improvement in laboratory rock deformation data, and more sophisticated analyses of exhumed fault rocks from appropriate depths, these geodetic studies provide a new opportunity to constrain the rheology of the lower crust and upper mantle. Such information not only has important implications for understanding the earthquake process, it can also provide rheological constraints that are applicable to many other problems in geology and geodynamics.

A review of some recent postseismic deformation studies by Handy et al. (Chapter 6) shows that, depending on fault type, earthquake magnitude, and the composition and environment of the surrounding lithosphere, different processes and rheological parameters fit the deformation data. However, these studies also show that different authors come to different conclusions for the same earthquake. This may, in part, reflect the variety of tectonic settings, which includes as examples the 1992 Landers and 1999 Hector Mine earthquakes in California, two strike-slip earthquakes in Iceland, the 1999 Chi-Chi earthquake in Taiwan, and the 2003 Tokachi-Oki earthquake in Japan. Depending on the method of observation and the earthquake studied, the observed postseismic deformation has been explained by afterslip below the rupture, distributed lower crustal and/or upper mantle flow, and poroelastic readjustments. Rheologies inferred for the lower crust and upper mantle include linear viscous, biviscous rheology, and power law creep. The strain associated with the postseismic deformation is very small, on the order of 0.02 to 0.2%, depending on the width of the deformation zone. Thus, depending on the ratio of the stress perturbation to the long-term stress state, the nonlinear response may largely reflect transient creep processes rather than steady-state deformation.

To clarify the issues raised by the current alternate interpretations, more work is needed to incorporate the the spatial and temporal information contained in all existing and future deformation data. Careful model testing and comparisons should carefully consider all candidate mechanisms. Even with recent improvements in the geodetic community's response following large earthquakes, we are still limited by the lack of sufficiently precise horizontal and vertical deformation data and would benefit from better and more complete spatial and temporal sampling of the relaxation signal. The small strain associated with postseismic deformation illustrates the need for further experimental and theoretical work on transient creep processes.

SEISMIC CYCLE: NUCLEATION LEADING TO UNSTABLE SLIP

Fault nucleation requires further study, not only because it is associated with precursors that could someday provide a means for short-term (weeks to minutes) prediction, but because it is fundamental to the onset of dynamic rupture and coseismic slip (Dieterich and Kilgore 1996; Tullis 1996). Nucleation occurs on a restricted portion of the fault, whereas dynamic rupture typically propagates over a much larger area of the fault (Figure 7.1). The time period of nucleation and the dimensions of the nucleation area are not well known and may be quite variable. Our current knowledge comes from seismic, strainmeter (Johnston and Linde 2002), and theoretical studies. These are consistent with laboratory-derived friction laws discussed below.

Rate and State Friction

Laboratory studies have shown that the frictional resistance of rocks involves two main dependences: a nonlinearly viscous dependence on slip velocity, in which an increase in stress causes an increase in slip rate, and a state dependence in which the state of the fault evolves with both time and displacement (Dieterich 1979; Dieterich and Kilgore 1996; Marone 1998; Ruina 1983; Tullis 1996). The slip velocities over which these experiments have been conducted span the range of slip rates that are expected to occur during earthquake nucleation. Theoretical studies of nucleation using empirically derived rate and state friction laws show interesting nucleation behavior and represent the basis for our best expectation of what may occur during nucleation (Dieterich 1986, 1992; Dieterich and Kilgore 1996). The rate and state friction laws are based on fitting experimental data to empirical functions over the range of the laboratory experimental variables and are not based on theoretical descriptions of the operative processes, although some progress has been made in this direction (Chapter 5). Neither of the two evolution laws in common use, the "slowness" or "aging" law and the "slip" law, fit the experimental data over the entire range of slip velocities from which they were derived (Kato and Tullis 2001; Marone 1998). The two laws make quite different predictions for behavior during earthquake nucleation. In particular, the aging law leads to much larger nucleation zones, which are more likely to be seismically detectable, than does the slip law.

Thus, constitutive laws are needed that fit laboratory data better and that are based on identifiable processes in fault rocks. Although rate and state friction is reasonably well understood as a thermally activated process (Chapter 5), we need to determine the processes responsible for this behavior. This would create a firmer basis for extrapolating the laws from experimental conditions to conditions in nature that are relevant to earthquake nucleation and propagation. There may be other processes operating in nature that are not well studied by laboratory experiments and which may produce behavior similar to that of the present rate and state laws, but with different evolution laws and different values of the controlling parameters. Consequently, theoretical studies of mechanisms that produce state evolution on faults should be undertaken to determine the form of their evolution laws. Such processes may include pressure solution, phase equilibrium-induced reactions, and even biologically enhanced mineral precipitation. The effect of pore-fluid-related processes on nucleation needs to be investigated. Although it appears that thermal pressurization (discussed below) alone cannot lead to earthquake nucleation, it may play an important role in accelerating slip at slip speeds exceeding 1 mm s^{-1} (Segall and Rice 2004).

Direct Studies of Nucleation

Direct measurements of behavior during earthquake nucleation should be undertaken. This can be done in a range of settings: from large laboratory experiments on rock, to small earthquakes induced by mining or by fluid injection in geothermic experiments, to standard earthquakes of a range of sizes. It is critical to place measurement devices, for example, seismometers and strainmeters, as close as possible to the nucleation site of events whose occurrence we can anticipate, ideally at distances comparable to the dimensions of the nucleation area itself. Although this spatial requirement is difficult to fulfill, borehole strainmeter installations are probably best suited for locating sufficiently close to be able to detect signals over the frequency band in which aseismic nucleation signals are expected. Studying the repeating earthquakes where the SAFOD experiment will hopefully penetrate in the summer of 2007 represents one possible opportunity, but better opportunities may exist in the case of mining-induced earthquakes.

The seismic nucleation stage may yield information about the aseismic nucleation process. One strategy for making progress in this area is to develop the capability for imaging earthquake slip using the waves generated during the beginning of seismic slip. Slip that occurs without much rupture propagation is indicative of nucleation over an area, rather than at a point. Therefore, we should try to identify microearthquakes that have stopped soon after nucleation, because much of the character of such earthquakes will be that of the nucleation stage. If such earthquakes can be identified, they may allow us to constrain the dimensions of the nucleation area. Knowledge of the nucleation area will help us evaluate the applicability of predictions from rate and state friction to understanding the nucleation process. Such microearthquakes might be identified by small ratios of radiated energy to seismic moment. One possibility is to seek microearthquakes that delimit the boundaries between areas on faults that creep and areas that slip in large earthquakes, such as those boundaries under Middle Mountain on the Parkfield segment of the San Andreas fault, or those located near the base of the seismogenic zone. The material properties in these areas are presumably marginal for unstable slip, in which case the nucleation zones will tend to be larger and thus more likely to produce detectable signals. A subset of the aftershocks of the Morgan Hill earthquake on the Calaveras fault, California, defines a truncated Omori-Law decay sequence, suggesting that they are just able to slip seismically. This renders them good candidates for study.

Another strategy would be to study foreshock sequences. There is evidence that foreshocks may involve dynamic rather than static triggering of the subsequent mainshock (Dodge et al. 1996). Alternatively, both the foreshocks and the mainshock are manifestations of an aseismic nucleation process.

SEISMIC CYCLE: DYNAMIC RUPTURE

Dynamic Weakening

Once the earthquake rupture propagates at speeds approaching the shear wave velocity or faster, slip velocities jump to rates of a few m s^{-1} behind the rupture tip, and continued slip occurs at about 1 m s^{-1}. It is likely that slip resistance drops to values much lower than the stress needed to initiate slip (Figure 7.1), but much of what occurs on the rupture surface is presently unknown. The frictional parameters measured in conventional friction experiments at slip rates less than 1 mm s^{-1} are probably no longer applicable because many other processes can also operate at high slip rates. These processes induce reductions in shear resistance to values much lower than those obtained from a typical frictional coefficient of ~0.6. A detailed constitutive description for many of these processes is not yet known, nor do we understand the relative importance of the potential mechanisms during earthquakes, despite some recent progress (see review by Rice and Cocco, Chapter 5). We focus our suggestions for future work on the dynamic weakening mechanisms of flash heating, wholesale frictional melting, gel formation, and thermal pressurization of pore fluids (Figure 7.2) because we feel that these processes (see Chapter 5, section: DYNAMIC WEAKENING PROCESSES (THERMAL, FLUID) DURING SEISMIC SLIP) are potentially important for better understanding the coseismic stage of the earthquake cycle.

Weakening due to flash heating appears to occur after a few mm of slip at slip speeds above about 100 mm s^{-1} in all rocks tested (quartzite, granite, gabbro) for which melting occurs at room pressure (Goldsby and Tullis 2003). It does not occur for calcite, which breaks down to CaO and CO_2. Due to the importance of clay minerals in fault zones, it is important to determine if weakening due to flash heating occurs for clay. The breakdown of clay minerals via dehydration may prevent weakening, although the very short times involved in flash heating (Figure 7.2) may not allow dehydration to occur.

With continued slip, the transition from local melting via flash heating to wholesale melting along the fault surface is a complex process involving a temporary increase in shear resistance (Hirose and Shimamoto 2004; Tsutsumi and Shimamoto 1997). Complicated negative feedback can occur with melting, as high heat generation leading to melting occurs when the frictional resistance is high; however, production of enough melt can reduce the resistance to the point that it freezes due to lower heat generation. Consequently, interesting cyclic behavior is possible where sufficient slip occurs. Unless other weakening processes reduce the shear stress to low values, melting is expected to occur on faults with any significant displacement below depths of about 5 km. A better understanding of the shear melting process may be obtained from experimental, theoretical, and field studies; pseudotachylite in natural fault rocks shows the importance of this process. It is commonly believed that melting may not be widespread on faults because pseudotachylite is quite rare in nature. Yet, pseudotachylite may be more common than its apparently rare occurrence would lead one to believe (Figure 7.1), especially given that later faulting may overprint it, or that slow cooling near the base of the seismogenic zone favor devitrification and/or static recrystallization.

Frictional sliding on smooth surfaces of a rock with silica contents of 70% or greater facilitates the generation of a layer of silica gel that results in large reductions in shear resistance (Di Toro et al. 2004; Goldsby and Tullis 2002). This occurs at lower speeds, 1 to 100 mm s^{-1}, than required for weakening caused by flash heating, but it requires displacements of about 1 m to attain steady state. The shear resistance can be extrapolated to zero at seismic slip speeds of 1 m s^{-1}. However, it is not known whether this weakening would occur at speeds above about 100 mm s^{-1} because the weakening caused by flash heating and thermal pressurization may inhibit or preclude the gel weakening. Large displacement experiments at slip speeds above 100 mm s^{-1} should be conducted to ascertain whether gel weakening operates in this velocity range. Related experiments at elevated ambient temperature would allow us to determine whether gel production is inhibited when the water required for the gel is driven off.

Weakening due to thermal pressurization of a pore fluid requires that slip be sufficiently localized for temperatures to rise rapidly, and that permeability be low enough that fluid pressure cannot decay rapidly. Thus, it is important to determine whether the narrow zones of slip found for fault zones exhumed from about 3 km depth (see Figure 7.1a and Chapter 5, Figure 5.1, and references therein) are characteristic of the fault structure at that depth and whether faults at deeper levels have the same degree of localization (Figure 7.1; see also Chapter 2). Finding suitable exposures in exhumed faults where overprinting from shallower brittle deformation did not occur is a formidable task and systematic searches for suitable large-displacement faults with structures preserved from the 5 to 10 km depth interval need to be undertaken (see Chapter 6, Figures 6.5 and 6.6 and Chapter 4, Figure 4.2,). It is also important to collect data on the permeability structure of the damaged rock immediately surrounding the narrow slip surfaces in such exhumed faults (Wibberley and Shimamoto 2003). The thermal pressurization mechanism also requires that fault zone dilatancy does not increase with progressive slip by as much as the increase in volume of the pore fluid due to its thermal expansion (Junger and Tullis 2003). Thus, it is important to evaluate the role of dilatancy in counteracting the thermal pressurization effect. Experiments designed to test the thermal pressurization mechanism should be undertaken, but they should be conducted in the absence of weakening by other mechanisms (Goldsby and Tullis 1998; Roig Silva et al. 2004) to ensure that sufficient heat is generated. Furthermore, the poroelastic and inelastic response to transient increases and decreases of stress during dynamic rupture needs to be included when considering pore-pressure changes.

Geological and seismological field studies are necessary to evaluate these various weakening mechanisms during earthquakes. Searches for thermal signatures immediately adjacent to localized slip surfaces in exhumed faults should be undertaken. Thermobarometric and geochronological studies of samples from exhumed fossil fault zones that were active at depths corresponding to large earthquakes may help to constrain the maximum heat released during earthquakes (d'Alessio et al. 2003). Models of heat generated during dynamic slip suggest that sampling would have to focus on a zone of only a few mm in width, rather than the 2–5 cm sampled by d'Alessio et al. (2003). The core of exhumed fault zones should be also examined for microstructural and physiochemical signatures that may be diagnostic of dynamic weakening mechanisms. Dating of such rocks would require applying geochronometers sensitive to short heat pulses (e.g., electron spin resonance, fission track dating, and U/Th-He methods).

Better seismological ways to determine fracture energy and slip weakening distances for earthquakes are needed to compare with model predictions from several of the high-speed weakening mechanisms discussed above. The current methods are subject to more uncertainty than is desirable for comparison with predictions.

Slip Distributions and Propagation during Dynamic Rupture

Kinematic inversions of slip during dynamic rupture yield very heterogeneous patterns, but their details may be incorrect. The character and significance of these heterogeneities should be further explored by conducting kinematic inversions that are dynamically consistent, in contrast to standard kinematic inversions. Such models could better define what combination of stress, constitutive properties, and fault geometry gives rise to an observed heterogeneity. In addition, we must understand the relative roles of geometry and stress heterogeneity in causing heterogeneity of slip, as well as in causing where ruptures initiate and terminate. It is also important to understand how any slip deficits are accommodated, whether by microseismicity, fault creep, or complementary behavior in successive earthquakes. Continued seismic and geodetic monitoring of strike-slip faults is an important way to do this, in particular at Parkfield, California, an area with many useful existing data. Evaluating the effect of lithology on the behavior of different parts of faults may be approached by analyzing the head waves traveling along faults (McGuire and Ben-Zion 2005).

One approach to understanding earthquake mechanics is to image the earthquake source using seismic waves. Currently, there is a large deficit in observational studies of large earthquakes. Excluding sites in Japan, we do not have instruments that can record large earthquakes on scale in relatively low-noise (and simple in terms of wave propagation effects) borehole environments. The recent Parkfield earthquake (September 28, 2004) was recorded by dozens of strong motion instruments; however, the data are very difficult to interpret for several reasons:

1. The instruments are old, and hence the analog signals must be digitized, which results in a loss of signal-to-noise ratio.
2. The instruments are at Earth's surface and not in solid rock, and hence nonlinear attenuation renders the records difficult to interpret. This also means that high-frequency ground motion, which might allow one to constrain the slip-weakening displacement, will be lost to seismic attenuation.
3. The instruments cannot record smaller earthquakes. Digital recordings of smaller earthquakes would allow us to calibrate wave propagation effects to estimate how much nonlinear attenuation might occur, and hence to isolate source effects in the data.

Neither EarthScope nor ANSS (the Advanced National Seismic System in the U.S.) will address this observational gap adequately: EarthScope because there is no strong motion component to that project, ANSS because it is focused on instrumenting urban environments rather than faults.

One possibility is to set up an array of monitoring instruments in boreholes along large faults. These might monitor ground motion, strain, and pore pressure, yet be designed to stay on scale during earthquake ruptures. This would amount to monitoring a large biaxial laboratory experiment, but conducted on the scale of Earth's crust in the field. The Plate Boundary Observatory component of EarthScope is doing this, but not with downhole seismometers that would stay on scale during strong shaking.

FAULT ZONE RESTRENGTHENING DURING CO-, POST-, AND INTER-SEISMIC PERIODS

Immediately following an earthquake (postseismic stage) and until the time of the next one (interseismic stage), the rock along and adjacent to the fault heals. The rate of healing likely occurs in a manner that is proportional to the log of the time since the earthquake. Studies, however, are needed to determine the kinds of healing processes that operate and the rate at which strength recovery actually occurs (Chapter 12). This involves learning about the same processes of strength evolution discussed earlier. Because interseismic times are typically long compared to times available for laboratory experiments, it is important to know which processes operate and their corresponding healing laws in order to extrapolate beyond the times of lab data (Figure 7.2). Geological evidence for healing in exhumed faults is needed to identify and understand the processes involved. Geochemical studies of pore-fluid compositions sampled *in situ* and over a period of time in borehole experiments might help constrain the processes and their rates (see Chapter 14). Seismological studies of changes in wave velocities as a function of time after earthquakes, together with magnetotelluric studies to estimate changes of fluid content, are desirable.

Concepts

Healing rates may be greatest immediately after the dynamic weakening, as slip velocities decrease during the coseismic rupture propagation. Rapid healing is required in a homogeneous fault to explain the finite duration of slip (see Figure 5.6, this volume). In the recent literature, several modifications to the constitutive relations governing fault friction have been proposed to model healing and to simulate dynamic traction evolution; they show fast and immediate restrengthening at the end of the dynamic weakening phase (Figure 7.1). Stress (and/or strength) heterogeneities might explain the finite slip (and slip velocity) durations without requiring such rapid restrengthening (Figure 7.1). Whether finite slip durations are due to heterogeneities or to high-velocity constitutive behavior is an unresolved issue requiring further investigation.

More generally, co-, post-, and interseismic restrengthening is a particular example of the time dependence of strength. In the well-known expression for strength, each term may be thought of as being time dependent:

$$\tau(t) = \mu(t)\,[\sigma_n(t) - P_f(t)] \ . \tag{7.1}$$

Rate- and state-dependent friction have been proposed to explain the time dependency of the frictional coefficient $\mu(t)$ under constant normal stress and at low slip rates:

$$\mu(t) = \mu_0 + a\ln\,(V/V_0) + b\ln\,(\theta V_0/D_c) \ , \tag{7.2}$$

where V is slip velocity, μ_0 is the frictional coefficient at steady state at slip velocity V_0, D_c is the e-folding displacement for evolution of friction with slip following a step in slip velocity, and θ is a state variable with units of time. Clearly it is not correct to describe friction as being solely time dependent, as might appear to be the case from writing $\mu(t)$. In fact, time dependency arises both from the dependence on slip velocity V, and the evolution of the state variable θ with time and slip according to several possible evolution laws (Marone 1998; Ruina 1983; Tullis 1996).

In this equation, the state variable θ represents the fault surface memory of previous slip episodes. For the short times and high slip velocities relevant to coseismic and postseismic restrengthening, the frictional coefficient may depend even more strongly on time and slip velocity than is the case for typical rate and state friction. In addition, both pore-pressure evolution and normal stress variations can contribute to this time dependency of fault strength over a wide range of time scales.

Observations

Evidence of fault restrengthening comes from several observations. Laboratory experiments illustrate the importance of restrengthening both during coseismic and interseismic periods (Dieterich 1979; Marone 1998; Tullis 1996; Chapter 12). Seismological evidence for healing is provided by duplets or multiple events at the same locations. It is sometimes observed that the stress drop or the magnitude of these repeating earthquakes depends upon the recurrence interval between them. Finally, analysis of fault zone-trapped waves allows imaging of the time variations of elastic properties of fault zones (Chapter 2). Many more of these types of observations are needed.

Processes

Several physical and chemical processes play a role in evolving fault zone strength (Figure 7.2). Testing the relative importance of each of these requires estimating the relative rates and amounts by which they operate and, in turn, necessitates an in-depth, theoretical understanding of the mechanisms. It is difficult to separate the physical from the chemical processes, because they both rely on the presence and behavior of a fluid phase—the physicochemical role of the fluid is of crucial importance. The evolution during the seismic cycle of porosity and permeability of the fault rock in both the fault core and the damage zone are fundamental in controlling the relative rates of these processes. Laboratory friction experiments in the presence and absence of chemically inert and active fluids have started to define the relative time evolutions for restrengthening but need to be complemented by field observations and laboratory porosity–permeability measurements of fault rocks to evaluate the potential importance of these different processes. The restrengthening shown by the increase of static friction with log of stationary contact time is one manifestation of the evolution of state in the formulation of rate and state friction. As discussed below, several physical and chemical processes offer plausible explanations for this evolution effect; experiments and theoretical modeling need to be undertaken to determine how many of them may play a role.

Physical Processes

One possible explanation for the experimentally observed increase of static friction with log of stationary contact time is that the real area of contacting asperities increases with time. The contact area could grow due to time-dependent creep of the contacts by a number of possible physical processes. Such growth might occur by thermally activated, and therefore time-dependent, dislocation motion. Any dislocation motion might occur either as dislocation creep, in which the strain rate is expected to be proportional to the local differential stress to a power of approximately three, or as dislocation glide in which the strain rate is expected to show an exponential dependence on the local differential stress. Two lines of evidence suggest that, if the evolution effect is due to dislocation-assisted growth of contact area, the process of dislocation glide is the more likely explanation. First, the increase of frictional strength depends on the log of the slip velocity, a functional form that is consistent with dislocation glide, not dislocation creep. Detailed modeling of the growth of contact area should be undertaken to verify this. Second, the stress magnitudes estimated to exist at contacting asperities approach the theoretical strength and, if dislocations are involved at all at such high stress levels, dislocation glide rather than dislocation creep is expected. In fact, the high estimated stress levels suggest that increases in contact area could occur by intracrystalline slip without the propagation of dislocations, that is, deformation at the theoretical strength. Increases in frictional contact area could also occur by small-scale time-dependent microfracturing. The process of subcritical crack growth, typically occurring by chemically assisted stress corrosion (see discussion on chemical processes in Chapter 14), could allow this to occur. It is important to make direct observations of the deformation processes at contacting frictional asperities to determine what role, if any, dislocations and cracking play in increasing asperity contact areas.

Fault strength is partly controlled by the evolution of fluid pressure. Coseismic overpressuring of the slip zone by thermal pressurization may be dissipated by outward fluid flow or by thermal contraction of the heated fluid due to conduction of heat away from the slip zone after frictional heating subsides. These mechanisms of fluid pressure reduction lead to a relatively rapid recovery of postseismic strength. On the other hand, fluid pressure increases may be generated by compaction creep and porosity reduction in parts of fault zones with a low permeability. Generation of an electrochemical potential gradient across the fault zone may also generate an equally large increase in fluid pressure. This could be stimulated by rapid, coseismic introduction of a more saline fluid along the fault from below, or by chemical exchange between the fluid and fresh, intragranular fracture surfaces in the fault rock, possibly with periodic (coseismic?) flushing of the surrounding fracture-damage-zone reservoir by fresh low-salinity fluids. The reader is referred to Chapter 14, Theme 2, for more discussion of these issues.

Chemical Processes

Chemical processes also possibly contribute to interseismic strength evolution, as discussed in detail by Gratier and Gueydan (Chapter 12). Chemical reactions at contacting surfaces assist in the restrengthening seen in the frictional evolution effect, both in situations involving the contact of bare fault surfaces and grain-grain contacts within gouge layers. These processes include pressure-solution compaction hardening of a gouge zone, healing of the fault by cementation, and increase in the asperity contact surface area by diffusive mass transfer and/or grain surface diffusion. Chemical reactions may strengthen contacts, even if no increase in contact area is involved. For example, gradual removal of contaminant water layers separating crystalline phases at a frictional contact can allow stronger bonding between the crystals to develop. Time-dependent polymerization of water-rich gel layers at frictional interfaces is a related way in which such contacts can restrengthen after sliding has ceased. As mentioned above, the relative importance of all of these processes can be established by determining the rate-limiting factor(s) for each process.

Although these processes have been proposed, in some cases with field, microstructural, and borehole evidence, we do not have a unified physical model of fluid effects in which to incorporate the constitutive properties of faults. As discussed by Yardley and Baumgartner (Chapter 11), faulting can affect fluid flow and availability just as fluid presence and pressure can affect faulting. Some of the important factors to consider are the open versus closed nature of the fluid supply and the ultimate source of the fluids involved (Chapter 11). More studies of the fluid from borehole samplings and from fluid inclusions in healed fault rocks in exhumed faults are needed, with attention paid to both the chemistry and the fluid pressure. In particular it would be useful to monitor fluid pressure evolution during the seismic cycle in the fracture damage network around the fault core to assess both the predictive capability of models of fluid pressure evolution during earthquakes and the relative rates of proposed physical processes operating in the fault zone in the interseismic period. In addition, it would be useful to monitor changes through time in the pore-water chemistry in the low-permeability fault core and surrounding fracture zone, coupled with geophysical estimates of porosity changes. This would allow estimation of the rates of dissolution and precipitation at different times during the interseismic period and the importance of electrochemical potential gradients for generating overpressures in the fault zone.

The Nojima Fault Probe drilling project provides an opportunity to conduct some of these studies. This project is unique in exploring the stress state and physical properties of the faulted crust soon after a major earthquake (M 7.2 Kobe 1995 earthquake). Borehole injection tests at 1.5–1.8 km depth around the fault zone only two years after the earthquake were followed by migration of fluid pressure-induced microseismicity parallel to the fault at 2–4 km depth. Permeability estimates of 10^{-15} to 10^{-14} m^2 obtained from velocities of the migrating microseismicity front generally correlate with the highest laboratory core-scale permeability measurements of very heterogeneous fractured damage zones around fault cores. Sealing of the poroelastic fracture damage zone network through the post- and interseismic stages is an important process in controlling fluid communication and pore-pressure distribution. Similar experiments undertaken through time may therefore provide us with valuable information on the rates of sealing. Also of possible help in estimating porosity evolution could be data from remote geophysical experiments, such as Vs/Vp and $Vs \times Vp$, and magnetotelluric studies.

FAULT GEOMETRY AND THE SIGNIFICANCE OF DISTRIBUTED DAMAGE

Faults idealized as in Figure 7.1 accommodate predominantly simple shear, but the boundary conditions applied in the Earth usually deviate from simple shear, being some form of general noncoaxial shear. Hence, no single fault or fault orientation is sufficient to accommodate bulk deformation compatibly. Faults with several orientations must coexist and be simultaneously active. In general, these must meet at bends or junctions. Unlike the triple junctions of oceanic plates, no lithospheric sources (ridges) or sinks (subduction zones) are present, and as a consequence these junctions are unstable. Owing to this, multiple fracturing must occur where faults change strike or faults of different orientations meet (King 1983). The result is the creation of regions of continuously evolving zones of fragmentation or damage (King and Yielding 1984; King 1986; King and Nabelek 1985). The existence of such regions has implications for the mechanics of earthquakes since narrow zones may not adequately represent faults. There is some evidence that earthquakes initiate and terminate in these complex regions. Systematic studies of such regions of deformation should be undertaken and may reveal features important to earthquake mechanics.

Even where boundary conditions are simple, large damage zones appear at the ends of individual faults. Like fault junctions, these zones are associated with multiple fracture directions (Scholz 2002). Information about these regions can be deduced from studies of the cumulative displacement distributions on geological faults (Manighetti et al. 2004, and references therein), which conclude that these process zones are much larger than encountered in engineering materials. Like fragmented regions around fault junctions, these zones evolve as a fault evolves, and because each fault is the product of numerous individual earthquakes, they necessarily play an important role in the earthquake cycle. Geological and seismic studies should examine how slip during individual events combines to create observed geological slip distributions. The damage zones around such evolving faults should also be studied in detail. Together, such data will provide information pertinent to the role of such structures in the dynamics of individual earthquakes.

SUBDUCTION THRUSTS

Convergent continental margins are the Earth's principal locus of important earthquake hazards. Some 90% of global seismicity and nearly all interplate megathrust earthquakes with magnitudes >8 occur in the seismogenic coupling zone between converging plates. Despite the societal, economic, and scientific importance associated with the interplate coupling zone, the processes that shape it and its mechanics are poorly understood. Drilling has not accessed seismogenic plate interfaces, nor are they exposed at the surface onshore.

Seismogenic coupling zones occupy a depth interval of convergent plate interfaces that range from 5 to 10 km depth at the up-dip end and 30 to 60 km at the down-dip end (e.g., Oleskevich et al. 1999, and references therein). The down-dip end tends to align with the coastline in the Circum-Pacific belt of convergent margins. The reason for this loose correlation is poorly understood, although it has significant implications for the potential of large seismic events to generate tsunamis as well as severe ground shaking, depending on the slip distribution on the rupture plane and its relation to the coastline. Features along the seismogenic coupling zone that affect the distribution of slip and ground shaking at the surface include large asperities on the down-going plate (seamounts, transform faults, or oceanic ridges) and lateral variations in fluid pressure. Both of these features affect the areal extent and the degree of seismic coupling and therefore play major roles in the generation of great interplate earthquakes. Because elastic strain prior to earthquake rupture does not accumulate uniformly in time and space, geodetic observations provide a valuable means of deducing the degree of coupling along the seismogenic zone. Questions about material behavior along the coupling surface can be addressed via laboratory experiments. Key questions include: How are rock strength and seismogenic coupling distributed at depth during the various stages of the seismic cycle? What are the constitutive properties of fault rocks along aseismic and seismic fault segments? In particular, what mechanisms are responsible for the upper and lower transitions from seismic to aseismic behavior? And how are all of these factors related to surface hazard? A more reliable assessment of hazard distribution is only possible if we understand why earthquakes with a similar rupture history and magnitude recur along the same fault segment.

The relatively shallow up-dip ends of seismogenic coupling zones are accessible to drilling and, indirectly, to sampling of rock and volatiles from faults that splay upward from the plate interface. Furthermore, the offshore location of these zones allows geophysical imaging at reasonably high resolution. On the other hand, this portion of the subduction interface is difficult to study geodetically. The down-dip transition zone is much less accessible to geophysical imaging and monitoring. Given the intriguing observations of deformation transients and tremor activity, as well as the first hints of temporal changes in geophysically imaged properties, intensified monitoring and imaging of subduction zones should be a high priority for future studies. This should be combined with detailed structural and thermobarometric studies of fault rocks from fossil subduction zones, including eclogitic rocks containing high-pressure and ultra-high-pressure metamorphic mineral phases. Only then can rock physical properties derived from geophysical studies be related to processes preserved *in situ* in disequilibrium structures of natural fault rocks.

The concomitant occurrence in several subduction zones of seismic tremors and transient slip events which seem to be located along the down-dip extension of the seismogenic zone (e.g., Rogers and Dragert 2003) suggests that they represent aseismic loading of the seismogenic zone. Alternatively, and less likely, they may be part of a long-term nucleation event. As precursors, they could be of value in predicting large underthrust earthquakes, as suggested by a $M_W = 6.7$, up-dip thrust event that followed one of the slow slip events observed at Guerrero, Mexico, on April 18, 2002 (Liu and Rice 2005). Understanding the origin of these transient slip events and the tremors sometimes associated with them could help us learn more about the nucleation process. The origin of the tremors could be constrained by looking carefully at the relative onset times of tremor occurrence and the slip transient. Does slip occur first and allow the tremors to occur? Or do the tremors alter the stress state on the plate interface and allow the slip to occur? Dense sampling of the seismic wavefield with arrays of closely spaced instruments would allow us to constrain the depth and propagation of tremors by tracking the phase velocity of locally coherent tremor waveforms across the array. The station spacings within current earthquake monitoring arrays are too broad to allow this. Using the waveforms to determine whether motions in the source region of the tremors involve shearing or dilatancy could help us to understand their origin. Models of how tremors might form should be developed and compared with observations. Magnetotelluric studies might help determine whether the motion of fluids causes tremors. Understanding the evolution of the fluid budget, especially the relative rates of fluid production, percolation, and sealing in the vicinity of faults during the seismic cycle, is key to quantifying the short-term mechanical evolution of the plate interface (Chapter 12). Here again, examination of exhumed parts of fossil seismogenic plate interfaces is required to understand the interaction between fluid flux, seismic failure, and aseismic slip (Chapter 6). In addition, high sensitivity monitoring to see if tremors exist in a variety of tectonic environments is worthwhile to ensure they represent a fault-related signal, since they have been found by careful re-examination of records in several places where they were not previously recognized.

STRESS ON FAULTS

The discussion above makes clear that although the strength of most fault rocks which are not slipping, or are slipping at rates less than 1 mm s^{-1}, is approximated by Byerlee's Law, faults that slip coseismically may have much lower shear strength, perhaps near zero strength (Figure 7.1, see also Chapter 6, Figure 6.1). This leads to a conceptual model (described below) in which the slip-integrated strength of fault zones in the seismogenic coupling zone is very low compared to their high time-integrated strength and the strength of the surrounding unfaulted crust. Therefore, there may be no heat flow paradox for large faults and no problem with the idea that large plate boundary faults like the San Andreas fault are weak (Zoback et al. 1987). We will return to the issue of the strength of the creeping section of the San Andreas fault below.

If we speculate that faults can be strong most of the time, but weak while slipping in earthquakes, what might be the values of stress on these faults at different times? It is possible that the stress is well below the Coulomb failure value most of the time and that it only reaches that value at unusual places or times. Only in earthquake nucleation zones would the static stress be high enough to overcome the Byerlee's Law static friction coefficient times the local effective normal stress. Elsewhere along a fault experiencing an unusual earthquake event, the stress would only very momentarily overcome the static strength, owing to the dynamic stress concentration at the tip of the propagating rupture (Figure 7.1). This means that the time average of the tectonic stress on the fault could be quite low and that the stress during most of the dynamic slip is even lower, possibly approaching zero. In fact, if the dynamic resistance were quite low as we presently expect (but need to verify), it would be difficult to argue that the ambient or tectonic stress on the fault could be too high. This is because, if it were, the dynamic stress drop (the difference between the initial stress and the dynamic friction) would be too large to fit seismological observations (see Chapter 6). The same would be true of the static stress drop, the difference between the initial and the final stress after slip ceased. Thus, by using seismological observations of stress drops and tying these delta values to a low absolute level equal to low dynamic friction, we obtain a model in which the stress on faults in most places is much lower than predicted by Byerlee's Law. We can have a strong crust that is able to deform seismically at low levels of tectonic stress! How can we test this model?

One way would be to focus on understanding what could cause stress on faults to be so heterogeneous such that it could statically overcome friction in the nucleation zone (Figure 7.1) and simultaneously be much lower elsewhere. This heterogeneity could be in the shear stress on the fault, perhaps due to stress concentrations left from previous slip events, or it could be in the effective normal stress. The nucleation regions could have elevated pore pressures or, perhaps as a result of geometric irregularities, could have reduced normal stress. Evidence that pore-fluid pressure can be involved in the triggering of slip in the nucleation zone and that stress levels must be essentially at the static failure or static friction stress in the nucleation zone comes from fluid injection experiments in deep boreholes, for example, the German deep drilling project, KTB. There, it was shown that pore-pressure perturbations of only 1–100 KPa can induce microseismic events with a magnitude of 2 and sometimes even greater (Shapiro et al. 1997). These data do not tell us how the stress varied over the area that slipped in such microearthquakes, but they do indicate that the preexisting stress at the hypocenter of those small events was high.

What other tests of this model can be devised? Improving our knowledge of dynamic weakening mechanisms is critical, as discussed above. In addition, we need to develop better methods with which to estimate dynamic and static stress drops (Figure 7.1). Measurements of the stress in the SAFOD and Nankai drilling projects will help, although the limited number of stress measurements may be insufficient to capture the heterogeneity of stress state along the faults. The results will have to be interpreted with caution.

Dynamic weakening processes do not operate on creeping faults; thus the stress level along creeping segments of faults (e.g., on San Andreas and North Anatolian faults) are worthy of careful attention. Brune (2002) suggested that the heat flow may be elevated only on the eastern side of the creeping segment of the San Andreas fault; this asymmetrical pattern is expected if rocks there underwent creep at high stresses, whereas rocks on the western side previously slipped dynamically at low stresses when they occupied a position further to the south. As the heat flow data along this creeping section are rare, more data are necessary to test this hypothesis, especially in light of the considerable scatter and attendant averaging of the heat flow values from such measurements.

SUMMARY

Much has been learned about the rheology and mechanics of faults and earthquakes in the past several years, and it is an exciting as well as societally important area of research. We have identified numerous areas where additional research efforts could be focused to fill in some of the critical gaps in our knowledge. This would bring us closer to the goals of creating computer simulations of dynamic rupture of earthquakes using what appear to be realistically low values of dynamic shear resistance. Through such models, we may then be able to predict strong ground motions from first principles. This can help us know where to build structures and how to design them to withstand damage. Studies of the earthquake nucleation process may even bring us closer to the elusive goal of reliable earthquake prediction.

REFERENCES

Brune, J. 2002. Heat flow on the creeping section of the San Andreas fault: A localized transient perspective. In: 2003 SCEC Annual Meeting Proc. and Abstr., pp. 58–59. Palm Springs, CA: Southern California Earthquake Center.

d'Alessio, M.A., A.E. Blythe, and R. Bürgmann. 2003. No frictional heat along the San Gabriel fault, California: Evidence from fission track thermochronology. *Geology* **31**:541–544.

Dieterich, J.H. 1979. Modeling of rock friction. 1. Experimental results and constitutive equations. *J. Geophys. Res.* **84**:2161–2168.

Dieterich, J.H. 1986. A model for the nucleation of earthquake slip. In: Earthquake Source Mechanics, Geophys. Monogr. Ser. 37, pp. 37–47. Washington, D.C.: Am. Geophys. Union.

Dieterich, J.H. 1992. Earthquake nucleation on faults with rate- and state-dependent strength, *Tectonophysics* **211**:115–134.

Dieterich, J.H., and B. Kilgore. 1996. Implications of fault constitutive properties for earthquake prediction. *PNAS* **93**:3787–3794.

Di Toro, G., D.L. Goldsby, and T.E. Tullis. 2004. Friction falls toward zero in quartz rock as slip velocity approaches seismic rates. *Nature* **427**:436–439.

Dodge, D.A., G.C. Beroza, and W.L. Ellsworth. 1996. Detailed observations of California foreshock sequences: Implications for the earthquake initiation process. *J. Geophys. Res.* **101**:22,371–22,392.

Goldsby, D., and T.E. Tullis. 1998. Experimental observations of frictional weakening during large and rapid slip. *EOS Trans. Am. Geophys. Union* **97**:F610.

Goldsby, D., and T.E. Tullis. 2002. Low frictional strength of quartz rocks at subseismic slip rates. *Geophys. Res. Lett.* **29**:1844, doi:10.1029/2002GL015240.

Goldsby, D.L., and T. Tullis. 2003. Flash heating/melting phenomena for crustal rocks at (nearly) seismic slip rates. In: 2003 SCEC Annual Meeting Proc. and Abstr., pp. 98–90. Palm Springs, CA: Southern California Earthquake Center.

Hirose, T., and T. Shimamoto. 2005. Growth of a molten zone as a mechanism of slip weakening of simulated faults in gabbro during frictional melting. *J. Geophys. Res* **110**:B05202, doi:10.1029/2004JB003207.

Johnston, M.J.S., and A.T. Linde. 2002. Implications of crustal strain during conventional, slow, and silent earthquakes. In: International Handbook of Earthquake and Engineering Seismology, ed. W. Lee et al., pp. 589–605. San Diego: Academic.

Junger, J.A., and T.E. Tullis. 2003. Fault roughness and matedness suggest significant fault-interface dilatancy with slip. *EOS Trans. Am. Geophys. Union* **84 Suppl.**: Abstr. S51B-03.

Kato, N., and T.E. Tullis. 2001. A composite rate- and state-dependent law for rock friction. *Geophys. Res. Lett.* **28**:1103–1106.

King, G.C.P. 1983. The accommodation of strain in the upper lithosphere of the Earth by self-similar fault systems: The geometrical origin of b-value. *Pure & Appl. Geophys.* **121**:761–815.

King, G.C.P. 1986. Speculations on the geometry of the initiation and termination of earthquake rupture and the evolution of morphology and geological structures. *Pure & Appl. Geophys.* **124**:567–585.

King, G.C.P., and J. Nabelek. 1985. The role of bends in faults in the initiation and termination of earthquake rupture: Implications for earthquake prediction. *Science* **228**:986–987.

King, G.C.P., and G. Yielding. 1984. The evolution of a thrust fault system: Processes of rupture initiation, propagation and termination in the 1980 El Asnam (Algeria) earthquake. *Geophys. J. Roy. Astron. Soc.* **77**:915–933.

Liu, Y., and J.R. Rice. 2005. Aseismic slip transients emerge spontaneously in 3D rate and state modeling of subduction earthquake sequences. *J. Geophys. Res.* **110**:B08307, doi:10.1029/2004JB003424.

Manighetti, I., G.C.P. King, and C. Sammis. 2004. The role of off-fault damage in the evolution of normal faults and the back-arc extension in the Aegean. *Earth Planet. Sci. Lett.* **217**:339–408.

Marone, C.J. 1998. Laboratory-derived friction constitutive laws and their application to seismic faulting. *Ann. Rev. Earth Planet. Sci.* **26**:643–696.

McGuire, J., and Y. Ben-Zion. 2005. High-resolution imaging of the Bear Valley section of the San Andreas fault at seismogenic depths with fault zone head waves and relocated seismicity. *Geophys. J. Intl.* **163**:152–164.

Oleskevich, D.A., R.D. Hyndman, and K. Wang. 1999. The updip and downdip limits to great subduction earthquakes: Thermal and structural models of Cascadia, South Alaska, SW Japan, and Chile. *J. Geophys. Res.* **104**:14,965–14,991.

Rogers, G., and H. Dragert. 2003. Episodic tremor and slip on the Cascadia subduction zone: The chatter of silent slip. *Science* **300**:1942–1943; ScienceExpress, 10.1126.

Roig Silva, C., D.L. Goldsby, G. Di Toro, and T.E. Tullis. 2004. The role of silica content in dynamic fault weakening due to gel lubrication. *EOS Trans. Am. Geophys. Union* **Suppl**. **85**:T21D-07.

Ruina, A.L. 1983. Slip instability and state variable friction laws. *J. Geophys. Res.* **88**:10,359–10,370.

Scholz, C.H. 2002. The Mechanics of Earthquakes and Faulting. Cambridge: Cambridge Univ. Press.

Segall, P., and J.R. Rice. 2004. Dilatancy, compaction, and slip instability of a fluid-infiltrated fault. *EOS Trans. Am. Geophys. Union* **75 Suppl**.:425.

Shapiro, S.A., E. Huenges, and G. Borm. 1997. Estimating the crust permeability from the fluid-injection induced seismic emission at the KTB site. *Geophys. J. Intl.* **131**:F15–F18.

Tsutsumi, A., and T. Shimamoto. 1997. High-velocity frictional properties of gabbro. *Geophys. Res. Lett.* **24**:699–702.

Tullis, T.E. 1996. Rock friction and its implications for earthquake prediction examined via models of Parkfield earthquakes. In: Earthquake Prediction: The Scientific Challenge, ed. L. Knopoff. *PNAS* **93**:3803–3810.

Wibberley, C.A.J., and T. Shimamoto. 2003. Internal structure and permeability of major strike-slip fault zones: The Median Tectonic Line in Mid Prefecture, Southwest Japan. *J. Struct. Geol.* **25**:59–78.

Zoback, M.D., M.L. Zoback, J. Mount et al. 1987. New evidence on the state of stress of the San Andreas fault system. *Science* **238**:1105–1111.

8

Topography, Denudation, and Deformation

The Role of Surface Processes in Fault Evolution

PETER O. KOONS[1] and ERIC KIRBY[2]

[1]Department of Earth Sciences, University of Maine, Bryand Global Science Center, Orono, ME 04469–5790, U.S.A.
[2]Department of Geosciences, Pennsylvania State University, 218 Deike Bldg., University Park, PA 16802, U.S.A.

ABSTRACT

Topography formed during active orogenesis and the elevated denudation rates established on that topography fundamentally influence the location and rates of interplate deformation. Stress concentration through slope-generated shear and normal stresses reduces the amount of stress of tectonic origin required to reach failure by variable amounts, depending upon topographic wavelengths, orientation, magnitude, and material property. In areas of extreme topography associated with plate convergence, this stress concentration can lead to significant strain concentration into valleys and away from loads generated by large mountains. This strain concentration then leads to crustal weakening via advection of hot lower crust into the upper crust resulting in significant reduction in integrated strength of the crust. Together, these mechanisms produce a highly heterogeneous lithosphere reflecting the distribution and intensity of surface processes that controls numerous crustal phenomena including the degree of metamorphism, decompression melting, and strain partitioning within oblique orogens over time frames of $\sim 10^6$ yr. Although three-dimensional topography produces stress states with large spatial variability, the general form of these stress states is readily approximated, and they should not be avoided when calculating the mechanical evolution of crustal fault zones. The magnitude of these stresses may be sufficient to influence earthquake nucleation or propagation.

INTRODUCTION

Of the many influences on crustal fault behavior, the surface of the Earth and those processes acting on the surface are some of the most accessible, simplest to evaluate, and often ignored. For mechanical settings where differential stresses on the order of 10^7–10^8 Pa can be sufficient to cause elastic deformation, frictional failure, and viscous flow, the surface and surface processes can make a significant contribution to stress states and crustal rheology. The role of tectonic processes in altering Earth's topography at the long wavelengths of plate boundaries has received a great deal of attention. In this presentation, however, we concentrate on the opposite side of tectonic-surface interaction: the influence that surface *state* (topography) and *process* (erosion) have upon Earth deformation. Although state and process are clearly linked, for discussion purposes we shall separate the two and correlate surface state with stress perturbations and surface process with strength perturbation.

Cooperation among tectonic and surface processes at orogen scale has been considered within the rather different frameworks of coupling between surface processes and crustal deformation due to isostasy and thermally activated viscous flow in the lower crust (e.g., Avouac and Burov 1996), gravity sliding (e.g., Ramberg 1973), and critical wedges (e.g., Koons 1990, 1994; Beaumont et al. 1992; Willet et al. 1993; Willet 1999). The effect of mass removal and precipitation regimes on orogen-scale mechanics has been thoroughly evaluated for two-sided orogens by Willett (1999), who investigated the competing influences of erosion and tectonics in two dimensions at long wavelengths. In addition, tectonic and erosional cooperation has been explored at the scale of individual river catchments (e.g., Norris and Cooper 1997; Pavlis et al. 1997; Thiede et al. 2004). Each of these studies shares a consideration of the effect that mass removal and redistribution has on the distribution and rates of deformation.

After briefly reviewing the basis of the surface influence on stress state, we examine three modern orogens and discuss the controlling surface processes. To save space, we contract "dominantly topographically generated stresses" to *topographic stresses,* "dominantly tectonically generated stresses" to *"tectonic stresses,"* and use "load" to refer *lithostatic load.*

INFLUENCE OF SURFACE *STATE* ON CRUSTAL STRESS

Shear and normal stresses related to topographic slopes and loads have long been recognized as a basic component of the force balance in natural Earth settings (e.g., McTigue and Mei 1981; Savage and Swolfs 1986; Liu and Zoback 1992). Stresses developed in an elastic medium are a function of the topographic slope, the amplitude and orientation of the topography, the wavelength of the topography, and the basal boundary condition (McTigue and Mei 1981;

Savage and Swolfs 1986; reviewed in Liu and Zoback 1992, who demonstrated the utility of the elastic half-space approximation). Topographic stresses are greatest near the free surface and are related to the changes in vertical normal stress due to loading (σ_{zz}), to the horizontal compressive stresses (σ_{xx}, σ_{yy}), and to the coordinate shear stresses (vertical shear stresses; τ_{xz}, τ_{yz}, and horizontal shear stress, τ_{xy}) as a function of the topographic slope (α).

As demonstrated in soil mechanics applications and in tectonic settings, limit states for wedges everywhere at the point of failure may be written that balance the tectonic stresses, basal and internal shear stresses with normal and shear stresses generated by topography. For example, in two dimensions:

$$\frac{\tau_{xz}}{\sigma_{zz}} = \tan \alpha \ , \tag{8.1}$$

(Terzaghi 1943; Chapple 1978; Davis et al. 1983; Dahlen 1984; Jaeger and Cook 1969; Mandl 1988). Several expressions can be written to demonstrate the balance of internal and basal shear properties when a basal weak zone is introduced, which can be simplified to:

$$N\alpha + (N - 1)\beta \approx \mu_b \ , \tag{8.2}$$

in which β is the slope of the basal weak plane; $\mu_b = \tan\phi_b$ is the basal coefficient of friction; ϕ is the internal angle of friction; and

$$N = \tan^2\left(\frac{\pi}{4} + \frac{\phi}{2}\right) = \frac{1 + \sin\phi}{1 - \sin\phi}$$

is a measure of the strength of the deforming material, i.e., $N_0 = \sigma_1 / \sigma_3$ for cohesionless materials, and is usually between 3 and 4 for natural Earth materials (Terzaghi 1943; Enlow and Koons 1998). While, the Eq. 8.2 applies for a cohesionless material, a similar form exists for uniform cohesion. The critical force balance argument can be extended to three dimensions without loss of generality (Koons 1994; Enlow and Koons 1998).

Simple natural wedges in the absence of erosion tend to approximate the topography predicted by the equilibrium arguments expressed above, with strongbase-critical wedges associated with relatively large topographical slopes that support relatively large vertical shear stresses in the critical state. However, in this chapter we are concerned with complex state and processes which may act at significant angles to natural orogen boundaries. Under these conditions, the requirements of constant boundary conditions and of incipient failure everywhere, as required by the critical wedge equilibrium formulation, are rarely approached. Natural topography can influence all components of the stress tensor, and we restrict the topic somewhat by examining the role of topography-induced stresses in localizing regional strain in natural, convergent settings. The net contribution of these topographic stresses to crustal principal stresses has been considered in two dimensions by Savage and Swolfs 1986) and McTigue and Mei (1981) and for small slopes in three dimensions by Liu and Zoback (1992). To gauge the role of topographic stresses on tectonic strain in natural settings, we have chosen to evaluate numerically the stresses generated by complex natural topography (Koons et al. 2002).

As natural examples of the influence of both stress and strength on deformation, we use numerical solutions for the stress state beneath the natural topography of Nanga Parbat region of the western syntaxis in northwest Pakistan, the Alpine fault of the South Island, New Zealand, and the Himalayan Front with its extreme relief and where deformation has been concentrated for >30 Ma (Lavé and Avouac 2000; Bollinger et al. 2004).

Topographic Stresses and Deformation in Nanga Parbat, Western Himalaya

Nanga Parbat, with its topographic relief of >5000 m, Pleistocene granites and metamorphism, high heat flow and high erosional fluxes (e.g., Zeitler et al. 2001a), provides an example of positive feedback among surface and tectonic processes that led to the concept of *tectonic aneurysm* (Koons et al. 2002). Geological and geophysical observations, summarized elsewhere (Zeitler et al. 2001b) indicate a region of concentrated contraction in which modern strain is partitioned into the Nanga Parbat massif to the extent that decompression melting is occurring beneath the massif and is rapidly exhumed in predominantly northward directed thrusting into the Indus Valley. The dominant structure is currently the Raikhot fault—the most obvious manifestation of the present plate boundary—but strain is occurring throughout the massif and along its margins.

We calculated possible contribution of topographic stresses on the present stress state by placing a GTOPO30 topographic model, smoothed to approximately 1 km^2 grids, onto a numerical crust with an elastic/Mohr Coulomb upper block extending to ~15 km below the surface and resting upon an elastic/viscoplastic block with yield stress inversely proportional to temperature similar to that of a feldspar rheological model (we refer to any two-layer, weak beneath strong, crust as *BK crust* after Brace and Kohlstedt 1980). In the reference frame used here, y = north, x = east, and z = vertical (Figure 8.1). The stress state is represented by ratios of stress tensor components on planes at approximately sea level and, for some ratios, on vertical sections oriented across the plate boundary through Nanga Parbat, across the Indus Valley, and into Kohistan. It must be noted that the natural stress field can differ from our calculated due to spatial and temporal variation in material density, cohesion, and frictional properties of the natural massif.

In addition to the vertical influence of the load, σ_{zz}, slope-generated normal stresses in the horizontal plane (σ_{xx}, σ_{yy}) also contribute to the state of stress beneath topography as seen in the contours of the maximum principal stress

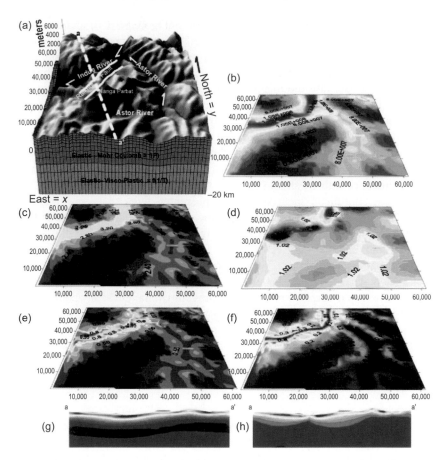

Figure 8.1 Stress state associated with Nanga Parbat present topography with Cartesian reference frame oriented x = east; y = north and z = up; grid in meters. (a) Surface of topography gridded at ~1 km intervals from GTOPO30 data onto a pressure- and temperature-dependent, 20 km thick block. a–a' indicate position of vertical section in (g) and (h) below. The following calculations are made for a horizontal plane at sea level: (b) Contours of absolute value of differential stress distribution (= $\sigma_{max} - \sigma_{min}$) in MPa. (c) Ratio of horizontal normal coordinate stresses to vertical normal stress. Horizontal stresses dominate except at high elevations. (d) Ratio of the two slope-generated horizontal normal stresses ($\sigma_{xx} / \sigma_{yy}$). (e) Topographic stress index (TSI) as a measure of proximity to failure due to topographic stresses. Failure at TSI = 1 is approached along the entire Indus Valley and elevated in the Astor and other drainages. (f) Vertical coordinate shear stresses normalized by the vertical normal stress $\sqrt{\tau_{xz}^2 + \tau_{yz}^2} / \sigma_{zz}$. The stress concentrations adjacent to the Indus Valley are augmented by the deep tributaries. (g) Vertical section of TSI through the crustal model showing location of stress concentration in the upper crust. Contours are the same as (e). (h) Vertical section of shear stress ratio in (f) illustrating the slope concentration.

(Figure 8.1a) and are also represented as ratios (Figure 8.1b–d) and incorporated in the topographic stress index developed below. Comparison of the load-generated, vertical normal stress (σ_{xx}) with the slope-generated contribution to the normal coordinate stresses (σ_{xx}, σ_{yy}) may be seen in the plot of their ratios (Figure 8.1c). Beneath the major massif, σ_{zz} is the dominant normal stress; however, horizontal normal stresses dominate throughout the remainder of the region, increasing toward the valleys, with the direction of the major drainage controlling the orientation of the dominant normal coordinate stress (Figure 8.1d) In the unstrained Nanga Parbat block, the major horizontal coordinate normal stress varies greatly and only in the Indus and Astor valleys is σ_{yy} consistently the dominant coordinate stress (i.e., $\sigma_{xx}/\sigma_{yy} < 1$ in Figure 8.1d).

Vertical coordinate shear stresses (represented by $\sqrt{\tau_{xz}^2 + \tau_{yz}^2}/\sigma_{zz}$; see Figure 8.1f, h) are concentrated beneath the slopes of the large massif and on the edges of the larger valleys with the highest values aligned parallel to the Raikhot fault and the Astor valley.

To illustrate the combined effects of load-generated normal stress patterns as well as slope-generated shear stresses produced by natural topography on the stress state of a natural orogen, we represent the proximity to failure at any point with a dimensionless topographic stress index (TSI), defined as the ratio of the maximum and minimum principle stresses at any point (= σ_{rat} to that predicted by the Mohr Coulomb failure criteria presented in terms of Terzaghi's (1943) flow value (N_0)

$$\text{TSI} = \sigma_{rat}/N_0 \ , \qquad\qquad\qquad (8.3)$$

in which $\sigma_{rat} = (\sigma_{max}/\sigma_{min})$.

The TSI, which varies between 1/N for no topographical influence and 1 at failure, includes therefore the different influences of load-generated normal stress patterns as well as slope-generated shear and normal stresses produced by natural topography. Large values of TSI within the crust, as observed to the west and below the Indus (Figure 8.1e, g) reflect significant stress concentration. Here, we ignore the effects of material cohesion but recognize that spatial variation in cohesion can significantly perturb the near surface stress regime.

Influence of Topography on Regional Strain

The effect of those stresses generated by spatially varying topography on tectonic stresses is a function of orientation of topographic features relative to the tectonic stress field (Savage and Swolfs 1986; McTigue and Mei 1981; Liu and Zoback 1992) as well as the other variables mentioned earlier. If the topographic stresses at high angles to the plate boundary are large, then dramatically different displacement patterns can coexist. A measure of one distinctive change in displacement patterns (i.e., from thrusting to strike-slip) is indicated when the following ratio approaches unity (Enlow and Koons 1998):

$$\tau_{xy} = \sqrt{\tau_{xz}^2 + (\sigma_{xx} - \sigma_{yy})(\sigma_{yy} - \sigma_{zz})} \ . \tag{8.4}$$

In a dominantly convergent margin, any orientation of topography that increases σ_{xx}, σ_{yy}, τ_{xz} will stabilize thrusting relative to strike-slip displacement, whereas increases in σ_{zz} or τ_{xy} favor strike-slip displacement. In collisional zones that are dominantly two dimensional without extreme relief, topographic stresses are unlikely to be sufficient to alter the displacement pattern from thrusting to strike-slip (Liu and Zoback 1992); however, minor horizontal shear stress associated with oblique collision can readily impose coexisting thrusting and strike-slip structural trend upon the topography (Koons 1994; Enlow and Koons 1998).

To investigate the influence of existing topography on tectonic strain associated with nearly orthogonal convergence, we applied a constant northern velocity of 0.01 Ma^{-1} to the southern edge and along the southern base of the same GTOPO30 NPHM block described above, with mirror conditions on either edge (Figure 8.2). This imposed velocity pattern is not intended to capture the larger-scale India/Eurasia strain pattern, but rather to illuminate the role of complex, extreme topography on a regional strain field.

Our applied velocity boundary conditions act in a north–south direction, oblique to the major plate structure, the Raikhot fault. Because the topographic stress pattern is dominantly one of horizontal compression with the major horizontal normal stress oriented toward the north, the applied velocity produces stress that augment the horizontal compressional pattern over much of the region. Therefore, in the strained model, σ_{yy} dominates the horizontal normal stress with particular concentrations associated with the northeast oriented Indus Gorge and upper Astor River to the south of Nanga Parbat. The effect of adding a tectonic stress is to increase spatial correlation of the principal stresses throughout the massif. The strain patterns are clearly influenced by the stress patterns generated by the natural, high-relief topography, as illustrated with strain concentrated in the Indus Valley near sea level, becoming less dominated by the topographic pattern at depth (Figure 8.2f). In the upper crust, strain is clearly diverted into valleys relative to the surrounding ridges and mountains. Investigations into similar relationships in the early stages of orogen growth demonstrate that topographic influence on regional strain distribution is not limited to regions of extreme topographic relief (Simpson 2004; Simon 2005).

In summary, topographic stresses by themselves partition strain away from the massif load and into the negative loads of the valleys in a manner compatible with a least work principle for a laterally homogeneous material. This predicted strain pattern differs distinctly from that observed in the Nanga Parbat region, where strain is diverted into the load of Nanga Parbat. The reasons for the departure of the predicted kinematics from those observed yield important information on the rheological evolution of the deforming region over time. In the following section we examine the influence of exhumation as a major agent of lithospheric strength modification.

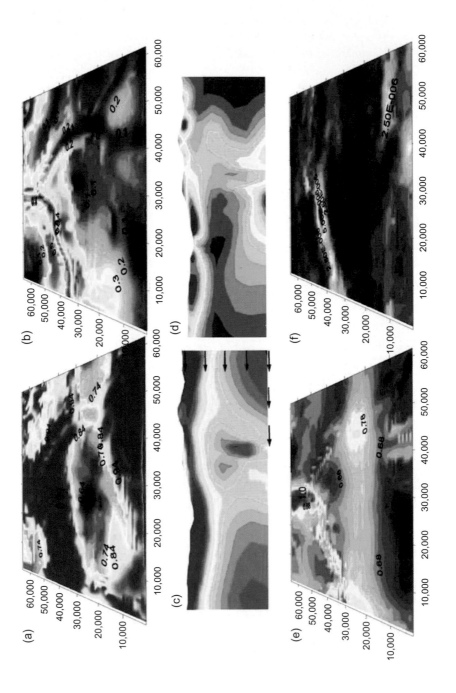

◄ **Figure 8.2** Stress and strain distribution of Nanga Parbat block in Figure 8.1 with velocity (V_y) conditions applied to the south and along the base (boundary velocities shown in (c); contours along the sea-level section except for the two vertical sections where the scale is the same as in the relevant horizontal sections. (a) TSI illustrates the shift of principle stresses toward failure due to tectonically generated stresses and the topographic influence on stress state. (b) Vertical shear stresses augmented by tectonic stresses. Pattern at sea level remains dominated by topographic stresses. (c) Vertical sections along a–a' (Figure 8.1a) showing the proximity to failure caused by convergence. TSI is elevated along the surface and along the oppositely dipping planes that emerge in the Astor and Indus valleys. (d) Concentration of vertical coordinate shear stresses with depth along a–a'. Tectonic stress pattern is significantly influenced by topography throughout the upper crust. (e) Ratio of horizontal normal coordinate stresses (σ_{xx}/σ_{yy}). Applied velocity conditions elevates σ_{yy}, acting in the direction of convergence, to the maximum horizontal normal stress over much of the region except in the topographic gap to the north where σ_{xx} generated by the valley slopes still dominates. (f) Strain rates (microstrain rate per year) along the ~ sea-level horizontal section. At this elevation, strain is distributed into the Indus Valley and to a lesser extent, the Astor Valley and away from the Nanga Parbat massif.

INFLUENCE OF SURFACE *PROCESSES* ON CRUSTAL STRENGTH

The static contribution of topography to regional strain described above leads by itself to a negative feedback situation in which valleys are preferentially filled by advecting mass and the stress heterogeneities generated by the original topography are reduced to insignificance. Consequently, continuing and efficient exhumation is a necessary condition for continued influence of topography on crustal stress states and can profoundly alter crustal rheology and therefore regional strain patterns (Koons et al. 2002).

For a pressure-dependent material underlain by a thermally activated lower crust (=BK crust; Brace and Kohlstedt 1980), the integrated strength of the crust is proportional to the square of the upper frictional layer (e.g., Sonder and England 1986) as

$$F_c \cong \int_0^{h_{BD}} \Delta\sigma \, dz = \frac{h_{BD}\Delta\sigma_{BD}}{2} \tag{8.5}$$

or

$$F_c \cong (N_0 - 1)\frac{\rho g h_{BD}^2}{2} \tag{8.6}$$

where $\Delta\sigma = (\sigma_{max} - \sigma_{min}) = \sigma_{min}(N_0 - 1)$, $\Delta\sigma_{BD} = \Delta\sigma$ at the base of the frictional layer. Because the upper region of the ductile material can contribute to the total crustal strength (Koons et al. 2002), F_c will generally lie between

$$F_c \cong \rho g(h_{BD}^2) \quad \text{and} \quad F_c \cong 2\rho g(h_{BD}^2) \ . \tag{8.7}$$

Although incision by itself into the upper crust can reduce the thickness of the upper plate, h_{BD}, incision that leads to rock uplift can result in much greater strength reduction (Koons et al. 2002). The enhanced role of uplift and the associated vertical advection of heat in crustal strength reduction are due to thinning, from the base, of the frictional crust (Figure 8.3). Because the greatest strength of a pressure-dependent material generally resides at the base of the frictional layer, which for quartz/feldspar-dominated crust is ~ 350–$400°C$, heating of that region by thermal advection leads to dramatic weakening of a standard, BK crust (e.g., Albarede 1976; Koons 1987).

The thickness of the frictional layer in a region of steady thermal state in an advecting medium is related to the inverse of the Péclet number (Carslaw and Jaeger 1959); (in 1D; $Pe = V_z L/\kappa$, where κ is the thermal diffusivity and L the advective distance, commonly the depth to detachment) as well as to the temperature at detachment. Consequently, for high-Péclet systems, the integrated strength of the crust is reduced at temporal and spatial scales defined by advection. A 50% reduction in integrated strength can occur after ~ 1 Ma of vertical displacement of lower crust at rates approaching 0.005 Ma^{-1} (Koons et al. 2002).

In Nanga Parbat, the thermal structure is highly perturbed, as indicated by the presence of hot springs at high elevations, steam-dominated fluid inclusions, elevated B/D transition and the formation of granite decompression melts at relatively shallow depths (Craw et al. 1994; Zeitler and Chamberlain 1991), in a manner compatible with that predicted from advection modeling (Craw et al. 1994). Incorporating advection-related weakening along exhumation structures and employing the same two-layer, BK crustal rheology discussed above in a three-dimensional numerical model, we have reproduced the pattern of strain concentration into the massif that one observes in the tectonic aneurysm of Nanga Parbat (Koons et al. 2002).

Figure 8.3 Strength reduction due to exhumation and resulting strain localization at Nanga ▶ Parbat. Dimensions and conditions are given in the figure. The three-dimensional model uses a standard initial BK rheology (curve 1 in b) that becomes perturbed through advection along a single river drainage. Resulting tectonic aneurysm results from concentration of strain into the massif rather than around the topographic load and can perturb thermal topographic and metamorphic fields. (a) Crustal strength profile for a cohesionless quartz-dominated material of $\varphi = 30°$ for crust with no topographic relief, for the region beneath a transverse topographic gap. (b) Strength profile beneath the same gap after thermal thinning of the high-strength brittle region by 5 km due to differential exhumation. h_{BD} and h_1 used in equations are shown in Figure 8.3a. Curve 1 in Figure 8.3b represents the calculated strength profile for a quartz-feldspar-dominated crust (Ranalli 1987) Curve 2 is taken from the un-thinned model crust (Figure 8.3c). Curve 3 comes from the thermally thinned central massif of the numerical model (Figure 8.3c). The topographic gap modifies the upper portion of the crustal strength envelope, whereas thermal thinning removes the highest strength part of the continental crust. (c) Oblique view of NPHM mechanical model with surface of vertical velocity and contraction rates. The dome in the vertical velocity is associated with rapid exhumation in the region of the incising river and advection hotspot.

(a)

Effect of transverse gap in topography

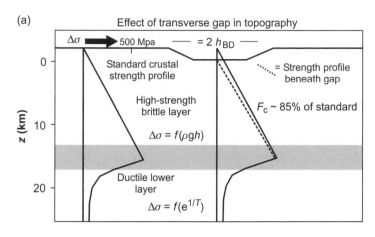

(b)

Effect of transverse gap and exhumation

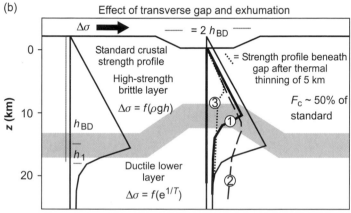

(c)

Vertical particle velocity (Vz)

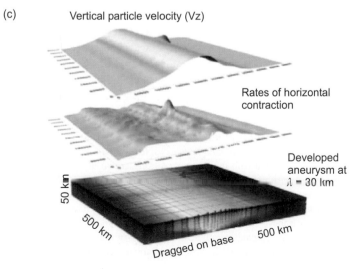

The spatial scale of this advection weakening is a function of the distribution of surface processes, as discussed below. The required condition that at least the main channel remains at approximately constant elevation during vertical rock displacement is met in the Indus (Burbank et al. 1996) and in other major transverse drainages. Nanga Parbat represents a river aneurysm where the thermal perturbation is spatially restricted to the region surrounding a major incising river, while the example from the Alpine fault, discussed below, represents the response of a three-dimensional oblique convergence zone deforming under a spatially extensive, asymmetric orographic condition.

It should be noted that while many variations on the thermally activated rheological model employed here will produce a generally similar strain concentration in the presence of thermal advection, the results are dependent upon two important components:

1. The upper crust is stronger than the lower crust; therefore, replacing the upper with the lower crust leads to weakening.
2. Although weaker than the upper crust, the lower crust must be capable of transmitting strain from the mantle and therefore aneurysm behavior is *not* compatible with a very weak lower crust.

These conditions leave open a means of reconciling recent interpretations about crustal rheology based upon earthquake locations in western Pakistan (Maggi et al. 2000; Handy and Brun 2004).

Topographically Generated Stresses along the Himalayan Front

The Himalayan Front, with its extreme relief parallel to and normal to the Eurasian–Indian plate boundary, large moment earthquakes and high erosion rates, provides a somewhat different topographic and strain situation than Nanga Parbat. Berger et al. (2004) have incorporated the along-strike-averaged topography in mechanical models of deformation along the Himalayan Front and here we examine the distribution of topographic stresses in three dimensions along the Himalayan Front in an attempt to identify regions of anomalous stress concentration. As above, we calculate the stress tensor as a function of topography for an elastic-plastic material. Topography is derived from GTOPO30 gridded at ~6 km; x is east; y is north (Figure 8.4).

Figure 8.4 Stress state at sea level associated with topography along the Himalayan Front. ▶ (a) Relief map showing location of area in degrees; the following stress maps are in meters and are calculated on a grid with the same conditions as that of Figure 8.1a. (b) Ratio of horizontal normal coordinate stresses to vertical normal stress. At sea level, the horizontal stresses dominate, rising to maxima along the front. (c) Ratio of the two slope-generated horizontal normal stresses (4_y). Values < 1 = dominated by main Himalayan Front; > 1 = dominated by transverse (north–south) gorges. (d) TSI indicates that topography along the southern front of the Himalayas and near the largest of the transverse gorges maintains the stress state at close to failure in the absence of any tectonic stresses.

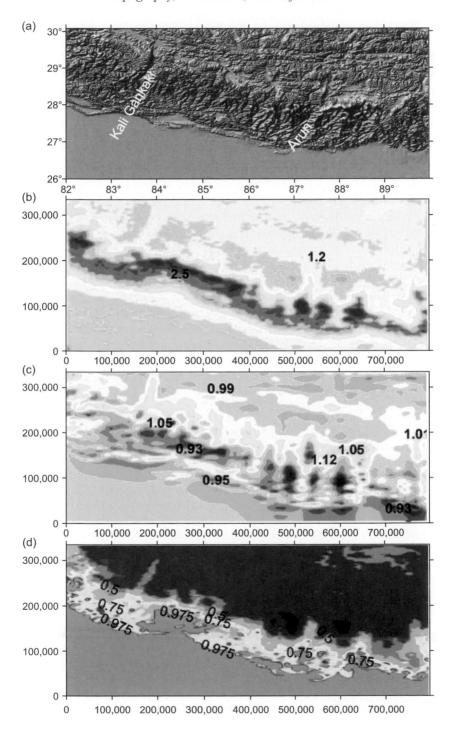

The largest perturbations to a uniform stress field are generated by the long wavelength Himalayan Front and by the transverse gorges that intersect the front at high angles as indicated in the ratios of the normal coordinate stresses (Figures 8.4b, c). At sea level, the horizontal normal stresses dominate over σ_{zz}, with the normal stress ratio varying from ~ 1.2 beneath the southern part of the plateau, to >2.5 along the front (Figure 8.4b). At higher elevations, σ_{zz} is the larger normal stress beneath the plateau. Along the southern part of the Himalayan Front, the horizontal stress in the north–south direction dominates ($\sigma_{xx}/\sigma_{yy} < 1$; Figure 8.4c), however, further to the north where the transverse gorges show the greatest relief, the horizontal stress is dominated by the east–west component ($\sigma_{xx}/\sigma_{yy} > 1$; Figure 8.4c).

The deep gorges generated by the major transverse Himalayan rivers normal to the Himalayan Front produce vertical shear stress concentrations that decay more rapidly with depth than does that generated by the Himalayan Front, but is of approximately equal magnitude on planes at sea level. Intersection of the Himalayan Front with the transverse gorges produces intense knots of vertical shear stress (Figure 8.4d). The map of TSI, incorporating the weakening effect of load reduction as well as the shear stress effect, shows the proximity to failure parallel to the Himalayan Front and also at the southern edges of the Transverse gorges. The transverse TSI maxima are probably responsible for focusing of strain into "River Anticlines" of the transverse rivers (e.g., Gansser 1964). In addition, we suggest that these knots may serve as nucleation sites of large earthquakes, or to delineate the rupture segments of large Himalayan events (e.g., Lavé and Avouac 2000).

Figure 8.5 Numerical model of oblique convergence illustrating the sensitivity of three- ▶ dimensional strain fields to rheological heterogeneities. All boundary conditions remain the same, as described below, but the model is allowed to evolve through exhumation of deeper ductile material along the exhumation structure (Koons et al. 2003). (a) *Incipient state:* Initial mechanical model of oblique compression of a two-layered crust dragged along its base at velocities of 40 mm yr^{-1} parallel to (along y-axis) and 10 mm yr^{-1} normal to (along x-axis) the plate boundary; z-axis is vertical downward. Boundary conditions on both models are identical, and approximate steady-state elevations are maintained adjacent to the plate boundary. In this initial model, two distinct structures accommodate strain through the upper crust. The initial model geometry and velocity contours are shown in section normal to plate boundary. Crustal rheology is approximately horizontal and not yet perturbed by advection. For contours of lateral strain rates ($= \partial V_y/\partial x + \partial V_y/\partial z$), dark shading represents highest rates. Vertical velocity (in mm yr^{-1}) is V_z. Convergent strain rates are represented as rotation rates in the x–z-plane about the y-axis ($= -0.5(\partial V_x/\partial z - \partial V_z/\partial x)$). (b) Results with velocity diagrams and strain-rate analysis for evolved model, which includes thermal weakening associated with exhumation in response to intense erosion along the western (left) side of the orogen. All variable identification and other conditions are the same as in the initial model (Figure 8.4a). At this stage, both lateral and convergent strains are concentrated along a single, east-dipping structure through the upper, brittle crust.

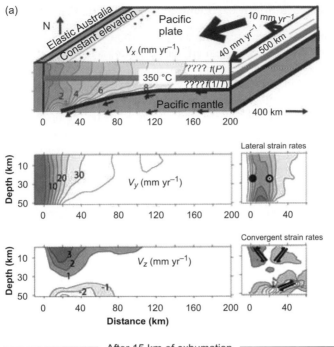

After 15 km of exhumation

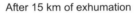

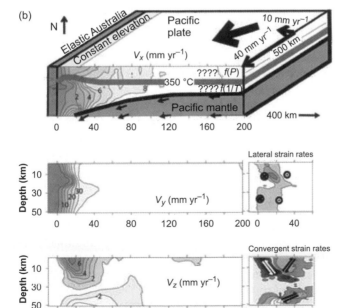

Process-controlled Strain Partitioning: Alpine Fault, New Zealand

In collisional zones that are dominantly two-dimensional with little oblique relative motion, as in the western Himalaya, topographic stresses are unlikely to be sufficient to alter the displacement pattern from thrusting to strike-slip (Liu and Zoback 1992). However, the displacement pattern in oblique collisional plate boundaries can be quite sensitive to stress and strength perturbations (Enlow and Koons 1998), and observations on the displacement patterns in three-dimensionally deforming regions can yield significant information on the rheological structure at depth. The displacement nature of the Alpine fault of the South Island, New Zealand, is an enigmatic three-dimensional structure in that it currently accommodates both lateral and contractional displacements on a single structure that dips at $\sim 50°$ to the southeast (Norris et al. 1990; Stern et al. 2000). For a laterally homogeneous rheology, the two strain components are predicted to separate into a vertical structure taking up the lateral strain and a dipping zone accommodating contractional strain (Figure 8.5a).

Elsewhere, we have argued that the concentration of both strain components onto a single dipping structure on the Alpine fault is related to thermal weakening by vigorous, orographically controlled exhumation (Koons et al. 2003). The highly asymmetric erosion pattern of the Southern Alps produces weakening along the exhumation structure that leads to a shift from the two structure plate boundary to the single oblique structure analogous to the Alpine fault (Figure 8.5b). The transition to a single structure is a function of the advective weakening and consequently this behavior should occur on the timescales characteristic of advective crustal heating (~ 1 Ma; Koons 1987; Batt and Braun 1999). Again, we emphasize that this model of fault behavior requires that the lower crust be weaker than the upper crust.

SURFACE PROCESSES AND TOPOGRAPHY: A CRITICAL LINK

In the previous sections, we have argued that topography at the Earth's surface influences the static stress state of the crust which, under certain conditions, can modify the distribution of strain and, in certain cases, lead to localization and faulting in the upper crust. Moreover, the distribution of erosion rate across a landscape modulates, via advection of heat and mass, regional thermal gradients and thus the integrated strength of crust. Although we have, thus far, considered them independent, erosion rates and topography are strongly correlated (e.g., Ahnert 1970); the efficacy of sediment transport on hillslopes, by debris flows, and in river systems depends largely on topographic gradients. Thus, the degree to which a geomorphic transport law (e.g., Dietrich et al. 2003) depends on topographic gradients plays a key role in the evolution of topography in active mountain belts (Whipple and Tucker 1999). Here we review what is currently known about the relationship between erosion rates and landscape topography with an intent to highlight key uncertainties.

The potential for a dynamic coupling between the efficiency of mass removal by erosion and the evolution of topography, rock uplift rate, and strain within active orogens has been well established by models of critical wedge mechanics (e.g., Davis et al. 1983; Koons 1989; Whipple and Meade 2004) as well as those which utilize temperature-dependent rheologies (e.g., Koons et al. 2002; Willett et al. 2001; Beaumont et al. 2001). If one seeks to isolate the response of surface processes to tectonic forcing, however, ideal field laboratories are small orogens where rates and patterns of rock uplift are well known at timescales appropriate to the erosional process, where climatic and lithologic variations are minimal, and where the erosional response is known (i.e., steady-state erosion versus a transient of known magnitude). Such sites are rare. Below we review the results from several case studies.

Landscape Response to Variations in Erosion Rate

Although it has been long recognized that many mountain ranges exhibit correlations between topographic relief and erosion rate (e.g., Ahnert 1970; Hurtrez et al. 1999; Montgomery and Brandon 2002), considerations of landscape scale (Montgomery and Dietrich 1992) suggest that individual geomorphic processes acting on separate portion of the landscape serve to dictate topographic relief. Thus, in considering topography in a landscape, it is useful to consider different process regimes.

Hillslopes

Although sediment liberated from hillslopes comprises the bulk of erosional mass flux from mountain belts, slope stability thresholds dictate that, once erosion rates outpace soil production (e.g., Heimsath et al. 1997), hillslope relief in active mountain belts becomes relatively insensitive to further increases in erosion rate (Schmidt and Montgomery 1995; Burbank et al. 1996). As hillslope gradients approach the mean angle for failure, topographic relief above the local base level (set by the channel network) is dictated primarily by the internal controls on rock strength (e.g., tensile strength, fracture density, degree of weathering) and by the spacing of channel systems. However, recent observations of a correlation between mean hillslope gradient and precipitation in the Nepalese Himalaya (Gabet et al. 2004) raises the interesting possibility that stability thresholds may also respond to variations in climate, either due to variations in pore pressures (Terzaghi 1943) and/or changes in rock mass quality due to enhanced weathering. Insofar as topographically generated stresses reflect local relief, such a climatic feedback could have important implications for fracturing of bedrock near valley bottoms (e.g., Miller and Dunne 1996; Molnar 2004).

Fluvial Channels

Despite the fact that fluvial channels occupy only a small part of the land-scape, river incision rates set the lower boundary condition on hillslopes, ef-fectively dictating denudation rates across a landscape. Moreover, in active mountain ranges, a large percentage of topographic relief is set by the eleva-tion drop on channel systems (Whipple et al. 1999). In addition, the channel network conveys tectonic and climatic signals throughout the landscape and controls landscape response to these perturbations (Whipple and Tucker 1999). Consequently, the controls on channel incision rates and longitudinal profiles have been the subject of much recent attention. For the purposes of this chap-ter, we focus on the relationship between the morphology of the channel pro-file and erosion rates; for a comprehensive discussion of bedrock channel pro-cesses and models used to represent those processes, the reader is referred to Whipple (2004).

In many active mountain belts, longitudinal profiles of bedrock rivers ex-hibit smooth, concave-up longitudinal profiles that can be described by a scal-ing between local channel gradient (S) and contributing drainage area (A):

$$S = k_s A^{-\theta} \tag{8.8}$$

where k_s and θ are referred to as the steepness and concavity indices, respec-tively. This slope-area scaling only holds downstream of a critical drainage area (e.g., Tarboton 1989). Although many longitudinal profiles exhibit a single scaling along their entire reach, segments of an individual profile may be char-acterized by different steepness and concavity indices and may depend on down-stream changes in bed state (alluvial vs. bedrock), substrate properties, and rock uplift rate (e.g., Whipple and Tucker 1999; Kirby and Whipple 2001; Kirby et al. 2003).

In a number of active mountain ranges, between 70% and 90% of topo-graphic relief occurs within the scaling regime described above (Whipple 2004). Thus, the controls on the steepness and concavity indices directly impact land-scape topography. At present, despite empirical support for a relationship be-tween channel steepness and erosion rate (e.g., Wobus et al. 2006), a number of complexities plague a quantitative understanding of this relation. Perhaps the most important of these are non-linearities in incision process (Hancock et al. 1998; Whipple and Tucker 1999), which may include incision thresholds (Tucker 2004). Other effects include: (a) adjustments in the state of the channel bed, including sediment grain size, extent of alluvial cover, and/or hydraulic rough-ness (Sklar and Dietrich 1998, 2004; Whipple and Tucker 2002); (b) changes in channel form, including sinuosity and hydraulic geometry (Lavé and Avouac 2000; Snyder et al. 2003; Duvall et al. 2004); (c) feedbacks between topogra-phy and orographically enhanced precipitation (Roe et al. 2002), and (d) changes in the erosive efficacy of debris-flows (Stock and Dietrich 2003). Convolved with all of these effects is the ever-present problem of lithologic heterogeneity; rock strength varies widely in mountain belts and can strongly influence k_s (e.g., Duvall et al. 2004).

We seek to illustrate some of these effects with two examples drawn from our previous work. The first case study comes from the Siwalik Hills where strong spatial gradients in deformation are developed across a growing fault-related fold above the Main Frontal Thrust in central Nepal (Lavé and Avouac 2000). The fold is developed in the relatively homogenous terrestrial sandstones and siltstones of the Lower and Middle Siwalik formations. Deformation rates inferred from the distribution of Holocene terraces vary from ~4 mm yr^{-1} north of the range, to ~16–17 mm yr^{-1} at the range crest, and back to near zero south of the range; multiple flights of terraces indicate that incision rates both balance rock uplift rates across the fold and have been steady over Holocene time (Lavé and Avouac 2000). Thus, this field site provides an opportunity to compare channel morphology developed in uniform lithology under widely varying erosion rates (Kirby and Whipple 2001).

In a study of channel topography derived from digital elevation data, Kirby and Whipple (2001) argued that systematic downstream changes in the concavity of channels crossing uplift rate gradients were consistent with an incision process that was linear in channel slope. Moreover, recent analysis of higher resolution data of tributary channels flowing parallel to uplift rate gradients (and thus, experiencing uniform uplift rates along their lengths) demonstrates that channel steepness and erosion/uplift rate are linearly related in this landscape (Figure 8.6), apparently confirming the earlier work. In addition, the steepness indices of small, first-order channels (debris-flow and colluvial channels) show a linear correlation to erosion rate (Lague and Davy 2003). Moreover, Hurtrez and colleagues (1999) demonstrated a significant linear relationship between local relief (measured within a 500 m radius, and thus probably capturing primarily hillslope relief) and erosion rate. Thus, in this relatively simple field site, it appears that steady-state topography scales linearly with erosion/uplift rate. The implications of this set of observations, that landscape morphology adjusts linearly with uplift/erosion rate, have not yet been fully realized: however, if validated elsewhere, they suggest a fairly simple means of relating topographic stresses and erosion rate.

Other studies of similar experimental design, however, have revealed nonlinear relationships between channel topography and erosion rate. Comparison of steady-state channel profiles under conditions of uniform lithology but differences in rock uplift rate in two separate sites in coastal California yielded steepness indices that display an apparent non-linear relation to erosion rate (Snyder et al. 2000; Duvall et al. 2004). In the King Range, Snyder and others (2003) argued that the apparent non-linearity was best explained as a consequence of a threshold for bedrock incision, while in the western Transverse

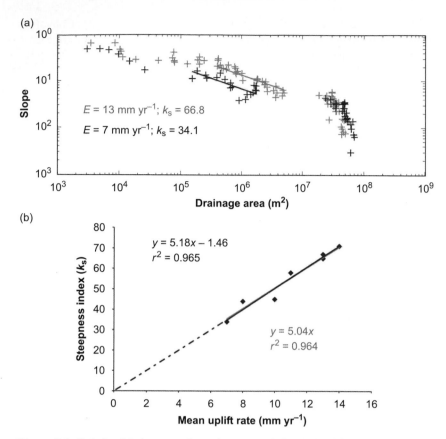

Figure 8.6 Relationship between channel steepness index and uplift rate in the Siwalik Hills, Nepal. All channels are developed in a single lithology, are oriented parallel to uplift rate gradients, and are thus experiencing uniform uplift rate along their lengths (Kirby and Whipple 2001). Uplift rates are taken from data of Lavé and Avouac (2000) and are presumed to be matched by channel erosion rates (e.g., Hurtrez et al. 1999). (A) Examples of topographic data (channel gradients) for two channels. Note that a two-fold difference in rock uplift rate engenders a doubling of channel gradients along the entire channel reach. (B) Channel steepness indices derived for 10 strike-parallel channels in the study region demonstrate a linear dependence of the steepness index on erosion rate in this landscape.

Ranges, Duvall et al. (2004) argued that adjustments of channel width accounted for the difference. These studies serve to highlight several of the uncertainties described above—the controls on bedrock channel hydraulic geometry, and on the incision processes themselves, remain major hurdles in our ability to quantitatively relate channel form to incision rate.

Glacial Valleys

Many of the world's active mountain ranges extend into the elevation range where glaciers become important agents of valley erosion. Measurements of sediment flux from glacial termini (e.g., Hallet et al. 1996) and classical geomorphic characteristics of glacial landscapes (low-gradient, U-shaped valleys) both indicate that glaciers can be an extremely effective erosive agent. Moreover, it has been argued that the efficiency of glacial erosion effectively maintains an upper limit on landscape elevations (Brozovic et al. 1997), near the climatically driven equilibrium line altitude (ELA). This contrasts, to some degree, with studies of glacial landscapes (Brocklehurst and Whipple 2002; Montgomery 2002), which suggest that relief production may be quite limited, restricted primarily to the development of high cirque headwalls and steep valley sidewalls.

A recent study of subglacial hydrology may hold the key to reconciling the erosive power of glacial ice with relatively limited relief. Alley et al. (2003) suggests that the thermodynamics of subglacial water can markedly affect glacial erosion. As water at the glacier bed flows downslope (or gradually upslope), heat production from the dissipation of viscous stresses maintains water at the pressure-melting temperature. However, when the bed angle is steep and facing upstream (as in an overdeepening or at a terminal moraine), subglacial water becomes supercooled, and begins to freeze, constricting subglacial channels and reducing sediment flux (Alley et al. 2003). The net effect decreases erosion of the bed allowing the accumulation of an armoring layer of till. Overall, this effect may regulate the shape of longitudinal profile of the glacial bed and maintain it in a "graded" condition. Once this condition is established, continued lowering of the glacial valley is dependent upon the lowering rate of the fluvial system downstream of the glacial margin.

CONCLUSIONS

Earth deformation is fundamentally influenced by surface processes via stress perturbations and strength perturbations. Both produce positive feedback that generates long term lithospheric weakness and transforms a homogeneous lithosphere to a highly heterogeneous lithosphere.

Contributions to the local stress state from natural topography in regions of extreme relief control the stress orientations in the upper 3–8 km below the surface, while stresses generated by boundary tectonic conditions dominate through the lower crust and the lower part of the upper frictional layer. Although this paper has not examined the interaction of topography and tectonics for incipient, subdued topography, recent studies indicate that cooperation occurs at the much higher frequencies of incipient fold-thrust belts (e.g., Simpson 2004; Simon 2005).

Because relief is generated in part by tectonic strain, there will always tend to be some correlation in the orientation of tectonic and topographic stresses at long wavelengths. As the topographic field becomes more complex at higher spatial frequencies, the kinematics, as indicated by structural fabric elements or seismic first motions, also undergoes reduced correlation.

The general tendency toward extensional features at high elevations and contractional features at lower elevations resulting entirely from topographic stresses requires that interpretation of regional strain from structures in high relief be done with great caution. Consequently, strain at higher elevations, in regions of extreme relief, reflects topographically generated stresses.

Where the topography results from tectonic strain at low frequencies, and topographic stresses correlate with tectonic stresses, a positive feedback situation is maintained that localizes both tectonic and topographic strain along a front which in turn leads to additional feedback through strength reduction resulting from advective thinning. The net effect of stress and strength weakening is the tendency to stabilize contractional strain along a single structure or family of structures that defines the plate boundary. Associated with this contractional structure will be a tendency to form extensional structures at high elevations, parallel to the plate boundary independently of channel-type behavior (e.g., Beaumont et al. 2001).

Evidence for rheological modification related to surface processes may be searched for in oblique orogens where the distribution of strain components (lateral and convergent) reflects the nature of rheological structure at depth. Distributions that differ from those predicted for a laterally homogeneous Earth carry information about rheological heterogeneity. Concentration of lateral strain on a single plate bounding structure in the Southern Alps of New Zealand, the Alpine fault, that dips ~50° (Norris et al. 1990; Stern et al. 2000) is compatible with significant departure from a homogeneous layered rheology caused by advective weakening along the dipping exhumation structure (Koons et al. 2003). Analogously, long-term exhumation on steeper than predicted oblique structures, such as the Karakoram fault, argues for an alternative mechanism of weakening on near vertical structures initially by igneous advection.

The intersection of large incising rivers into topographic fronts, such as the Arun into the Himalayan front, produces stress knots that concentrate long-term strain and may influence the distribution of large earthquakes.

ACKNOWLEDGMENTS

Thanks to P. Upton, A. Densmore, and M. Handy for comments and discussion. We are grateful to J.-P. Avouac and G. Simpson for help in clarifying some of the concepts.

REFERENCES

Ahnert, F. 1970. Functional relationships between denudation, relief, and uplift in large mid-latitude drainage basins. *Am. J. Sci.* **268**:243–263.

Albarede, F. 1976. Thermal models of post-tectonic decompression as exemplified by the Haut Allier granulites (Massif Central, France). *Bull. Soc. Géol. France* **18**:1023–1032.

Alley, R.B., D.E. Lawson, G.J. Larson, E.B. Evenson, and G.S. Baker. 2003. Stabilizing feedbacks in glacier-bed erosion. *Nature* **424**:758–760.

Avouac, J.-P., and E.B. Burov. 1996. Erosion as a driving mechanism of intracontinental growth. *J. Geophys. Res.* **101**:17,747–17,769,

Batt, G.E., and J. Braun. 1999. The tectonic evolution of the Southern Alps, New Zealand: Insights from fully thermally coupled dynamical modeling. *Geophys. J. Intl.* **136**:403–420.

Beaumont, C., P. Fullsack, and J. Hamilton. 1992. Erosional control of active compressional orogens. In: Thrust Tectonics, ed. K.R. McClay, pp. 1–18. New York: Chapman & Hall.

Beaumont, C., R.A. Jamieson, M.H. Nguyen, and B. Lee. 2001. Himalayan tectonics explained by extrusion of a low-viscosity crustal channel coupled to focused surface denudation. *Nature* **414**:738–742.

Berger, A., J. Jouanne, R.Hassani, and J.L. Mugnier. 2004. Modelling the spatial distribution of present-day deformation in Nepal: How cylindrical is the Main Himalayan Thrust in Nepal? *Geophys. J. Intl.* **156**:94–114.

Bollinger, L., J.-P. Avouac, R. Cattin, and M.R. Pandey. 2004. Stress buildup in the Himalaya. *J. Geophys. Res.* **109**:B11405, doi:10.129/2003JB002911.

Brace, W.F., and D.L. Kohlstedt. 1980. Limits on lithospheric stress imposed by laboratory experiments. *J. Geophys. Res.* **85**:6248–6252.

Brocklehurst, S.H., and K.X. Whipple. 2002. Glacial erosion and relief production in the Eastern Sierra Nevada, California. *Geomorphology* **42**:1–24.

Brozovic, N., D.W. Burbank, and A.J. Meigs. 1997. Climatic limits on landscape development in the northwestern Himalaya. *Science* **276**:571–574.

Burbank, D.W., J. Leland, E. Fielding et al. 1996. Bedrock incision, rock uplift and threshold hillslopes in the northwestern Himalayas. *Nature* **379**:505–510.

Carslaw, H.S., and J.C. Jaeger. 1959. Conduction of Heat in Solids, 2nd ed. New York: Oxford Univ. Press.

Chapple, W.M. 1978. Mechanics of thin-skinned fold and thrust belts. *Geol. Soc. Am. Bull.* **89**:1189–1198.

Craw, D., P.O. Koons, D. Winslow, C.P. Chamberlain, and P.K. Zeitler. 1994. Boiling fluids in a region of rapid uplift, Nanga Parbat Massif, Pakistan. *Earth Planet. Sci. Lett.* **128**:169–182.

Dahlen, F.A. 1984. Noncohesive critical Coulomb wedges: An exact solution. *J. Geophys. Res.* **89**:10,125–10,133.

Davis, D., J. Suppe, and F.A. Dahlen. 1983. Mechanics of fold-and-thrust belts and accretionary wedges. *J. Geophys. Res.* **88**:1153–1172.

Dietrich, W.E., D.G. Bellugi, L.S. Sklar et al. 2003. Geomorphic transport laws for predicting landscape form and dynamics. In: Prediction in Geomorphology, ed. P.R. Wilcock and R.M. Iverson, pp. 103–132. Washington, D.C.: Am. Geophys. Union.

Duvall, A., E. Kirby, and D. Burbank. 2004. Tectonic and lithologic controls on bedrock channel profiles in coastal California. *J. Geophys. Res.* **109**:F03002, doi:10.1029/2003JF000086.

Enlow, R.L., and P.O. Koons. 1998. Critical wedges in three dimensions: Analytical expressions from Mohr-Coulomb constrained perturbation analysis. *J. Geophys. Res.* **103**:4897–4914.

Gabet, E.J., B.A. Pratt-Sitaula, and D.W. Burbank. 2004. Climatic controls on hillslope angle and relief in the Himalayas. *Geology* **32**:629–632.

Gansser, A. 1964. Geology of the Himalaya, p. 289. New York: Wiley.

Hallet, B., L. Hunter, and J. Bogen. 1996. Rates of erosion and sediment evacuation by glaciers: A review of field data and their implications. *Global Planet. Change* **12**:213–235.

Hancock, G.S., R.S. Anderson, and K.X. Whipple. 1998. Beyond power: Bedrock river incision process and form. In: Rivers Over Rock: Fluvial Processes in Bedrock Channels, ed. K.J. Tinkler and E.E. Wohl, pp. 35–60. Washington, D.C.: Am. Geophys. Union.

Handy, M.R., and J.P. Brun. 2004. Seismicity, structure, and strength of the continental lithosphere. *Earth Planet. Sci. Lett.* **223**:427–441.

Heimsath, A., W.E. Dietrich, K. Nishiizumi, and R.C. Finkel. 1997. The soil production function and landscape equilibrium. *Nature* **388**:358–361.

Hurtrez, J.-E., F. Lucazeau, J. Lavé, and J.-P. Avouac. 1999. Investigation of the relationship between basin morphology, tectonic uplift, and denudation from the study of an active fold belt in the Siwalik Hills, central Nepal. *J. Geophys. Res.* **104**:12,779–12,796.

Jaeger, J.C., and N.G.W. Cook. 1969. Fundamentals of Rock Mechanics. New York: Chapman and Hall.

Kirby, E., and K. Whipple. 2001. Quantifying differential rock-uplift rates via stream profile analysis. *Geology* **29**:415–418.

Kirby, E., K.X. Whipple, W. Tang, and Z. Chen. 2003. Distribution of active rock uplift along the eastern margin of the Tibetan Plateau: Inferences from bedrock channel longitudinal profiles. *J. Geophys. Res.* **108**:B4, doi:10.1029/2001JB000861.

Koons, P.O. 1987. Some thermal and mechanical consequences of rapid uplift: An example from the Southern Alps. *Earth Planet. Sci. Lett.* **86**:307–319.

Koons, P.O. 1989. The topographic evolution of collisional mountain belts: A numerical look at the Southern Alps, New Zealand. *Am. J. Sci.* **289**:1041–1069.

Koons, P.O. 1990. The two-sided wedge in orogeny: Erosion and collision from the sand box to the Southern Alps, New Zealand. *Geology* **18**:679–682.

Koons, P.O. 1994. Three-dimensional critical wedges: Tectonics and topography in oblique collisional orogens. *J. Geophys. Res.* **99** (Spec. Iss. Tectonics and Topography):12,301–12,315.

Koons, P.O., R.J. Norris, D. Craw, and A.F. Cooper. 2003. Influence of exhumation on the structural evolution of transpressional plate boundaries: An example from the Southern Alps, New Zealand. *Geology* **31**:3–6.

Koons, P.O., P.K. Zeitler, C.P. Chamberlain, D. Craw, and A.S. Melzer. 2002. Mechanical links between erosion and metamorphism in Nanga Parbat, Pakistan Himalaya. *Am. J. Sci.* **302**:749–773.

Lague, D., and P. Davy. 2003. Constraints on the long-term colluvial erosion law by analyzing slope-area relationships at various tectonic uplift rates in the Siwalik Hills, Nepal. *J. Geophys. Res.* **108**:doi:10.1029/2002JB001893.

Lavé, J., and J.-P. Avouac. 2000. Active folding of fluvial terraces across the Siwalik Hills, Himalayas, of central Nepal. *J. Geophys. Res.* **105**:5735–5770.

Liu, L., and M.D. Zoback. 1992. The effect of topography on the state of the crust: Application to the site of the Cajon Pass Scientific Drilling Project. *J. Geophys. Res.* **97**:5095–5108.

Maggi, A., J.A. Jackson, D. McKenzie, and K. Priestley. 2000. Earthquake focal depths, effective elastic thickness, and the strength of the continental lithosphere. *Geology* **28**:495–498.

Mandl, G. 1988. Mechanics of Tectonic Faulting: Models and Basic Concepts. New York: Elsevier.

McTigue, D.F., and C.C. Mei. 1981. Gravity-induced stresses near topography of small slope. *J. Geophys. Res.* **86**:9268–9278.

Miller, D.J., and T. Dunne. 1996. Topographic perturbations of regional stresses and consequent bedrock fracturing. *J. Geophys. Res.* **101**:B11, 25,523–25,536.

Molnar, P. 2004. Interactions among topographically induced elastic stress, static fatigue, and valley incision, *J. Geophys. Res.* **109**:F02010, doi: 10.1029/2003JF000097.

Montgomery, D.R. 2002. Valley formation by fluvial and glacial erosion. *Geology* **30**: 1047–1050.

Montgomery, D.R., and M.T. Brandon. 2002. Topographic controls on erosion rates in tectonically active mountain ranges. *Earth Planet. Sci. Lett.* **201**:481–489.

Montgomery, D.R., and W.E. Dietrich. 1992. Channel initiation and the problem of landscape scale. *Science* **255**:826–830.

Norris, R.J., and A.F. Cooper. 1997. Erosional control on the structural evolution of a transpressional thrust complex on the Alpine fault, New Zealand. *J. Struct. Geol.* **19**:1323–1342.

Norris, R.J., P.O. Koons, and A.F. Cooper. 1990. The obliquely-convergent plate boundary in the South Island of New Zealand: Implications for ancient collision zones. *J. Struct. Geol.* **12**:715–725.

Pavlis, T.L., M.W. Hamburger, and G.L. Pavlis. 1997. Erosional processes as a control on the structural evolution of an actively deforming fold and thrust belt: An example from the Pamir-Tien Shan region, central Asia. *Tectonics* **16**:810–822.

Ramberg, H. 1973. Model studies of gravity controlled tectonics by the centrifuge technique. In: Gravity and Tectonics, ed. K.A. DeJong and R. Scholten, pp. 49–77. New York: Wiley.

Ranalli, G. 1987. Rheology of the Earth: Deformation and Flow Processes in Geophysics and Geodynamics. New York: Allen & Unwin.

Roe, G.H., D.R. Montgomery, and B. Hallet. 2002. Effects of orographic precipitation variations on the concavity of steady-state river profiles. *Geology* **30**:143–146.

Savage, W.Z., and H.S. Swolfs. 1986. Tectonic and gravitational stress in long symmetric ridges and valleys. *J. Geophys. Res.* **91**:3677–3685.

Schmidt, K.M., and D.R. Montgomery. 1995. Limits to relief. *Science* **270**:617–619.

Simon, J.L. 2005. Erosion-controlled geometry of buckle fold interference. *Geology* **33**:561–564, doi: 10.1130/G21468.1.

Simpson, G. 2004. Role of river incision in enhancing deformation. *Geology* **32**: 341–344, doi: 10.1130/G20190.2.

Sklar, L., and W.E. Dietrich. 1998. River longitudinal profiles and bedrock incision models: Stream power and the influence of sediment supply. In: Rivers over Rock: Fluvial Processes in Bedrock Channels, ed. K.J. Tinkler and E.E. Wohl, pp. 237–260. Washington, D.C.: Am. Geophys. Union.

Sklar, L.S., and W.E. Dietrich. 2004. A mechanistic model for river incision into bedrock by saltating bed load. *Water Resour. Res.* **40**:W06301, doi: 10.1029/2003WR002496.

Snyder, N.P., K.X. Whipple, G.E. Tucker, and D.J. Merritts. 2000. Landscape response to tectonic forcing: Digital elevation model analysis of stream profiles in the Mendocino triple junction region, northern California. *Geol. Soc. Am. Bull.* **112**:1250–1263.

Snyder, N.P., K.X. Whipple, G.E. Tucker, and D.J. Merritts. 2003. Importance of a stochastic distribution of floods and erosion thresholds in the bedrock river incision problem. *J. Geophys. Res.* **108**:B2, doi: 10.1029/2001JB001655.

Sonder, L.J., and P. England. 1986. Vertical averages of rheology of the continental lithosphere: Relation to thin sheet parameters. *Earth Planet. Sci. Lett.* **77**:81–90.

Stern, T., P. Molnar, D. Okaya, and D. Eberhart-Philip. 2000. Teleseismic P-wave delays and modes of shortening the mantle lithosphere beneath South Island, New Zealand. *J. Geophys. Res.* **105**:21,615–21,631.

Stock, J., and W.E. Dietrich. 2003. Valley incision by debris flows: Evidence of a topographic signature. *Water Resour. Res.* **39**:doi: 10.1029/2001WR001057.

Tarboton, D.G., R.L. Bras, and I. Rodriguez-Iturbe. 1989. Scaling and elevation in river networks. *Water Resour. Res.* **25**:2037–2051.

Terzaghi, K. 1943. Theoretical Soil Mechanics. New York: Wiley.

Thiede, R.C., B. Bookhagen, J.R. Arrowsmith, E.R. Sobel, and M.R. Strecker. 2004. Climatic control on rapid exhumation along the Southern Himalayan Front. *Earth Planet. Sci. Lett.* **222**:791–806.

Tucker, G.E. 2004. Drainage basin sensitivity to tectonic and climatic forcing: Implications of a stochastic model for the role of entrainment and erosion thresholds. *Earth Surf. Proc. & Landforms* **29**:185–205.

Whipple, K.X. 2004. Bedrock rivers and the geomorphology of active orogens. *Ann. Rev. Earth & Planet. Sci.* **32**:151–185.

Whipple, K.X., E. Kirby, and S.H. Brocklehurst. 1999. Geomorphic limits to climate-induced increases in topographic relief. *Nature* **401**:39–43.

Whipple, K.X., and B.J. Meade. 2004. Controls on the strength of coupling among climate, erosion, and deformation in two-sided, frictional orogenic wedges at steady state. *J. Geophys. Res.* **109**:F01011, doi:10.1029/2003JF000019.

Whipple, K.X., and G.E. Tucker. 1999. Dynamics of the stream-power river incision model: Implications for height limits of mountain ranges, landscape response timescales, and research needs. *J. Geophys. Res.* **104**:17,661–17,674.

Whipple, K.X., and G.E. Tucker. 2002. Implications of sediment flux-dependent river incision models for landscape evolution. *J. Geophys. Res.* **107**:3–20.

Willett, S.D. 1999. Orogeny and orography: The effects of erosion on the structure of mountain belts. *J. Geophys. Res.* **104**:28,957–28,982.

Willett, S.D., C. Beaumont, and P. Fullsack. 1993. Mechanical model for the tectonics of doubly vergent compressional orogens. *Geology* **21**:371–374.

Willett, S.D., R. Slingerland, and N. Hovius. 2001. Uplift, shortening, and steady state topography inactive mountain belts. *Am. J. Sci.* **301**:455–485.

Wobus, C.W., K.X. Whipple, E. Kirby et al. Tectonics from topography: Procedures, promise, and pitfalls. In: Tectonics, Climate, and Landscape Evolution, ed. S.D. Willett, N. Hovius, M.T. Brandon, and D. Fisher. *GSA Spec. Paper* 398:55–74.

Zeitler, P.K., and C.P. Chamberlain. 1991. Petrogenetic and tectonic significance of young leucogranites from the northwestern Himalaya, Pakistan. *Tectonics* **10**:729–741.

Zeitler, P.K., P.O. Koons, M.P. Bishop et al. 2001b. Crustal reworking at Nanga Parbat: Evidence for erosional focusing of crustal strain. *Tectonics* **20**:712–728.

Zeitler, P.K., A.S. Meltzer, P.O. Koons et al. 2001a. Erosion, Himalayan tectonics and the geomorphology of metamorphism. *GSA Today* **11**:4–8.

9

Constraining the Denudational Response to Faulting

NIELS HOVIUS[1] and FRIEDHELM VON BLANCKENBURG[2]

[1]Department of Earth Sciences, University of Cambridge, Downing Street,
Cambridge CB2 3EQ, U.K.
[2]Institute of Mineralogy, University of Hanover, Callinstr. 3,
30167 Hannover, Germany

ABSTRACT

Denudation links tectonics with climate by changing topographic loads and promoting the drawdown of CO_2. Measurements of denudation are a key to understanding this link. In particular, they are required to test and calibrate geodynamic models, to evaluate the tectonic control on landscape evolution, to quantify the geomorphic impact of faulting and seismicity, and to assess the role of tectonically driven denudation in stabilizing Earth's climate. We review techniques used to measure denudation, and weathering on timescales relevant to faulting and the dynamics of fault zones, with particular attention paid to the use of hydrometric data and cosmogenic isotopes. Using selected examples, we illustrate the application of these techniques to problems ranging from soil formation and coseismic erosion of earthquake epicentral areas to the erosion of orogens and estimation of catchment-scale erosion and weathering fluxes. The examples show that faulting is the Earth's premier erosion and weathering engine. Globally, erosion scales with tectonic forcing. Locally, fluvial incision and landscape lowering are correlated with faulting and seismic activity. Thus, tectonically active areas yield disproportionate amounts of sediment. Erosion refreshes rock surfaces in these areas, thereby enhancing chemical weathering rates and CO_2 consumption. The effects of climate variability and change are evident in the patterns and rates of erosion and weathering. However, they are almost always superimposed on a stronger tectonic signal. We highlight the potential of cosmogenic nuclides to quantify present and past rates and patterns of denudation associated with faulting. Finally, we identify outstanding challenges for future work: (a) to characterize crustal deformation, climate, and denudation over their full range of time and length scales; (b) to analyze the geomorphic impact and stratigraphic record of recent earthquakes; (c) to identify the processes, thresholds, and feedback mechanisms that control global weathering and regulate the long-term climate; and (d) to provide constraints that help to mitigate the risks associated with geomorphic processes triggered by earthquakes. Constraining the denudational response to faulting will help to meet these challenges.

INTRODUCTION

Fault dynamics are controlled not only by the rheology of the lithosphere but also by the distribution and redistribution of rock and soil masses at Earth's surface via denudation and deposition (Koons 1989; Beaumont et al. 1991). Climate exerts a fundamental control on crustal deformation by driving localized denudation and changing topographic load (Molnar and England 1990; Willett 1999). In turn, climate is affected by several factors, including topography which can alter atmospheric circulation patterns (Kutzbach et al. 1989), rapid weathering of fresh silicate crust exposed in areas with rapid rock uplift which promotes drawdown of CO_2 (Raymo and Ruddiman 1992), and burial of organic carbon in sedimentary basins (Palaca et al. 2001; Lyons et al. 2002). Thus, denudation provides a first-order, two-way link between crustal deformation and atmospheric processes.

This link is now widely recognized and has been incorporated in both geodynamic and landscape evolution models (e.g., Whipple and Tucker 1999; Willett et al. 2001; Simpson 2004; Chapter 8, this volume). For given tectonic and climatic boundary conditions, these models predict that landscape and the deformation of its substrate will develop towards a steady state (Willett and Brandon 2002), that rivers adjust their longitudinal and cross-sectional geometry to ambient rock uplift rates (Snyder et al. 2000; Lague et al. 2000; Lavé and Avouac 2001), and that crustal deformation is focused in areas with exceptionally fast denudation (Beaumont et al. 2001; Zeitler et al. 2001). However, many model predictions lack a quantitative, observational basis and therefore are merely plausible rather than proven. It is fair to say that geodynamic modeling capabilities have outpaced our ability to quantify crustal deformation and denudation. This is partly because denudation has proven difficult to measure, especially on timescales of crustal deformation and faulting.

Upper crustal deformation is often accompanied by seismicity. The short-lived, unpredictable nature of large earthquakes has so far precluded a systematic documentation of coseismic denudation in all but a handful of cases (e.g., Harp and Jibson 1996; Keefer 1994; Malamud et al. 2004). If indeed a relationship exists between earthquake magnitude and denudational impact, then the form of this relationship is not yet clear. Similarly, few observations exist that allow us to constrain the minimum time over which denudation forces crustal deformation. Conceivably, such observations are best acquired where denudation rates are high.

The same does not hold for coupled faulting and erosion. Large fault systems evolve on timescales equal to or longer than their seismic cycle (10^2–10^4 yr) (Figure 4.1 in Chapter 4; see also Chapter 10). This time is at and beyond the limit of direct measurement of denudation, and too short for the use of established thermochronometric methods to estimate exhumation. Although the sedimentary and geomorphic records can contain information on these intermediate timescales (Amorosi et al. 1996; Lavé and Avouac 2001), they do not necessarily provide the comprehensive spatial coverage and temporal resolution required to evaluate the relationship between denudation and faulting (Figures 10.2 and 10.3). New techniques using cosmogenic nuclides to measure rock exposure ages as well as past and present denudation rates have the potential to fill this gap.

Coupling within the tectonics–denudation–climate system stems primarily from the drawdown of CO_2 by weathering of silicate rocks and the sequestration of particulate organic carbon. This counters the input of CO_2 by outgassing and stabilizes Earth's climate within a narrow range of livable conditions. Faulting may be crucial to the efficiency of these mechanisms because it exposes fresh rock surfaces to weathering and enhances topographic relief, thereby mobilizing biomass by mass wasting. However, without quantitative constraints on the rates of erosion and weathering as well as on the yield of organic matter, it is difficult to assess the interdependency of these processes, their impact on the global carbon cycle, and the importance of faulting in this context. Here, too, the quantification of rates and patterns of surface processes sets the pace of progress.

The first part of this chapter contains a method primer, in which we review some techniques used to measure present and past rates of denudation. We focus on the use of fluvial sediment transport data to constrain short-term erosion rates, and the use of cosmogenic isotopes to estimate denudation rates on longer timescales of crustal deformation and faulting. These two approaches have become important tools for understanding the coupling between tectonics, climate, and denudation. They are complemented by the use of low-temperature thermochronometers, such as apatite fission tracks (Reiners and Ehlers 2005) and the U/Th-He system (Ehlers and Farley 2003) to assess the exhumation of rocks from shallow (<5 km) depths and the evolution of local relief. The characteristic timescales of these techniques are 10^6 yr in all but the fastest exhuming domains. We therefore limit our attention to the shorter timescales on which many individual fault systems operate, and do not consider thermochronometric techniques and their results in detail.

After the method primer, we present several case studies of linked weathering, erosion, and denudation on a range of spatial scales of crustal deformation and faulting. We argue that faulting is Earth's premier engine of erosion and weathering, and that through the mediational effects of these processes, faulting has a profound impact on the composition of the atmosphere and thus on climate.

Throughout the chapter, we refer to denudation as the removal of mass from locations *at or near* the Earth's surface due to combined effect of *physical and chemical* processes. Erosion is more narrowly defined as the removal of rock mass and/or weathering products *at* the Earth's surface by *mechanical* processes. Weathering is the *in situ* dissolution by chemical processes (chemical weathering) or mechanical disintegration (mechanical weathering) of rock mass at or near the surface. Other definitions of geomorphological terms are contained in the appendix.

QUANTIFYING DENUDATION: A BRIEF METHOD PRIMER

Measurements of Fluvial Bedrock Erosion

Of all erosion processes, fluvial bedrock incision is the most important because it lowers river base level and drives mass wasting of adjacent hillslopes. Patterns and rates of erosion in bedrock river channels can be constrained by repeatedly surveying a channel reach from a benchmark with known location and measuring the change in values over a known time interval. This requires repeated measurement at identical locations with precision that allows the resolution of incision rates over relatively short intervals. Maximum incision rates are on the order of $1–10$ mm yr^{-1}. Sub-mm precision of measurements is therefore required even in the most quickly eroding channels. A common way of quantifying channel erosion is to measure the length of metal pins fixed in the channel bed, or holes drilled into the bed. Local perturbations of the flow are inherent in both methods. Such perturbations may give rise to anomalous erosion rates at the measurement site, and these methods are therefore unlikely to yield optimal results. Better methods avoid or minimize the introduction of artificial flow perturbations. Electronic range-finding devices with sufficient precision over length scales relevant to channel bed surveying are now available but not yet in general use. A recently developed alternative, if time-intensive, method is the use of a sliding point gauge mounted on a bar between recessed benchmarks (Hartshorn et al. 2002). None of the available methods permits easy measurement of wear below the water line, and surveys are normally limited to those sections of a channel that are dry at low water level. Regardless of the survey method, it is important to make observations at representative sites across an active channel, from the low flow line up to the level of the maximum flood, as wear rates may vary significantly (Hartshorn et al. 2002).

Measurements of River Sediment Load

Away from areas where extreme aridity and low temperatures cause aeolian and glacial processes to dominate, the products of weathering and erosion are removed by rivers. Measurements of river sediment load are therefore fundamental to constraining catchment-scale erosion rates. Rivers transport sediment in suspension or as bedload in regular or constant contact with the channel bed.

In most cases, suspended sediment transport dominates, and it is on this process that we focus our attention first.

Suspended sediment concentrations have been measured systematically in many of the world's large rivers (Figure 9.1), and global compilations of these measurements form the basis for calculating patterns and rates of continental erosion and sediment supply to the oceans (Pinet and Souriau 1988; Milliman and Syvitski 1992; Summerfield and Hulton 1994; Hay 1998; Hovius 1998). The accuracy of sediment load and erosion rate estimates depends on the quality of the suspended sediment sampling and discharge record, the length of the sampling record relative to the return time of the maximum discharge event, the interval between samples, the method used to turn water discharge and sediment concentration measurements into estimates of suspended sediment transport, and the availability and accuracy of bedload transport measurements. We will address each of these issues in turn.

Suspended sediment is not uniformly concentrated in the water column and across a channel. It is therefore necessary to collect samples that integrate over the full depth of flow, at representative locations along a cross section of the channel. Most modern hydrometric stations operate according to a well-defined sampling protocol, but published estimates of the sediment load of some rivers are based on samples collected by dipping a bucket in the stream. Sediment concentration measurements are combined with water discharge measurements to obtain an estimate of the river load. Water discharge measurements are commonly collected by

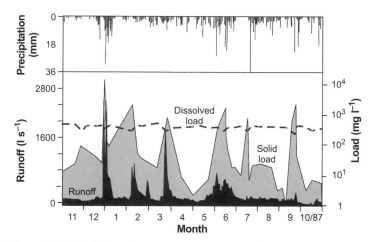

Figure 9.1 Example of a hydrometric record for the Wendebach River, a small tributary of the Leine River, near Göttingen, Germany. The suspended sediment concentration is highly dependent on precipitation and runoff, with most of the suspended flux occurring over the few days with high water discharge. In contrast, the concentration of dissolved material is virtually invariant with time and shows only minor dissolution effects during high discharge (modified from Pörtge 1995).

automatic stage recorders. The recorded stage, or water level of the river, is then converted to an estimate of water discharge in a channel with a known cross section by means of an empirically constrained stage-flow velocity relation.

Any record of hydrometric measurements is necessarily limited to human timescales, but most rivers have variable flow and sediment transport conditions in which the return time of the maximum event (flood) is typically much longer than the total length of the record. Where the maximum event is also the formative event (i.e., the event that dominates the erosional dynamics of a catchment) the limited length of the hydrometric record yields chronically unstable transport and erosion estimates. If no maximum event occurs during the monitoring interval, then the transport and erosion estimates based on measurements are lower than the rates on timescales longer than the return time of the maximum event. On the other hand, if a maximum event is perchance included in the monitoring interval, then it is statistically overrepresented and gives rise to disproportionately high transport and erosion estimates. Unless one is interested in quantifying the actual transport and erosion rates during the interval covered by the available measurements, all measurements should be prorated for the return time of the hydrometric conditions during sampling. Moreover, the maximum possible event size and its return time should be estimated in order to define the full range of system states. Extrapolation of the scaling of event size and frequency over the full dynamic range of events would then permit estimation of true transport and erosion rates. Where the formative event is not the largest event, but is rather an event of intermediate size and frequency, inclusion of the largest floods is less important. Estimates of transport and erosion based on hydrometric records are then much more likely to be close to the true values.

Although river stage tends to be recorded continuously, suspended sediment concentrations are measured at intervals, sometimes only during times of high flow rather that at peak flow (Figure 9.2). Thus, there are several procedures for estimating suspended sediment discharge rates from intermittent measurements (for a detailed review, see Cohn 1995). These methods can be divided into two groups: those that are based on an average of the available observations, and those that combine available sediment observations with other information (usually water discharge data) to obtain a model, or rating curve, which is then used to interpolate or predict sediment concentration in the absence of direct measurements. Averaging estimators use only measured data and therefore rely on the fewest assumptions.

Rating curve estimators employ a quantitative relation between water discharge and sediment concentration or load. The most straightforward relation is an empirically calibrated power-law relation of the standard form:

$$Q_s / aQ^c \ , \tag{9.1}$$

where Q_s is suspended sediment discharge and Q is water discharge. Rating curves are useful only when a robust model can be defined. They are therefore

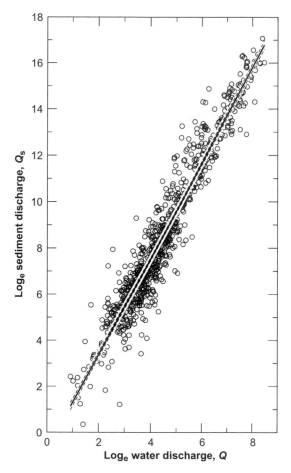

Figure 9.2 Suspended sediment rating curve for the Peinan River (station number 2200H011), Taiwan. This is a mountain river with a catchment area of ~1600 km^2 and a runoff of 1.8 m yr^{-1}. Observations are shown as open circles with a least-squares regression fit (solid line) and confidence interval of regression (dashed lines). Note that for any given water discharge, sediment discharge varies by more than two orders of magnitude. This is the effect of variable sediment supply from hillslopes.

appropriate in transport-limited settings, that is, where the suspended sediment load is close to the carrying capacity of the river. In supply-limited settings, where sediment transport is limited by sediment inputs from hillslopes and alluvial stores, definition of a robust rating curve is difficult because water discharge is only one of many influences on sediment transport rate. There, use of averaging estimators is appropriate. Most rivers in erosional landscapes that are affected by faulting fall into this latter category.

Measurements of bedload transport are rare, especially for small, steep rivers which drain an area affected by recent faulting. Where available, such measurements show significant spatial and temporal variability. This makes it hard to derive a general relationship between channel discharge characteristics and bedload transport. Where no bedload measurements are available, bedload is commonly assumed to be 10% of the total sediment load of a river. Although this may be an appropriate value for large, low-gradient rivers, most mountain rivers probably transport a much larger percentage of their sediment load as bedload. For example in the mountain rivers of Taiwan, bedload is generally 30% of the total sediment load (Dadson et al. 2003).

Measurement of Dissolved Loads

In contrast to clastic sediment, concentrations of dissolved matter are more or less independent of discharge (Figure 9.1). They are, however, often subject to pollution and must be corrected for industrial inputs (Meybeck 1986, 1987). The contribution from cyclic salts (e.g., in sea spray) is usually removed by estimating Cl content, which in turn is corrected for evaporite weathering by using sulfate concentrations (Gaillardet et al. 1999). If the main interest is the contribution of silicate weathering (due to its importance for CO_2 drawdown) then the carbonate contribution of Ca and Mg must be subtracted by titrating for the HCO_3 concentrations. These in turn must be corrected for atmospheric exchange. A global database of dissolved loads in the largest rivers (White and Blum 1995; Gaillardet 1999; Oliva et al. 2003; Dupré et al. 2003) contains fluxes that have been carefully corrected for these effects.

Measurements of Hillslope Mass Wasting

Rivers occupy only a small part of the landscape. Most sediment is produced on hillslopes and transferred downslope by several mass wasting processes. Although many processes can operate on a hillslope, one or a few are normally dominant. Each process and site requires a tailored measurement method and protocol. For example, where mass wasting is dominated by overland flow, erosion can be estimated by monitoring the fill rate of sediment traps placed below rills or at slope inflections (Morgan 2005). Slowly moving landslide complexes can be gauged with creep meters installed along depth profiles through the slide, electronic distance measurements, remote sensing of the deforming surface, or monitoring of embedded GPS receivers or permanent scatterers (Hilley et al. 2004). Volumes of sediment displaced by debris flows can be estimated from their deposits or from the fill of sediment traps in the debris flow channel, whereas information on the frequency of debris flows and some other mass wasting processes can be extracted from the tree ring record (Scott et al. 1995; Lavé and Burbank 2004).

A full assessment of the long-term erosional impact of mass wasting requires information on the frequency and size of events. We illustrate this with the example of landslide populations mapped from time series of aerial photographs or other remote sensing images, but point out that similar considerations apply to many other erosion processes. The size–frequency distribution of landslides is characterized by a maximum at small to intermediate size events (10^3 m^2) and a broad, negative power-law tail for larger landslides (Figure 9.3). This power-law scaling holds true whether the landslide size is defined as the scar area (Hovius et al. 1997) or the total area disturbed (Pelletier et al. 1997), and whether landslides are triggered over a long period of time (Guzzetti et al. 2002) or almost instantaneously (Harp and Jibson 1996). It also holds true if landslide volume is considered instead of area (Brardinoni and Church 2004), although volume is typically much more difficult to measure (both in the field

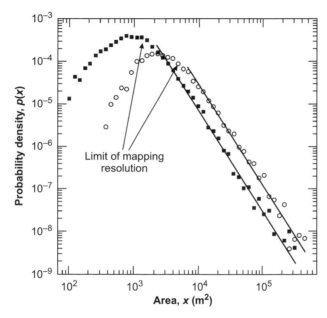

Figure 9.3 Examples of landslide size distributions, from the western Southern Alps, New Zealand, plotted as a probability density function $p(x)$ in log[$p(x)$] versus log(x) coordinates. Solid squares show the probability density of landslides in the Whataroa catchment mapped at 1 : 25,000, $N = 3986$. Open circles show the probability density of landslides in a larger part of the western Southern Alps mapped at 1 : 50,000, $N = 5086$. The data sets show similar scaling of landslide magnitude and frequency. Above a cutoff size (which is proportional to the rollover in this diagram and related to the resolution of the mapping and/or a break in the failure mechanism), the data scale as a power law. This portion of the data is the tail end of the distribution and represents about a quarter of the observed landslides (after Stark and Hovius 2001).

and in aerial photographs). For an idealized landslide size distribution with a power-law distribution across the size range $x \in [c, \infty)$, the size probability density is defined as:

$$p(x) \equiv \alpha c^\alpha x^{-\alpha-1} \ , \quad c > 0 \ , \quad \alpha > 0 \tag{9.2}$$

where α is the power-law scaling exponent (Stark and Hovius 2001), x is usually defined as planform area, and c is a constant. The scaling exponent explicitly determines the impact of large versus small landslides on integrated measures, such as the total area disturbed or the volume of material yielded. Power-law scaling is typically observed for areas greater than 1000–5000 m^2 up to the largest landslide areas for which a distribution can be reliably estimated (of the order of 10^5 m^2). Over this size range, size–frequency distributions are consistently steep for a variety of data sets, notwithstanding some exceptions. The scaling exponent α expresses this steepness and generally varies between 1.3 and 1.5. This indicates that, in the long term, the area affected by landslides is dominated by small- to medium-scale failures with areas up to 10^3–10^4 m^2. Assessing the sediment yield from an area dominated by landslide mass wasting requires a well-defined relation between landslide thickness and planform. This scaling is important because it determines the transformation between the area–frequency and the volume–frequency distributions. For strictly soil/regolith failures, it can be argued that the depth of landslide failure is approximately constant. For failures that involve bedrock, however, the depth of failure likely correlates with the length scale of the landslide, giving a volume to area relation of $V \sim A^{3/2}$ (Hovius et al. 1997). At present, neither model has been verified by field data, although such verification is requisite for any reliable estimate of total landslide sediment flux. If the constant thickness model applies, then the volume eroded by landslides is set by the frequency of the average area landslide and is weakly dependent on the power-law scaling. In contrast, if the scaling thickness model applies, then the total eroded volume is determined in more equal measures by the input from small and large landslides, with a weighting that is a sensitive function of the power-law scaling exponent, α.

Denudation Rates from In Situ-produced Cosmogenic Nuclides in Bedrock

The techniques discussed above measure erosion and denudation rates on timescales of local fault evolution that are significantly shorter than the time for a large fault system to develop (10^4–10^6 yr; see Chapter 4). Measurements of *in situ*-produced cosmogenic nuclides can be used to constrain erosion and denudation over timescales that are meaningful for faulting, weathering, and landscape change (Lal 1991; Gosse and Phillips 2001; Bierman and Nichols 2004). For example, [10]Be and [26]Al have been widely used to estimate denudation rates of bedrock surfaces. Assume that a surface has been denuded for a period that is long compared to the denudational timescale (as defined below). Then, the surface nuclide concentration of a steadily denuding bedrock (C, atoms g^{-1}) is inversely proportional to the rate of denudation (ε, mm ka^{-1}) (Lal 1991):

$$C = P_0 / (\lambda + \varepsilon/z^*) , \tag{9.3}$$

where P_0 (atoms g^{-1} yr^{-1}; scaled for latitude and altitude) is the cosmogenic production rate at the surface in a mineral of known composition, λ (yr^{-1}) is the decay constant and z^* (cm) is the absorption depth scale. The latter depends mainly on the density ρ (g cm^{-3}) of the material irradiated; it can be calculated from Λ/ρ, where Λ is the cosmic ray absorption mean free path (typically 150 g cm^{-2} on Earth). With Eq. 9.3 a bedrock denudation rate can be estimated from the measured surface nuclide concentration. The rate reflects the total denudation of the bedrock, including both erosion and chemical weathering, and integrates approximately over the time required to remove the upper meter of regolith or 60 cm of bedrock (about 10^3 to 10^5 yr in most cases). Cosmogenic nuclide abundancies are usually measured in quartz, a very common mineral with simple chemical composition, well-known cosmogenic nuclide production rate, and high resistance to weathering.

Denudation Rates from In situ-produced Cosmogenic Nuclides in Soils

Soil denudation rates can be obtained with a similar approach from samples collected at the top of the soil profile. The denudation rates can be ambiguous, especially if it cannot be ruled out that the sampled soil section was affected by sudden removal of a surface layer (thereby exhuming soil with lower concentrations) or by addition of allochtonous material (loess, ash falls). Later addition of material can usually be identified by soil characterization. Removal of a surface layer is not pertinent if the regolith has been thoroughly mixed by creep, bioturbation, or tree fall. In a mixed layer, the cosmogenic nuclide concentrations in quartz at all depths are equal to the concentration that would be observed at the surface if mixing had not occurred (Braucher et al. 1998; Granger et al. 1996; Schaller et al. 2003; Small et al. 1999). As a result, samples taken from any depth in the mixed layer faithfully record the denudation of the soil section. Downslope advection of soil material (e.g., by periglacial processes) would result in nuclide concentrations that are not necessarily representative of the denudation rate of the sampled soil section. Instead, the estimated denudation rate is the average of rates upslope of the sample location (Schaller et al. 2003).

Heimsath et al. (1997) developed a method to determine the rate of conversion of bedrock to soil. This involves measuring the cosmogenic nuclide concentration in a rock sample from immediately below the rock–saprolite interface and correcting this measured value for shielding by the overlying soil. This estimated rate is insensitive to downslope advection of slope sediments and mixing of the regolith and soil.

Catchment-wide Denudation Rates from Cosmogenic Nuclides in River Sediment

Quantification of the erosion rates of a faulted landscape would require analysis of a large number of representative soil and bedrock samples from the various geomorphic components of that landscape (Figure 9.4, e.g., bedrock surfaces, soil-mantled plateaus, hillslopes, river valleys, gorges, landslides). Alternatively, we can let nature do the averaging. Spatially averaged erosion rates can be estimated using cosmogenic nuclide concentrations in river sediments (Bierman and Steig 1996; Brown et al. 1995; Granger et al. 1996). A sample of sediment collected from the active channel at the outlet of a catchment is an aggregate of grains that originate from different upstream locations (Figure 9.4). These sediments have been eroded at different rates from different locations within the catchment and have therefore inherited different nuclide concentrations. Mixing of sediment during transport down hillslopes and in river channels

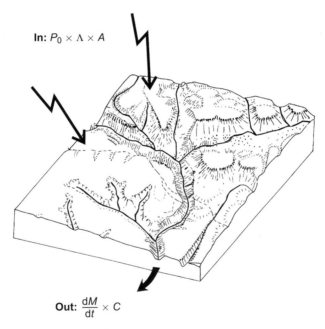

In: $P_0 \times \Lambda \times A$

Out: $\frac{dM}{dt} \times C$

Figure 9.4 Illustration of the "let nature do the averaging" principles of the cosmogenic nuclide-based approach to calculating catchment-wide denudation rates. This approach averages all geomorphic processes operating within the catchment. Mathematically, cosmogenic nuclides are produced in a catchment with an area A ("In") and are exported with the river sediment ("Out"). If the inbound flux by production equals the outbound flux by erosion, then the basin is at cosmogenic steady state and the mass flux dM/dt (t yr^{-1}) can be calculated. Dividing the mass flux by the catchment area and the rock density results in the catchment-wide erosion rate, ε (mm ka^{-1}). From von Blanckenburg (2005); reproduced with permission from Elsevier.

is assumed to cause sufficient homogenization of the sediment that the nuclide concentration measured in the sample is representative of the average erosion of the catchment. This technique ensures that the answer to the much-sung question, "How many years can a mountain exist, before it's washed to the sea?" is no longer "blowing in the wind" (Dylan 1963). Bierman and Nichols (2004) and von Blanckenburg (2005) review this method, and the latter paper offers a detailed account of the assumptions and limitations of the method.

Weathering, Erosion, Denudation, Faulting, and Climate

The availability of detailed digitally based topographic information and technical advances, which allow the measurement of erosion and denudation rates on scales from the single soil section through incising rivers to whole drainage networks and even entire mountain belts, have revolutionized our approach to and understanding of the denudational response to faulting. In this section we present examples that show how the methods above can be used to determine rates on the full range of spatial scales. Taken together, these examples provide strong evidence of the control exerted by crustal deformation and faulting on landscape denudation.

Soil Production in the Vicinity of Faults

The rate of bedrock conversion to soil has long been predicted to be a function of soil thickness (Gilbert 1877). If this is the case, then the dependency between soil thickness and soil production rate is a sensitive test for recent perturbations of a landscape. Heimsath et al. (1999) measured cosmogenic ^{10}Be in bedrock immediately beneath the weathering front and determined a smooth linear function which shows that the soil production rate indeed declines with increasing regolith thickness (Figure 9.5). This relationship seems to be a general rule for convex parts of a landscape (Heimsath et al. 2005). In the case of the Tennessee Valley, California, catchment-wide denudation rates have also been observed to be higher than the average of soil samples. This means that erosion occurs mostly along hill crests in steeper parts of the catchment. This observation has been interpreted to indicate that the landscape is out of equilibrium and is adjusting to recent tectonic change. Indeed, the Tennessee Valley site is in the vicinity of the San Andreas fault, with major active strike-slip faults to the east and the west of the area. Tectonically induced incision can have caused the perturbation of the landscape.

Fluvial Bedrock Incision and Gorge Formation Associated with Faulting

Fluvial bedrock incision results in the formation of valleys and gorges. It forms the local base level in erosional landscapes and carves the conduits for the

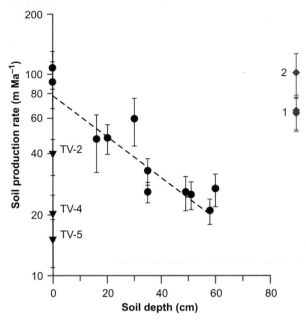

Figure 9.5 The "soil production function" in the Tennessee Valley (TV), California (Heimsath et al. 1999). Denudation rates, or soil production rates, decrease with increasing soil depth. Catchment-wide denudation rates (data 1, 2) are higher than most soil denudation rates, indicating that denudation occurs mostly along hill crests and in steeper parts of the catchment. If true, this observation suggests that the landscape is changing, possibly in response to a recent tectonic change. Reprinted with permission from Elsevier.

transport of erosional products. Here, we consider examples from the Himalayas, Taiwan, and the eastern U.S.A. to illustrate the tectonic and climatic controls on fluvial incision.

We first consider the Himalayas, the Earth's highest mountain belt. Burbank et al. (1996) measured incision of the Indus River with [10]Be exposure ages of Holocene and late Pleistocene bedrock channel remnants above the active Indus channel (Figure 9.6a). Incision rates were found to range from 2 to 8 mm yr[-1] along the study reach (Figure 9.6b) and were highest where the river crosses a horst bounded by the Stak and Raikhot faults. This incision pattern is a long-term feature of the Indus River, as demonstrated by apatite fission track ages that date the cooling of rocks currently exposed in the river bed to below ~110°C at between 1.3 and 8.2 Ma upstream of the fault block, but at only 0.4 Ma between the two faults. (Figure 9.6c). The long-term incision rates implied by these ages are similar to the cosmogenically determined, late Quaternary rates. They show that faulting and fluvial incision along the Indus River are intimately linked.

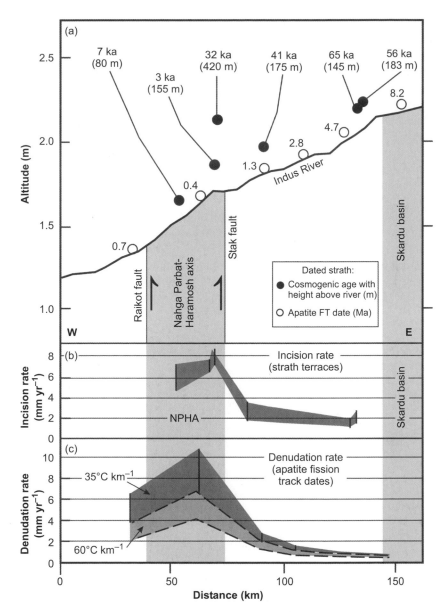

Figure 9.6 Bedrock incision along the Indus River, Pakistani Himalayas: (a) Strath terraces are shown with their cosmogenic nuclide exposure ages and heights above the modern river level. Apatite fission track ages are shown at river level. (b) Incision rates versus distance along the Indus, calculated from cosmogenic nuclide exposure ages. (c) Rates of denudation based on apatite fission track ages. Denudation rates were calculated for geothermal gradients of 35°C km^{-1} and 60°C km^{-1}. Modified from Burbank and Anderson (2001).

A sharp increase in denudation rates of catchments contributing to the Burhi Gandaki River of the Nepalese Himalayas has been attributed to thrusting on a hitherto unidentified north-dipping fault located somewhere between the Main Himalayan fault in the south and the Main Central Thrust in the north (Figure 9.7; Wobus et al. 2005). Catchment-wide [10]Be denudation rates from tributaries draining into the trunk of the Burhi Gandaki River show a pronounced change in denudation rate across a narrow zone where this fault is presumably located. [40]Ar-[39]Ar cooling ages on white mica (recording the time when rocks cooled to ca. 350°C) are younger to the north of this zone. This is interpreted in terms of sustained faulting leading to progressive exhumation of deeper rocks in the hangingwall of this thrust. The authors speculate that the development of this fault was aided by erosion associated with strong monsoonal precipitation focused along the southern flank of the Himalayas. Similar mechanisms are discussed in Koons and Kirby and Buck et al. (both this volume).

Next, we consider Taiwan, where the mechanisms, patterns, and rates of fluvial bedrock wear have been measured over a range of timescales in the Liwu River. These measurements reveal a strong climatic control on river incision rates in a region with high tectonic rates. By measuring incremental changes of channel

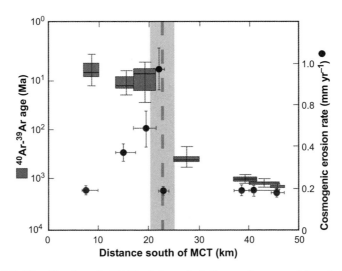

Figure 9.7 Identification of a "hidden" thrust fault from an increase of denudation rates in tributaries to the Burhi Gandaki River of the Nepal Himalayas. This hitherto unidentified fault may lie between the Main Himalayan Fault in the south and the Main Central Thrust (MCT) in the north. Catchment-wide [10]Be denudation rates (solid circles) from tributaries draining into the Burhi Gandaki trunk stream show a pronounced increase across the presumed fault; [40]Ar-[39]Ar cooling ages (gray bars) on white mica (that record cooling to ca. 350°C) are younger to the south of the N-dipping thrust fault, suggesting that it has exhumed originally warmer, deeper rocks in its hangingwall in the north. Modified from Wobus et al. (2005).

bed topography across the active river channel, Hartshorn et al. (2002) established that between February 2000 and December 2001 the spatially averaged wear rate in schists—the dominant lithology in the central part of the catchment—was 3.4 mm yr^{-1}. The maximal local incision rate was 69 mm yr^{-1}. A wide range of discharges was recorded during this interval, including a 25-year return time flood. This flood caused a disproportionate amount of erosion on the sides of the channel at intermediate flow depths, but did not significantly deepen the thalweg (central line of the channel floor). Instead, the thalweg was lowered during more frequent, smaller floods.

Water discharge and suspended sediment concentrations have been measured since 1970 at the same station where Hartshorn et al. conducted their study (Water Resources Agency 1970–2003). Using a bias-corrected average of the measurements, Dadson et al. (2003) obtained an average annual suspended sediment load of 14.4 megatonnes (Mt) for an area of some 435 km^2 upstream of the station. This is equivalent to a catchment-wide average surface lowering rate of 12.5 mm yr^{-1} (using the density of quartz).

In a study of river terraces and perched paleochannels along the Liwu River, Liew (1988) reported ages of two terrace deposits dated with ^{14}C of 2.4 ka and 2.5 ka. Using the height of the bedrock-fill interface of these features above the current channel, Liew calculated river incision rates of 6 mm yr^{-1} and 11 mm yr^{-1}. These are considered to be minimum incision rates because in some cases the river has shifted its course after aggradation and incised a new bedrock channel adjacent to the filled passage. Therefore, the thickness of the fill should be added to incision rate calculations.

Schaller et al. (2005) employed cosmogenic nuclides to show that channel lowering in the Liwu catchment occurred at a maximum average rate of 9 mm yr^{-1} over the last 130 years. Using samples from a 200 m high, fluvially sculpted gorge wall, and assuming progressive channel lowering without temporary aggradation, these authors calculated a maximum average rate of river incision of 26 ±4 mm yr^{-1} since 6.5 ka.

Viewed in the context of a long-term exhumation rate of 3–5 mm yr^{-1} from fission track analysis (Willett et al. 2003), the average Holocene river incision rates of the Liwu River must be higher than the average Quaternary exhumation rate of the catchment by at least a factor two and possibly much more (Figure 9.8). If this value is representative of Quaternary climate cycles, then one may infer that in Taiwan warm and cold stages are associated, respectively, with high and low erosion rates. If sustained over 10^4 years, these differences in climate could give rise to episodic erosional unloading of the Taiwan orogen and clustering of erosionally driven faulting events and seismic activity (see Chapter 10). However, the dramatic changes in erosion and incision rates observed in the Liwu catchment could also be due to increased rates of tectonically driven rock uplift in the Holocene. Although the current rate of relative plate motion in the Taiwan region is close to the Quaternary average rate, this second possibility cannot be excluded.

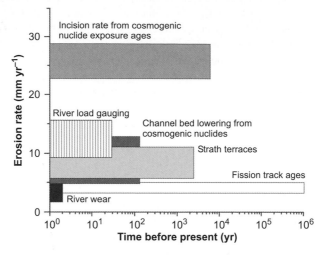

Figure 9.8 Compilation of incision and erosion rates over different timescales from the Taroko Gorge area. Cosmogenic nuclide-derived rates (Schaller et al. 2005), rates from dated fill terraces (Liew 1988), and bedrock lowering rates (Hartshorn et al. 2002) are incision rates. Rates derived from river load gauging (Dadson et al. 2003) and from fission track dating (Liu et al. 2001; Willett et al. 2003) reflect erosion rates. The Holocene incision rates are higher than the long-term erosion rate determined by fission track dating. Modified from Schaller et al. (2005).

Away from regions of active compression, the impact of climate change can also be seen in the dissection of the passive Atlantic margin of North America, which has not been affected by major recent faulting (Reusser et al. 2004). The Susquehanna and Potomac Rivers have incised this margin at rates of 0.6 and 0.8 mm yr^{-1} (Figure 9.9) between 40 and 10 ka according to measurements of cosmogenic ^{10}Be abundances in fluvially eroded bedrock surfaces. Minimum exposure ages of 50–100 ka for weathered but undissected surfaces above the river gorges indicate that gorge formation is a recent, possibly episodic phenomenon. These data demonstrate that rivers draining passive margins are capable of quick, periodic cutting through bedrock. In the absence of faulting, the most likely causes of gorge formation are related to climate change, for example, increased storm frequency and/or magnitude during the Last Glacial Maximum, pulses of meltwater discharge, upwarping of a forebulge in front of a glacial hinterland, or eustatic sea-level change (Reusser et al. 2004).

Catchment Denudation

Faulting can generate relief and relief facilitates erosion. In a much-cited publication, Ahnert (1970) reported a correlation between catchment relief and erosion rate. This correlation has been confirmed and refined in several studies

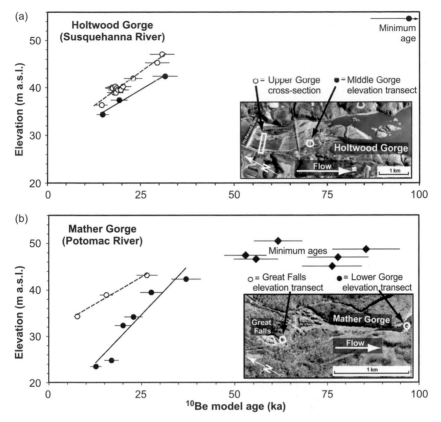

Figure 9.9 Plots of [10]Be exposure age and elevation of fluvially eroded bedrock surfaces along river gorges atop the Atlantic passive margin (Susquehanna and Potomac Rivers, eastern U.S.A.). "Minimum Ages" are from weathered outcrops above the gorges and represent the minimum age of the highlands above the gorges. From Reusser et al. (2004); reproduced with permission from AAAS.

that relate coarse topographic parameters with erosion rates derived from river loads (Pinet and Souriau 1988; Summerfield and Hulton 1994; Hovius 1998) (Figure 9.10). In these studies, relief is used as a proxy for the parameters and processes that actually control erosion, that is, local topographic slope and curvature, and rock uplift. Erosion rates usually scale linearly with relief in tectonically inactive areas, such that areas with greater local relief have higher erosion rates (Figure 9.10). However, erosion rates in active orogens vary by more than an order of magnitude (see next section below), even where mean local relief over a distance of 10 km only varies between 1.0 km and 1.5 km (Figure 9.10). This indicates that topographic relief is not a first-order control on the rate of hillslope mass wasting in active mountain belts. Instead, erosion

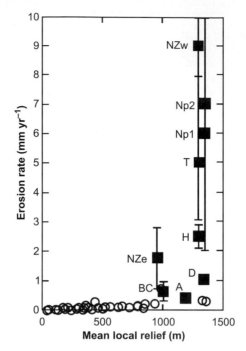

Figure 9.10 Erosion rate versus mean local relief measured over 10 km from mostly tectonically inactive orogens (open circles) and tectonically active, convergent orogens (solid squares). NZ: Southern Alps, New Zealand; NP: Nanga Parbat region, western Himalaya; T: Taiwan; H: central Himalaya; D: Denali portion of the Alaska Range; A: European Alps; BC: British Columbia (after Montgomery and Brandon 2002).

rates are set by tectonic forcing and there is a limit to local relief imposed by the mechanical properties of the near-surface rock mass (Montgomery and Brandon 2002).

These global considerations show general trends but do not give clear insight into the actual processes operating on landscapes. The combination of sediment yield measurements or cosmogenic nuclide-derived erosion and denudation rates with digital elevation models allows for more detailed testing of hypotheses on the processes involved in the link between erosion and tectonics, at least in well-defined tectonic settings and geomorphic provinces.

Denudation rates in the uplands of Germany, France, and Belgium show a strong linear dependency on basin relief (cf. Ahnert 1970), with high-relief catchments denuding up to 5 times more quickly than catchments with low relief (Schaller et al. 2001) (Figure 9.11a). Using a linear relationship of denudation rate to relief, one can calculate that the characteristic timescale of relief reduction (i.e., the time required to reduce the local relief at a given length scale to 1/e of its initial value) in the German uplands is about 10 Myr. If this is

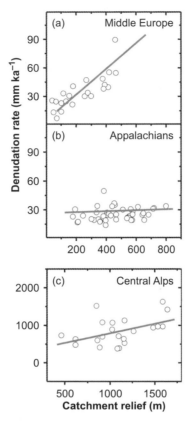

Figure 9.11 Cosmogenic nuclide-derived denudation rate as a function of relief (calculated here as mean catchment altitude minus minimum altitude. The data from middle European Highlands (Massif Central, Ardennes, Bayerischer Wald) is from Schaller et al. (2001); denudation rates recalculated by Schaller et al. (2002). Data from the Appalachians (Great Smoky Mountains) is from Matmon et al. (2003a, b). Data from the European Central Alps is preliminary, unpublished data (H. Wittmann, T. Krüsmann, F. von Blanckenburg, and P. Kubik, in prep.). From von Blanckenburg (2005); reproduced with permission from Elsevier.

correct, then the presence of high local relief (up to 600 m over a distance of 10 km) indicates that the landscape has experienced a relative base level change in the recent geological past. Indeed, parts of this region have been affected by Neogene faulting, uplift and subsidence, volcanic activity, and epirogeny, possibly driven by basaltic underplating concentrated in the vicinity of the Rhine Graben (Ritter et al. 2001).

An entirely different picture emerges from an analysis of the Great Smoky Mountains of the Appalachians (Figure 9.11b) (Matmon et al. 2003a, b). There, denudation rates are uniform and independent of local relief. The values are mostly lower than in Germany, France, and Belgium. From this, Matmon et al. inferred that the Appalachians are in topographic steady state; that is, for geologically substantial lengths of time, all parts of the landscape have been denuding at the same rate, regardless of relief, topographic slope, or substrate (Gilbert 1877; Hack 1960). Where denudation occurs, topographic steady state

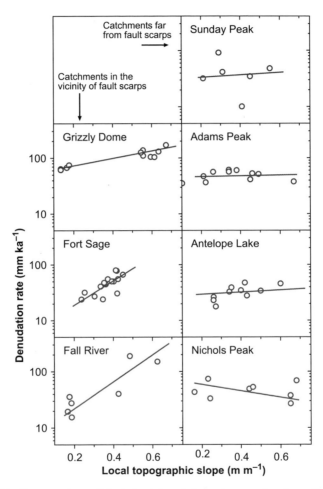

Figure 9.12 Cosmogenic nuclide-derived denudation rates as a function of local topographic gradient in seven catchments in the Sierra Nevada, U.S.A. (Riebe et al. 2000: recalculated by Riebe, pers. comm.). In the three catchments in the left column, denudation and local topographic slope are strongly correlated and denudation increases with proximity to active fault scarps or canyons. In the four catchments in the right column, denudation rates are more uniform and independent of the local topographic slope. These catchments are far from any fault. From von Blanckenburg (2005); reproduced with permission from Elsevier.

requires a tectonic input of rock mass that is equal to the denudational flux. It is difficult to see how this is the case in the Appalachian mountain belt, most of which formed during Variscan or earlier orogenic events. Topographic steady state is more likely in younger, active mountain belts (Willett and Brandon 2002). For example, in the Central Alps of Switzerland denudation rates determined with cosmogenic isotopes are high and similar to geodetic uplift rates, but surprisingly uniform and only weakly correlated with local relief (Figure 9.11c, unpublished data). This implies that the Central Alps are at present close to topographic steady state, even though glacial overprinting of the topography during the Quaternary may have perturbed a long-term balance. At present, shortening across the European Alps is minor. Present rock uplift there may be largely isostatically driven. At the ambient rates observed on short timescales, it is impossible to distinguish between topographic steady state and isostatic compensation of denudation fluxes.

Topographic gradient is a more meaningful parameter for quantifying erosion than local relief, because it relates directly and predictably to the rate of erosion (Montgomery and Brandon 2002). Measurements of local topographic slope depend highly on the resolution of the topographic data at hand and on the length scale of the measurement (Montgomery and Brandon 2002). Unless hillslopes are absolutely straight, the highest resolution topographic data treated at the shortest possible length scale yield the most useful and reproducible slope estimates. In a study of seven catchments in the Sierra Nevada, U.S.A., Riebe et al. (2000, 2001) investigated the relation between local topographic slope and denudation rates derived from cosmogenic nuclides. Four catchments located away from active faults were found to have approximately uniform denudation rates that were independent of the local topographic gradient. In these four catchments all sections of the landscape are denuded at the same rate. Three other catchments located close to active faults or young canyons have denudation rates up to 15 times higher that correlate positively with the local topographic gradient (Figure 9.12). This was attributed to the fact that faulting lowers the local base level, adjusts the drainage network, and increases erosion of hillslopes and associated chemical weathering (Riebe et al. 2001). All of this is in line with the concepts outlined above and shows that in the Sierra Nevada, active faulting exerts a dominant control on denudation.

Erosion of an Active Orogen

Direct measurements of erosion on the scale of an orogen are rare. The size of some orogens precludes comprehensive monitoring of erosional fluxes; in other orogens, there has been no economic incentive to measure erosion. When this incentive exists, human activity has often pushed the surface system far from its natural state, so that present-day erosional fluxes are not naturally forced. The Taiwan orogen is an exception on both counts. It is sufficiently small such that a detailed picture of erosion can be obtained from measurements at a limited number of sites. Furthermore, the Taiwanese have been driven to collect river statistics in their pursuit of hydroelectricity and the island's mountainous interior has thus far escaped wholesale development.

Since 1970, river-suspended sediment discharge has been recorded at over 150 stations with an average spacing of 20 km across Taiwan. Dadson et al. (2003) estimated that the 30-year average suspended sediment erosion rate of Taiwan is 3.9 mm yr^{-1} (Dadson et al. 2003), and that Taiwan supplied 384 Mt yr^{-1} of suspended sediment to the ocean. This amounts to 1.9% of the estimated global suspended sediment discharge (Milliman and Syvitski 1992) and yet is derived from only 0.024% of Earth's subaerial surface. Accounting for 30% bedload (as indicated by the fill rate of mountain reservoirs) would increase the estimated erosion rate to 5.2 mm yr^{-1} and the average annual sediment yield to 500 Mt yr^{-1}, although much bedload is trapped in floodplains before reaching the sea. Notably, this total sediment yield is approximately equal to the estimated accretionary mass flux into the orogen (around 480 Mt yr^{-1}) due to convergence of the Philippine Sea Plate and the Asian continental plate. This implies that the Taiwan orogen is close to steady state in terms of tectonic and erosional fluxes (Willett and Brandon 2002).

Between 1970 and 1999, decadal-scale erosion rates were high in the eastern Central Range and southwestern Taiwan, but low in the north and west. The metamorphic core was eroding at ~6 mm yr^{-1}. Erosion rates of up to 60 mm yr^{-1} were found near active thrusts located south of the Western Foothills thrust belt. Northern and western Taiwan had lower erosion rates (1–4 mm yr^{-1}). The spatial pattern of average erosion rates does not correlate with local relief, slope, precipitation runoff, and stream power. These are all factors that are traditionally seen as important controls on erosion rate. However, the erosional pattern can be explained to a statistically significant degree by cumulative seismic moment release ($r^2 = 0.13$; Figure 9.13a) and temporal variability in runoff ($r^2 = 0.27$; Figure 9.13b). Most of the remaining variation in erosion rate probably originates in the natural variability in mountain drainage basins. This variability dominates on the decadal timescale of this study, but may be less important on timescales longer than the erosional response to very large earthquakes (see below). Nevertheless, the results above indicate that modern erosion rates are strongly influenced by large earthquakes and typhoons, both of which augment widespread sediment supply to channels. Moreover, the data indicate that in the active Taiwan orogen there is no strong link between erosion rates and the topographic attributes considered in the previous section. This should not come as a surprise, because erosion is driven by processes, not landforms. Landforms develop in response to tectonic and erosional forcing in a way that is determined by the geomechanical properties of the substrate. Without due consideration of substrate properties, analyses of erosion and topographic attributes are bound to yield ambiguous conclusions.

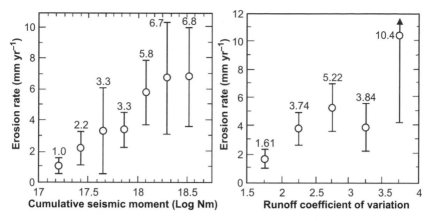

Figure 9.13 Controls on erosion rates in the Taiwan orogen. (a) Suspended sediment erosion rate as a function of cumulative scalar seismic moment, M_0. A total of 128 drainage basins are shown, binned by mean cumulative seismic moment release within their boundaries. The figure shows mean erosion rate within these bins and its 95% confidence interval. (b) Suspended sediment erosion rate as a function of coefficient of runoff variation. All 128 drainage basins shown are binned by the coefficient of runoff variation defined by their water discharge record. The figure shows mean erosion rate within these bins and its 95% confidence interval (after Dadson et al. 2003).

Across Taiwan, cumulative seismic moment release correlates linearly with decadal erosion rate over a sixfold range of erosion rate. Sediment production resulting from earthquakes is therefore an important factor in the overall sediment budget of the Taiwan orogen. However, the strong north–south gradient in the seismically driven erosion pattern may be relatively short-lived compared to the long history of strike-parallel uniform shortening across the mountain belt. The southwestern section of the western Taiwan thrust belt experienced eleven $M_w > 6.0$ earthquakes from 1900–1998, whereas only three such earthquakes occurred in the north. The 1999 Chi-Chi M_w 7.6 earthquake caused a considerable increase of erosion rates in the north of Taiwan (Dadson et al. 2004). This earthquake was the largest in Taiwan in the previous fifty years and the largest on the Chelungpu thrust fault in the previous 300–620 years (Shin and Teng 2001; Chen et al. 2001). The earthquake had a focal depth of 8 km and resulted in rupture of an approximately 100 km segment of the north–south trending, east-dipping ($\sim 30°$) Chelungpu thrust. It produced measured ground accelerations of up to 1 g. Geodetic observations show that coseismic displacement of up to 3 m on the fault decreased to zero at a distance of 20 km away from the fault (Yu et al. 2001). Sediment discharge to the ocean increased following the Chi-Chi earthquake. The average annual suspended sediment discharge from Choshui River draining the epicentral area increased by a factor of 2.6 to 143 Mt yr^{-1}. Most of this discharge occurred during brief floods associated with the passage of large typhoons. 67% of the annual suspended sediment discharge from Choshui River following the Chi-Chi earthquake involved suspended sediment concentrations greater than 40 g l^{-1}. River water with this turbidity can underflow seawater and initiate marine turbidity currents. Hydrometric data indicate that this may have happened on at least four occasions since the earthquake and before 2004, each time during a typhoon. Thus, earthquakes are recorded in the stratigraphic record, not as single pulses of sediment (Beattie and Dade 1996) but as episodes of increased magnitude and frequency of delivery of coarse clastic material to depocenters (Dadson et al. 2005). During the 2004 typhoon floods, the characteristic sediment concentrations were still anomalously high in rivers draining the epicentral area of the Chi-Chi earthquake. It is not yet clear how and on what timescale this effect decays. However, it is likely that a simple relationship exists between the seismic moment and the magnitude and duration of the geomorphic (erosional) response. Quantification of this relationship will provide a key for inverting the stratigraphic record to yield detailed information about the spatial and temporal patterns of large, prehistoric earthquakes. This is an outstanding research challenge.

Chemical Weathering on the Catchment Scale

Experiments suggest that mineral dissolution rates increase with temperature (Blum and Stillings 1995). Meanwhile, precipitation provides the medium for dissolution and transport of dissolved products and drives erosion. Of course, erosion and chemical weathering are intrinsically linked: erosion creates fresh mineral surfaces on which reaction rates related to weathering are enhanced. Erosion, therefore, links climate and weathering to tectonics. A substantial body of literature has focused on the relative importance of climatic and tectonic drivers of silicate weathering (Berner and Berner 1997; Edmond and Huh 1997).

This issue is confounded by the fact that catchment studies of weathering typically yield rates of mineral dissolution that are orders of magnitude lower than those predicted from laboratory experiments (White and Blum 1995; White and Brantley 2003) (Figure 9.14). Mineral dissolution rates have been measured on timescales ranging from less than four days for the shortest experiment to more than a million years in the oldest natural weathering environment. Each mineral exhibits a dynamic range of weathering rates that spans about six orders of magnitude from the highest experimental rate to the lowest natural rate. The fact that natural rates are consistently the lowest may be due partly to increased available mineral surface area with time (note that mineral dissolution rates are normalized by the available surface area). However, the available mineral surface area measured in natural weathering environments increases by only two orders of magnitude over a million years while the weathering rates decrease by close to six orders of magnitude (White and Brantley 2003). This implies that the bulk of the decrease in weathering rate with observation

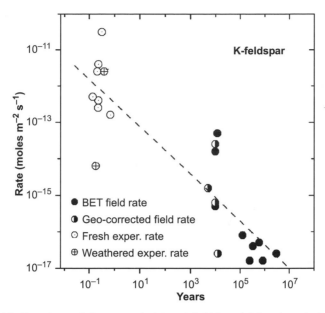

Figure 9.14 Experimental (open symbols) and field-based (closed symbols) rates of mineral (K-feldspar) weathering as a function of timescale. Laboratory rates reflect the short duration of the experiments. Field rates reflect the estimated age of the weathering environment. The surface area of minerals was measured by BET (gas absorption isotherms) or by geometric estimates (SEM). Watershed studies yield rates that are consistently several orders of magnitude slower than laboratory rates. This suggests that external controls (e.g., pore fluid saturation) set weathering rates, which are in turn determined by physical hillslope processes. From White and Brantley (2003); reproduced with permission from Elsevier.

time must be related to other time-dependent factors: (a) decrease in reactive surface area due to diminishing compositional and structural heterogeneities; (b) physical occlusion of the substrate by secondary precipitates and leached layers; and (c) extrinsic weathering properties (e.g., the nature of the solution species or the saturation states of the surrounding fluids). For dilute conditions in the experiments, reaction rates are interface limited and independent of fluid concentrations and rates of solute transport. As solute concentrations approach equilibrium, reaction rates are transport limited. For weathering in soil columns, reaction rates increase with increasing percolation rate of the solution. This is the classic transport-controlled environment. Therefore, for mineral dissolution under conditions that are close to equilibrium (the case in many field settings), transport control is important for weathering (White and Brantley 2003). Groundwater flow rates tend to decrease with increasing depth in a soil and regolith profile, such that the progressive accumulation of weathering products would cause a gradual decrease of weathering rates at the rock–regolith interface. Erosion of weathering products might prevent this decrease. For this reason, it is desirable to separate the chemical and physical components of catchment denudation.

Riebe and coworkers (2003, 2004, 2001) have developed a technique that allows the acquisition of both rates separately in homogeneous granitic lithologies. Their approach is to measure the total denudation rate in a given catchment, using cosmogenic nuclides, and then to determine a "chemical depletion fraction" (CDF) by measuring the bulk chemical composition of representative soil samples and unweathered bedrock within the catchment. Both sets of compositions are normalized over zirconium, a highly refractive element during weathering, and from this the fraction of solutes lost from soil relative to

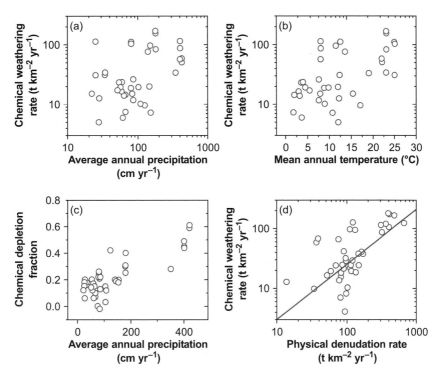

Figure 9.15 Chemical weathering rates obtained from the combination of cosmogenic nuclide-derived total denudation rates and the calculation of chemical depletion using zirconium-normalized soil and bedrock chemical compositions: (a) chemical weathering rate versus mean annual precipitation; (b) chemical weathering rate versus mean annual temperature; (c) chemical depletion fraction (ratio of chemical weathering rate to total denudation rate) versus mean annual precipitation; (d) chemical weathering rate versus physical denudation rate (from Riebe et al. 2004), showing that chemical weathering is closely related to physical erosion. From von Blanckenburg (2005); reproduced with permission from Elsevier.

unweathered bedrock is obtained. The CDF is also equal to the ratio of the rate of chemical weathering to the rate of total denudation. Multiplying the CDF by the total (cosmogenic nuclide-derived) denudation rate yields the chemical weathering rate. This chemical rate subtracted from the total denudation rate is the erosion rate. In this way, the entire denudational system is constrained. Because chemical weathering operates on the same timescale as total denudation at steady state, the rates obtained in this way all integrate over the same cosmogenic timescale. This weathering timescale is much more useful for assessing long-term silicate weathering than the short timescale constraints from measurements of solute river loads.

Results of a global survey of chemical weathering rates in granitic catchments are shown in Figure 9.15. At first glance, neither precipitation (Figure 9.15a) nor temperature (Figure 9.15b) appear to exert any control over silicate weathering, but erosion and chemical weathering are interlinked (Figure 9.15d). The best fit to the data in Figure 9.15d corresponds to a mean global CDF of about 0.2. Superimposed on this trend, however, is some nonrandom scatter due to climate dependency of the CDF. This is exemplified in Figure 9.15c by correlation of the chemical depletion fraction with mean annual precipitation. The same is true for temperature (Riebe et al. 2004). We note that the climate dependence of weathering becomes apparent only when the weathering rate is normalized with respect to physical erosion, as for the CDF.

The examples of Riebe and coworkers show that neither the denudation rates nor the chemical weathering rates are in any straightforward way controlled by climate. Once erosional effects on weathering are removed from the data, weak trends become apparent that support the notion that both temperature and precipitation enhance the rate of weathering. However, the dominant external, erosional control on weathering is tectonic processes such as faulting.

Very similar conclusions were reached from global surveys of dissolved fluxes from granitic catchments (Dupré et al. 2003; Gaillardet et al. 1999; Oliva et al. 2003; White and Blum 1995). These compilations show a relationship between chemical weathering fluxes and several factors, including runoff, a proxy for precipitation (Figure 9.16), physical erosion rate (Figure 9.17), and, to a lesser extent, temperature. Throughout, silicate weathering rates are ca. 10% of erosion rates. This confirms that silicate weathering rates are affected by tectonic activity and associated erosion, but depend only weakly on climate. The link to climate is mainly through precipitation-driven groundwater flow.

The strength of the tectonic control on silicate weathering is perhaps best illustrated in Sri Lanka, where pronounced relief, high elevation, hot climate, and high precipitation would be expected to result in high rates of weathering and erosion. Instead, Sri Lanka has the lowest weathering rates of all mountain sites studied to date (von Blanckenburg et al. 2004). We argue that this is due to the absence of tectonic forcing. The last major phase of compressional tectonic activity in Sri Lanka occurred 500 Ma ago and rifting in the region ceased 130 Ma

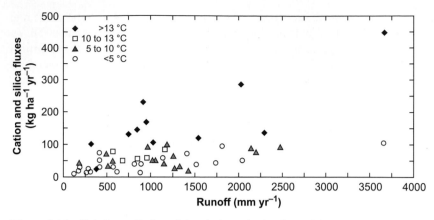

Figure 9.16 Global compilation of chemical weathering fluxes (Si + Na + Mg + K + Ca) and runoff from catchments with granitic substrates (modified from Oliva et al. 2003). Chemical fluxes are corrected for atmospheric inputs. Catchments within the envelope have high temperature and humidity. The Nsimi River carries in part lower cation loads as compared to other rivers of warm regions. Presumably, this is due to the thick soils prevailing in that catchment (Oliva et al. 2003). Diamonds: $T > 13°C$; squares: $T = 10°–13°C$; triangles: $T = 5°–10°C$; circles: $T < 5°C$.

ago. Since then, the absence of any tectonic activity has impeded weathering and denudation. Thick, clay-rich weathered layers cover most of the landscape, shielding the bedrock from corrosive fluids. Weathering rates are therefore extremely low. Nevertheless, it is in settings like this that the temperature and precipitation control of silicate weathering is best constrained because the much stronger, physical controls have been removed (Riebe et al. 2004).

The examples described above show that the driving forces of silicate weathering and CO_2 drawdown can now be distinguished, the relative importance of the contributing processes quantified, and the feedback mechanisms relevant to climate identified. It is not premature to state that faulting is the Earth's premier erosion and weathering engine. It acts along two chains of coupled processes:

1. *Mechanical chain:* Faulting → fall in relative base level fall → incision of the drainage network → enhancement of soil erosion and of the rates of physical denudation and sediment production.
2. *Chemical chain:* Faulting → fall in relative base level → incision of the drainage network → development of new groundwater flow paths → stronger dilution and less saturation of soil pore waters → enhanced rates of mineral dissolution → higher rates of chemical weathering and soil production → increased CO_2 drawdown.

These processes are more active in highly faulted continental areas, for example, mountain belts. It will be interesting to see if faulting in non-orogenic areas exerts a similarly important control on global silicate weathering cycles.

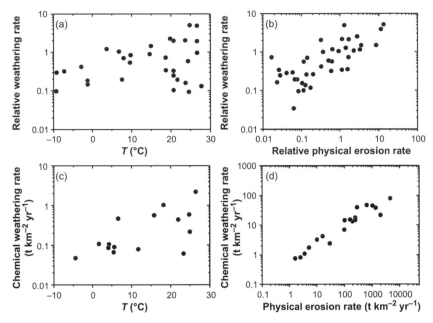

Figure 9.17 Silicate weathering plotted against temperature and specific sediment yield (erosion rate) for a global set of large catchments (a, b), and for a selection of smaller catchments with granitic substrates (c, d). For large catchments, observed weathering and erosion values have been normalized to the Amazon catchment. Note the clear correlation between physical erosion and chemical weathering. Modified from Dupré et al. (2003).

Reconstructing Denudation Histories

At this point, it is appropriate to highlight the opportunities offered by novel applications of cosmogenic nuclides. These nuclides can be used to determine paleodenudation rates from sediments of known age. The principle is that sediment which was deposited and buried at some stage in the past contains an inherited nuclide inventory accumulated in the hinterland during erosional exhumation and transfer of this material (Anderson et al. 1996). If the transfer of sediment from source to sink occurred over a short amount of time compared to its residence in the near-surface zone where cosmogenic nuclides accumulate, and if the sediment was buried quickly upon deposition, then the measured cosmogenic nuclide concentration in fluvial deposits can be corrected for radioactive decay after deposition and converted into a catchment-wide paleodenudation estimate.

For example, Schaller et al. (2004) presented a paleodenudation sequence reaching back 1.3 Ma (Figure 9.18). This sequence was obtained by sampling a series of fill terraces in the Lower Meuse Valley near Maastricht, the Netherlands. Each terrace represents a cold stage deposit, and the terrace chronology

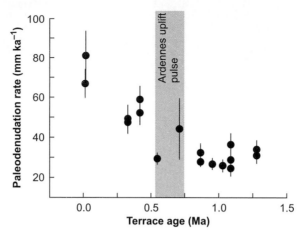

Figure 9.18 Paleodenudation rates derived from cosmogenic nuclides and from independently dated terrace deposits of the Meuse River, the Netherlands. The terrace sediments were deposited during cold stages; however, the Ardennes Mountains, from which much of the material was eroded, were not glaciated during the Quaternary. The gray bar indicates the timing of increased incision of the Meuse trunk stream, possibly associated with rapid regional uplift of the Ardennes. This is followed by a progressive increase of catchment denudation rates, possibly with a time lag of about 200 ka (Schaller et al. 2004). Reprinted from von Blanckenburg (2005), with permission from Elsevier.

is known independently from pollen records, geomagnetic reversals, tephra-chronology, and consideration of orbital parameters. Most of the sediment in the terraces comes from the Ardennes uplands situated directly to the south of the Lower Meuse Valley. Cosmogenic nuclide measurements indicate that denudation rates in the Ardennes from 1.3 Ma to 0.5 Ma were approximately uniform at ca. 30 mm ka^{-1}, but have since increased progressively to 80 mm ka^{-1}. The resolution of this record is not very high, but the change in erosion rate at 0.5 Ma and possibly as early as 0.8 Ma is a robust feature of the data. One possible explanation for this increase in denudation rate is the change in climate cycling around 0.7 Ma. At this time, climate cycles increased both in duration (from 41 ka to 100 ka) and in amplitude due to the strengthening of the 100 ka eccentricity. If this is the principal cause of the increased denudation of the Meuse hinterland since the mid-Pleistocene, then one can conclude that the amplitude of climate cycles is a stronger control on denudation than the frequency of climate change.

However, other processes played in the Meuse catchment at the time of denudation change. The nearby Eifel volcanic field became more active between 0.7 Ma and 0.5 Ma. This has been attributed to a pulse of the Eifel mantle plume, possibly accompanied by basaltic underplating of the Ardennes and Rhenish Massif. The geomorphic record indicates that the Meuse trunk stream and other larger rivers in the region incised rapidly during Eifel volcanic activity (Garcia-Castellanos et al. 2000), perhaps in response to rapid regional uplift. An erosional pulse propagated into the Ardennes uplands from the trunk streams, causing a progressive increase of landscape denudation rates. Central parts of the Ardennes have not yet been affected by this erosional wave, and it is reasonable to expect that catchment denudation rates will continue to increase during the next few climate cycles. The cause of this increase may not have anything to do with climate change, but may instead be related to mantle processes and associated crustal movements. Regardless of the cause of mid-Pliocene elevated denudation rates in the Meuse catchment, it is clear that applying cosmogenic nuclide analysis to well-dated sedimentary sequences potentially yields detailed, quantitative histories of landscape denudation on timescales of faulting and crustal deformation. Through a systematic pursuit of this approach in settings where faulting is accompanied by continuous or episodic basin filling, we may learn more about the denudational response to faulting.

OUTLOOK

We have argued that faulting and seismic activity are first-order controls on the stochastic nature of erosion and sediment routing in active mountain belts. Similarly, it is apparent that erosion alters stress fields and can trigger earthquakes on timescales shorter than or comparable to that of the seismogenic cycle (Dadson et al. 2004). However, it is over longer intervals that the interactions between erosion and tectonics are most important. On such timescales, climate is no longer a constant, but changes both in terms of its average conditions and the variability of these conditions. It is therefore imperative that the history of erosional forcing of faulting is regarded in the context of a changing climate. Too often a link is sought between observed patterns of erosional exhumation, usually constrained on the Ma timescale, and some attributes of the present climate (weather). Alternatively, if one starts from the perspective of faulting, then it is clear that temporal variations in the rate and style of faulting will affect the denudational response. Faulting and denudation must be constrained on matching scales in order to find out how they are linked. The quantitative characterization of crustal deformation, climate, and denudation over their full dynamic ranges and on matching scales is a key requirement for progress (Anderson et al. 1996; Schaller et al. 2004).

Determining the characteristic length scale of the erosional response to faulting and its sedimentary derivatives remains a challenge. Although it is evident that larger earthquakes cause more erosion, it is not yet clear that a global relationship exists between earthquake magnitude and erosional impact. The relation between peak ground acceleration and density of landsliding in the epicentral area of the Chi-Chi earthquake in Taiwan hints at the existence of such a relationship (Dadson et al. 2004). This relationship is expected to scale with distance from the epicenter due to the dispersion and attenuation of seismic waves. Seismically induced erosion is best quantified within this scaling distance; beyond it, other factors unrelated to the earthquake obviously control erosion. Preliminary observations indicate that for M_w 7 earthquakes this distance is of the order of 10^2 km. However, in the absence of well-calibrated models, one measurement of erosion within an epicentral area is insufficient. Instead, erosion should be measured along an array aligned in the direction of seismic wave propagation, and with a sampling density proportional to the length scale of attenuation of peak ground acceleration.

Similar considerations apply to the sedimentary record of earthquakes. If space is created within the epicentral area where erosion is dominated by the seismic event, then the near-field sedimentary record of this event is expected to consist of a pulse of hillslope debris, reworked by rivers and transferred over distances of no more than a few tens of kilometers. This pulse results in enhanced deposition rates of relatively coarse material, possibly sustained for some years after the earthquake. In contrast, aseismic intervals are expected to have lower deposition rates and sediments with finer average grain sizes. Paleodeposition rates are therefore a key tool in the hunt for old earthquakes near seismogenic or formerly seismogenic faults. Outside the epicentral area and beyond the characteristic distance of the erosional response to an earthquake, an aseismic sedimentary signal is admixed with the seismic sediment pulse. Therefore, in depocenters located in the far field of seismogenic faults, changes in deposition rate alone are not enough to recognize earthquakes. Instead, information on paleoearthquakes is contained in the mode of delivery of sediment. An example of this is the increased hyperpycnal sediment delivery from the Chi-Chi epicentral area (Dadson et al. 2005). The potential for using the sedimentary record to identify episodes of faulting in the hinterland remains largely unexplored. Progress in this line of research must involve systematic analyses of the geomorphic impact and stratigraphic record of well-documented, recent earthquakes (e.g., California, Alaska, Japan, Taiwan, Papua New Guinea, Honduras, and Kashmir).

Finally, the evidence that we have reviewed indicate that erosion and chemical weathering are tightly interlinked. Chemical weathering is strongest in landscapes that have been rejuvenated by faulting. Rejuvenation ensures that fluids containing carbonic acid have permanent access to bedrock and that solutes loaded with dissolved cations and anions are flushed out from hillslopes as they are produced by weathering. Mountain belts have been identified as premier sites for this process, rendering them the primary sinks of atmospheric CO_2. The combination of accurate measurements over matching timescales of the rates of tectonic deformation, denudation, and indices of chemical weathering could facilitate identification of the exact settings, processes, thresholds, and feedback mechanisms that control global weathering and regulate the long-term climate.

ACKNOWLEDGMENTS

We are grateful to Mark Handy, Arjun Heimsath, and an anonymous reviewer for their comments on earlier versions of this paper. The thoughts reflected in this contribution have arisen from work supported by the NASA, the U.S. National Science Foundation (NSF), the U.K. Natural Environment Research Council, the Swiss National Science Foundation (SNF), the German Science Foundation (DFG), the Royal Society, the Leverhulme Trust, and the European Commission.

APPENDIX: GLOSSARY OF TERMS

Denudation: Removal of mass from locations at or near Earth's surface due to the combined effect of physical and chemical processes acting to break up the rock mass, and remove the debris and solutes.

Weathering: *In situ* dissolution by chemical processes (chemical weathering) or mechanical disintegration (mechanical weathering) of rock mass at or near the surface.

Erosion: Removal of rock mass and weathering products at the Earth's surface by mechanical processes.

Tectonic denudation: Removal of rock mass by tectonic processes such as faulting.

Runoff: Water discharged from a given surface area, normally a drainage basin. The principal source of runoff is precipitation, but advected deep water may contribution to runoff. Surface runoff is the component of total runoff that is discharged along Earth's surface. It is commonly expressed as the height of water column accumulated at the surface in an imaginary column within a unit of time (e.g., mm yr^{-1}). Broadly speaking: runoff = precipitation – evapotranspiration.

Water discharge: Volume of water displaced per unit time (e.g., $km^3 \, yr^{-1}$). For surface areas greater than ~1 km^2, water discharge tends to be concentrated in well-defined channels.

Sediment load: Granular material (organic and inorganic) in suspension and moving along the river or stream bed. The sediment load is a proxy for the amount of material eroded mechanically from a drainage area, and is commonly recorded per unit time (e.g., tonnes per year, t yr^{-1}). Note that sediment loads are often calculated from measurements of *suspended* sediment only. The bedload is more difficult to measure, and is often assumed to be 10% of the total load.

   Calculation: sediment load = (suspended sediment concentration × water discharge) + bedload transport rate.

Dissolved load: All dissolved components discharged in a river from a drainage area (units: t yr^{-1}). Typically, the major constituents of the dissolved load are: $Ca^{2+} + Mg^{2+} + Na^+ + K^+ + SiO_2 + Cl^- + SO_4^{2-} + HCO_3^-$. In order to obtain an estimate of the weathering input, HCO_3^- must be adjusted for rainfall input, and Na^+, C^-, and SO_4^{2-} for pollution and cyclic salts (e.g., sea spray). *Calculation:* dissolved load = concentration of total dissolved loads × water discharge.

Sediment yield: Amount of material eroded mechanically from a given surface area. When expressed per unit time and area, it is the *specific sediment yield* (units: t km^{-2} yr^{-1}). It is commonly calculated as the ratio of the total annual sediment load of a river, and the area upstream of the point of measurement of the river sediment load. A similar measure can be defined for the *dissolved yield.*

Erosion rate: Rate of lowering of Earth's surface due to mechanical processes (units: e.g., m Myr^{-1} = mm ka^{-1} = μm yr^{-1}). Note that "soil erosion" as used by many geographers and agronomists describes erosion caused by human action only.

Calculation: erosion rate = specific sediment yield / bedrock or regolith density.

Total denudation rate: Total rate of removal of material from an area by solid and solute transport (units: e.g., m Myr^{-1} = mm ka^{-1} = μm yr^{-1}). Part of the denuded material may be derived from below Earth's surface, but subsurface material should be excluded.

Calculation: denudation rate = (specific sediment yield + specific dissolved yield) / bedrock density.

REFERENCES

Ahnert, F. 1970. Functional relationships between denudation, relief, and uplift in large mid-latitude drainage basins. *Am. J. Sci.* **268**:243–263.

Amorosi A., M. Farina, P. Severi et al. 1996. Genetically related alluvial deposits across active fault zones: An example of alluvial fan-terrace correlation from the upper Quaternary of the southern Po Basin, Italy. Sediment. Geol. **102**:274–295.

Anderson, R.S., J.L. Repka, and G.S. Dick. 1996. Explicit treatment of inheritance in dating depositional surfaces using *in situ* [10]Be and [26]Al. *Geology* 24:47–51.

Beattie, P.D., and W.B. Dade. 1996. Is scaling in turbidite deposition consistent with forcing by earthquakes? *J. Sediment. Res.* **A66**:909–915.

Beaumont, C., P. Fullsack, and J. Hamilton. 1991. Erosional control of active compressional orogens. In: Thrust Tectonics., ed. K.R. McClay, pp. 1–18. New York: Chapman and Hall.

Beaumont, C., R.A. Jamieson, M.H. Nguyen, and B. Lee. 2001. Himalayan tectonics explained by extrusion of a low-viscosity crustal channel coupled to focused surface denudation. *Nature* **414**:738–742.

Berner, R.A., and E.K. Berner. 1997. Silicate weathering and climate. In: Tectonic Uplift and Climate Change, ed. W.F. Ruddiman and W. Prell, pp. 353–365. New York: Plenum.

Bierman, P., and K.K. Nichols. 2004. Rock to sediment—slope to sea with [10]Be—Rates of landscape change. *Ann. Rev. Earth Planet. Sci.* **32**:215–255.

Bierman, P.R., and E.J. Steig. 1996. Estimating rates of denudation using cosmogenic isotope abundances in sediment. *Earth Surf. Proc. & Landforms* **21**:125–139.

Blum, A.E., and L.L. Stillings. 1995. Feldspar dissolution kinetics. *Mineral. Soc. Am. Rev. Mineral.* **31**:291–351.

Brardinoni, F., and M. Church. 2004. Representing the landslide magnitude–frequency relation: Capilano River basin, British Columbia. *Earth Surf. Proc. & Landforms* **29**:115–124.

Braucher, R., F. Colin, E.T. Brown et al. 1998. African laterite dynamics using *in situ*-produced [10]Be. *Geochim. Cosmochim. Acta* **62**:1501–1507.

Brown, E.T., R.F. Stallard, M.C. Larsen, G.M. Raisbeck, and F. Yiou. 1995. Denudation rates determined from the accumulation of *in situ*-produced [10]Be in the Luquillo Experimental Forest, Puerto Rico. *Earth Planet. Sci. Lett.* **129**:193–202.

Burbank, D.W., and R.S. Anderson. 2001. Tectonic Geomorphology. Malden, MA: Blackwell.

Burbank, D.W., J. Leland, E. Fielding et al. 1996. Bedrock incision, rock uplift, and threshold hillslopes in the northwestern Himalayas. *Nature* **379**:505–510.

Chen, W.S., Y.G. Chen, H.C. Chang, Y.H. Lee, and J.C. Lee. 2001. Palaeoseismic study of the Chelungpu fault in the Wanfung area. *Western Pacific Earth Sci.* **1**:499–506.

Cohn, T.A. 1995. Recent advances in statistical methods for the estimation of sediment and nutrient transport in rivers. *Rev. Geophys.* **33 Suppl.**:1–18.

Dadson, S., N. Hovius, H. Chen et al. 2004. Earthquake-triggered increase in sediment delivery from an active mountain belt. *Geology* **32**:733–736.

Dadson, S.J., N. Hovius, W.B. Dade et al. 2003. Erosion of the Taiwan orogen. *Nature* **426**:648–651.

Dadson, S., N. Hovius, S. Pegg, W.B. Dade, and M.J. Horng. 2005. Hyperpycnal river flows from an active mountain belt. *J. Geophys. Res.* **110**:F04016, doi:10.1029/2004JF000244.

Dupré, B., C. Dessert, P. Oliva et al. 2003. Rivers, chemical weathering, and Earth's climate. *C.R. Geoscience* **335**:1141–1160.

Dylan, B. 1963. The Freewheelin' Bob Dylan. New York: Columbia Records.

Edmond, J.M., and Y. Huh. 1997. Chemical weathering yields from basement and orogenic terrains in hot and cold climates. In: Tectonic Uplift and Climate Change, ed. W.F. Ruddiman and W. Prell, pp. 329–351. New York: Plenum.

Ehlers, T.A., and K.A. Farley. 2003. Apatite (U-Th)/He thermochronometry: Methods and applications to problems in tectonic and surface processes. *Earth Planet. Sci. Lett.* **206**:1–14.

Gaillardet, J., B. Dupré, P. Louvat, and C.J. Allègre. 1999. Global silicate weathering and CO_2 consumption rates deduced from the chemistry of large rivers. *Chem. Geol.* **159**:3–30.

Garcia-Castellanos, D., S. Cloetingh, and R. Van Balen. 2000. Modelling the Middle Pleistocene uplift in the Ardennes-Rhenish Massif: Thermo-mechanical weakening under the Eifel? *Glob. Planet. Change* **27**:39–52.

Gilbert, G.K. 1877. Report on the Geology of the Henry Mountains, p. 160. Washington, D.C.: U.S. Geol. Survey.

Gosse, J.C., and, F.M. Phillips. 2001. Terrestrial *in situ* cosmogenic nuclides: Theory and application. *Quat. Sci. Rev.* **20**:1475–1560.

Granger, D.E., J.W. Kirchner, and R. Finkel. 1996. Spatially averaged long-term erosion rates measured from *in situ*-produced cosmogenic nuclides in alluvial sediment. *J. Geol.* **104**:249–257.

Guzzetti, F., B.D. Malamud, D.L. Turcotte, and P. Reichenbach. 2002. Power-law correlations of landslide areas in central Italy. *Earth Planet. Sci. Lett.* **195**:169–183.

Hack, J.T. 1960. Interpretation of erosional topography in humid temperate regions. *Am. J. Sci.* **258**:80–97.

Harp, E.L., and R.L. Jibson. 1996. Landslides triggered by the 1994 Northridge, California earthquake. *Bull. Seismol. Soc. Am.* **86**:319–332.

Hartshorn, K., N. Hovius, D. Brian, and R.L. Slingerland. 2002. Climate-driven bedrock incision in an active mountain belt. *Science* **297**:2036–2038.

Hay, W. 1998. Detrital sediment fluxes from continents to oceans. *Chem. Geol.* **145**: 287–323.

Heimsath, A.M. 1999. Cosmogenic nuclides, topography, and the spatial variation of soil depth. *Geomorphology* **27**:151–172.

Heimsath, A.M., J. Chappell, R.C. Finkel, K. Fifield, and A. Alimanovic. 2005. Escarpment erosion and landscape evolution in southeastern Australia. In: Tectonics, Climate, and Landscape Evolution, Geol. Soc. Am. Bull. Spec. Publ., in press

Heimsath, A.M., W.E. Dietrich, K. Nishiizumi, and R.C.Finkel. 1997. The soil production function and landscape equilibrium. *Nature* **388**:358–361.

Heimsath, A.M., W.E. Dietrich, K. Nishiizumi, and R.C. Finkel. 1999. Cosmogenic nuclides, topography, and the spatial variation of soil depth. *Geomorphology* **27(1-2)**: 151–172.

Hilley, G.E., R. Bürgmann, A. Ferretti, F. Novali, and F. Rocca. 2004. Dynamics of slow-moving landslides from permanent scatterer analysis. *Science* **304**:1952–1955.

Hovius, N. 1998. Controls on sediment supply by large rivers. In: Relative Role of Eustacy, Climate and Tectonics in Continental Rocks, ed. K.W. Shanley and P.J. McCabe, Special Publication 59, pp. 3–16. Tulsa, OK: Society for Sedimentary Geology (SEPM).

Hovius, N., C.P Stark., and P.A. Allen. 1997. Sediment flux from a mountain belt derived by landslide mapping. *Geology* **25**:231–234.

Keefer, D.K. 1994. The importance of earthquake-induced landslides to long-term slope erosion and slope failure hazards in seismically active regions. *Geomorphology* **10**: 265–284.

Koons, P.O. 1989. The topographic evolution of collisional mountain belts: A numerical look at the Southern Alps, New Zealand. *Am. J. Sci.* **289**:1041–1069.

Kutzbach, J.E., P.J. Guetter, W.F. Ruddiman, and W.L. Prell. 1989. The sensitivity of climate to Late Cenozoic uplift in southern Asia and the American west: Numerical experiments. *J. Geophys. Res.* **94**:18,393–18,407.

Lague, D., P. Davy, and A. Crave. 2000. Estimating uplift rate and erodibility from the area–slope relationship: Examples from Brittany (France) and numerical modelling. *Phys. & Chem. Earth* **25**:543–548.

Lal, D. 1991. Cosmic ray labeling of erosion surfaces: *In situ* nuclide production rates and erosion models. *Earth Planet. Sci. Lett.* **104**:424–439.

Lavé, J., and J.P. Avouac. 2001. Fluvial incision and tectonic uplift across the Himalayas of Central Nepal. *J. Geophys. Res.* **106**:26,561–26,591.

Lavé, J., and, D.W. Burbank. 2004. Denudation processes and rates in the Transverse Ranges, Southern California: Erosional response of a transitional landscape to external and anthropogenic forcing. *J. Geophys. Res.* **109**:F01006.

Liew, P.M. 1988. Sedimentology and River Terrace Correlation of the Liwu River. Taipei: National Taiwan University (in Chinese).

Liu, T.-K., S. Hseih, Y.-G. Chen, and W.-S. Chen 2001. Thermo-kinematic evolution of the Taiwan oblique-collision mountain belt as revealed by zircon fission track dating. *Earth Planet. Sci. Lett.* **186**:45–56.

Lyons, W.B., C.A. Nezat, A.E. Carey, and D.M. Hicks. 2002. Organic carbon fluxes to the ocean from high-standing islands. *Geology* **30**:443–446.

Malamud, B.D., D.L. Turcotte, F. Guzzetti, and P. Reichenbach. 2004. Landslides, earthquakes, and erosion. *Earth Planet. Sci. Lett.* **229**:45–59.

Matmon, A., P.R. Bierman, J. Larsen et al. 2003a. Temporally and spatially uniform rates of erosion in the southern Appalachian Great Smoky Mountains. *Geology* **31**:155–158.

Matmon, A., P.R. Bierman, J. Larsen et al. 2003b. Erosion of an ancient mountain range, the Great Smoky Mountains, North Carolina and Tennessee. *Am. J. Sci.* **303**:817–855.

Meybeck, M. 1986. Composition chimique des ruisseaux non pollués de France. *Sci. Géol. Bull. Strasbourg* **39**:3–77.

Meybeck, M. 1987. Global chemical weathering of surficial rocks estimated from river dissolved loads. *Am. J. Sci.* **287**:401–428.

Milliman, J.D., and J.P.M. Syvitski. 1992. Geomorphic/tectonic control of sediment discharge to the ocean: The importance of small mountainous rivers. *J. Geol.* **100**:525–544.

Molnar, P., and P. England. 1990. Late Cenozoic uplift of mountain ranges and global climate change. *Nature* **346**:29–34.

Montgomery, D.R., and M.T. Brandon. 2002. Topographic controls on erosion rates in tectonically active mountain ranges. *Earth Planet. Sci. Lett.* **201**:481–489.

Morgan, R.P.C. 2005. Soil Erosion and Conservation. Oxford: Blackwell.

Oliva, P., J. Viers, and B. Dupré. 2003. Chemical weathering in granitic environments. *Chem. Geology* **202**:225–256.

Palaca, S.W., G.C. Hurt, D. Baker et al. 2001. Consistent land and atmosphere-based U.S. carbon sink estimates. *Science* **292**:2316–2320.

Pelletier, J.D., B.D. Malamud, T.A. Blodgett, and D.L. Turcotte. 1997. Scale-invariance of soil moisture variability and its implications for the frequency-size distribution of landslides. *Eng. Geol.* **48**:254–268.

Pinet, P., and M. Souriau. 1988. Continental erosion and large-scale relief. *Tectonics* **7**:563–582.

Pörtge, K.H. 1995. Temporal and spatial variation in dissolved and solid load yields in partial catchment areas of the upper Leine River (southern Lower Saxony). *Z. Geomorph.* **100**:167–179.

Raymo, M.E., and W.F. Ruddiman. 1992. Tectonic forcing of late Cenozoic climate. *Nature* **359**:117–122.

Reiners, P.W., and T.A. Ehlers, eds. 2005. Low-temperature Thermochronology: Techniques, Interpretations, and Applications, Reviews in Mineralogy and Geochemistry 58. Blacksburg: Mineral. Soc. Am.

Reusser, L.J., P.R. Bierman, M.J. Pavich et al. 2004. Rapid Late Pleistocene incision of Atlantic passive-margin river gorges. *Science* **305**:499–502.

Riebe, C.S. 2004. Erosional and climatic effects in long-term chemical weathering rates in granitic landscapes spanning diverse climate regimes. *Earth Planet. Sci. Lett.* **224**:547–562.

Riebe, C.S., J.W. Kirchner, and R.C Finkel. 2003. Long-term rates of chemical weathering and physical erosion from cosmogenic nuclides and geochemical mass balance. *Geochim. Cosmochim. Acta* **67**:4411–4427.

Riebe, C.S., J.W. Kirchner, D.E. Granger, and R.C Finkel. 2000. Erosional equilibrium and disequilibrium in the Sierra Nevada, inferred from cosmogenic ^{26}Al and ^{10}Be in alluvial sediment. *Geology* **28**:803–806.

Riebe, C.S., J.W. Kirchner, D.E. Granger, and R.C Finkel. 2001. Strong tectonic and weak climatic control of long-term chemical weathering rates. *Geology* **29**:511–514.

Riebe, C.S., J.W. Kirchner, and R.C. Finkel. 2004. Sharp decrease in long-term chemical weathering rates along an altitudinal transect. *Earth Planet. Sci. Lett.***218**:421–434.

Ritter, J.R.R., M. Jordan, U.R Christensen, and U. Achauer. 2001. A mantle plume below the Eifel volcanic fields, Germany. *Earth Planet. Sci. Lett.* **186**:7–14.

Schaller, M., F. von Blanckenburg, N. Hovius, and P.W. Kubik. 2001. Large-scale erosion rates from in situ-produced cosmogenic nuclides in European river sediments. *Earth Planet. Sci. Lett.* **188**:441–458.

Schaller, M., F. von Blanckenburg, N. Hovius et al. 2004. Paleoerosion rates from cosmogenic ^{10}Be in a 1.3 Ma terrace sequence: Response of the River Meuse to changes in climate and rock uplift rate. *J. Geol.* **112**:127–144.

Schaller, M., F. von Blanckenburg, H. Veit, and P.W. Kubik. 2003. Influence of periglacial cover-beds on *in situ*-produced cosmogenic ^{10}Be in soil sections. *Geomorphology* **49**:255–267.

Schaller, M., F. von Blanckenburg, A. Veldkamp et al. 2002. A 30,000-year record of erosion rates from cosmogenic ^{10}Be in Middle European river terraces. *Earth Planet. Sci. Lett.* **204**:307–320.

Schaller, M., N. Hovius, S.D. Willett et al. 2005. Fluvial bedrock incision in the active mountain belt of Taiwan from *in situ*-produced cosmogenic nuclides. *Earth Surf. Proc. & Landforms* **30**:955–971.

Scott, K.M., J.W. Vallance, and P.T. Pringle. 1995. Sedimentology, behavior, and hazards of debris flows at Mount Rainier, Prof. Paper 1547. Washington, D.C.: U.S. Geol. Survey.

Shin, T.C., and T.L. Teng. 2001. An overview of the 1999 Chi-Chi, Taiwan, earthquake. *Bull. Seismol. Soc. Am.* **91**:895–913.

Simpson, G. 2004. A dynamic model to investigate coupling between erosion, deposition, and three-dimensional (thin-plate) deformation. *J. Geophys. Res.* **109**:F02006, doi: 10,1029/2003JF000078.

Small, E.E., R.S. Anderson, and G.S. Hancock. 1999. Estimates of the rate of regolith production using ^{10}Be and ^{26}Al from an alpine hillslope. *Geomorphology* **27**:131–150.

Snyder, N.P., K.X. Whipple, G.E. Tucker, and D.J. Merritts. 2000. Landscape response to tectonic forcing: Digital elevation model analysis of stream profiles in the Mendocino triple junction region, northern California. *Geol. Soc. Am. Bull.* **112**:1250–1263.

Stark, C.P., and N. Hovius. 2001. The characterization of landslide size distributions. *Geophys. Res. Lett.* **28**:1091–1094.

Summerfield, M.A., and, N.J. Hulton. 1994. Natural controls of fluvial denudation rate in major world drainage basins. *J. Geophys. Res.* **99**:13,871–13,883.

von Blanckenburg, F. 2005. The control mechanisms of erosion and weathering at basin scale from cosmogenic nuclides in river sediment. *Earth Planet. Sci. Lett.* **237**: 462–479.

von Blanckenburg, F., T. Hewawasam, and P. Kubik. 2004. Cosmogenic nuclide evidence for low weathering and denudation in the wet tropical Highlands of Sri Lanka. *J. Geophys. Res.* **109**:F03008, doi: 10.1029/2003JF000049.

Water Resources Agency. 1970–2003. Hydrological Yearbook of Taiwan, Republic of China: Taipei, Water Resources Agency; also available at http://gweb.wra.gov.tw/wrweb.

Whipple, K.X., and G.E. Tucker 1999. Dynamics of the stream-power river incision model: Implications for height limits of mountain ranges, landscape response times-cales, and research needs. *J. Geophys. Res.* **104**:17,647–17,661.

White, A.F., and A.E. Blum. 1995. Effects of climate on chemical weathering in water-sheds. *Geochim. Cosmochim. Acta* **59**:1729–1747.

White, A.F., and S.L. Brantley. 2003. The effect of time on the weathering of silicate minerals: Why do weathering rates differ in the laboratory and field? *Chem. Geol.* **202**:479–506.

Willett, S.D. 1999. Orogeny and orography: The effects of erosion on the structure of mountain belts. *J. Geophys. Res.* **104**:28,957–28,981.

Willett, S.D., and M.T. Brandon. 2002. On steady states in mountain belts. *Geology* **30**:175–178.

Willett, S.D., D. Fisher, C.W. Fuller, E.C. Yeh, and C.Y. Lu. 2003. Erosion rates and orogenic wedge kinematics in Taiwan inferred from apatite fission track thermo-chronometry. *Geology* **31**:945–948.

Willett, S.D., R.L. Slingerland, and N. Hovius. 2001. Uplift, shortening, and steady state topography in active mountain belts. *Am. J. Sci.* **301**:455–485.

Wobus, C., A.M. Heimsath, K. Whipple, and K. Hodges. 2005. Active out-of-sequence thrust faulting in the central Nepalese Himalaya. *Nature* **434**:1008–1011.

Yu, S.B., Y.-J. Hsu, H.-H. Su et al. 2001. Preseismic deformation and coseismic dis-placements associated with the 1999 Chi-Chi, Taiwan, earthquake. *Bull. Seismol. Soc. Am.* **91**:995–1012.

From left to right: Peter Koons, Niels Hovius, Anke Friedrich, Manfred Strecker, Eric Kirby, Roger Buck, Alexander Densmore, Fritz Schlunegger, Thorsten Nagel, Friedhelm von Blanckenburg

10

Group Report:
Surface Environmental Effects
on and of Faulting

W. ROGER BUCK, Rapporteur

ALEXANDER L. DENSMORE, ANKE M. FRIEDRICH, NIELS HOVIUS,
ERIC KIRBY, PETER O. KOONS, THORSTEN J. NAGEL,
FRITZ SCHLUNEGGER, MANFRED R. STRECKER,
and FRIEDHELM VON BLANCKENBURG

OVERVIEW

One of the most exciting developments in the geosciences over the past two decades has been the recognition that mass redistribution by surface processes modulates the geodynamic evolution of mountain ranges (e.g., Koons 1990; Beaumont et al. 1992). Our present understanding of the nature of this coupled system has emerged from several convergent developments: the proposition of provocative hypotheses for the manner in which climate impacts mountain building, progress in numerical modeling of lithospheric deformation, quantitative description of surface processes, and the application of new (thermo)chronologic techniques to the analysis of landscape processes. Despite significant progress in elucidating the feedbacks between climate, erosion, and tectonics, many avenues for research remain only superficially explored.

We see specific value in focusing future research on characterizing simple systems down to the length scales of landscapes associated with individual faults in the upper crust (e.g., lengths of ~ 10–100 km). This approach would represent a marked change from the current emphasis on analyzing interactive surface and tectonic processes on the scale of entire active orogens. These simple,

small-scale settings offer several crucial advantages in designing natural experiments to relate deformation, climate, topography, and mass transfer. First,
such settings offer the best opportunity to constrain the spatial and temporal
patterns of tectonic displacements on timescales ranging from the earthquake
cycle up to the life span of the structure, typically a few million years. Second,
interactions between atmospheric circulation, the hydrologic cycle, erosion,
and sediment transport are more easily observed and constrained in simple topographic settings. The power of simple systems lies in our ability to regard a
relatively large number of variables constant and to use the areal variation of
fault geometry and topography as a proxy for the evolution of faults and landscapes in time. Using space-for-time substitution may allow us better to identify the response of both landscapes and faults to perturbations in climate and
other surface processes.

In this chapter, we have attempted to highlight important gaps in our understanding and to identify promising new research directions. We have broken
down the issue of faulting and surface processes into five key topic areas: Topics 1 and 2 consider the interplay between erosion, deposition, and faulting.
Underlying these topics is the realization that redistribution of land mass can
affect the lithospheric stress field and that lithospheric deformation is a first-
order control on topography. Topic 3 addresses secular and cyclical changes in
climate that drive erosion. Topic 4 explores the fidelity of the surface records
of deformation, climate, and erosion. Finally, Topic 5 discusses how faulting
affects the human environment.

For the sake of brevity, we use the term "frequency", unless otherwise specified, to refer to temporal and spatial frequencies. Likewise, "erosion" means
the patterns and rates of erosion, "deposition" the patterns and rates of deposition, and "resolution" refers to temporal and spatial resolution.

TOPIC 1:
SURFACE PROCESSES AND STRAIN LOCALIZATION

The term "surface processes" is used to denote such diverse processes as weathering, erosion, and deposition that are triggered by changes in atmospheric dynamics, and/or variations in local base level or sea level. Geologists have long
realized that surface processes affect large-scale tectonic structures on Earth
(Hutton 1788), for example, by wearing down mountain ranges. Over the last
few decades, there has been increased interest in how surface processes affect
deformation on shorter length and timescales, down to the scale of individual
active faults (e.g., Merritts and Ellis 1994).

The pattern of flexure around volcanic ocean islands shows that the Earth's
lithosphere (i.e., its relatively cold outer shell) behaves elastically on timescales

of many millions of years (Watts and Burov 2003). Still, rocks yield and eventually break where the differential stress exceeds their fracture strength. The subsequent decrease in rock strength is associated with the localization of deformation along fault zones. At greater depths, however, high temperatures, low strain rates, and/or high confining pressures cause rocks to flow viscously. Viscous behavior at depth is suggested by post- and interseismic changes in displacement rates along active faults (see, e.g., Chapter 6) or by the slow, time-dependent rebound of Earth's surface in response to melting of the most recent continental ice sheets about 10^4 years ago (e.g., Peltier 2004). Therefore, the bulk mechanical response, or rheology, of the crust and mantle reflects a combination of elastic, brittle, and viscous behavior.

Interactions of tectonic and surface processes at the orogenic scale have been considered within the rather different frameworks of isostatic compensation involving viscoelastic flow (Holmes 1944; Avouac and Burov 1996), gravity sliding on the lithospheric scale (Ramberg 1973), and critical Coulomb behavior in orogenic wedges (Dahlen and Suppe 1988; Koons 1990, 1994; Beaumont et al. 1992; Willett et al. 1993). One possible effect of erosion is that unloading induces the upward advection of warm asthenospheric material, causing the overlying crust to become hotter. Due to the temperature dependence of viscosity, this hot, rapidly eroding area becomes weaker, enabling deformation to become more concentrated there.

Topography formed during orogenesis and the increased denudation rates in regions of high relief fundamentally influence the location and rates of interplate deformation at various temporal scales. A basic issue that we seek to address is the lower limit of temporal and spatial scales at which surface processes can influence tectonic kinematics.

Hypothesis: Surface processes affect tectonic responses on length scales shorter than the thickness of the seismogenic layer and over time intervals less than the seismic recurrence time.

Apparent first-order links between geomorphology and tectonics occur through stress effects due to topographic loads, slopes, and groundwater perturbations as well as through strength effects due to strain-dependent changes in fault–rock structure and rheology. The former linkage can be explored through evaluation of timing and magnitude of erosional perturbations and the patterns of tectonic response at different frequencies. As our observational resolution increases (see discussion under TOPIC 4) we can test the response of individual extreme events, such as cyclonic storms, or the effect of large-scale decadal atmospheric oscillations (e.g., El Niño Southern Oscillations (ENSOs) discussed under TOPIC 5) on the rate of seismic moment release. The influence of major precipitation events depends on several factors: (a) the tectonic setting, (b) the distribution, frequency, and intensity of precipitation, and (c) the permeability of the fault system[1]. For example, it is well established that an influx of water increases pore-fluid pressure, reduces effective normal stress and therefore leads to a reduction in shear stress (Law of Effective Stress or Terzaghi's Law).

Changes in fault strength can arise from fluid/fault or fluid/slope interactions in which groundwater perturbations influence the frictional properties of fault zones. Rate- and state-dependent fault stability is sensitive to perturbations in frictional coefficient and cohesion (see Chapter 5) and, consequently, may be sensitive to mineral solution–precipitation events (see Chapters 7 and 12). Thus, near-surface fault strength is expected to vary in space and time, depending on the length scale of groundwater diffusion and the kinetics of the reactions that alter the near-surface mineralogy and permeability of the fault rock. This is an important area of future research, as pointed out in the by Person et al. (Chapter 13). A related hypothesis is that the spatial scale of the response to surface processes, like erosion, depends on the tectonic setting. Here, "tectonic setting" refers to a variety of factors that affect the response of the lithosphere to loads. The most basic factor is whether the region is undergoing predominantly compressional, extensional, or strike-slip deformation. Much interest has focused on surface processes in compressional mountain belts, but erosion may have a major impact on how faults develop in extensional rifts.

Another potentially important factor is residual fault strength. The stress state around a fault depends partly on the strength of that fault. If the fault is extremely weak, then the fault reactivates and differential stress in the surrounding intact rock never reaches the level needed for brittle failure. Thus, a small erosion-related change in load may produce regional elastic deformation rather than permanent deformation of the intact rock. Where the active fault is relatively strong, the same erosion-related load change might induce brittle failure of the wall rock.

Progress can be made in natural laboratories where it may be shown that there is either constant tectonic forcing and variable erosion or constant erosional forcing. Can we identify simple systems where the effects of surface processes and deep-seated tectonic processes can be isolated and distinguished?

The western escarpment of the Andes has great potential to record high-frequency feedback mechanisms between crustal deformation, seismicity, and surface erosion for three reasons. First, this part of the Andes is one of the most active mountain belts (Farías et al. 2005). Second, it comprises several folds and faults with well-preserved growth strata, which allow a high-resolution reconstruction of the chronology of crustal deformation and strain accumulation. Third, the Andes of northern Chile and Peru are characterized by a distinct pattern of surface erosion rates that vary at high frequencies in both space and time.

---

[1] Fault system is defined as the fault core, damage zone, and adjacent wall rock.

In a recent integrated study in the Andes, Zeilinger et al. (2005) showed how surface erosion has caused localization of strain accumulation on the Oxaya Anticline in northern Chile. The power of this study lies in the detailed reconstruction of the deformation mechanisms and of the chronology of strain accumulation in relation to the spatial and temporal pattern of surface erosion rates of the rivers that cross this fold. This study emphasizes the potential of simple systems for exploring possible feedback mechanisms between surface erosion and strain accumulation; it also highlights the limits set by the resolution of the data from natural systems.

Another region with potential for work in this field is the Basin and Range Province of the western United States. For several reasons, this area comprises a large number of distinct, fault-bounded ranges that are attractive for elucidating the smaller-scale relationships between faulting and surface processes: First, the length, displacement, and displacement rates of the faults vary broadly, so that landscape response to different magnitudes of tectonic perturbation and over different spatial scales can be isolated. Second, the activity of individual faults initiated and ceased at various times during regional Cenozoic extension. This allows straightforward space-for-time substitutions to be made, giving us a means of characterizing landscape development through all stages of fault evolution, from initiation and growth to death. Finally, footwall erosion and mass transfer to adjacent basins have occurred by a variety of processes that differ in rate and importance from range to range. For example, glaciers were important geomorphic agents in some ranges but were absent from others. With a better understanding of the spatial and temporal distribution of strain accumulation in this region, these range-to-range differences should allow us to isolate and quantify important feedbacks between tectonics and surface processes.

In these respects, the Basin and Range Province is no different than any large continental region containing faults that are active over a variety of spatial and temporal scales under varying climatic conditions and surface process regimes. The following information is needed in the Basin and Range Province to make it a first-rate field laboratory for studying the effects of surface processes on tectonics: (a) a better understanding of the distribution of strain accumulation on individual faults, particularly in terms of spatial variations in displacement and temporal clustering of earthquakes; (b) an improved chronology of fault initiation and death to help constrain the timescales over which faults grow and erosion can affect deformation; and (c) better quantification of the spatial patterns and rates of denudation and mass transport.

In addition to smaller, focused studies, we see a substantial challenge for future research in exploring feedback mechanisms between surface erosion and crustal deformation at the orogenic scale. Acknowledging the research that has already been carried out in this field, we emphasize that thus far studies have been hampered by limits in the resolution of the structural, geomorphic, and

stratigraphic data sets. In this regard, the European Alps represent a particularly promising mountain belt for improving our knowledge in this field. There
is a wealth of information about the geodynamic, structural, and topographic
evolution of this orogen (Schmid et al. 1996). Similarly, the architecture of its
neighboring basins has been explored in great detail (Kuhlemann and Kempf
2001), resulting in a high-resolution reconstruction of erosional mass flux from
the Alps (Kuhlemann et al. 2001). There is also sufficient evidence that the
evolution of the Alps occurred in distinct geodynamic phases that were related
potentially to variations in climate-driven erosion rates (Schlunegger and
Simpson 2002; Willett et al. 2005). Nevertheless, the precision and resolution
of existing data do not yet allow for an unambiguous identification of feedback
mechanisms between surface erosion and structural deformation at the scale of
the Alpine orogeny. We highlight improvements in our ability to read and resolve structural and stratigraphic recorders as one of the outstanding challenges,
and address this more fully under TOPIC 4.

TOPIC 2:
TOPOGRAPHIC RESPONSE TO COUPLED EROSION,
WEATHERING, AND TECTONICS

Earth's topography reflects the integrated effects of rock velocity, material properties, and erosion and transport processes. Each of these components of the
coupled erosional–tectonic system is directly and indirectly influenced by the
dynamics of faulting. We identify three primary, yet overlapping, feedback loops
where significant potential for new insight exists.

Hypothesis: There is a predictable and systematic relationship between erosion rates and differential rock velocity fields imparted by active deformation
of the crust.

Deformation gradients within the upper portion of the crust often lead to changes
in the elevation of the Earth's surface. Surface processes acting on the resultant
topographic gradients give rise to the rugged, mountainous topography commonly
associated with regions of active faulting. Geomorphologists have long recognized that erosion and sediment transport increase with increasing topographic
gradients. Recent studies are beginning to yield insight into the relationship between crustal strain and landscape form, particularly as expressed in bedrock channel
gradients (see Chapter 8), in settings where erosion rates appear to balance rock
velocity fields. However, both the manner in which and rates at which these
systems respond to changes in the rock displacement field—and by implication, the sensitivity of erosion rates to active deformation—remain uncertain.

Continued development in our ability to read the record of active deformation in the landscape requires well-constrained field sites where we can isolate

the response of geomorphic systems to known perturbations in base level (Van der Beek and Bishop 2003). By examining the response timescale of both fluvial channels and hillslopes to transient forcing, we can begin to place bounds on the rates at which information on crustal deformation is recorded in, and lost from the geomorphic system (see discussion under TOPIC 4). Of critical importance to this effort are studies which combine measurement of erosion rates in both the pre- and post-transient portions of the landscape. The use of cosmogenic radionuclides shows tremendous promise in this effort and will form the basis of a new generation of landscape evolution models (von Blanckenburg 2005; see also Chapter 9, this volume).

Hypothesis: Topography reflects the finite strain field of the Earth's crust.

The correlation length of topography, a measure of the simplicity (predictability) of Earth's surface, is a function of interaction between the surface velocity field and the distribution of crustal strain. The latter obviously depends on *in situ* rock strength. For a homogeneous material, erosion generally causes roughening (i.e., increasing complexity) of the topography. However, surface simplicity arises if erosional heterogeneities are introduced, for example, by faulting and weathering. Fault zones that have been brittly overprinted are particularly susceptible to weathering, because grain-size reduction increases surface to volume ratios and renders them open to fluid influx. Increased erodability of cataclastic fault zones promotes localized fluvial incision. As a result, the higher-order components of topography reflect deformational patterns. This notion is not entirely new, but the advent of digital elevation models has improved our ability to identify erosional topography, in turn facilitating detailed, quantitative investigation of the topographic fingerprint of deformation.

Hypothesis: Localization of crustal strain promotes significant silicate weathering and CO_2 drawdown.

Silicate weathering by carbonic acid provides the most important long-term CO_2 sink at the Earth's surface, maintaining a comfortable greenhouse effect and stabilizing the climate at habitable conditions. A major debate has persisted for years about the actual mechanisms of such weathering in rocks and soils (Chapter 9). When landscapes are stable and not subjected to significant physical erosion, groundwater pathways to unweathered rock mass lengthen, the water becomes saturated with cations dissolved from silicate minerals, and weathering all but ceases. In contrast, weathering is efficient in well-drained landscapes where fresh mineral surfaces are continuously exposed by rapid physical erosion. This situation occurs in mountain belts, which have been identified as primary sinks of atmospheric CO_2. From the perspective of a global budget, the open question is whether surface deformation at smaller spatial scales (e.g., that of an individual fault) drives a weathering flux that is also significant. Given the widespread occurrence of faulting, this might be the case if faulted topography has a specific weathering yield that is close to that of orogenic regions. Along many active faults, new topography and advection of fresh material into the zone of chemical weathering combine with relatively prolonged fluid residence to make faults efficient and potentially important weathering "factories."

To address this issue, we require measurements of physical and chemical erosion rates and fluxes in areas of active deformation over a wide range of climatic conditions. In particular, the following activities are likely to be fruitful: (a) measurement of rates of hillslope denudation and soil production using cosmogenic nuclides; (b) measurement of chemical weathering indices at the hillslope scale, through chemical analyses of soils and unweathered bedrock; and (c) measurement of denudation rates at the catchment scale by monitoring solid and dissolved river loads and cosmogenic nuclide analysis of sediment (von Blanckenburg 2005). These analyses must be done in field settings where the rate of tectonically driven base-level fall can be carefully constrained. These constraints are best derived by mapping active structures and determining surface velocity fields with geodetic techniques and displacement of geologic and geomorphic markers.

TOPIC 3:
IMPACT OF CLIMATE AND CLIMATE CHANGE ON SURFACE DEFORMATION

As outlined in TOPIC 1, surface processes are likely to affect the localization of strain at the Earth's surface. This is generally attributed to changes in surface loads and upper crustal stresses due to mass redistribution at the surface. However, it seems likely that other effects, including transient loads caused by shifting glaciers, lakes, and the distribution of fluids in the upper crust, may also affect fault deformation. As these variables are modulated by climate, an obvious corollary is that climate change could cause observable temporal variations in the patterns and/or rates of surface deformation. We identify the following testable hypotheses to explore these relationships.

Hypothesis: The temporal variation in slip exhibited by fault systems on orbital timescales (10^3–10^5 yr) is due to climate-driven surface effects. These effects may include glacial or lacustrine loading and unloading, pore-fluid pressure and hydrological changes during climate cycles, or base-level change. Conversely, observed temporal variations in slip rate may be used to determine changes in these hydrological parameters.

Displacement accumulation on many faults is episodic, with periods of enhanced slip rate followed by periods of quiescence. This episodicity has been attributed to a host of factors operating at depth (see Chapter 4–6), but in some cases it may also be linked to variations in surficial loading of the fault. So far, this relationship has not been quantified and its mechanisms remain ambiguous. In particular, climate may influence deformation in a number of ways, including precipitation-driven changes in pore-fluid pressure, constructive or destructive interference between climatic and seismic cycles, and changes of surface erosion patterns and rates due to base-level changes. There may be time lags associated with the redistribution of mass and surface loads that are important; however, these cannot be resolved without high-resolution records of seismicity.

Field tests of the relationship between climate and faulting will require better understanding of the statistics of precipitation events and the patterns of precipitation averages as well as of the changes of these statistics and patterns during climate cycles. In addition, we must understand how the patterns are related to changes in groundwater infiltration rates and pore-fluid pressure in fault zones. Simultaneously, it is important to identify and measure the sources and magnitudes of near-surface loading and unloading of fault systems. These may include hillslope saturation, mass redistribution by erosion and deposition, and growth and decay of lakes or ice fields. Finally, a key requirement is a detailed, high-resolution record of earthquake frequency, magnitude and locations (down to the level of the earthquake cycle), and measurement of strain accumulation on individual faults.

One promising approach to this problem may be to focus on regions with a large number of individual fault-bounded ranges to take advantage of regional variations in glacial or lacustrine loading and unloading as well as hydrological changes during climatic cycles. We may be able to compare these variations with the histories of strain accumulation on individual faults in order to deduce the effects of both loading/unloading cycles and changes in groundwater flow and fluid pressure on fault displacement. This can be done, for example, in the Basin and Range Province, where temporal clustering of earthquakes and variations in strain accumulation have been recently documented.

In the Great Basin region, which includes the Basin and Range Province, changing surface loads induced by Quaternary climatic oscillations have resulted in lithospheric deformation and strain localization on at least two timescales. During cold and pluvial periods, at least four major lake cycles have occurred in the past ~600 ka, the largest of which is the Bonneville Lake cycle with the highest lake levels reached between 25 and 17 ka. By 10 ka, the water level had dropped by over 200 m. On the longest spatial and temporal wavelengths (several hundred kilometers and a few thousands of years), the lake Bonneville load caused a deflection of the Earth's surface of up to 70 m near the center of the lake (Gilbert 1890; Bills et al. 1994). Modeling of the shape of Earth's rebounding surface implies that the load was compensated at a depth near the base of the lithosphere (Bills et al. 1994). On a shorter timescale, a rapid drop of the lake level between 17 and 10 ka appears to have affected the rate of strain localization along the Wasatch and neighboring fault systems. This is best documented for the Salt Lake fault segment, where the vertical displacement rate accelerated from <0.6 mm yr^{-1} before 10 ka to >1.5 mm yr^{-1} in the time since. The rate change is apparent from a decrease in recurrence intervals of large (~2 m) fault ruptures from >5 ka to ~1.3 ±0.3 ka between 14 and 10 ka (McCalpin 2000). The inference that changing surface loads have affected the rate of strain release is based on the twin assumptions that the far-field (tectonic) strain accumulation rate has remained constant and that a change in loading rate may cause a change in the rate of strain release on the timescale of the earthquake recurrence. Therefore, a fundamental challenge is to determine on what timescales a change in the far-field loading rate may affect strain release on a single fault, and how strain release is coupled to strain accumulation (e.g., Wallace- versus Reid-type behavior; Friedrich et al. 2003).

Hypothesis: Long-term climatic changes can alter the efficiency of the sediment transport system enough to cause observable differences in surface deformation.

The complex interrelationships between climate change and tectonic activity have been partly addressed on a global scale; however, the dynamics and implications of these relationships are largely unresolved at the scale of individual structures or orogens. In particular, we do not sufficiently understand the relative importance of different climatic variables (e.g., mean precipitation and variability of precipitation) in modulating erosion and mass transport. There is a strong theoretical basis for the idea that mass redistribution leads to changes in near-surface crustal stress sufficient large to influence the patterns and/or rates of rock mass and surface deformation (see TOPIC 1). We lack unambiguous field evidence of this effect because the resolution of the available structural and stratigraphic data is not sufficiently high.

To address this issue, we need independent information on the amplitude and frequency of climate variations in a given region of faulting or orogeny, as outlined above. In addition, we require a high-resolution record to establish how climatic attributes control erosion and sediment routing. Ideally, the temporal resolution should be such that we can record changes over times that are shorter than the response time of the major river systems in an orogen (10^3–10^6 yr, depending on the size of the catchment and the rates at which the river systems operate on the surface). The final component of such a study is the chronology of surface deformation, particularly the timing of fault displacement and the cumulative spatial pattern of slip at different intervals during deformation.

All of these records must be sufficiently continuous to represent the important climatic variations over orbital timescales. An ideal field test of this hypothesis would be a single fault-basin pair within a fold and thrust belt, or even (at a larger spatial scale) the belt itself. Such settings generally offer the possibility that sediments preserved in footwall or thrust-top basins may record progressive deformation through growth strata or propagation of faulting into the footwall. At present, the greatest obstacles to overcome are in recovering the chronology of deformation, that is, in relating the sedimentary record to rates and patterns of tectonic activity and in obtaining an independent characterization of past climate variability and change. We currently lack the ability to reconstruct patterns and rates of displacement with independent data sets.

TOPIC 4:
THE FIDELITY OF THE GEOMORPHIC AND GEOLOGIC RECORDS OF CHANGES IN RATES OF SURFACE DEFORMATION AND EROSION/SEDIMENTATION

To address the questions posed above, it is necessary to explore the extent to which the geomorphic and geologic records may be used to quantify the rates of landscape or surface evolution in response to tectonic or climatic events. A related, fundamental issue is the degree to which the Earth's surface is evolving as the result of catastrophic events versus gradual change. On a human timescale, near-surface crustal deformation occurs mainly in rare, episodic events, for example, expressed as ground-rupturing earthquakes. Most large faults have not yet ruptured twice in historic times. This would imply that rare, catastrophic events play a major role in shaping Earth's surface. However, instrumentally recorded seismic events reveal that for any given active fault system, the number of earthquakes of a given magnitude decreases linearly with every tenfold increase in magnitude, that is, the size–frequency distribution follows a power-law distribution. Most natural events, such as volcanic eruptions, landslides, and storms, exhibit similar size–frequency behavior. In such cases, the relative number of events of different sizes is determined by the exponent in the power-law relation of event size and frequency. In a general sense, large events are nothing but the rare equivalents of smaller events.

How can catastrophic events, such as the recurrence of ground-rupturing seismic events or the secular change in the tectonic loading rate of a region, be recognized in the geologic record? On the human timescale, the largest possible faulting events may not yet have been observed because recurrence times are commonly on a scale of several thousand years. On the geologic timescale these events may not be recognized due to limited resolution of the sedimentary and geomorphic records. Both landscapes and basin fills have the potential to evolve on timescales of earthquake recurrence. However, three challenges must be addressed before this can be done. First, the recurrence time of ground-rupturing earthquakes is not regular, as noted above. Second, the sedimentary record of earthquake activity may be biased by changes in the sedimentation rate. Third, the geomorphic response to fault slip may vary in time due to secular

Figure 10.1   Three-dimensional perspective view of the Lost River Range, a simple fault-bounded mountain range in Idaho, western U.S.A. Image produced by draping Landsat 7 ETM+ imagery (bands 7, 4, and 2) on USGS digital topographic data with 30 m resolution. Black lines show surface trace of the active normal-slip Lost River fault (Janecke 1995), which separates the Lost River Range footwall from the adjacent hangingwall basin of the Big Lost River Valley. Fault segments in foreground last ruptured in the 1983 M_s 7.3 Borah Peak earthquake. Maximum throw on the Lost River fault is approximately 4.5–6 km (Janecke et al. 1991). Note high relief and evidence for glacial erosion in the footwall, and coalesced alluvial and debris-flow fans in the hangingwall.

changes in climate. For typical continental deformation rates, the time constant of secular change of tectonic loading may exceed tens of thousands of years. Thus, the challenge is to identify field areas that expose a high-resolution sedimentary record or landforms that may be directly related to individual events to characterize the tectonic history of a region better.

Hypothesis: Individual tectonic events can be identified using combinations of geologic markers and geomorphic recorders of erosion and deposition with overlapping response frequencies, at timescales shorter than the time constant of the characteristic event (e.g., seismic cycle). However, geologic and geomorphic signals with a relatively low temporal resolution are too susceptible to aliasing[2] to be useful in reconstructing high-frequency tectonic deformation far back in time.

---

[2] Aliasing is an effect that causes different continuous signals to become indistinguishable when sampled.

A key requirement in inferring fault evolution is to be able to relate fault displacement to a distinctive displacement field at the surface. We therefore require geologic or geomorphic markers that are capable of faithfully recording that displacement field, either incrementally or over a given time interval. A number of methodological and geological factors limit our ability to resolve individual fault displacement events in the geologic record (Figure 10.1). For example, our ability to separate events in time is limited by the precision and decay constant of the available geochronometers, given a sufficiently high recording rate (e.g., sedimentation rate, erosion rate). Moreover, tectonic elements in the landscape are continuously attacked by weathering and erosion, and their persistence depends on the geomorphic response time to a perturbation and the rate of erosion (Figure 10.2) during response. Whereas erosional processes constantly reduce the expression of tectonic events in the landscape, the sedimentary record may preserve high-frequency signals for a very long

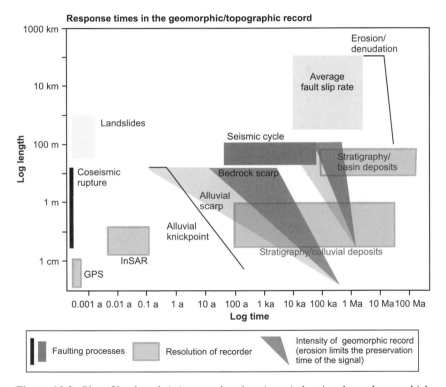

Figure 10.2 Plot of log length (m) versus log time (years) showing the scales on which geomorphic and tectonic processes can be recorded and preserved in the rock record. A single earthquake (black bar; coseismic slip ranging from 10 cm to 10 m) results in geologic, geomorphic and sedimentary responses (yellow and pink boxes). Size of boxes indicates limits of resolution; color intensity within the boxes indicates the strength of the signal in the geologic record. Figure from A. Friedrich.

time. Realistically, the available length of this record also depends on logistical limits such as depth of excavation or natural incision of a stratigraphic section.

Coseismic displacement on a dip-slip fault typically results in a vertical surface offset that may be described by a step function. This surface rupture also offsets marker horizons across the fault, erosion of the up-thrown side near the fault, and deposition of colluvial sediments on the down-thrown side. Depending on the rates of geomorphic and sedimentary processes affecting the faulted region after the event, different strategies may be applied to recover the age and displacement records (Figure 10.3). On the down-thrown side, sedimentary deposits allow the identification of event horizons. The precision with which this record can be resolved is limited by the sedimentation rate, and the availability of an independent reference frame to which the preserved strata can be tied. An episode of faulting may be dated by determining the age of the associated sedimentary deposit; the resolution of this age is limited by the bracketing stratigraphic markers and the precision of the geochronometer. The reliability and resolution of this stratigraphic record remain constant as long as the sediments are preserved and the isotopic system was open to diffusion for the duration of faulting. The best results using this method have been obtained where

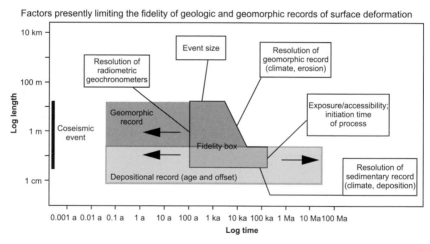

Figure 10.3 Plot of log length (m) versus log time (years) showing the spatiotemporal limits on the fidelity (outlined area) of preserving and measuring a "high-frequency" event (e.g., a seismic rupture) in the geologic record. In contrast to real-time measurements of seismic events, geologic signals thereof can only rarely be dated directly. However, these signals leave traces in the geomorphic and sedimentary records. Our ability to measure and date such an event depends on event size, the degree to which traces of the event are preserved, and the constraints on the age of the geomorphic or sedimentary record (colored boxes). Note that both geologic recorders and measurement tools are restricted to lower-frequency bands, making it difficult and sometimes impossible to recover high-frequency event time series. Figure from A. Friedrich.

sediments were deposited at sea level. For example, the record of formation and growth of deltas in the Gulf of California (Dorsey et al. 1995; Dorsey and Umhoefer 2000) can be inverted for individual earthquakes or earthquake clusters, due to the proximity of the controlling fault to the coast line (and base level). In this way, the postseismic increase in sediment flux is transmitted into the basin without significant loss of signal strength.

If the seismic event does not repeat, geomorphic recorders (fault scarps) will decay in intensity, that is, they leave no significant geologic signal. However, if large earthquakes repeat at a rate (seismic cycle, Figure 10.3, dark gray box for 100 m of slip) higher than the geomorphic decay rate but lower than the sedimentation rate, distinct signals are left behind in the geologic, geomorphic, and sedimentary records, for example, fault systems, mountain ranges, and basins, respectively. Note the limited overlap of windows for geologic processes (gray boxes) and geologic recorders (green boxes). In particular, notice that a tectonic signal recorded in the landscape will not be preserved for very long (intensity of recorder, red lines).

On the hangingwall of the fault, the free surface of the scarp has eroded with time, and information about fault slip events is progressively lost. The decay in this geomorphic record of faulting is often considered as a function of system-scale diffusivity (Andrews and Hanks 1985). However, it is possible to derive constraints on rock uplift rates and patterns from the erosional topography if hydrometric parameters and substrate properties are known (Snyder et al. 2003). Most commonly used geomorphic displacement indicators, such as marine and river terraces and coseismic scarps developed along mountain fronts, do not record the full three-dimensional surface displacement field. Therefore, the challenge is to find geological or geomorphic datums with the greatest possible areal extent. As an example, shallow marine limestones have undergone surface uplift without significant erosion in the Finisterre Mountains of Papua New Guinea. This has allowed measurement of the cumulative displacement field across an entire mountain range (Abbott et al. 1997). Superimposed on this deformed pre-orogenic surface are sets of uplifted coral terraces whose lateral extent can be used to derive patterns and rates of surface and rock uplift along the northeast tip of the mountain belt. This requires knowledge of sea-level changes for the relevant interval (Chappell 1974). In turn, these terraces contain meter-scale topographic steps associated with individual large earthquakes (Ota and Chappell 1996). Composite records like this should allow a reliable estimate of earthquake activity back into geological time. Future work should employ this approach at other sites. In addition, efforts should be directed towards closed sedimentary systems, such as internally drained, fault-bounded basins, where the sedimentary record of faulting is likely to be most complete. Where both the geomorphic and sedimentary records are accessible, deformation rates may also be obtained by cross-calibrating geomorphic and sedimentary processes.

A key and hitherto underexplored issue in the relationship between tectonic activity and surface evolution is that various elements of the surface (e.g., hillslopes, bedrock channels, and alluvial rivers) respond to tectonic perturbations in different ways and over different timescales. Thus, we argue that the response of these surface systems is best considered in terms of characteristic response times, that is, the time required for an element of the landscape or geologic record to adjust to a tectonic event, such as an earthquake or the inception of displacement on a fault. These response times are particularly useful in guiding our search for geomorphic recorders of the dynamics of tectonic deformation. For example, we should not hope to find a record of short-period (10 ka) variations in fault slip rate in a large alluvial catchment, in which the response to fault slip and base-level change is strongly buffered by changes in floodplain sediment storage possibly on the order of millions of years. Instead, we should seek other elements of the landscape or record which are less likely to filter out this high-frequency signal. At present, our understanding of landscape response times is based largely on theory and numerical modeling. An important research challenge is to design field experiments that allow us to observe and quantify the response times of various parts of the landscape and geologic record to well-constrained tectonic perturbations.

TOPIC 5:
IMPACT OF FAULTING ON THE HUMAN ENVIRONMENT

The risk of fault activity has again become all too clear with the December 26, 2004 earthquake off the coast of Sumatra in the Indian Ocean. The damage and terrible loss of life due to such events makes it easy to overlook the many beneficial effects of fault activity. Although posing a threat to life and livelihood, crustal deformation has contributed to the origin, wealth, and well-being of humankind. Therefore, public awareness of the benefits and risks of faulting should be raised in equal measure, as outlined below.

Benefits

Crustal deformation in general and faulting in particular have many beneficial effects. First and foremost they are form and maintain topography above sea level and have provided the specific topographic and environmental niches within which hominid species have evolved and developed complex societies (King and Bailey 2006). Further, they condition and stabilize Earth's atmosphere and climate by promoting CO_2 drawdown by rapid weathering of freshly exposed silicate rocks (Chapter 9). At present, humankind benefits from fault-controlled hydrothermal activity, groundwater flow, generation and preservation of hydrocarbons, production and accumulation of soils, routing of surface runoff, and the ecological diversity that derives from topographic and climatic gradients.

Hazards and Risks

A hazard is a (natural) process or phenomenon that has the potential to do damage. Risk is the impact of a hazardous event on society and can be thought of as the product of hazard and probability of occurrence for a given area of Earth's surface. The hazards associated with faulting include ground shaking and related dislocation and deformation of the Earth's surface, as well as a variety of surface processes: soil liquefaction, slope failure, avalanches, ponding and rerouting of surface and subsurface drainage, breaching of flow barriers, volcanic activity, accelerated glacier flow, and tsunamis. Often, most damage associated with earthquakes is caused by seismically triggered processes that affect areas far beyond an earthquake's epicenter. Documented patterns of seismically triggered surface processes can be cast as simple mathematical expressions.

A case in point is the pattern of landsliding associated with the 1999 M_W 7.6 Chi-Chi earthquake in Taiwan. This earthquake was the largest of its size in Taiwan over the last fifty years, and the largest on the Chelungpu thrust fault in the previous 300–400 years. The earthquake had a focal depth of 8 km, ruptured an approximately 100 km segment of the north–south trending, east-dipping (~30°) Chelungpu thrust, and produced ground accelerations of up to 1 g. More than 20,000 landslides covering a total area of 150 km^2 were triggered by the earthquake within a ~3000 km^2 region broadly bounded by the 0.2 g peak vertical ground acceleration contour. The area disturbed by landsliding exceeded two percent of a ~20 km zone around the Chelungpu fault and decreased away from the epicentral area in an exponential fashion, with a decay constant of ~30 km. Subsequent typhoons have triggered >30,000 landslides with a total area of 500 km^2 in the epicentral region. The intensity of typhoon-triggered landsliding decreased systematically with distance from the Chelungpu fault, suggesting either that typhoon Toraji remobilized landslide debris generated during the Chi-Chi earthquake (Figure 10.4), or that the substrate was preconditioned to fail through damage of hillslope rock mass during strong seismic ground motion.

Another dramatic example of a well-quantified hazard is the catastrophic increase in sediment flux related to ENSO events (Schneider et al. 2005). For instance, high-resolution pluviometric data from the Piura drainage basin of northern Peru reveal that ENSOs cause precipitation rates in the Sechura desert (located adjacent to the Pacific coast) to increase to 4000 $mmyr^{-1}$. In the Cordillera at ca. 2000 m a.s.l., ENSOs cause a threefold increase in precipitation rates (ca. 3000 $mmyr^{-1}$). Gauging stations measured a tremendous increase in water and sediment discharge due to ENSOs. During a "normal" year, water discharge is ca. 15 m^3s^{-1} in Piura (located in the Sechura desert), and sediment discharge measures ca. 179 $ktyr^{-1}$. ENSOs, however, have resulted in a ca. 24- and 807-fold increase in water and sediment discharge, respectively. Hence, denudation rates increase to ca. 8.8 $mmyr^{-1}$ during ENSO events. It is unclear

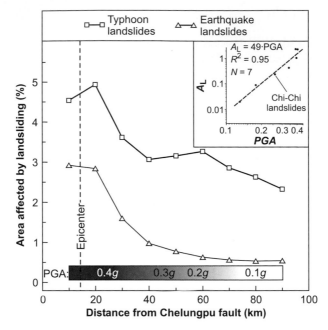

Figure 10.4 Geomorphic response to the 1999 M_w 7.6 Chi-Chi earthquake, Taiwan. The density of earthquake-triggered landslides decays exponentially with distance from the epicenter. The density of landslides triggered by typhoons between 2000 and 2002 is higher everywhere and decays in a similar fashion away from the epicenter. The intensity and distribution of postseismic slope failure is therefore strongly controlled by proximity to earthquakes. The inset shows a simple power-law relation between peak ground acceleration (PGA) during the earthquake, and the percent area affected by earthquake-triggered landslides (A_L).

whether episodic erosion and sediment transport related to ENSOs have affected the accumulation of strain and the distribution of seismic hazard in the Andes. This issue should be explored in more detail.

Mitigation

Scientists have a responsibility to convey information and understanding of natural hazards to society. Possibly the most effective way that earth scientists can aid the mitigation of earthquake hazard risk is by advising policy makers and architects on better building codes and practices. Construction or reinforcement of buildings to specifications proportional to maximum expected earthquake-generated stresses can greatly reduce the loss of life during earthquakes. This is illustrated by the fact that the 1999 Izmit earthquake caused about ten times more fatalities in a densely populated part of Turkey than the

1999 Chi-Chi earthquake in an equally densely populated part of Taiwan, even though the latter was several times stronger. In Taiwan most recently constructed buildings have been built to withstand strong earthquakes. Crucial to the success of this practice is that potential seismic loads on buildings are known with confidence and that overspecification is avoided. This requires sound knowledge of regional earthquake magnitude-frequency distributions and of possible site effects due to substrate properties and topography. Site effects can enhance seismic strong ground motion by over an order of magnitude and deserve special attention when planning and building.

A second way to mitigate the impact of faulting and earthquakes is general education. Fast processes, such as earthquakes, do not leave time to inform people about life-saving behavior. The general public in earthquake-prone areas must be made aware of the existence of a hazard, the nature of that hazard, the likelihood of its impact, and the best way to behave during an event. Scientists can help by preparing educational material for wide circulation: What are earthquakes? Where and why do they occur? What happens when they strike? What should one do in the event of an earthquake? These aspects can and should be explained at a level comprehensible to a normal twelve year old. Information should be made available in a way that takes into account the level and mode of local education, and may take the form of printed text and/or figures, film, oral presentations, or web pages. The principal objective should be to enable people to make informed decisions about their own conduct and to prepare their communities. As part of this effort, misinformation about natural hazards for commercial gain (e.g., disaster movies) should be countered. Embedding background information in media reports on natural catastrophes should be encouraged.

A third way to mitigate earthquake hazard is for scientists to collaborate with civil defense organizations. The efficiency of these organizations depends strongly on scientific information of two kinds: (a) warning of the possible natural catastrophes in areas of jurisdiction and (b) information on what happens during a catastrophic event. Scientists can quantify hazards and their historic spatial and temporal patterns. They can point out that hazards are often based on a sequence of events, with the earthquake being only the first event that triggers an entire series of events. In the event of a catastrophe, scientists can make available information from observatories and monitoring networks. Information should also flow the other way: civil defense organizations should explain to scientists their organizational structure, mode of operation, and the type and format of information they require to function optimally. By their nature, civil defense organizations are a portal through which the scientific community can influence natural hazard management.

A fourth and final way in which natural scientists can mitigate earthquake-related hazards is to continue to inform policy makers. Decision makers need to be aware of patterns of natural hazards and associated risks in order to control land use, to plan infrastructure, to judge requirements for contingency funding and insurance, and to regulate civil defense and educational efforts. Scientific briefs on fault-related phenomena might include summaries of seismogenic structures, patterns of historic seismicity, statistics of earthquake magnitude and frequency, patterns of ground shaking and seismically triggered processes, characteristic timescales of these processes, as well as spatial and temporal trends in seismicity and triggered processes, with careful anticipation of future activity. It is important to state the level of (un)certainty of this information and to ensure that (un)certainties remain visible when information is propagated through communication networks.

It is important for the scientific community to be proactive about mitigating natural hazards and risks, seeking interaction with potential partners in government, industry, education, and civil defense, and understanding the requirements of these institutions and explaining its own capabilities. In doing so, scientists should strive to be interpreters of nature for the public, to communicate what is known or new and to raise awareness of what is not yet known. A side effect of this action is that it lays the foundation for support of future research.

SUMMARY

In this report, we have described the promise and problems of characterizing links between climate, surface processes, faulting, and hazards. The field is in a state of flux because new techniques to measure and date surfaces and processes, such as cosmogenic nuclides, are not yet fully developed and advancing rapidly. Application of new measurement techniques to active fault systems are yielding startling results that call into question long-held assumptions. For example, fault activity is recognized as a major control on the rate of rock weathering and erosion and may be as (or more) important than surface relief or rainfall. Numerical models of interactive tectonic and surface processes are now able to reproduce the effects of smaller-scale features like faults and rivers on landscape and climate evolution. The model predictions have the potential to guide both the formulation of new hypotheses and the search for large-scale field studies to test these hypotheses.

Among the major challenges in this field, we assign special importance to improving the resolution and coverage of measurements used to constrain the history of fault activity, regional strain, as well as of climate and erosion. A related challenge is to find natural laboratories where changes of only one or a few variables that affect tectonic and surface processes can be identified and quantified. Recent progress in identifying such areas and in demonstrating the controls on their evolution give us confidence that new insights into the processes that shape Earth's surface and shallow subsurface will be forthcoming.

REFERENCES

Abbott, L.D., E.A. Silver, R.S. Anderson et al. 1997. Measurement of tectonic surface uplift rate in a young collisional mountain belt. *Nature* **385**:501–507.

Andrews, D.J., and T.C. Hanks. 1985. Scarp degraded by linear diffusion: Inverse solution for age. *J. Geophys. Res.* **90**:10,193–10,208.

Avouac, J.P., and E.B. Burov. 1996. Erosion as a driving mechanism of intracontinental growth. *J. Geophys. Res.* **101**:17,747–17,769.

Beaumont, C., P. Fullsack, and J. Hamilton. 1992. Erosional control of active compressional orogens. In: Thrust Tectonics., ed. K.R. McClay, pp. 1–18. New York: Chapman & Hall.

Bills, B.G., D.R. Currey, and G.A. Marshall. 1994. Viscosity estimates for the crust and upper mantle from patterns of lacustrine shoreline deformation in the Eastern Great Basin. *J. Geophys. Res.* **99**:22,059–22,086.

Chappell, J. 1974. Geology of coral terraces, Huon Peninsula, New Guinea: A study of Quaternary tectonic movements and sea level changes. *Geol. Soc. Am. Bull.* **85**: 553–570.

Dahlen, F.A., and J. Suppe. 1988. Mechanics, growth, and erosion of mountain belts. Spec. Paper 218, pp. 161–178. Boulder, CO: Geol. Soc. Am.

Dorsey, R.J., and P.J. Umhoefer. 2000. Tectonic and eustatic controls on sequence stratigraphy of the Pliocene Loreto basin, Baja California Sur, Mexico. *Geol. Soc. Am. Bull.* **112**:177–199.

Dorsey, R.J., P.J. Umhoefer, and P.R. Renne. 1995. Rapid subsidence and stacked Gilbert-type fan deltas, Pliocene Loreto Basin, Baja California Sur, Mexico. *Sediment. Geol.* **98**:181–204.

Farías, M., R. Charrier, D. Comte, J. Martinod, and G. Hérail. 2005. Late Cenozoic uplift of the western flank of the Altiplano: Evidence from the depositional, tectonic, and geomorphologic evolution and shallow seismic activity (northern Chile at 19°30' S). *Tectonics* **24**:10.1029.

Friedrich, A.M., B. Wernicke, N.A. Niemi, R.A. Bennett, and J.L. Davis. 2003. Comparison of geodetic and geologic data from the Wasatch region, Utah, and implications for the spectral character of Earth deformation at periods of 10 to 10 million years. *J. Geophys. Res.* **108**:doi:10.1029/2001JB000682.

Gilbert, G.K. 1890. Lake Bonneville. Monograph 1. Reston, VA: U.S. Geol. Survey.

Holmes, A. 1944. Principles of Physical Geology. London: Thomas Nelson & Sons.

Hutton, J. 1788. Theory of the Earth; or an Investigation of the Laws Observable in the Composition, Dissolution, and Restoration of Land upon the Globe. *Trans. Roy. Soc. Edinburgh* Volume 1.

Janecke, S.U. 1995. Eocene to Oligocene half grabens of east-central Idaho: Structure, stratigraphy, age, and tectonics. *Northwest Geol.* **24**:159–199.

Janecke, S.U., J.W. Geissman, and R.L. Bruhn. 1991. Localized rotation during Paleogene extension in east-central Idaho: Paleomagnetic and geologic evidence. *Tectonics* **10**:403–432.

King, G.C.P., and G. Bailey. 2006. Tectonics and human evolution. *Antiquity* **80**: 265–286.

Koons, P.O. 1990. The two-sided wedge in orogeny: Erosion and collision from the sand box to the Southern Alps, New Zealand. *Geology* **18**:679–682.

Koons, P.O. 1994. Three-dimensional critical wedges: Tectonics and topography in oblique collisional orogens. (Special issue: Tectonics and Topography). *J. Geophys. Res.* **99**:12,301–12,315

Kuhlemann, A., and O. Kempf. 2001. Foreland basin geometry and facies: Response to the tectonic setting of the Alpine orogen. *Sediment. Geol.* **152**:45–78.

Kuhlemann, J., W. Frisch, I. Dunkl, and B. Szekely. 2001. Quantifying tectonic versus erosive denudation by the sediment budget: The Miocene core complexes of the Alps. *Tectonophysics* **330**:1–23

McCalpin, J.P. 2000. The Wasatch fault megatrench of 1999: Evidence for temporal clustering of post-Bonneville paleoearthquakes. In: Geol. Soc. Am. Rocky Mountain Section, 52nd annual meeting, Abstracts with Programs, 32, 5, 16. Washington, D.C.: Geol. Soc. Am.

Merritts, D.J., and M. Ellis. 1994. Introduction to special section on tectonics and topography. *J. Geophys. Res.* **12**:135–141.

Ota, Y., and J. Chappell. 1996. Late Quaternary coseismic uplift events on the Huon peninsula, Papua New Guinea, deduced from coral terrace data. *J. Geophys. Res.* **101**:6071–6082.

Peltier, W.R. 2004. Global glacial isostasy and the surface of the Ice Age Earth: The ICE-5G (VM2) model and GRACE. *Ann. Rev. Earth & Planet. Sci.* **32**:111–149.

Ramberg, H. 1973. Model studies of gravity-controlled tectonics by the centrifuge technique. In: Gravity and Tectonics, ed. K.A. DeJong and R. Scholten, pp. 49–77. New York: Wiley.

Schlunegger, F., and G. Simpson. 2002. Possible erosional control on lateral growth of the European Central Alps. *Geology* **30**:907–910.

Schmid, S.M., O.A. Pfiffner, N. Froitzheim, G. Schönborn, and E. Kissling. 1996. Geophysical-geological transect and tectonic evolution of the Swiss-Italian Alps. *Tectonics* **15**:1036–1064.

Schneider, H., F. Schlunegger, D. Rieke-Zapp et al. 2005. Landscape response to high-episodic precipitation (ENSO); Piura area, northern Peru. *EGU Abstracts* EGU05-A-02049.

Snyder, N.P., K.X. Whipple, G.E. Tucker, and D.J. Merritts. 2003. Channel response to tectonic forcing: Analysis of stream morphology and hydrology in the Mendocino triple junction region, northern California. *Geomorphology* **53**:97–127.

Van der Beek, P., and P. Bishop. 2003. Cenozoic river profile development in the upper Lachlan catchment (SE Australia) as a test of quantitative fluvial incision models. *J. Geophys. Res.* **108**:6 10.1029/2002JB002125

von Blanckenburg, F. 2005. The control mechanisms of erosion and weathering at basin scale from cosmogenic nuclides in river sediment. *Earth Planet. Sci. Lett.* **237**:462–479.

Watts, A.B., and E.B. Burov. 2003. Lithospheric strength and its relationship to the elastic and seismogenic layer thickness. *Earth Planet. Sci. Lett.* **213**:113–131.

Willett, S.D., C. Beaumont, and P. Fullsack. 1993. A mechanical model for the tectonics of doubly vergent compressional orogens. *Geology* **21**:371–374.

Willett, S.D., D.B. Stolar, F. Schlunegger, and V. Picotti. 2005. Evidence for late Miocene climate change from the tectonics of the Alps. *EGU Abstracts* EGU05-A-09891.

Zeilinger, G., F. Schlunegger, and G. Simpson 2005. The Oxaya anticlines (northern Chile): A buckle enhanced by river incision? *Terra Nova* **17**:368–375.

11

Fluid Processes in Deep Crustal Fault Zones

BRUCE W. D. YARDLEY[1] and LUKAS P. BAUMGARTNER[2]

[1]School of Earth and Environment, Earth Sciences, University of Leeds,
Leeds LS2 9JT, U.K.
[2]Institut de Minéralogie et Géochimie, Université de Lausanne,
1015 Lausanne, Switzerland

ABSTRACT

Fluid as a C-O-H dominated phase is widespread, but not ubiquitous, in the Earth's crust. The presence or absence of fluid is in large part a function of thermal history, at least up to the onset of melting. Rocks containing relatively low-temperature assemblages that are subject to further heating release fluid and so are commonly saturated, while rocks undergoing cooling resorb fluid into hydrous minerals and so are dry. Fluid may be introduced from external sources during faulting or magmatic activity, and the degree to which it persists depends on the interplay between injection rates and reaction rates. Where fluids do occur in the crust, fluxes are generally low, so that many aspects of fluid chemistry are dictated by saturation with rock-forming minerals. These mineralogical controls on fluid chemistry and activities of volatile species further affect the rheology of the crust by determining whether or not deformation can be fluxed by fluid processes. It is argued that rocks undergoing progressive metamorphism are wet and experience widespread deformation, while rocks that are cooling are strong and deformation is localized into zones, particularly during times of fluid infiltration. The transition between brittle and ductile behavior may therefore reflect changes in the availability of water rather than changes in temperature. Faults themselves are important loci of fluid flow, but it is often difficult to identify the sources of fluid, because geochemical tracers are mainly reset in a rock-dominated environment. Nevertheless, it is unlikely that faults are commonly effective drains of fluid being released by prograde metamorphism, because the very low permeability of such rocks (inferred from evidence for strong overpressuring) means that fluid cannot easily drain into fractures, even where a strong gradient in hydraulic head exists.

INTRODUCTION

Evidence for fluid involvement in deep crustal fault zones comes from a variety of sources, including fluid fluxes encountered in deep drilling or at the surface after earthquakes, geophysical characteristics of deep faults, and the chemical, isotopic, and mineralogical characteristics of rocks from exhumed faults and shear zones. In this chapter, we will be concerned primarily with evidence from exhumed rocks, since these provide an insight into long-term, deep crustal processes in a wide range of settings. The objective of this chapter is to highlight evidence from the geological, geochemical, and petrological record for the involvement of fluids in deep crustal deformation, and to examine the implications for the geochemical environment of the deep crust. In particular, we will explore the implications for crustal deformation mechanisms of the patterns of fluid involvement that we deduce. Chapter 12 addresses many of the companion issues of the specific interactions between fluids and minerals during faulting processes.

THE NATURE OF CRUSTAL FLUIDS

By the term *crustal fluids* we mean here mobile phases that occupy pore spaces or fractures within rocks, which can move independently of the rocks that host them, and which have a sufficiently long residence time in the subsurface that they cannot be considered to be part of the shallow hydrological cycle. Crustal fluids have compositions that are dominated by combinations of water, carbon dioxide, hydrocarbon, chloride salt, or silicate melt. These are all components that we are familiar with from surface processes and from drilling, but their mutual solubility, their physical properties, and their interactions with rocks can vary greatly with pressure and temperature in the crust. The crustal fluids discussed in this chapter are C-H-O-salt fluids rather than silicate melts, and despite the large compositional variations that are possible, they are distinguished from melts by being significantly less dense than their host rocks.

C-H-O fluids are typically water-rich, and are supercritical at temperatures above around 350°C, although the presence of salt can extend two-phase behavior to magmatic temperatures. Therefore, silicate melt can coexist with a salt-rich aqueous fluid and a C-rich vapor phase simultaneously (Shmulovich et al. 1995).

The composition of water-rich fluids is variable in both its solvent and solute composition. Fluid species found in lower crustal fluids are mainly H_2O and CO_2, with minor CH_4, nitrogen, and sulfur. Aqueous fluids are often rich in salts, with chloride being almost invariably the major anion. The most common cation is normally Na, with Ca and K also important in most fluids. At higher temperatures, transition metals become progressively more important components of saline fluids, with Fe and even Mn reaching percent concentrations in some concentrated magmatic brines. Many aspects of the solute composition are dominated by equilibration with the rock mineral assemblage, and vary systematically with temperature even at sedimentary basin temperatures. The systematic variation in Fe-content of fluids is illustrated in Figure 11.1a (shown normalized to Cl), from Yardley (2005). Sedimentary brines and other low temperature fluids have relatively low Fe/Cl ratios, whereas for magmatic fluids this ratio is orders of magnitude higher. Metamorphic and geothermal fluids formed at intermediate temperatures have intermediate Fe/Cl values. The systematic trend from sedimentary basin fluids to magmatic brines demonstrates the importance for most crustal fluids of buffering of fluid composition by silicate minerals. In other words, fluids throughout the crust are saturated solutions of a relatively restricted range of rock types.

In addition to dissolved salts, crustal fluids contain silica (and probably also Al) at concentrations that are approximately independent of the salt chemistry, except at extreme salinities (Shmulovich et al. 2006). Their concentrations are almost invariably buffered by the adjacent rocks, and in the case of silica, quartz solubility in water provides a reasonable approximation of the concentrations likely in many natural crustal fluids (Figure 11.1b). Redistribution of quartz by solution-precipitation mechanisms is an important process throughout the crust, forming veins or cements in many rock types. The driving forces for such redistribution have been controversial, especially in the literature of diagenesis, with pressure solution, mineral reactions, and fluid migration down temperature (i.e., with decreasing temperature) all proposed. Integrated studies considering the behavior of all fluid components suggest that down-temperature flow is important for veining in many ore deposits, but metamorphic and diagenetic veins form predominantly by segregation processes and are driven by small, local chemical potential gradients without extensive fluid flux.

Although quartz solubility increases at high pressures and temperatures, there remains a distinct miscibility gap between silicate melts and simple aqueous solutions for many crustal conditions. Recent results show, however, that there are compositions for which this gap is reduced or absent. In fact, continuous liquid solutions exist for F- and B-rich pegmatitic fluids at temperatures as low as 700°C at 1 kbar (Thomas et al. 2000), and the critical curve descends to below 600°C for strongly peralkaline melt compositions rich in fluorine (Sowerby and Keppler 2002). The majority of melts are silicate melts, though carbonate-, chloride-, and sulfide-rich melts have also been reported (Kamenetsky et al. 2004; Campbell et al. 1995; Frost et al. 2002).

(a)

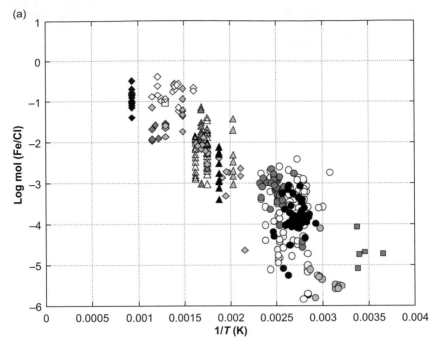

(b)

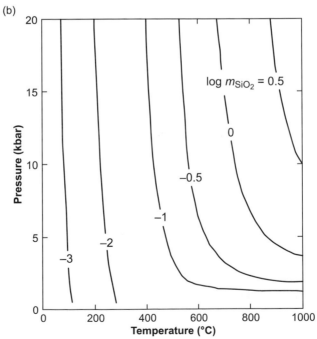

◄ **Figure 11.1** (a) The effect of temperature on the Fe-content of crustal fluids, normalized to Cl. The data ranges from low-temperature shield brines (square symbols) and oilfield formation waters (circles) through metamorphic fluids (triangles) to geothermal brines and high-temperature magmatic fluid inclusions (diamonds). The systematic trend with temperature is interpreted to indicate that the Fe-content of crustal fluids is buffered within distinct limits by interactions with silicate rocks (from Yardley 2005). (b) Quartz solubility is a good approximation of SiO_2 content in nearly all crustal fluids since quartz is present in most rocks of the continental crust. SiO_2 content is nearly independent of solution composition, though salting out and salting in behavior has been observed. Note the strong increases with depth (pressure) and temperature in the crust leading to silica-rich solution in lower crustal fluids. Decompression and cooling of crustal fluids lead to quartz precipitation. Modified from Manning (1997).

MODELS OF CRUSTAL FLUID DISTRIBUTION

There are two fundamentally opposed views of the occurrence of fluids in the continental crust. On the one hand, geophysicists have pointed to the absence of lower crustal earthquakes as evidence for a weak, water-saturated lower crust with the rheology of "wet," hydrolytically weakened quartz (Chen and Molnar 1983). Likewise, deep fluids have been invoked as giving rise to deep seismic reflectors and electric conductors (Jones 1987). On the other hand, petrologists point to the preservation of anhydrous, high-temperature mineral assemblages through subsequent exhumation as evidence of a dry crust at depth, since these would otherwise react with any available water to form retrograde assemblages once they had begun to cool from their peak assemblages (Yardley 1981; Frost and Bucher 1994; Yardley and Valley 1997). Handy and Brun (2004) argued that seismicity is an ambiguous indicator of crustal strength, such that the lack of seismicity in a given depth interval may be interpreted to indicate either that the crust there is weak (has yielded) or strong (remains largely undeformed). Experimental flow laws for quartz and feldspar aggregates suggest that dry continental crust is strong (Kohlstedt et al. 1995), providing an alternative explanation for the scarcity of deep crustal earthquakes to the weak and wet model of Chen and Molnar (1983). Our view of a strong, dry lower crust is supported by a recent reinvestigation of deep continental earthquakes (Maggi et al. 2000), in which a significant number of seismic events previously thought to have occurred in the upper mantle were relocated within the lower crust. This implies the existence of far greater differential stresses than would be possible if the rheology of the lower continental crust were dominated by hydrous quartz (e.g., Chapter 1).

This divergence of views arises partly from differences in the interpretation of seismic data, but partly also from a lack of resolution of crustal processes in time and space. Few studies clearly distinguish between tectonically active parts of the crust, especially where thrusting and subduction cause rocks to undergo progressive heating and dehydration (and hence where water is most likely present today), and formerly active, but now stable, continental regions. Arguments for a dry crust mainly apply to the latter, since they concern rocks that have cooled below their original temperature of formation. Some of the crustal sections interpreted as water saturated from conductivity profiles are not contentious, since they are areas of active tectonism (e.g., the Southern Alps of New Zealand, Upton et al. 2003), but beneath large areas of the continents the petrological and geophysical approaches lead to diametrically opposite conclusions.

THE EVIDENCE FOR FLUIDS AND THEIR AMOUNT IN THE CRUST

Upper Crust

Crystalline Basement

Faulting in basement affects low-porosity, low-permeability rocks (Brace 1980) with very low water contents and probably with very low partial water pressures (Stober and Bucher 2004). Fluid activity is localized in faults or joints which act as permeable pathways, at least transiently, and this in turn influences the faulting process (Miller 2002; Miller et al. 2003). While the existence of fluids in fractures intersected by deep continental drill holes has been widely reported, it is perhaps less well known that in the best documented example, the KTB (*Kontinentales Tiefbohrprogramm der Bundesrepublik Deutschland) borehole,* fluid pockets are only very weakly interconnected so that the electrical conductivity is dominated by graphite smears, not brine (ELEKTB 1997). These rocks are fluid-poor, with water restricted to occasional open fractures where it resides at near-hydrostatic pressure. Although fluxes through fractures may be high at very shallow crustal levels, deeper fluids are isolated from the hydrological cycle, and their high salinities may arise in part from their progressive dehydration due to formation of hydrous silicates in the altered wall rocks adjacent to the fractures. At some deep crustal drilling sites, notably Saatly, the crystalline midcrust is a sink for fluid and this has prevented further drilling (Borevsky et al. 1995; see also Stober and Bucher 2004). The geological record of upper crustal fluid penetration into crystalline rocks is ubiquitous, although often passed over. Joints and fault planes are often lined with hydrous minerals such as chlorite or clays, or with hydrothermal quartz, carbonates, or sulfates. Detailed studies linked to the investigation of possible nuclear waste repository sites have demonstrated that these mineral linings can form down to temperatures below 100°C (Dublyansky et al. 2001).

Sedimentary Basins

Sedimentary basins expel fluids due to both compaction and mineral break-down. In contrast to crystalline basement, the host rocks are often porous, permeable, and water-rich. Fault rock permeability is complex; fault rocks can be rather impermeable, for example, as a result of grain crushing or clay smearing (Fisher and Knipe 2001), or highly permeable due to the creation of microfracture networks. Thus fault rocks act as seals in some circumstances, but as permeable pathways in others (e.g., Miller and Nur 2000). At higher temperatures, as metamorphic conditions are attained during continued heating, it is less clear that faulting or strain partitioning into shear zones plays such an important role in deformation. Instead, the length scale of strain localization increases, and rocks experience pervasive ductile deformation accompanied by folding. Fractures formed under metamorphic conditions may result from embrittlement due to high fluid pressures. The transition to pervasive deformation with folding and cleavage development probably begins at temperatures as low as 200–250°C in deeply buried sedimentary sequences. It is clear that the fluid regime of a deep sedimentary basin undergoing burial and heating differs fundamentally from that of basement rocks at similar crustal depths, because the sediments are experiencing prograde dewatering rather than retrogression. Fluids are present pervasively, and are replenished by mineral dehydration rather than depleted by mineral hydration. Furthermore, fluid pressures are commonly in excess of hydrostatic values, and may approach lithostatic pressures.

The Basin–Basement Interface

In recent years, studies of sediment-hosted ore deposits in particular have high-lighted the interaction of basinal fluids with underlying basement. Lead isotope studies of the Irish limestone-hosted Pb-Zn deposits showed that they changed progressively from southeast to northwest, reflecting variations in the nature of the underlying Caledonian basement (Caulfield et al. 1986). This pattern is believed to arise because the metal in the ore bodies was leached by brines from the local underlying basement; where this has been drilled or is exposed at the surface, the rocks show evidence for extensive retrograde alteration focused on faults and joints. At Modum, Norway, Munz et al. (2002) and Gleeson et al. (2003) document both hydrocarbon fluids and basinal brines trapped in quartz veins hosted by Precambrian basement rocks beneath the Caledonian foreland basin. Fluid inclusions document very variable temperatures and pressures of trapping within the same quartz crystal. This was a dynamic system and both cool immature fluids drawn down from the basin, and heated and dehydrated fluids expelled back up from depth have been preserved. The basin–basement interface provides a clear demonstration of the very different behavior of fluids in the two settings.

Middle to Lower Crust

Field evidence of deformed, metamorphic, and igneous rocks is the main source of information about fluid regimes deeper in the crust. Here, we will separate evidence about the prograde part of the metamorphic cycle, that is, when fluids are progressively released by devolatilization reactions, and the retrograde stage, when the reaction direction is reversed.

Prograde metamorphism, and indeed deep diagenesis of sedimentary sequences, is frequently accompanied by the formation of veins whose morphology and mineralogy is indicative of fluid pressures close to lithostatic values (Yardley 1986). Likewise, equilibria between hydrated and less hydrated assemblages support a similar conclusion. These types of arguments have led metamorphic petrologists to agree that prograde metamorphism occurs under conditions of near-lithostatic fluid pressure, with fluid loss being balanced by release of further fluid from mineral lattices. Thus the fluid is present as a distinct phase, present pervasively through at least those rocks that contain hydrous or carbonate minerals, although the interplay of rates of dehydration and fluid loss results in rhythmic fluctuations of fluid pressure (Connolly and Podladchikov 1998).

The nature of fluids present during high-temperature processes in the lower crust can be investigated directly by studying the composition of fluid and melt inclusions in minerals (subject to the proviso that care is often necessary to distinguish true deep fluids from those trapped during uplift). Alternatively, the presence or absence of fluids and melts can be deduced through geochemical and phase petrological studies, and their composition constrained through experiments and thermodynamic modeling.

Petrology collections abound with evidence for an absence of fluids in cooled crust. The onset of cooling of high-temperature rocks results in the absorption of any remaining pore fluid to form trace amounts of secondary retrograde phases. The extremely fresh character of many high-grade metamorphic rocks therefore attests to the very low porosity of rocks undergoing metamorphism. The main evidence for fluid activity during cooling is localized hydration, often accompanied by deformation. Infiltration of water from an external source is a prerequisite for the formation of detectable quantities of retrograde minerals. Veins are a rarer phenomenon at this stage, except in large shear zones where rocks have been fully hydrated, and form at discrete stages (e.g., Hay et al. 1988). Large masses of cooled crystalline rocks survive over geological timescales in the deep crust without interacting with fluids at all, as evidenced by the preservation of granulite facies mineral assemblages in Precambrian rocks even where, as in West Norway, they can be shown to have spent much of the intervening time residing in the lower crust, not near the surface (Jamtveit et al. 1990). They are effectively completely impermeable; in part because anhydrous minerals react with the fluid phase to provide mineral sinks for any fluid that penetrates.

The most obvious evidence for retrograde fluid activity in basement rocks is in shear zones, where hydrated assemblages commonly develop in the deformed rocks. The precise nature of the minerals and their composition may be a guide to the depth and temperature when shearing and fluid influx occurred. In the upper crust, similar retrograde alteration is associated with major thrusts that interleave basement with cover sequences (Badertscher et al. 2002). The association of retrograde reaction products and deformation in shear zones is a close one, with the retrograde minerals typically contributing to the deformation fabrics. McCaig (1997) has reviewed examples of fluid flow in shear zones from a range of structural settings. While not all shear zones are retrograde, hydration reactions appear to occur in deformation zones over a very wide range of crustal conditions. Most of the shear zones which are cited by McCaig (1997) as examples have formed under greenschist facies conditions, but there are more extreme cases, notably the retrograde alteration of granulites (sometimes also eclogites) to amphibolites. The formation of eclogite shear zones after granulite also involves hydration. Common themes that emerge from deep fault zones in crystalline rocks are that deformation has very commonly taken place at lower temperatures than the original formation of the host rocks, so that the deformation may be considered retrograde, and that there is a correlation between the length scale of strain localization and degree of hydration.

CRUSTAL FLUID DISTRIBUTION ON THE GRAIN SCALE

The literature abounds with discussions of the importance of fluid–mineral wetting angles for fluid and melt segregation and flow, and much effort has been spent to determine this property (Holness 1997). Wetting angles influence the localization of fluids in porous media at textural equilibrium. Wetting fluids, that is, fluids with low wetting angles, will be distributed in a rock along all types of grain boundaries, forming a permanent film of fluid. With increasing wetting angle, fluid is first concentrated along triple grain junctions and eventually fluid forms isolated pockets (Figure 11.2a). In silicates, aqueous fluids tend to form angles larger than 60°, only partially wetting the grains (Holness 1997). This has the potential to reduce the reactivity of minerals in systems where only small amounts of fluids are present, although because reactions themselves lead to a deviation from textural equilibrium, it is not clear how important this effect actually is. Melts form lower wetting angles with silicate grains and wetting is greater if they are volatile-rich (Figure 11.2b).

It is clear that fluid significantly affects rock strength, usually by decreasing effective normal stress within pore spaces, or by enhancing diffusional mass transport of material along grain boundaries. Investigations of naturally and experimentally deformed, partially melted aggregates have shown that melt volume is the most important parameter affecting rock strength; above a critical

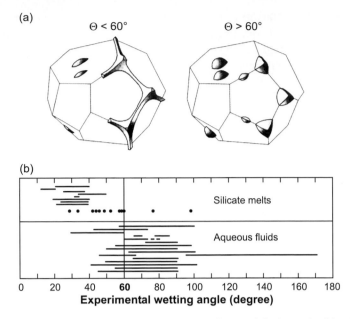

Figure 11.2 (a) The equilibrium fluid distribution in a rock is determined by the wetting properties between fluid and minerals, as well as the amount of fluid present. The smaller the wetting angle the larger the contact area between fluid and minerals, and the smaller the fluid amount to obtain a pervasive fluid film on mineral grains. After Watson and Brenan (1987). (b) Aqueous fluids have larger wetting angles than siliceous melts (from Holness 1997).

melt fraction (ca. 0.07), experimentally deformed aggregates experience a strength drop of several orders of magnitude. In nature, this strength drop may be much less, depending on the mechanism (fracture versus viscous flow) by which melt/fluid pockets interconnect (Rosenberg and Handy 2005). Deformation tends to decrease fluid wetting angles and to increase fluid (and melt) interconnectivity (e.g., Jin et al. 1994). Thus, the presence of even minor amounts of melt can lead to large-scale strain localization, as discussed in Chapter 13.

KINETICS OF FLUID PROCESSES IN THE CRUST

Reactions that involve interactions between rocks and fluids are of three types: (a) when fluid components are released from mineral lattices, (b) when a fluid phase is consumed by incorporation of fluid species into mineral lattices, and (c) mineral surface processes that are catalyzed by fluid species. The rate-controlling steps for these different types of reactions are likely to be different and so we will consider them separately.

Prograde Metamorphism

Case studies on the growth of porphyroblasts in natural rocks (e.g., Müller et al. 2004; Carlson et al. 1995) have quantitatively described their nucleation and growth in relation to each other and to the rock matrix. They suggest that non-equilibrium processes can significantly affect the evolution of a typical metamorphic rock. Three end-member models for mineral growth have been proposed for metamorphic rocks (see review in Kerrick et al. 1991; Fisher 1978): transport limited (TL), interface controlled (IC), and external rate controlled. The latter is the case if, for example, the heat flux reaching the system limits the rate of crystal growth. Experimental data is available to constrain the interface-controlled reaction kinetics (e.g., Kerrick et al. 1991; Lüttge et al. 2004), and results show that interface reactions are fast, so that at moderate overstepping of the equilibrium, significant reaction progress occurs in hours to a few weeks.

Overall, dehydration reactions are probably limited by the rate of heat supply, if sufficient fluid is present to allow diffusion along the grain-boundary structure, connecting the reactant phases. This is because such reactions are strongly endothermic, but also tend to take place continuously over a range of temperatures, so that there is often no nucleation step. If heat supply is faster than the reaction rate, the degree of overstepping increases, speeding up the reaction rate until the two balance. The existence of corona textures in metamorphic rocks indicates that despite this, reactions can be limited by transport, even at the slow heating rates (tens to hundreds of degrees per million years) experienced in regional metamorphic terrains. This implies very small amounts of fluid on grain boundaries.

Retrograde Metamorphism

Hydration reactions in the crust are likely to be constrained by the rate of fluid supply. From the low-grade metamorphism of pyroclastic sediments, we know that where water is available, high-temperature phases react rather rapidly because extensive alteration accompanies active geothermal circulation in fields that have been active for perhaps thousands of years. In many parts of the crust where retrograde hydration is incomplete, it is likely that fluid was not available in sufficient amounts. Initiation of cracking will potentially advect fluids and spur new mineral reactions, or accelerate reactions to alleviate pre-existing disequilibrium (see Chapter 12). At the same time, syntectonic grain size reduction, due to either cataclasis or dynamic recrystallization, will enhance surface reaction kinetics. At the grain scale, these reactions can be considered to be controlled by the rate of mineral surface reactions involving the breakdown of a silicate lattice and the growth of a new one (Putnis 2002). At a larger scale, however, they may be controlled by the rate of fluid supply and the time over which fluid continues to be present after it has infiltrated dry rocks. It is interesting to speculate on whether this is directly coupled to large-scale tectonic strains.

CONTROLS ON FLUID PRESSURES
AND THE PERMEABILITY OF CRYSTALLINE CRUST

In the upper crust, permeability can be considered to be an intrinsic rock property that can be measured or predicted, and used to model how fluids flow through the rock mass. However, even under upper crustal conditions, fluid–rock interactions provide a feedback that can modify permeability through time. In contrast, for much of the remaining crust, it is questionable whether permeability is an intrinsic rock property. For example, during prograde metamorphism fluid pressure is at approximately lithostatic pressure (Yardley 1986), and is limited by the tensile strength of the rock. It has been proposed that as fluid pressure approaches lithostatic pressure due to progressive dehydration, rock permeability increases through inflation of pores and/or hydrofracturing, until it is in balance with fluid production; as a result permeability is coupled to the rate of fluid production, rather than being an intrinsic rock property (Yardley 1986; Hanson 1992). The rate of fluid production is itself driven by the rate of heat supply or limited by the other reaction mechanisms discussed above. Hence, heat supply can be the ultimate control on the permeability of rocks undergoing progressive devolatilization, while the mechanical properties of rocks, their depth of burial, and to a lesser extent their stress state dictate the fluid pressure.

For crystalline rocks that have begun to cool, the mineral assemblage provides the dominant constraint on fluid pressure. This is because the chemical potentials of fluid species are controlled by equilibria among minerals, and so are essentially dictated by their host rocks (Yardley 1981; Yardley and Valley 1997). These chemical potentials are directly related to partial pressures of the fluid species. Once rocks have cooled by 50°C or more below their peak temperatures, activities of fluid species drop to values orders of magnitude below the value for a free fluid phase, so it is not possible for a fluid to be present, except in isolated fluid inclusions where it is unable to interact with the grain-boundary network. Figure 11.3 is an activity diagram from Yardley and Valley (1997), showing the activities of the volatile components H_2O and CO_2 that occur in equilibrium with high-grade metamorphic assemblages under lower crustal conditions, together with the activities that are present in a free fluid ranging between these two end-member compositions, with or without coexisting salt. These diagrams demonstrate that no free fluid phase can coexist in cooled rocks which contain higher grade assemblages, since they will react rapidly to form retrograde minerals. This will drive the activities of volatile species far below those possible when a fluid phase is present. Where free fluids do occur in such rocks in the upper crust, they evidently persist only in poorly connected fractures whose host rocks have been fully retrogressed under the ambient *P-T* conditions, so that their buffer capacity has been exhausted. Rimming of these cavities with retrograde minerals effectively seals off the fluid phase, isolating it from the rest of the rocks. Permeabilities quoted for crystalline crust at the retrograde stage are essentially dominated by late stage

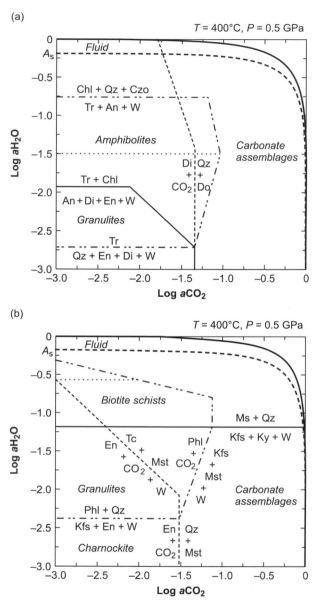

Figure 11.3 Log activity diagrams for (a) basic and (b) pelitic bulk compositions at 400°C and 0.5 GPa. Straight lines represent the volatile activities buffered by the coexistence of the reactants and products of the reactions indicated, all common assemblages in rocks. The curved lines define the activities of H_2O and CO_2 in fluids ranging between the pure fluid end members. Activities in salt-saturated solutions are shown by the dashed curve. Conditions inside these curves cannot be obtained if a free fluid phase is present, hence partially retrograded rocks must absorb any fluid phase completely (from Yardley and Valley 1997).

fractures (Brace 1980). Although laboratory measurements can be made of the permeability of fresh crystalline rocks, it is questionable whether the concept is valid for natural settings. This is because any initial infiltration of fluid results in hydration reactions that decrease permeability further.

FLUIDS AND THE MEANING OF THE BRITTLE-TO-DUCTILE TRANSITION (BDT)

The significance of contrasting fluid regimes in rocks undergoing heating as opposed to cooling is that they have different effects on crustal rheology. Modern work, discussed in Chapter 6, has demonstrated that the transition from brittle to ductile deformation (defined there more rigorously as the fictional-viscous transition or FVT) is not a single plane corresponding to an isotherm, but occurs within a depth interval that fluctuates with time; it is a function primarily (but not exclusively) of temperature, strain rate, effective pressure (lithostatic pressure minus fluid pressure), and grain size. Their work emphasizes the physical role of fluids, but in this section we consider the relation between fluid availability and thermal history in order to understand additional complexities of this zone.

During the prograde part of the metamorphic cycle, rocks are saturated with fluid at near-lithostatic pressure and can therefore deform pervasively by a variety of mechanisms that are not operative in the absence of fluid, including pressure solution and viscous granular flow at low differential stresses, as well as intracrystalline deformation by dislocation creep. During cooling, rocks are dry and therefore very strong. The frictional sliding behavior described by Byerlee's Law will persist to much greater depths and temperatures than is commonly assumed, because dislocation creep, pressure solution, and other deformation mechanisms that are facilitated by water, are not available. That crystalline rocks can rupture in a brittle way in response to very high differential stresses in the deep crust is demonstrated by the formation of eclogite facies pseudotachylite (Austrheim and Boundy 1994). The formation of these eclogites is accompanied by minor hydration, but because the surrounding granulites were dry and contain metastable plagioclase that breaks down under eclogite facies conditions if water is available to catalyze the reactions, the water associated with the pseudotachylites must have been introduced as an integral part of the fracturing that led to frictional melting.

In volumes of crust that are water-saturated, some crystalline quartz-bearing rocks exhibit ductile deformation at temperatures partly below those of the greenschist facies, that is, of the order of 300–400°C (e.g., the Haast Schists of New Zealand; Landis and Coombs 1967). For example, metamorphosed fine-grained mudstones deform ductilely at much lower temperatures, possibly also by viscous granular flow. Under these conditions, the role of water in enabling ductile deformation (by enhancing grain boundary diffusion) may supersede its role in lowering effective normal stress to promote brittle deformation. The brittle-to-ductile transition (BDT) can only be considered to be a response to temperature alone when comparing rocks of similar composition and grain size, that are either uniformly dry or uniformly wet, and that deform at constant strain rate and effective pressure. Field evidence would suggest that, up to melting, dry rocks are relatively strong and sometimes even brittle, irrespective of the conditions of metamorphism. In some cases, high-grade metamorphic rocks show little evidence of deformation after peak metamorphic temperatures were attained, irrespective of the absolute value of the peak temperature.

Hence, fluid-depleted rocks (e.g., granulites from the lower crust) may remain strong throughout their exhumation history, except where and when infiltrated by water during initial embrittlement (Figure 11.4). Since mineral reactions in these environments are expected to be fast, the rate of weakening will likely depend on the rate of fluid infiltration. Furthermore, hydration may leave a legacy of permanent weakness in the form of weak, fine-grained reaction

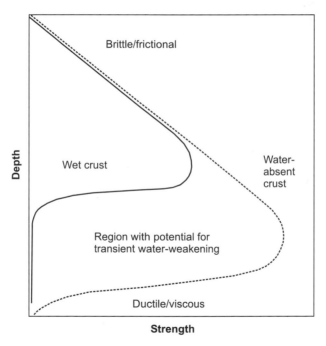

Figure 11.4 Schematic illustration of possible changes in the strength of the crust with depth. Solid curve represents the rheology for wet quartz-bearing rocks, or for a sequence of progressively metamorphosing sediments, for which strength drops rapidly by temperatures ca. 350°C. The broken curve represents dry crust which remains strong to temperatures that may range from 600–750°C according to lithology. For quartz-bearing rocks, large fluctuations in strength are possible at constant depth and temperature, in response to the infiltration or removal of a free water phase.

products (e.g., phyllosilicates, Bos et al. 2000). Therefore, the depth of the BDT is a function especially of fluid activity, as well as of temperature, strain rate and effective pressure, and grain size. Water-weakening may occur transiently in response to fluid infiltration accompanying episodic fracturing, and therefore may be tied to the earthquake cycle (Chapter 1).

Field studies of granulite shield terranes confirm the petrological model of a largely dry and strong lower continental crust. Characteristically, final-stage veins and other late metamorphic features of such granulites display no evidence of ductile deformation, even where (as in western Norway) they have demonstrably experienced prolonged residence in the lower crust before the onset of exhumation (Jamtveit et al. 1990). In other granulite terranes where post-peak ductile deformation is well known (e.g., Ivrea Zone, northern Italy), the fine dynamically recrystallized grain size in dynamically recrystallized quartz and feldspars indicate relatively high differential stress during creep (Zingg et al. 1990). This too is consistent with the notion of a relatively strong lower continental crust, even if stresses there did not attain the fracture strength. Most high-grade shields form impermeable blocks with primary assemblages that are cut by anastomozing fractures and/or shear zones exhibiting coupled deformation and hydration, sometimes also with carbonation. However, these formed during specific orogenic events and the persistence of remnant anhydrous phases away from major conduits suggests that the water supply was limited and transient, with the fluid pockets slowly drying out by absorbing water into minerals of the rock mass between episodes of fluid infiltration.

In summary, whereas rocks are weak during heating and dehydration, they are stronger during cooling, even at high metamorphic temperatures. This difference in rheology can be attributed to the effect of free fluid, which is present in the former case, but not in the latter. During cooling, fluid is restricted locally to fractures, and so only the immediately adjacent rocks are able to experience pressure solution, the growth of fine-grained phyllosilicates, lower effective normal stress, and other water-enhanced effects that are ubiquitous in rocks undergoing continuous dehydration such that fluid is continuously generated throughout the rock mass.

We conclude that the rheological properties of much of the lower continental crust vary considerably in time and space, depending on the transient presence of aqueous fluids. Scale is clearly important. Locally, an individual rock may be dry or wet, and hence strong or weak at different times, but on a larger scale it is the presence or absence of fractures or shear zones containing fluid that dictates the response of the crust. It may be appropriate to consider the effect of hydrated fractures and shear zones on the strength of dry crust as analogous to the effect of dislocations on the strength of an otherwise perfect crystal. If this is the case, large parts of the stable continental crust may contain planes of weakness that have the potential to yield in a ductile manner, despite the bulk of the crust at any time being dry and strong.

FLUID PRESSURE AND FLUID FLUXES IN FAULTS

Although there is ubiquitous evidence of fluid movement in fault zones in the upper crust, it was noted by Sibson et al. (1988) that, if the development of economic gold concentrations is a useful criterion, the largest fluid fluxes are not associated with the largest fault movements but with minor faults, joints, and fractures, including hydrofractures. It may be that the permeability of major fault planes is simply too low to permit extended fluid flow, whereas small-displacement fractures retain unhealed asperities that permit localized flow. The clear implication is that, although fluid flow may be associated with the development of a fault rupture, it continues for an extended period after fault movement. This is in accord with monitoring of stream flows in the vicinity of major earthquakes (Muir Wood and King 1993). Fluid inclusion studies have often documented evidence for transient overpressuring from minerals in veins related to faulting (Parry and Bruhn 1990). This is inconsistent with the extensive circulation of fluid that the existence of the Au-vein itself implies. Fluid inclusions that record evidence for near-lithostatic fluid pressures in veins may have been trapped during intervals between periods of fault-induced fluid flow, where stagnant fluid resided within fractures.

Some of the best-studied examples of extensive fluid flow in association with large-scale faults and shear zones are mesothermal gold-bearing quartz vein systems (Sibson et al. 1988). In summarizing the characteristics of many of these systems, Cox (1999) calculated that fluxes of the order of $10^6 \text{ m}^3 \text{ m}^{-2}$ are required to form deposits of the Eastern Goldfields Province of Western Australia, assuming the transportation of 1–10 ppb Au in solution. It is indicative of the large uncertainties in these types of estimates that recent LA-ICP-MS analyses at Leeds have detected Au at about the 1–3 ppm level in mesothermal gold fluids from Brusson, Italy, in the Western Alps (Yardley, unpublished data). Cox argues that fluid flow requires focusing from a large volume of surrounding rock; at the Golden Mile deposit, he estimates that fluid leached gold from $10,000 \text{ km}^3$ rock into a 2 km^3 volume, and that flow was driven by a near-lithostatic gradient in fluid pressure. This analysis raises serious questions about the sustainability of fluid overpressures in fractured and permeable media. In essence, why should fluid that resides in very low-permeability metamorphic rock with a fluid pressure approximately equal to lithostatic pressure, migrate more than 10 km laterally toward a fracture zone within which similar fluid overpressuring is maintained? Why should it not instead migrate vertically 10 km to the surface? In fact, numeric simulations indicate that rock draining times would be immense for this mechanism to be feasible (Matthäi and Roberts 1996). In the following section, we explore whether the admittedly large fluxes necessary for the formation of these fault-bound vein systems can be generated by the release of metamorphic fluid, as is commonly assumed, and is indeed implicit in the modeling of the fluid pressure as near-lithostatic. It may be necessary to revisit many of the assumptions about these major vein systems.

THE SOURCES OF FLUIDS THAT INFILTRATE FAULTS AND SHEAR ZONES

The identification of the sources of fluids that may play a role in weakening the deep crust is problematic because common geochemical tracers of fluid sources gradually equilibrate with wall rocks. This is because they are present, albeit in small amounts, in the wall rocks that dominate the mass of the fluid–rock interaction system. Only highly conservative tracers are able to retain a memory of the fluid source in a system that is so extensively rock-buffered. Some of these, the noble gases in particular, are of dubious value for the purpose because they may move independently of the aqueous phase, and therefore point to a different source. In addition, once a signature is obtained by a rock, it can be maintained during subsequent metamorphism. For example, an eclogite facies (450–550°C, 20 kbar) shear zone containing whiteschists in a Variscan granite from the Western Alps documents abundant fluid flow with stable isotope signatures indicating the involvement of surface fluids during granite emplacement (Pawlig and Baumgartner 2001). Subsequent metamorphism maintained this geochemical signature even during formation of the shear zone, indicating that, although there may have been a fluid phase present, the flux of water through the shear zone during Tertiary exhumation was too small to reset the oxygen isotope signature of the rocks.

Among the sources of fluids implicated in major fault and shear zone systems are those released by metamorphic reactions deeper in the crust, magmatic fluids, fluids expelled from overthrust lower grade rocks, and surface-derived fluids pumped down into the crust. While in some specific settings the case for a particular origin may be quite clear, in most cases there is only circumstantial evidence for the fluid source.

In the case of the mesothermal gold-bearing quartz veins discussed above, an origin from metamorphic fluids has been widely accepted. These deposits formed late in the metamorphic cycle, when the immediate host rocks were cooling, although it could be argued that metamorphism may have continued at depth. In the absence of magma emplacement, however, regional metamorphic heat fluxes give rise to heating rates of the order of $10°C\ Ma^{-1}$, whereas the dehydration reactions themselves are strongly endothermic. Rates of fluid generation by metamorphism are therefore very slow, in contrast to the remarkably fast rates of formation of these deposits. Ar-dating of the extensive veins systems of the Juneau belt, Alaska, has led to the remarkable result that they formed over a span of ca. 1.5 Ma (Goldfarb et al. 1991). Only a few percent of the total metamorphic fluid production could occur in the period for the deposits to form. Therefore, large fluxes of metamorphic fluid through the fault zone would imply massive and rapid focusing of fluids through very impermeable rock masses into the fault zone. Either the fracture system accessed some kind of deep metamorphic aquifer, or the fluid is not of direct metamorphic origin. The metamorphic fluid paradigm has not been strongly challenged in recent years, but the alternatives at most of these deposits are that either water from a magmatic origin or water from shallow crustal levels (either deep circulating groundwaters, or formation waters from an overlying and possibly overpressured basin) migrated down into the underlying hot basement and were pumped through the evolving fracture network. Such a model removes the constraints on the rate of fluid supply which inhibit the metamorphic model. On the other hand, the hydrogen isotope composition of the fluid at Juneau and many similar deposits appears to point strongly to a metamorphic origin. This is an important controversy that has yet to be resolved.

There are few cases of deep shear zones where a deep fluid origin can be demonstrated unequivocally. Equally, unequivocal examples of a shallow fluid origin, such as the oil-bearing quartz veins described by Munz et al. (2002) are not surprisingly restricted to rather shallow crustal levels. While it is usually assumed that major faults and shear zones act as pathways for the escape of deep fluids, there is no reason why this must be so if the surrounding rocks are dry and act as a fluid sink (Munz et al. 2002; Stober and Bucher 2004). It is possible that the fluids that hydrate major, deep crustal shear zones are derived in part from shallower levels. In detail, it is likely that deep crustal shear zones will prove to be sites of entrainment of fluid from more than one source. Noble gases and carbon dioxide are more likely to be derived from greater depth and at higher temperatures, whereas water, because it fluxes melting so readily, is more likely to have a shallow origin.

It would be especially valuable to be able to trace the origin of fluids associated with processes at the base of thickened continents, but their source is particularly difficult to constrain. Austrheim and Engvik (1997), following Jamtveit et al. (1990), demonstrated through meticulous field and petrological studies that the formation of Caledonian eclogites from Precambrian granulite basement in western Norway was linked to the introduction of aqueous fluids along fractures. Most of the granulites failed to react despite being subjected to very high pressures in the root of a continental collision zone, where eclogite assemblages are more stable. Jackson et al. (2004) argue that this process is occurring beneath the present-day Himalayas, and suggest that fluids derived from the underthrust mantle are fluxing eclogite facies recrystallization in response to earthquake failures in the basal crust today. In practice, however, though the fluids in the example from western Norway are known to be brines, their ultimate origin is not traceable at present.

CONCLUSIONS

We have attempted to summarize what is known about the involvement of fluids in the deep crust and deep crustal processes, and to highlight what remains uncertain or controversial. The rheology of crustal rocks depends strongly (but not exclusively) on a combination of temperature, strain rate, effective pressure, and the availability of aqueous fluids. To a good approximation, however, we can simplify these apparently independent effects. Mineral assemblages that are progressively heated beyond any previous temperature they have experienced will emit fluid pervasively. Fluid pressure will rise concurrently to near-lithostatic values, allowing the fluid to escape. In contrast, mineral assemblages at temperatures lower than those of their formation are a sink for fluid; they will react with any infiltrating water to form hydrous assemblages, driving fluid fugacities down to a few bars. Given the strong dependence of dislocation mobility on the presence of intracrystalline water, it follows that the bulk of the dry lower crust is very strong and may not deform by creep except where water can infiltrate, that is, into fractures and shear zones. What is far from clear is the origin of the water. Extensive flow systems, such as gold-quartz vein systems, are a characteristic of fault systems in the upper crust, and form too quickly to be fed by fluid released during simultaneous metamorphic reactions. Yet, some fluid characteristics point to a deep origin. In this case, pumping of fluid down into dry, deep rocks along faults may play an important role in weakening the crust. There is evidence of fluid involvement in seismic faulting throughout the continents, and a deep origin for the fluid involved in eclogite facies fracturing during subduction in the root of a continent collision also seems likely.

ACKNOWLEDGMENTS

We would like to thank Rainer Abart and Mark Handy for their in-depth reviews of the manuscript. LPB acknowledges the funding of the Swiss National Foundation, which supported part of the above research. A special thanks to Mark Handy for his efforts in organizing the Dahlem Workshop in January, 2005, and to all of the Dahlem staff.

REFERENCES

Austrheim, H., and T.M. Boundy. 1994. Pseudotachylites generated during seismic faulting and eclogitization of the deep crust. *Science* **265**:82–83.

Austrheim, H., and A.K. Engvik. 1997. Fluid transport, deformation, and metamorphism at depth in a collision zone. In: Fluid Flow and Transport in Rocks, ed. B. Jamtveit and B.W.D. Yardley, pp. 123–137. London: Chapman & Hall.

Badertscher, N.P., R. Abart, and M. Burkhard. 2002. Fluid flow pathways along the Glarus overthrust derived from stable and Sr-isotope patterns. *Am. J. Sci.* **302**:517–547.

Borevsky, L., S. Milanovsky, and L. Yakovlev. 1995. Fluid-thermal regime in the crust: Superdeep drilling data. In: Proc. 1995 World Geothermal Congress, ed. E. Barbier, G. Frye, E. Iglesias, and G. Palmason, pp. 975–981. Auckland: Intl. Geothermal Assoc.

Bos, B., C.J. Peach, and C.J. Spiers. 2000. Frictional-viscous flow of simulated fault gouge caused by the combined effects of phyllosilicates and pressure solution. *Tectonophysics* **327**:173–194.

Brace, W.F. 1980. Permeability of crystalline and argillaceous rocks. *Intl. J. Rock Mech. & Mineral Sci.* **17**:241–251.

Campbell, A.R., D.A. Banks, R.S. Phillips, and B.W.D. Yardley. 1995. Geochemistry of Th-U-REE mineralizing magmatic fluids, Capitan Mountains, New Mexico. *Econ. Geol.* **90**:1271–1287.

Carlson, W.D., C. Denison, and R.A. Ketcham. 1995. Controls on the nucleation and growth of porphyroblasts: Kinetics from natural textures and numerical models. *Geol. J.* **30**:207–225.

Carter, N.L., and M.C. Tsenn. 1987. Flow properties of continental lithosphere. *Tectonophysics* **136**:1–26.

Caulfield, J.B.D., A.P. LeHuray, and D.M. Rye. 1986. A review of lead and sulphur isotope investigations of Irish sediment-hosted base metal deposits with new data from the Keel, Ballinalack, Moyvoughly, and Tatestown deposits. In: Geology and Genesis of Mineral Deposits in Ireland, ed. C.J. Andrew, R.W.A. Crowe, S. Finlay, W.M. Pennell, and J.F. Pyne, pp. 591–615. Dublin: Irish Assoc. Econ. Geol.

Chen, W.P., and P. Molnar. 1983. Focal depths of intracontinental and intraplate earthquakes and their implications for the thermal and mechanical properties of the lithosphere. *J. Geophys. Res.* **88**:4183–4214.

Connolly, J.A.D., and Y.Y. Podladchikov. 1998. Compaction-driven fluid flow in viscoelastic rock. *Geodiniamica Acta (Paris)* **11**:55–84.

Cox, S.F. 1999. Deformational controls on the dynamics of fluid flow in mesothermal gold systems. In: Fractures, fluid flow and mineralization, ed. K.J.W. McCaffrey, L. Lonergan, and J.J. Wilkinson, Spec. Publ. 155, pp. 123–140. London: Geol. Soc.

Dell'Angelo, L., and J. Tullis. 1988. Experimental deformation of partially melted granitic aggregates. *J. Metamorphic Geol.* **6**:495–515.

Dublyansky, Y., D. Ford, and V. Reutski. 2001. Traces of epigenetic hydrothermal activity at Yucca Mountain, Nevada: Preliminary data on the fluid inclusion and stable isotope evidence. *Chem. Geol.* **173**:125–149.

ELEKTB Group. 1997. KTB and the electrical conductivity of the crust. *J. Geophys. Res.* **102**:18,289–18,306.

Fisher, D.W. 1978. Rate laws in metamorphism. *Geochim. Cosmochim. Acta* **42**: 1035–1050.

Fisher, Q.J., and R.J. Knipe. 2001. The permeability of faults within siliciclastic petroleum reservoirs of the North Sea and Norwegian Continental Shelf. *Marine & Petrol. Geol.* **18**:1063–1081.

Frost, B.R., and K. Bucher. 1994. Is water responsible for geophysical anomalies in the deep continental crust? A petrological perspective. *Tectonophysics* **231**:293–309.

Frost, B.R., J.A. Mavrogenes, and A.G. Tomkins. 2002. Partial melting of sulfide ore deposits during medium- and high-grade metamorphism. *Canad. Mineral.* **40**:1–18.

Gleeson, S.A., B.W.D. Yardley, I.A. Munz, and A.J. Boyce. 2003. Infiltration of basinal fluids into high-grade basement, South Norway: Sources and behaviour of waters and brine. *Geofluids* **3**:33–48.

Goldfarb, R.J., L.W. Snee, L.D. Miller, and R.J. Newberry. 1991. Rapid dewatering of the crust deduced from ages of mesothermal gold deposits. *Nature* **354**:296–298.

Handy M.R. and J.P. Brun. 2004. Seismicity, structure and strength of the continental lithosphere. *Earth Planet. Sci. Lett.* **223**:427–441.

Hanson, R.B. 1992. Effects of fluid production on fluid flow during regional and contact metamorphism. *J. Metamorphic Geol.* **10**:87–97.

Hay, S.J., J. Hall, G. Simmons, and M.J. Russell. 1988. Sealed microcracks in the Lewisian of NW Scotland: A record of 2 billion years of fluid circulation. *J. Geol. Soc. Lond.* **145**:819–830.

Holness, M.B. 1997. The permeability of non-deforming rock. In: Deformation-enhanced Fluid Transport in the Earth's Crust and Mantle, ed. M.B. Holness, pp. 9–39. London: The Mineralogical Society.

Jackson, J.A., H. Austrheim, D. McKenzie, and K. Priestley. 2004. Metastability, mechanical strength, and the support of mountain belts. *Geology* **32**:625–628.

Jamtveit, B., K. Bucher-Nurminen, and H. Austrheim. 1990. Fluid-controlled eclogitization of granulites in deep crustal shear zones, Bergen Arcs, western Norway. *Contrib. Mineral. & Petrol.* **104**:184–193.

Jin, Z.M., Q. Bai, and D.L. Kohlstedt. 1994. High-temperature creep of olivine crystals from four localities. *Phys. Earth & Planet. Inter.* **82**:55–64.

Jones, A.G. 1987. MT and reflection: An essential combination. *Geophys. J. R. Astron. Soc.* **89**:7–18.

Kamenetsky, M.B., A.V. Sobolev, V.S. Kamenetsky et al. 2004. Kimberlite melts rich in alkali chlorides and carbonates: A potent metasomatic agent in the mantle. *Geology* **32**:845–848.

Kerrick, D.M., A.C. Lasaga, and S.P. Raeburn. 1991. Kinetics of heterogeneous reactions. *Rev. Mineral.* **26**:583–671.

Kohlstedt, D.L.B. Evans, and S.J. Mackwell. 1995. Strength of the lithosphere: Constraints imposed by laboratory experiments. *J. Geophys. Res.* **100**:17,587–17,602.

Landis, C.A, and D.S. Coombs. 1967. Metamorphic belts and orogenesis in southern New Zealand. *Tectonophysics* **4**:501–518.

Lüttge, A., E.W. Bolton, and D.M. Rye. 2004. A kinetic model of metamorphism: An application to siliceous dolomites. *Contrib. Mineral. Petrol.* **146**:546–565.

Maggi, A., J.A. Jackson, D. McKenzie, and K. Priestley. 2000. Earthquake focal depths, effective elastic thickness, and the strength of the continental lithosphere. *Geology* **28**:495–498.

Manning, C.E. 1997. Coupled reaction and flow in subduction zones: Silica metasomatism in the mantle wedge. In: Fluid Flow and Transport in Rocks, ed. B. Jamveit and B.W.D. Yardley, pp. 139–148. New York: Chapman and Hall.

Matthäi, S.K., and S.G. Roberts. 1996. The influence of fault permeability on single-phase fluid flow near fault-sand intersections: Results from steady-state high-resolution models of pressure-driven fluid flow. *AAPG Bulletin* **80**:1763–1779.

McCaig, A.M. 1997. The geochemistry of volatile fluid flow in shear zones. In: Deformation-enhanced Fluid Transport in the Earth's Crust and Mantle, ed. M.B. Holness, pp. 227–266. London: The Mineralogical Society.

Miller, S.A. 2002. Properties of large ruptures and the dynamical influence of fluids on earthquakes and faulting, *J. Geophys. Res.* **107**:536–548.

Miller, S.A., and A. Nur. 2000. Permeability as a toggle switch in fluid-controlled crustal processes. *Earth Planet. Sci. Lett.* **183**:133–146.

Miller S.A., W. van der Zee, and D.L. Olgaard. 2003. A fluid-pressure feedback model of dehydration reactions: Experiments, modelling, and application to subduction zones. *Tectonophysics* **370**:241–251.

Müller, T., L.P. Baumgartner, C.T. Foster, Jr. and T.W. Vennemann. 2004Metastable prograde mineral reactions in contact aureoles. *Geology* **32**:821–824.

Muir-Wood, R., and G.C.P. King. 1993. Hydrological signatures of earthquake strain. *J. Geophys. Res.* **98**:22,035–22,068.

Munz I.A., B.W.D. Yardley, and S.A. Gleeson. 2002. Petroleum infiltration of high-grade basement, South Norway: Pressure-temperature-time-composition (*P-T-t-X*) constraints. *Geofluids* **2**:41–55.

Newton, R.C., and G.E. Manning. 2000. Quartz solubility in H_2O-NaCl and H_2O-CO_2 solutions at deep crust-upper mantle pressures and temperatures: 2–15 kbar and 500–900°C. *Geochim. Cosmochim. Acta* **64**:2993–3005.

Parry, W.T, and R.L. Bruhn. 1990. Fluid pressure transients on seismogenic normal faults. *Tectonophysics* **179**:335–344.

Pawlig, S., and L.P. Baumgartner. 2001. Geochemistry of a talc-kyanite-chloritoid shear zone within the Monte Rosa granite, Val d'Ayas, Italy. *Schweiz. Min. Pet. Mitt.* **81**:329–346.

Putnis, A. 2002. Mineral replacement reactions: from macroscopic observations to microscopic mechanisms. *Mineral. Mag.* **66**:689–708.

Rosenberg, C.L., and M.R. Handy. 2005. Experimental deformation of partially melted granite revisited: Implications for the continental crust. *J. Metamorphic Geol.* **23**:19–28.

Shmulovich, K.I., S.I. Tkachenko, and N.V. Plyaunova. 1995. Phase equilibria in fluid systems at high pressures and temperatures. In: Fluids in the Earth's Crust, ed. K.I. Shmulovich, B.W.D. Yardley, and G. Gonchar, pp. 193–214. London: Chapman & Hall.

Shmulovich, K.I., B.W.D. Yardley, and C.M. Graham. 2006. Solubility of quartz in crustal fluids: Experiments and general equations for salt solutions and H_2O–CO_2 mixtures at 400°–800°C and 0.1–0.9° GPaK. *Geofluids* **6**:154–167, doi:10.1111/j.1468-8123.2006.00140.x.

Sibson, R.H., F. Robert, and K.H. Poulsen. 1988. High-angle reversed faults, fluid pressure cycling, and mesothermal gold deposits. *Geology* **16**:551–555.

Sowerby, J.R., and H. Keppler. 2002. The effect of fluorine, boron, and excess sodium on the critical curve in the albite–H_2O system. *Contrib. Mineral. & Petrol.* **143**:32–37.

Stober, I., and K. Bucher. 2004. Fluid sinks within the Earth's crust. *Geofluids* **4**:143–151.

Thomas, R., J.D. Webster, and W. Heinrich. 2000. Melt inclusions in pegmatite quartz: Complete miscibility between silicate melts and hydrous fluids at low pressure. *Contrib. Mineral. & Petrol.* **139**:394–401.

Upton, P., D. Craw, T.G. Caldwell et al. 2003. Upper crustal fluid flow in the outboard region of the Southern Alps, New Zealand. *Geofluids* **3**:1–12.

Watson, E.B., and J.M. Brenan. 1987. Fluids in the lithosphere. 1. Experimentally-determined wetting characteristics of CO_2-H_2O fluids and their implications for fluid transport, host-rock physical properties, and fluid inclusion formation. *Earth Plan. Sci. Lett.* **85**:497–515.

Yardley, B.W.D. 1981. Effect of cooling on the water content and mechanical behaviour of metamorphosed rocks. *Geology* **9**:405–408.

Yardley, B.W.D. 1986. Fluid migration and veining in the Connemara Schists, Ireland. In: Advances in Physical Geochemistry 5, ed. J.V. Walther and B.J. Wood, pp. 109–131, Berlin: Springer.

Yardley, B.W.D. 2005. Metal concentrations in crustal fluids and their relationship to ore formation. *Econ. Geol.* **100**:613–632.

Yardley B.W.D., and J.W. Valley. 1997. The petrologic case for a dry lower crust. *J. Geophys. Res.* **102**:12,173–12,185.

Zingg, A., M.R. Handy, J.C. Hunziker, and S.M. Schmid. 1990. Tectonometamorphic history of the Ivrea Zone and its relation to the crustal evolution of the Southern Alps. *Tectonophysics* **182**:169–192.

12

Deformation in the Presence of Fluids and Mineral Reactions

Effect of Fracturing and Fluid–Rock Interaction on Seismic Cycles

JEAN-PIERRE GRATIER[1] and FRÉDÉRIC GUEYDAN[2]

[1]LGIT, CNRS-Observatoire, Geosciences, Rue de la Piscine,
Université Joseph Fourier, 38041 Grenoble, France
[2]Geosciences Rennes, Université Rennes 1, Bat. 15, Campus de Beaulieu,
35042 Rennes, France

ABSTRACT

Natural and experimental deformation of fault rocks show that fluid flow and mineral reactions are linked to fracturing in a nonlinear feedback relationship that potentially affects the displacement and stress histories of large faults. These interactions spawn instabilities that are expressed as episodic seismic events involving cataclasis, which alternate with slow, aseismic deformation involving pressure-solution creep, as well as healing and sealing by fluid-assisted mass transfer. This chapter focuses on the timescale of these processes during the earthquake cycle, with special emphasis on the evolution of rheological and transport properties of fault rock during the interseismic period. Fracturing weakens faults dramatically by enhancing the kinetics of pressure-solution creep and of mineral reactions. Therefore, during the postseismic period and initial part of the interseismic period, weakening is faster than fault strengthening by healing and sealing of fractures. During the interseismic period, mass transfer associated with fluid-assisted chemical reactions smoothes asperities on fault surfaces, heals fractures and enhances the formation of a foliation parallel to the fault plane, and decreases permeability. If advective fluid inflow is significant, this can increase pore-fluid pressure and reduce effective shear strength, at least locally within the fault. In the long term, however, the combined effect of fracturing, pressure-solution creep, and sealing is to restore the rheological and transport properties of the fault during the interseismic period, setting the

stage for renewed stress build-up and seismicity. We demonstrate the salient characteristics of fluid-assisted fault weakening and strengthening with a one-dimensional model
of an idealized fault zone undergoing simple shear at constant velocity. The model shows
that the kinetics of the weakening and strengthening processes determine the relative
rates of shear stress decrease and increase during the interseismic period. The kinetics
of dissolution precipitation and mineral reactions are therefore expected to exert an important control on the recurrence time of earthquakes.

INTRODUCTION

Several lines of evidence show that fluids play a key role in the dynamics of
faulting. Fluids are preserved within inclusions in minerals that grew in fault
zones. The isotopic composition of such fluids reveals various origins: meteoric fluids and metamorphic fluids produced by chemical reactions in the lower
crust or mantle (Hickman et al. 1995). In some cases, cyclical fault cementation (e.g., Stel 1981) betrays transient flow along paths which episodically opened
by critical fracturing and closed by crystal growth and compaction processes.
The presence of reactive fluids is also crucial in accommodating deformation
by mass transfer in solution. Pressure solution is the main mechanism competing with cataclasis in the upper crust (Figure 12.1a).

Both mechanisms are well documented in exhumed natural fault rocks. In
the upper crust, deformation accommodated by mass transfer in a solution does
not require high differential stresses. However, chemical reactions in a
nonhydrostatic stress field are generally very slow, resulting in strain rates ranging from 10^{-11} to 10^{-15} s^{-1} (Rutter 1976; Pfiffner and Ramsay 1982). Repro-
ducing such low strain rates in the presence of a fluid in the laboratory is therefore a challenge (Paterson 2001). In contrast, cataclasis can be very fast (strain
rates of 10^{-2} to 10^0 s^{-1}). Such rates are reproducible in the lab, but require
much higher differential stresses. Neither mechanism is very sensitive to temperature; they tend to become subordinate to thermally activated creep at depths
greater than 10–15 km (Figure 12.1a).

Cataclasis and deformation by diffusive mass transfer are not mutually independent processes. Their interaction leads to complex behavior both in time
and in space. Both mechanisms affect weakening and strengthening, but their
characteristic times are very different (Chapter 4). It is convenient to distinguish three characteristic times for the evolution of the mechanical properties
of fault rocks:

Seconds to minutes are characteristic of cataclastic failure associated with
earthquakes. The role of fluids during such short intervals is mostly mechanical (due to change of fluid pressure associated with fluid advection) or catalytic
(melting).

Tens to thousands of years is characteristic for the evolution of the rheological and transport properties of rock during the interseismic period (Chapter 7).

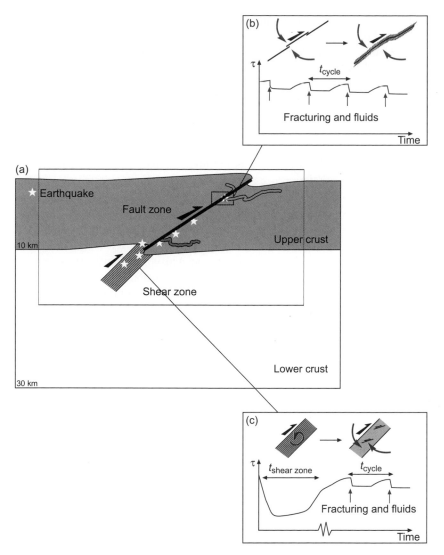

Figure 12.1 Schematic view of seismic and aseismic interactions within the crust. (a) Cataclastic, seismic faulting in the upper crust competes with aseismic deformation (folding, cleavage formation) associated with diffusive mass transfer (pressure solution, mineral reactions). In the lower crust, mostly aseismic deformation (predominantly by dislocation creep, diffusion creep) may be associated with rare earthquakes due to subordinate cataclasis. (b) Individual fault strands in the upper crust coalesce to form mature fault zone networks during successive cycles (stars) and associated inflow of fluids and stress changes (τ). (c) Shear zone networks in the lower crust evolve over longer times, initially in a closed system for fluids. Coseismic ruptures propagate downward into the lower crust, creating pathways for the downward flow of fluids.

In this chapter, we focus on the role of fracturing and fluid–rock interactions within this time interval because these processes are key in the weakening and strengthening of faults in the upper crust. In the following, we show how fracturing speeds up the kinetics of pressure-solution creep and mineral reaction, leading to dramatic weakening as well as to progressive healing and strengthening of faults. To demonstrate these effects we present some simple models which explain how the duration of the interseismic period, and thus the recurrence time of large earthquakes, is directly related to the kinetics of fluid–rock interaction.

Thousand to millions of years is the characteristic time of chemical and mechanical differentiation processes associated with mass transfer (e.g., Robin 1979). The deformation of natural polymineralic rocks leads to chemical and mechanical differentiation at all scales, from a few microns to several kilometers (Gratier 1987). For example, fault zones develop into narrow weak gouge between progressively strengthened damage zones (Figure 12.1b; e.g., Evans and Chester 1995). Long-term weakening associated with reactive mass transfer (Wintsch et al. 1995) is also responsible for the formation of ductile shear zones (Figure 12.1c). The duration of strain localization in ductile shear zones is crudely estimated to range anywhere from a few thousand years (Handy 1989) to tens of millions years (Muller et al. 2000). Placing tighter constraints on the rates of weakening and localization awaits the development of new analytical techniques to date fabric associated with syn-tectonic mineralization events (Chapter 14).

FRACTURING AND FLUID–ROCK INTERACTIONS

Basic Concepts

Thermodynamics: The Driving Forces

Stress affects the chemical potential of a solid by increasing both its molar free energy and surface chemical potential compared to that at zero stress (Gibbs 1877; Paterson 1973). Though Gibbs free energy is not defined for a nonhydrostatically stressed solid, dissolution and precipitation of a solid is commonly described by the surface chemical potential:

$$\Delta\mu = \Delta f + V_s \Delta\sigma_n + \Delta E_s \qquad (12.1)$$

where μ is the chemical potential of the dissolved component, σ_n is the normal stress or the fluid pressure P on the solid, f is the molar Helmholtz free energy, V_s is the molar volume of the solid, E_s is the surface energy $\gamma V_s (1/r + 1R)$, with γ interfacial energy, and r and R are the principal radii of curvature (Kingery et al. 1976). Equation 12.1 characterizes the chemical potential of the solid at each point on the mineral surface. This potential is a function of the normal

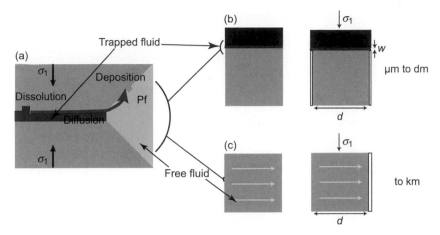

Figure 12.2 Basic concepts of pressure-solution creep: (a) Dissolution at contact between continuous or discontinuous fluid phases and stressed solids, oriented normal to σ_1; the dissolved species are transferred by diffusion within this trapped fluid and redeposited on the free surfaces of the solid at fluid pressure, P_f, or possibly on any surfaces in contact with the fluid and at normal stress less than σ_1. Various creep models are possible depending on the rate-limiting step of this process and on the mechanism of mass transfer. (b) Diffusion along the boundaries of the solid driven by a chemical potential gradient; (c) fluid advection through the solid driven by head gradient $\Delta h / \Delta x$.

stress, which varies from the contact surface (σ_n) to the pore surface (P_f) in Figure 12.2a. The driving force for material transfer along a grain surface is therefore the difference in chemical potentials at the stressed and un- or less-stressed surfaces. The term Δf contains various contributions (Paterson 1973):

$$\Delta f = \Delta E_e + \Delta E_p \tag{12.2}$$

where E_e is the elastic strain energy, and E_p is the plastic strain energy due to dislocations.

The driving force linked to the difference of surface curvature ΔE_s, sometimes called the coarsening potential (Ostwall ripening), includes both the effect of the small particles trapped in the crack and the change in surface curvature of the crack (including changes induced by crack healing). The effect of the difference in normal stresses is most often one or two orders of magnitude greater that of other effects (Paterson 1973; Lehner 1995; Shimizu 1995). However, in cases when the normal stress does not vary along the solid surface (e.g., in cavities under pressure), each of the contributions to Δf and ΔE_s may drive mass transfer. For example, ΔE_s is the driving force associated with self-healing of the crack (Brantley et al. 1990), ΔE_e and/or ΔE_p are responsible for local dissolution near the boundary of an indenter (Tada and Siever 1986).

Kinetics: The Limiting Processes

Pressure solution is a sequence of three processes: dissolution, mass transfer, and deposition. The slowest of these processes imposes its kinetics on the whole deformation process and is termed the rate-limiting step or process. Various creep equations may be written according to which process is rate-limiting, an approach which we have summarized briefly below. However, when the kinetics of the three processes is very similar, the driving force is partitioned subequally between these processes, leading to a relatively complex equation (Raj 1982).

Mass transfer in nature may occur both by diffusion and by advection, although generally advective fluid flow is more competitive at longer length scales (m to km), as discussed in Chapter 11. In this chapter we do not consider in detail the various creep laws based on the assumptions above (e.g., Paterson 2001). If the only driving force for pressure solution is the difference in normal stress between contacting and free surfaces, then creep laws for the deformation of a rock cube in a closed system take the form of the following equations:

(R) $$\dot{\varepsilon} = \alpha k c V_{s} (\Delta\sigma_{n})^{n} / R T d \qquad (12.3)$$

if dissolution or crystallization are the rate-limiting processes (Raj 1982)

(D) $$\dot{\varepsilon} = \beta D w c V_{s} (\Delta\sigma_{n})^{n} / R T d^{3} \qquad (12.4)$$

if diffusion through a stressed fluid phase is the rate-limiting process (Rutter 1976) and

(I) $$\dot{\varepsilon} = \lambda K \Delta c \Delta h / \Delta x d \qquad (12.5)$$

if advection through a porous aggregate is the rate-limiting process (Gratier and Gamond 1990).

In all three equations above, $\dot{\varepsilon}$ is the axial strain rate ($\Delta d / d\Delta t$), α, β, and λ are dimensionless constants that depend on the geometry of the interface, d is the length of one side of the deforming cube and corresponds to the size of the closed system for mass transfer (Figure 12.2b, model R and D; Figure 12.2c, model R and I), t is time, k is the rate of reaction, c is the concentration of solid in solution (models R and D, volumetric ratio), V_{s} is the molar volume of the stressed solid, and Δc (model I and Figure 12.2c) is the difference in concentration of the solid in solution between the zone of dissolution (fluid–mineral interaction through the entire cube of rock) and the zone of deposition (sealed fracture). R is the gas constant, T is temperature in K, K is the permeability coefficient, and $\Delta h / \Delta x$ is the head gradient that drives advective transport.

Three principal models can describe the stressed fluid phase: (a) the thin-fluid film model (Weyl 1959; Rutter 1976); (b) the island-channel model (Raj 1982; Spiers and Schutjens 1990); and (c) the microcracking model (Gratz 1991;

den Brok 1998). Here, we consider a thin-fluid model with diffusion coefficient, D, and thickness of the fluid phase, w, subjected to normal stress, σ_n. The stress exponent, n, is 1 in most pressure-solution creep laws. However, for several reasons this is only true to a first approximation:

- The difference in solubility of the solid between the zone of dissolution and the zone of deposition depends exponentially on the chemical potential (Dewers and Ortoleva 1990).
- When diffusion is rate-controlling and maintains the balance of the diffusive flux out of the fluid film, the normal stress at the center of the contact must be higher than the average stress across the contact (Weyl 1959; Rutter 1976). In this situation, a stress exponent of 1 is only predicted at differential stress values of $\Delta\sigma_n$ less than 30 MPa.
- When dissolution or precipitation is rate-controlling, n reflects the magnitude of the interface velocity versus the driving force relation and is typically 2 for spiral growth/dissolution (two-dimensional nucleation or dissolution on surfaces that are smooth and flat on the atomic scale) and 1 for a rough interface (evolution without a layer-to-layer mechanism of growth or dissolution; Niemeijer and Spiers 2002). Indentation experiments on quartz (Figure 12.3a) confirm this complex nonlinear behavior. For example, an experimental curve $\Delta l/\Delta t = f(\Delta\sigma_n)^n$ can be fitted by a power law with n ranging from 1.3 to 2 (Gratier, unpublished results).

Other creep laws are possible, with each step potentially rate-limiting. Crystallization must also be considered as a possible rate-limiting process, either due to coating of the depositional surface or to the small surface area available for precipitation. In the latter case, crystallization must occur under stress. This is the so-called "force of crystallization" concept that was verified experimentally by Taber (1916) and has rarely been discussed since (Weyl 1959; Means and Li 2001; Hilgers et al. 2004).

The creep laws in Eqs. 12.3, 12.4, and 12.5 operate over a large range of conditions. The most important parameters controlling the relative activity of the mechanisms in these models are strain rate, size of the closed system, temperature and nature of solids and fluids. For example, pressure solution of quartz is controlled by reaction rate at low temperature and by diffusion rate at high temperature (Oelkers et al. 1996; Renard et al. 1997). The nature of solids and fluids is also a key factor. For example, impurity ions in the pore fluid can slow the reaction rate (Zhang et al. 2002) and consequently control mass transfer rates in some natural systems. Special attention should be paid to the relationship between strain rate, $\dot{\varepsilon}$ and the size of the closed system, d, which varies from $1/d$ for the I and R laws, to $1/d^3$ for the D law. The size of the closed system also varies with the mass transfer process: micrometer to decimeter for diffusion transfer (model D) compared to hundreds to thousands of meters for advective transfer (model I, see also Chapter 11).

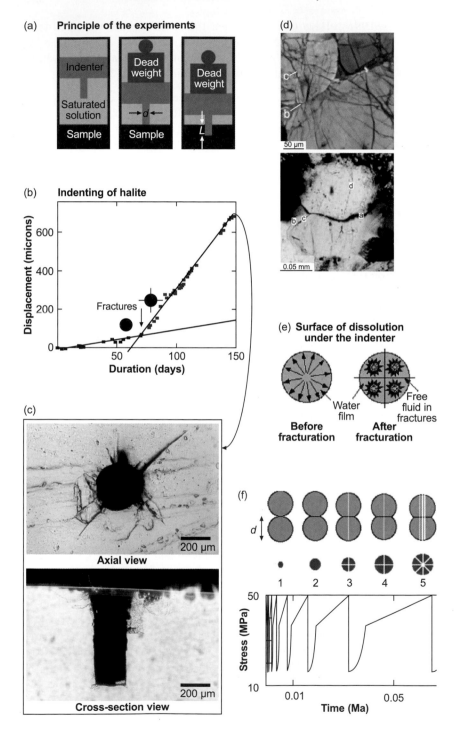

(a) **Principle of the experiments**

Indenter

Saturated solution

Sample

Dead weight

Sample

d

Dead weight

L

(b) **Indenting of halite**

Displacement (microns)

Duration (days)

Fractures

(c)

Axial view

200 µm

Cross-section view

200 µm

(d)

50 µm

0.05 mm

(e) **Surface of dissolution under the indenter**

Water film

Before fracturation

Free fluid in fractures

After fracturation

(f)

d

1 2 3 4 5

Stress (MPa)

Time (Ma)

Evaluating the validity of the creep laws above under natural conditions is not easy. The characteristic times for diffusion, transport, and precipitation on a given length scale are several orders of magnitude longer than the human life span. Therefore, the kinetics of these processes in the laboratory must be accelerated so that experimentalists can outlive their experiments. This can be achieved in one of several ways, each with its attendant problems:

- Solubility can be increased, but not without inducing possible chemical side effects.
- Temperature can be increased but only at the risk of favoring solid-state creep mechanisms with a higher activation energy than that of pressure solution (e.g., Nabarro-Herring creep, a mechanism that is otherwise unlikely to operate in fault rocks, Rutter 1983).
- Stress can be increased at the risk of inducing intracrystalline plasticity (dislocation glide plus climb) or cataclasis.

Taken together, these problems raise the question of how to model the rheological behavior of large volumes of rock with constitutive equations that are derived only from laboratory specimens at unnaturally high strain rates. Possible solutions to the problem of scaling are discussed by Paterson (2001) and touched on in Chapter 1.

Observations from both experiments and nature show that microfracturing of minerals can drastically increase the kinetics of pressure solution. Because microfracturing occurs suddenly, pressure solution favors transient creep.

Evidence from Experiments

In the experimental set-up in Figure 12.3a, a stainless steel cylindrical indenter is mounted beneath a Teflon piston in contact with a crystal of halite in its brine. Before the indenter is placed in contact with the sample, the crystal is immersed in brine that was previously saturated with halite powder at the temperature of the experiment. This is done to trap a saturated fluid phase beneath

---

◄ **Figure 12.3** Effect of fracturing on the kinetics of pressure-solution creep. (a) Configuration of the indenter experiment; (b) Displacement of indenter in halite with time (note the drastic increase of displacement rate associated with radial fracturing at day 71). (c) Axial view (top) and cross-sectional view (bottom) of the radial fractures that developed on day 71. (d) Natural fractures at the contact between two stressed solids that underwent pressure-solution indentation; (e) Axial view showing reduction of mean distance of diffusive mass transfer within trapped fluids from (d) (before fracturing) to less than $d/2$ (after fracturing). (f) Unstable behavior caused by competition of fracturing and fracture-induced increase in dissolution surface when diffusion is rate-limiting at constant displacement rate (adapted from Gratier et al. 1999).

the indenter before the indenter exerts stress. In this example, the dead weight induces a stress of about 16 MPa on the indenter. The device was maintained at a temperature of 25°C. For the 150-day duration of the experiment, the displacement of the indenter was registered by following a reference line on the piston under a microscope (Figure 12.3b). The result was rather surprising: the displacement rate of the indenter was constant for the first 71 days (less than 1 micron per day) before suddenly accelerating by a factor of 8. This increased displacement rate persisted for the remaining 77 days of the experiment. After the sample was polished, the depth of the hole created by dissolution of the halite beneath the indenter matched the measured displacement (Figure 12.3c). Other than this cylindrical hole, the only change in the structure of the sample occurred on day 71, when the displacement rate suddenly increased and radial fractures developed in the halite beneath the indenter, as seen in Figure 12.3c. Previous experiments conducted with a similar set-up (Hickman and Evans 1991, 1995; de Meer et al. 2002; Dysthe et al. 2003) did not show any fractures developed during indentation.

The effect of fracturing on the displacement rate in our experiments is obvious and irreversible. We interpreted fracturing to have augmented the rate of diffusive mass transfer along the contact between the indenter and halite (Gratier et al. 1999). Without fracturing, the displacement rate ($\Delta d / \Delta t$) is controlled by the rate of mass transfer out of the thin-fluid film trapped beneath the indenter, and is inversely proportional to the square of the diameter, d, of the indenter (model D, Eq. 12.4). Within this trapped fluid phase (Figure 12.3b, before fracturing), the product Dw (m^3 s^{-1}) is about 5×10^{-19} m^3 s^{-1}, which is in good agreement with values from previous work (Hickman and Evans 1995; de Meer et al. 2002). Radial fractures that are longer than the diameter of the indenter and several microns wide form paths filled with a free fluid that facilitates the fast removal of material away from the contact area. There, the product of the diffusion coefficient (2×10^{-9} m^2 s^{-1}) and the width of the mass transfer path (2×10^{-6} m) is about 4×10^{-15} m^3 s^{-1}, that is, about 8×10^3 higher than within the thin-fluid phase otherwise trapped beneath the indenter. After fracturing, the initially thin-fluid film is distributed among several smaller domains, with each domain bounded by fast diffusive paths (Figure 12.3e). Therefore, fracturing renders the displacement rate inversely proportional to the square of the mean size of the small domains bounded by the radial fractures. This explains the sudden increase of the displacement rate, as pressure-solution indentation is diffusion controlled. The ratio between the indenter diameter and the mean size of the fractured domains is about 2.8.

A significant increase in the rate of pressure solution was previously attributed to subcritical microcrack growth (Gratz 1991; den Brok 1998) at grain boundaries, where channels along closely spaced microcracks intersect the dissolution surface. The location of these channels changes continuously as microcracking progresses (dynamic channel island model). Therefore, microcracking and its effects on diffusivity were considered to be integral parts of steady-state pressure-solution creep.

In contrast, we propose that creep is unsteady, because fracturing is able to connect the fluid phase under stress with the free fluid located around the contact. This short cut in the diffusional mass transfer path is responsible for the sudden increase of the creep rate. We suggest that transgranular fracturing during earthquakes drastically enhances pressure solution and weakens rocks after seismic events.

Fluid-filled microfractures that potentially act as shortcuts in the diffusional path have also been proposed to account for high rates of measured diffusion (Farver and Yund 1998). However, fast healing may alleviate such shortcuts. Note that at the conditions of our experiment the fractures do not heal, but remain open paths of fast diffusion throughout the experiment. However, healing and sealing are very common processes in both experimental and natural deformation, and strength recovery is a key process that competes with dissolution. This will be discussed in the following section in the context of fault healing and strengthening.

Applications to Nature

Fractured Grains or Pebbles

The idea that fracturing enhances pressure solution has several applications in natural deformation. Pitted pebbles have long been attributed to pressure solution (Sorby 1865; McEwen 1978, Figure 12.3d). However, the large size of the dissolution areas (cm) is incompatible with the known duration of deformation and is certainly inconsistent with experimentally derived parameters for pressure solution. Only the development of fractures at stressed contacts explains the relatively high strain rate inferred for the large pits. A numerical model of the complex interaction between pressure solution and fracturing (Gratier et al. 1999) explains why these two mechanisms are so often associated in nature. For example, a cyclic stress-time function at constant displacement rate is derived by taking into account the ratio of stress to the diameter of dissolution (σ/d^2). Without any fracture, the contact area increases and stress must increase in order to maintain a constant displacement rate. However, stress cannot increase infinitely, otherwise the pebble will fracture. If fracturing does occur, then the dissolution contact area is broken up into smaller domains, reducing the mean distance of diffusion that controls the kinetics of mass transfer (Figure 12.3f). Therefore, the stress needed to maintain a constant displacement rate is drastically reduced. This model yields a mean viscosity of 3×10^{21} Pa s, but the viscosity and the stress values evolve with time (Figure 12.3f).

Fracturing and Reaction Weakening

Reaction weakening involves the nucleation and growth of reaction products that are weaker than the reactants. Reaction weakening assisted by micro-fracturing and diffusive mass transport in fluids is common during shear zone formation in the lithosphere (Wintsch et al. 1995; Handy and Stunitz 2002). Fracturing creates the space necessary for the infiltration of fluids that are required for the reaction to take place. Figure 12.4 shows three examples described below:

1. Midcrustal shear zones often consist of fine-grained mylonites within granitoid host rocks, typical of continental crust (FitzGerald and Stünitz 1993). At greenschist-facies conditions (depths of 10–20 km), quartz deforms predominantly by dislocation creep while feldspar fractures (Simpson 1985). The transformation of feldspar to very fine-grained white mica, albite, and clinozoisite is inferred to weaken the rocks (e.g., Mitra 1978; White and Knipe 1978; Dixon and Williams 1983; Gapais 1989; Wibberley 1999). Coeval syntectonic feldspar fracturing and reaction weakening within the Tenda Shear Zone of eastern Corsica (Gueydan et al. 2003; Figure 12.4) affected granitoids that can be considered as homogeneous at the regional scale. The modal amount of white mica increases with strain and reaches a maximum value of 50%. At midcrustal depths, the transformation of feldspar to white mica requires the addition of water. The formation of fine-grained white mica weakens the material and so favors strain localization, ultimately within meter- to kilometer-wide shear zones.

2. Handy and Stünitz (2002) showed that fracturing and reaction weakening are responsible for the formation of shear zones in the upper part of the lithospheric mantle during Early Mesozoic rifting of the Adriatic continental margin. During extensional unroofing of these mantle rocks, fracturing triggered the progressive replacement of olivine, clinopyroxene, orthopyroxene, and spinel by a hydrous, lower-pressure assemblage of olivine, plagioclase, and hornblende. This reaction, which required fracturing and fluids, weakened the material by at least an order of magnitude (stage 1, Figure 12.4) as inferred from microstructural evidence for viscous granular flow probably accommodated by diffusional mass transfer along grain boundaries of the very fine-grained syntectonic reaction products. The upper mantle was thus inferred to evolve from a high-strength layer into a low-viscosity detachment layer that accommodated significant extensional deformation within the rifted margin.

3. The transformation of granulite to eclogite also involves fracturing, fluid flux, and syntectonic reaction, and is inferred to have induced pronounced weakening and strain localization. Klaper (1990), Austrheim and Boundy (1994), and Jolivet et al. (2005) showed that during Caledonian subduction of nominally dry granulites rock in the Western Gneiss Region (Norway), metamorphic reactions were delayed until fracturing allowed fluid infiltration. The occurrence of pseudotachylites in some of these eclogitized fractures

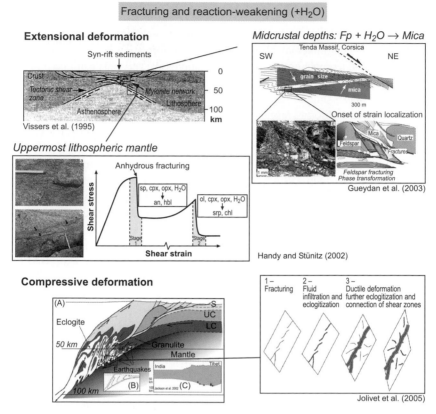

Figure 12.4   Three examples of fracturing followed by fluid influx that triggered reaction-weakening and strain localization within the lithosphere. At midcrustal depths, feldspar fracturing and reaction-softening (feldspar to mica transformation) lead to the formation of kilometer-scale extensional shear zone in the Tenda Massif, Corsica (Gueydan et al. 2003). In the uppermost mantle, fracturing and phase transformations are coeval with the formation of extensional shear zones in ultramafic rock (Handy and Stünitz 2002). Localized deformation zones (a) and (b) define a mylonitic shear zone network that accommodates extension at depths, as depicted by Vissers et al. (1995). During continental subduction (lower part), seismic fracturing of granulites followed by fluid influx and eclogitization facilitate strain localization (Austrheim and Boundy 1994; Jolivet et al. 2005). The exhumation of deep crustal units within the subduction channel may have occurred along eclogitic shear zones.

indicates that embrittlement was seismic and localized subsequent ductile deformation. In fact eclogitization was spurred by further deformation and fluid infiltration that resulted in the formation of isolated blocks of granulite separated by eclogitic shear zones. These blocks are inferred to have behaved like rigid inclusions between the shear zones.

Dissolution around a Fault

To accommodate large displacements, the country rock between and adjacent to faults must undergo internal deformation. This deformation is a geometrical necessity near the termination of a fault or at compressive and tensile bridges linking fault arrays. On a much smaller scale, deformation also occurs around asperities on the fault planes (Figure 12.1b). Exhumed fault segments of various seismically active faults in California reveal that pressure solution is clearly associated with fracturing through the entire thickness of the upper crust: at shallow depths, calcite is more mobile than quartz, whereas the reverse is true at greater depth (Gratier et al. 2003). These authors suggested that coseismic microfracturing enhances pressure-solution creep and may explain postseismic creep. Transient values of viscosity are expected to be high during postseismic creep; certainly greater than the mean viscosity values typical of interseismic creep (Hickman et al. 1995; Chapter 14, this volume). For example, in the same region, carbonate-rich rocks deformed at about 2 km depth within the Little Pine fault zone (California) yield strain values of 30%, and reasonable estimates of the differential stress (20 MPa) and the duration of deformation (4 Ma) indicate that the mean viscosity was about 8×10^{21} Pa s. Modeled values of effective postseismic viscosity have been evaluated assuming pressure solution of calcite in the upper part of the crust and of quartz in the lower part (see Figure 12.8f below). Postseismic viscosity is strongly dependent on the fracture density, which is not easy to evaluate. However, for a mean fracture spacing associated with each seismic episode of 200 microns, and using the creep equations above (Renard et al. 2000) the effective viscosity just after the earthquake may have been about 6×10^{18} Pa s (Gratier et al. 2003).

The evolution of viscosity is more complex as cracks and fluid advection paths evolve with time, and as fractures are progressively sealed. In fact, viscosity probably decreases exponentially with time according to the relation of Renard et al. (2000), as explained below. Strain rates in active fault zones vary due to cyclic fracturing, fluid fluxes, and reaction (Knipe and Wintsch 1985). Pressure-solution creep enhanced by microfracturing may account for some examples of aseismic displacement registered by continuously operated GPS stations arrayed on active faults (Bokelmann and Kovach 2003).

Finally, we note that dissolution in polymineralic rocks can lead to selective dissolution and removal of soluble phases, resulting in the development of tectonic layering (e.g., solution cleavage) that is initially oriented perpendicular to the direction of maximum stress (Cosgrove 1976; de Boer 1977; Robin 1978; Gratier 1987). Fletcher and Pollard (1981) have modeled this evolution as the propagation of a zone of negative dilation (a so-called "anti-crack") around weak or depleted inclusions. The nucleation and growth of platy minerals (e.g., micas) perpendicular to the direction of maximum stress forms a schistosity. The progressive development of a mechanical anisotropy that can rotate during shearing affects the rheological and transportational properties of fault rocks.

Dissolution of Faults' Asperities

Asperities on irregular fault surfaces hinder displacement, but can also change their shape by stress-induced dissolution (Figures 12.5a, b) thereby easing deformation. In Figure 12.5d, dissolution occurs on the surface that prevents sliding (for simplicity the surface is oriented perpendicular to the maximum

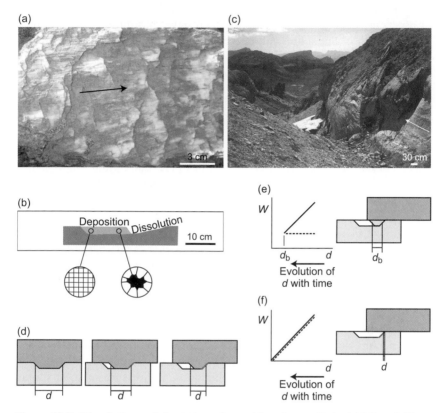

Figure 12.5 Dissolution and fracturing of asperities along a fault. (a) Mineral fibers within voids created by slow dissolution of asperities on a fault. (b) Detail of the process based on natural measurements showing the dissolution surface with directional roughness opposite to fault displacement. Mineral fibers formed by successive crack-seal events (left) attest to aseismic sliding controlled by the kinetics of pressure solution. Alternatively, cataclastic sliding was episodic, as evidenced by some large openings sealed by euhedral crystals (right). (c) Exhumed fault surface showing threading of fault parallel to fault displacement (arrows) and the various asperities at all scales. (d) Schematic evolution of dissolution deposition at the scale of a single asperity. (e) Evolution of energy (W) needed to break (continuous line) or to dissolve (dotted lines, dissolution models I or R) an asperity of decreasing size d. (f) Energy (W) needed to break (continuous line) or dissolve (dotted lines, dissolution model D) an asperity of decreasing size d. Adapted from Gratier and Gamond (1990).

stress), whereas deposition occurs in the void created by aseismic fault displacement. This in turn leads to the formation of mineral fibers on fault surfaces (Figures 12.5a, b; Gratier and Gamond 1990). The dissolution of asperities may reduce fault strength. The frictional shear strength of the fault is considered here to be the yield stress needed to fracture a large part of the fault. Its shear strength may be considered to be proportional to the cumulative length of the asperities along the fault. If the total length of asperities decreases due to their progressive dissolution, the frictional shear strength of the entire fault decreases with time. The rate of this weakening can be expressed as the relationship between the rate of dissolution and the length of the asperities, according to the theoretical relations in Eqs. 12.3–12.5 for the different rate-limiting processes. For the simple geometry adopted in Figure 12.5d, the sliding rate, d, is related to the displacement rate, $\Delta d / \Delta t$, by a numerical coefficient. In some relations (Eqs. 12.3 and 12.5), respectively, for models I and R), the sliding rate does not depend on the distance of mass transfer, d, from dissolution to deposition zones. In the other relation (Eq. 12.4, model D), the sliding rate depends on the inverse of the square of the distance of mass transfer, d. This distance, d, is the asperity length for dissolution of only one side of the fault. Because d also depends on the crystal growth mechanism, it is assumed here that each growth increment occurs at the vein-wall contact. The relation between stress and the asperity length is related to the change in energy consumed during sliding to an extent dependent on the creep law. On the other hand, the energy needed to break asperities always depends on the asperity length. Thus, the energies needed to accommodate cataclastic flow and pressure solution sliding are comparable (Figures 12.5e and 12.5f). At the scale of a single asperity, stable sliding is expected when the two energies vary similarly (e.g., pressure-solution sliding), whereas unstable sliding is expected when the two energies vary differently with successive mechanisms, thereby minimizing overall strain energy dissipation (e.g., pressure-solution sliding then cataclasis). The behavior of all asperities on a fault is more complex (Bos et al. 2000) and needs to be modeled numerically. In any case, considering all asperities, pressure solution reduces the shear strength of the fault by progressively reducing the total length of the asperities.

A main objective for future work will be to establish the geometrical evolution of asperities and their sliding mechanism (Nadeau and Johnson 1998; Sammis and Rice 1998). One way to do this is to compare the true geometry of a fault surface (Figure 12.5c) with that of threaded surfaces (Thibaut et al. 1996). A threaded surface is an uneven surface that allows the sliding of two rigid blocks without any deformation (like the surfaces of matching nuts and bolts). The deviation of real fault surfaces from that of ideally threaded surfaces may reveal the true geometry of fault asperities.

FAULT STRENGTHENING

Pervasive Strengthening

Numerous observations of natural deformation show that deposition selectively strengthens rocks. For example, quartz and calcite mineral deposits in pressure shadows of stronger minerals indent their slaty matrix by developing dissolution haloes (Figure 12.6a). Another example is "bamboo-like structures" created by boudins that selectively strengthened the boudinaged layer (Figure 12.6b). The same behavior was observed in experimentally deformed polymineralic aggregates with soluble and insoluble minerals (Zubtsov et al. 2004). At room temperature, the compaction rate of a polymineralic aggregate was significantly greater than that of a monomineralic rock comprising only a single soluble species. This may reflect the fact that the grains of monomineralic rocks grew together rather than dissolving under stress. The presence of another mineral phase favors dissolution along mutual contacts between phases with contrasting solubilities. Hickmann and Evans (1991) also found that dissolution at halite–silica interfaces (silica being insoluble) was much faster than at halite–halite interfaces, which tended to grow together. In all the natural examples (Figures 12.4 and 12.6), strengthening linked to mineral precipitation competed with weakening linked to the concentration of less-soluble micas. Deformation with mass transfer leads to a progressive segregation of rocks that induces rock heterogeneities (mass transfer from weakened zones of dissolution to strengthened zones of deposition). Therefore, rock viscosity evolves with space and time, with viscosity contrasts generally increasing to an extent dependent on the size of the closed system. The size of this closed system and the characteristic size of the induced heterogeneities mostly depend on the mechanism and scale of mass transfer: diffusion on length scales of microns to decimeters (e.g., tectonic layering) or advection on scales of hundreds to thousands of meters (e.g., ore deposits).

An atypical example where the degrees of strengthening and weakening are comparable comes from silicic volcanics of Vendée (France) affected by fluid-assisted deformation (Le Hébel et al. 2002). Deformation of the quartz-K-feldspar-phengite volcanic rock (Figure 12.6c) occurred at about 400°C and 4–5 kb. The deformation involved diffusion-driven dissolution–crystallization of quartz and feldspar, with the phengite behaving as a residual, relatively insoluble phase. Increasing strain is marked by the development of alternating mica-enriched layers (sources) and quartz-feldspar (sinks), as shown in Figure 12.6d. The former are expected to have weakened as mica content increased, whereas the latter underwent microcracking (Figure 12.6e). This process partitions strain, with very limited localization in the micaceous layers. Isotopic analysis revealed limited fluid–rock ratios, showing that the fluid originated locally and that the system remained closed during ductile deformation. No transient fluid

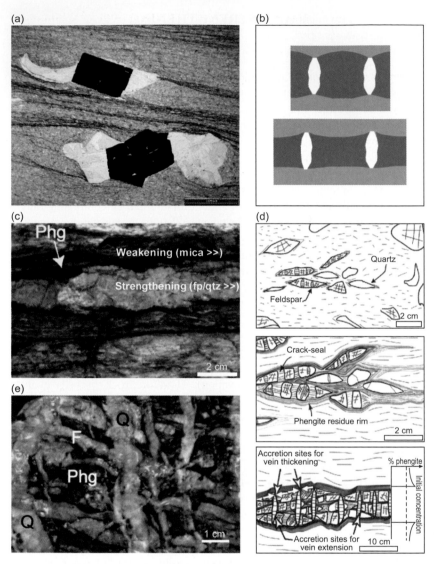

Figure 12.6 Three examples of inferred strengthening during high-strain deformation: (a) pressure shadow stronger than the slaty matrix; (b) quartz and calcite deposit between boudins (top) stronger than the boudins themselves (bottom), as sketched from a natural example. (c) to (e) Examples taken from Porphyroïds units (Vendée, France) of fluid-assisted crack-seal deformation (Le Hebel et al. 2002). (c) Porphyroïds marked by alternating quartzo-feldspathic layers rimmed by phengitic layers; (d) layering resulting from crack sealing (Le Hebel et al. 2002); pervasive fracturing of strong quartzo-feldspathic layers is coeval with dissolution of feldspar and quartz in the matrix, leading to the development of phengitic residues. K-rich fluids seal the numerous cracks, inducing growth of quartzo-feldspathic layers. (e) Detailed view of the pervasive fracturing of the quartzo-feldspathic veins.

fluxes induced by microcracking were discerned. The stability of K-feldspar is consistent with the absence of external fluids under greenschist-facies conditions. In this case, the absence of transient fluid fluxes precludes weakening at a regional scale and favors slight strengthening, as marked by numerous cracks. If the system was indeed closed, then the amount of dissolved material (quartz and K-feldspar) is equal to the amount of precipitated material.

This behavior contrasts with the large-scale evolution of zones of quartz deposition within ore deposits. Where large amounts of quartz are transported from depth by fluid flow along active faults (Sibson et al. 1988), the zone of quartz deposition appears be significantly stronger than the surrounding metamorphic country rocks.

Fault Healing and Sealing

Evidence for postseismic strength recovery comes from geophysical studies. For example, Li et al. (2003) described fault rocks from the Hector Mine rupture zone where P- and S-wave velocities increased, respectively, by 0.7%–1.4% and 0.5%–1.0% between 2000 and 2001. In contrast, velocities in surrounding rocks increased much less, indicating that the Hector Mine rupture zone healed and strengthened after the main shock, most likely as cracks that had opened during the 1999 earthquake closed up again and were sealed. The recovery of fault zone strength is consistent with a decrease in apparent crack density of 1.5% within the rupture zone. The ratio of travel time decreases for P- and S-waves was 0.72, suggesting that the cracks near the fault were partially filled with fluids. This restrengthening is similar to that observed after the 1992 M_L 7.4 Landers earthquake some 25 km to the west (Li and Vidale 2001).

The same observation can be made in experiments on faulted sandstones that were heated and subsequently re-deformed (Blanpied et al. 1995; Tenthorey et al. 2003). They show a strength recovery of 75% after 6 h of heating at 927°C. In the most extreme case, hydrothermally induced gouge compaction, cementation and crack healing resulted in 75% strength recovery after treatment at 6 h at this very high temperature. Isostatic hydrothermal treatment also resulted in dramatic reductions in porosity and permeability.

The question is thus how, and at what rate, do fractured rocks recover their strength after an earthquake? The nature of the minerals that constitute the country rocks is very important because fluid–rock interactions play a crucial role in the rheology of fault zones. We do not consider all the possibilities here, but just cite the examples of quartz and calcite, both of which are common mobile minerals in Earth's crust. The relative mobility of quartz and calcite clearly changes with depth due to differing pressure and temperature dependences of their solubilities. At shallow depths, calcite is more abundant and mobile than quartz, whereas the reverse is true at greater depth. This variation is related to two effects: first, the solubility of quartz increases with increasing temperature whereas calcite solubility decreases with increasing temperature; second, at low temperature, the kinetics of quartz dissolution is very slow, precluding significant pressure solution (Oelkers et al. 1996; Renard et al. 1997). Note that the situation is probably much more complex because temperature may have opposite effects on pressure-solution kinetics and solubility, at least in the case of calcite (Rutter 1976). Also, reaction kinetics is very sensitive to impurities in the pore fluid (Zhang et al. 2002) and this effect may outweigh the effect of temperature in nature.

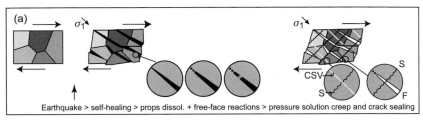

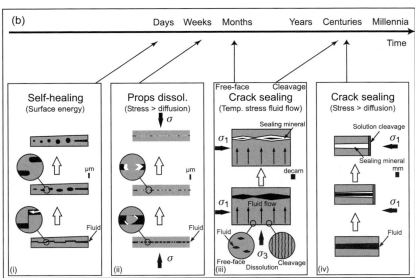

Figure 12.7 (a) Schematic view of postseismic sealing of a fault zone. The porosity (black) linked to the fracturing process is progressively sealed by a mineral deposit (white). Dissolution along stylolites or solution cleavage (S) is triggered by fracturing (F) and is associated with aseismic postseismic deformation that relaxes stress. Solution cleavage (S) and associated postseismic crack-sealed veins (CSV) also accommodate part of the strain aseismically for the duration of the interseismic period. (b) The characteristic times of sealing vary from days to millennia, with self healing driven by surface energy (i), dissolution of props driven by stress and a decrease in fluid pressure (ii), crack sealing with fluid infiltration controlled either by reactions on free faces or by diffusion along solution cleavage (iii), and crack sealing driven by stress with mass transfer from solution cleavage to fracture (iv). Figure adapted from Gratier et al. (2003).

Various mechanisms can lead to healing or sealing of rocks, as summarized in Figure 12.7. Here, we consider only the mechanisms that occur at the characteristic time of interseismic periods. Changes of fluid pressure that occur during, or just after, the main rupture with their mechanical consequences (i.e., dilational hardening, triggering of earthquakes) are discussed elsewhere (e.g., Miller 2002; Miller et al. 2003; Chapter 14, this volume).

Based on observations of natural structures (Gratier et al. 2003), the following succession of crack-sealing processes is expected in active fault zones during interseismic periods (Figure 12.7a). Initially, seismic rupture increases the overall permeability and reduces fluid pressures to near-hydrostatic values within the fault zone. This favors an increase in the reaction rates for the reasons outlined above. Self healing of the fractured minerals and some metamorphic reactions are relevant to this stage. Crack healing is driven by a decrease in microcrack surface energy (Figure 12.7b, i). No input of external material is required at this point. Solid–solid contacts are required along the fracture in order to facilitate mass transfer from sites with maximum wall curvature to sites with minimum curvature (Figure 12.7b, i). Because diffusive transfer occurs in a free (unstressed) fluid, this process is usually probably controlled by the kinetics of the interface reaction and may be rather fast (days to weeks; Brantley et al. 1990). Evidence of this process is common in naturally deformed rocks. Also relevant to this stage are free-face metamorphic reactions that are activated by the advective inflow of fluids in disequilibrium with the minerals lining the fault. This could lead either to dissolution or to deposition that is potentially rather fast (days to years) in the case of reaction on free faces around pores or voids (Figure 12.7b, iii). On the other hand, if dissolution occurs on a stressed surface, reaction kinetics is usually controlled by diffusion (i.e., sealing of large aperture cracks, Figure 12.7b, iv).

Other reactions may progress at a slower rate. Beeler and Hickman (2004) proposed that crack closure may be controlled by the dissolution of asperities and microfractured grains that prop the fracture open (Figure 12.7b, ii). This type of self-healing process is assisted by stress and requires a decrease in fluid pressure to operate effectively. A fluid pressure drop may explain the observation that dissolution pits occur in all the directions during gouge compaction (see Figure 12.8b). If this drop is sufficiently rapid, for example, during an earthquake, the rocks will tend to collapse. The experiments of Elias and Hajash (1992) in which grains or fractured rock are embedded in a "soft tube" show such pressure-solution compaction. A decrease of the compaction rate is expected with time when diffusion is rate-controlling (Figure 12.3).

The sealing of large aperture cracks (10 μm to mm–cm, Figure 12.7b, iv) combined with the large opening of the separated walls of the fracture (i.e., no contact of walls across the fracture) requires an influx of material into the cracks from outside sources. As pointed out above, long-range advective transport of dissolved solid (e.g., involving flushing by strongly oversaturated solution as

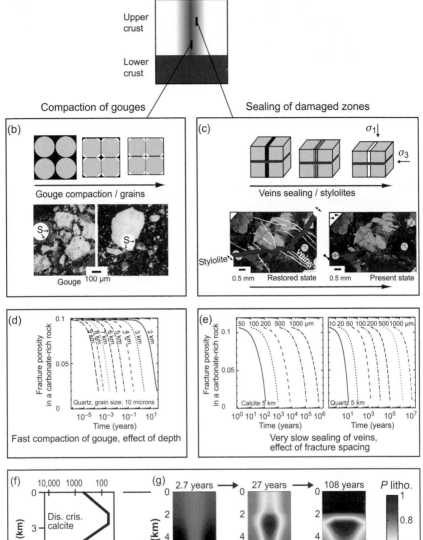

(a) Fluids in fault zones

Upper crust

Lower crust

Compaction of gouges

Sealing of damaged zones

(b)

Gouge compaction / grains

S→

S→

Gouge 100 µm

(c)

σ₁

σ₃

Veins sealing / stylolites

Stylolite

veins

0.5 mm   Restored state   0.5 mm   Present state

(d) Fast compaction of gouge, effect of depth

Fracture porosity in a carbonate-rich rock

0.1

0.05

0

9 km 8 km 7 km 6 km 5 km 4 km 3 km 2 km

Quartz, grain size: 10 microns

10^{-5} 10^{-3} 10^{-1} 10^{1}

Time (years)

(e) Very slow sealing of veins, effect of fracture spacing

Fracture porosity in a carbonate-rich rock

0.1

0.05

0

50 100 200 500 1000 µm

Calcite 5 km

10^0 10^1 10^2 10^3 10^4 10^5 10^6

Time (years)

10 20 50 100 200 500 1000 µm

Quartz 5 km

10^1 10^3 10^5 10^7

Time (years)

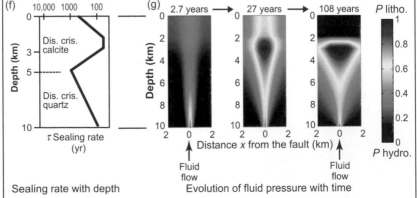

(f) Sealing rate with depth

10,000 1000 100

Depth (km)

0

3

5

10

Dis. cris. calcite

Dis. cris. quartz

τ Sealing rate (yr)

(g) Evolution of fluid pressure with time

2.7 years → 27 years → 108 years

Depth (km)

0 2 4 6 8 10

2 0 2   2 0 2   2 0 2

Distance x from the fault (km)

Fluid flow   Fluid flow

P litho.

1
0.8
0.6
0.4
0.2
0

P hydro.

shown in Figure 12.7b, iii) may contribute to fracture sealing (Fyfe et al. 1978; Etheridge et al. 1984). Inflow of supersaturated fluid is documented both by stable isotopes studies (Kennedy et al. 1997) and by observations of typical mineral reactions in fault gouge (Evans and Chester 1995). This contributes to a decrease in porosity during the interseismic period, mainly in the gouge. However, it has minor effects on the sealing of nearby fractured country rocks (Gratier et al. 2003).

Following initial fast sealing (e.g., involving self-healing [Figure 12.7b, i] or by free-face metamorphic reactions [Figure 12.7b, iii]), which most often does not eliminate porosity (Figure 12.7a), crack sealing involves diffusive mass transfer (e.g., Ramsay 1980; Gratier and Gamond 1990). Mechanism of dissolution and redeposition may be recognized directly by optical microscope, by chemical analysis, or by cathodoluminescence studies. Mass transfer is proportional to the difference in chemical potential between the zones of dissolution and deposition. Most often, soluble species are removed from grain–grain contacts (stylolites, solution cleavage) and reprecipitated in veins or voids (Weyl 1959; Rutter 1976; Gratier 1987). In this case, mass transfer involves diffusion through the fluid phase trapped along the stressed contact, and can lead to the complete sealing of veins and voids (Figure 12.7b, iv, upper part). Note that solution cleavage (S) and associated postseismic crack-sealed veins (CSV, Figure 12.7a), possibly in combination with subcritical fracturing (Atkinson 1982), may accommodate part of the stress aseismically over the entire interseismic period (see also Chapter 14).

Compaction and crack sealing involves two mechanisms (Figures 12.7, 12.8): pervasive pressure solution at the grain scale in the gouge (Figure 12.8b) and vein cementation associated with dissolution along stylolites in the damaged zone around the active faults (Figure 12.8c, Gratier et al. 2003). Compaction modeling shows that these two mechanisms have different time scales (Renard et al. 2000). Pervasive pressure solution at the grain scale in fine-grained gouge is much faster than pressure solution along stylolites and associated precipitation in veins. For example, the time required to reduce permeability from 10% to near 0 at five km depth is about a month for a fine-grained quartz gouge (10 microns) and a thousand years in a damaged zone with fractures spaced 100 microns. Therefore, slow stress-driven crack sealing and compaction

---

◄ **Figure 12.8** (a) Fluids in the upper crust rising along active faults as deduced from stable isotopes studies (Pili et al. 1998); (b) model and microstructure of gouge compaction; (c) geometry and microstructure of crack-seal veins, adapted from Gratier et al. (2003); (d) effect of depth on gouge compaction; (e) effect of fracture spacing on slow fracture sealing, adapted from Renard et al. (2000). (f) Postseismic sealing rate with depth; (g) evolution of fluid pressure within fault for predominantly calcite dissolution in the uppermost 5 km and dominantly quartz dissolution at 5 to 10 km depth, adapted from Gratier et al. (2003).

processes control the change of porosity and permeability at depth after an earthquake. The kinetics of the sealing process during advective fluid flow may also be very slow if diffusive mass transfer of dissolved rock through the stressed intergranular fluid is the rate-limiting step (Figure 12.7b, iii, right).

Modeling of fluid transfer along active faults is possible if pressure-solution fracture sealing is combined with advective inflow of lower crust fluids (Figure 12.8a). The compaction rate in the gouge (Figure 12.8b) and the sealing rate in the damaged zone (Figure 12.8c) can be calculated (Figures 12.8d and 12.8e, respectively), the latter being the rate-controlling process (Gratier et al. 2003). The rate of sealing is modeled to decay exponentially with time. An example is given in Figure 12.8g which shows the progressive change in fluid pressure from hydrostatic (just after the earthquake) to locally near-lithostatic. Lithostatic pressure develops at two different depth intervals: at depth due to the inflow of fluid from the lower crust, and in the upper crust where calcite is available for mass transfer and relatively fast sealing of the veins (Figure 12.8f). Note that hydrothermal reactions in fault zones may lead to two competing time-dependent effects: fault strengthening due to crack sealing and fault weakening that arises from elevated pore pressures within a well-cemented, low-permeability gouge layer (Figure 12.8g).

In summary, healing and sealing processes in faults may develop over a broad range of time scales, from several days (self healing) to a thousand years (complete sealing of the veins by pressure solution and deposition). Strengthening therefore competes with weakening due to an increase in fluid pressure during progressive healing. This could help to localize the main rupture along the healed gouge and scatter the aftershocks within the strengthened, damaged fault zones. Sealing restores and even increases the cohesion of the rock. Muhuri et al. (2003) argued that sealing does not necessarily change the frictional coefficient.

A MODEL OF UPPER CRUSTAL FAULT STRENGTH

Rheological Model

As explained above, the strength of fault zones can either decrease or increase with time. The difference in the characteristic timescales of weakening and strengthening can control the rheology of active faults and thus influence the duration of seismic cycles. The simple rheological model proposed below accounts for these two effects, as shown schematically in Figure 12.9a. The fault zone is modeled here as ductile material undergoing pressure-solution creep, healing, and reaction softening. These mechanisms are assumed to operate within the seismogenic crust, as suggested by evidence for pressure solution at all depths along the San Andreas fault (Gratier et al. 2003). An alternative approach has been to model fault weakening and strengthening as due to changes in fault friction, cohesion, or pore fluid pressure (Miller 2002; Miller et al. 2003;

(a) 1D layered structure

(b) Schematic evolution of weakening, strengthening, and shear stress

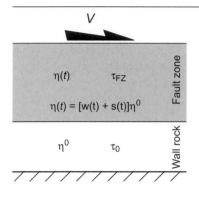

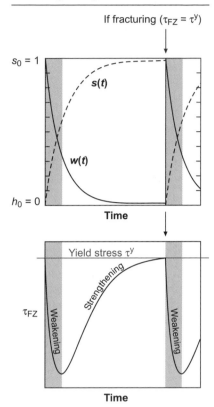

Figure 12.9 (a) Depiction of the one-dimensional-layered model of strengthening and weakening during faulting. The viscosity of the wall rock is invariant with time, while the fault zone viscosity weakens $w(t)$ and strengthens $s(t)$ with time. (b) Schematic evolution of weakening and strengthening (top) and of shear stress (bottom). Initial weakening is followed by strengthening, loading of the fault zone, leading to renewed fracturing at $\tau_{FZ} \geq \tau^y$.

Sibson and Rowland 2003). Since we are interested in quantifying the role of weakening and strengthening processes during ductile deformation, we assume that the viscosity of the fault varies with time, with respect to a constant, initial viscosity of the wall rock, η_0. Starting from this initial value, the viscosity of the fault rock, $\eta(t)$, varies with time as follows:

$$\eta(t) = [w(t) + s(t)]\eta_0 \ , \ \text{with} \ 0 < w(t) < 1 \ \text{and} \ 0 < s(t) < 1 \quad (12.6)$$

where $w(t)$ and $s(t)$ are the amounts of weakening and strengthening, respectively. The initial and final states are defined as:

Inital state: $w_0 = 1$, $s_0 = 0$, $\eta = \eta_0$ (12.7)

Final state: $\eta_\infty = [w_\infty + s_\infty]\eta_0$ (12.8)

The final value of weakening, w_∞, is a free parameter that is set here at $w_\infty = 10^{-2}$. The final amount of strengthening is such that $w_\infty + s_\infty = 1$ (i.e., no final change in the viscosity). s_∞ can thus be expressed as $s_\infty = 1 - w_\infty$. Note that although $w_\infty + s_\infty = 1$, $w(t) + s(t) \neq 1$ and depends on the relative amounts of weakening and strengthening. The viscosity of the fault zone thus changes with time according to relative rates of weakening and strengthening. The weakening rate, $\partial w/\partial t$, and strengthening rate, $\partial s/\partial t$, are proportional to $(w(t) - w_\infty)$ and $(s(t) - s_\infty)$, respectively. We have shown in the sections above that the strain rate, $\dot{\varepsilon}$, within the fault zone controls the rates of weakening and strengthening:

$$\frac{\partial w}{\partial t} = \varphi_w \dot{\varepsilon}[w_\infty - w(t)] \quad \text{and} \quad \frac{\partial s}{\partial t} = \varphi_s \dot{\varepsilon}[s_\infty - s(t)] \qquad (12.9)$$

The scalars φ_w and φ_s define the kinetics of weakening and strengthening and are free parameters. As discussed above, weakening may be triggered by microfracturing and fluid flow, leading to post-seismic stress relaxation. Subsequent strengthening is related to interseismic sealing of these fractures. In our rheological description, initially predominant weakening yields to the following condition: $\varphi_w > \varphi_s$.

For a constant strain rate, $\dot{\varepsilon}$, the weakening and strengthening derived from Figure 12.9 become

$$w(t) = [w_0 - w_\infty]\exp(-\varphi_w \dot{\varepsilon}t) + w_\infty \ ;$$
$$s(t) = [s_0 - s_\infty]\exp(-\varphi_s \dot{\varepsilon}t) + s_\infty \quad \text{for} \quad \dot{\varepsilon} = c^{te} \ . \qquad (12.10)$$

The characteristic time for weakening, t_w, and strengthening, t_s, can thus be defined as

$$t_w = 1/(\varphi_w \dot{\varepsilon}) \quad \text{and} \quad t_s = 1/(\varphi_s \dot{\varepsilon}) \ . \qquad (12.11)$$

Increasing φ_w at constant strain rate reduces the weakening time and leads to a sharper decrease in viscosity. An increase of strain rate (e.g., during strain localization) also enhances the viscosity reduction (decreases t_w). This corresponds to a positive feedback between weakening and strain localization. In the same way, strengthening tends to delocalize strain.

Note that at time $5t_w$, the term $\exp(-\varphi_w \dot{\varepsilon}t)$ in Eq. 12.10 is equal to 0.007, indicating that by then, weakening has almost reached its final value, s_∞. The same holds for strengthening at time $5t_s$. Weakening and strengthening thus reach their final values at the following times:

$$t_{w\,\text{end}} = 5t_w = 5/(\varphi_w \dot{\varepsilon}) \quad \text{and} \quad t_{s\,\text{end}} = 5t_s = 5/(\varphi_s \dot{\varepsilon}) \ . \quad (12.12)$$

After $t_{\text{w\_end}}$, weakening ceases unless a new fracture resets weakening and strengthening to their initial values, as explained below. Fracturing occurs if the shear stress in the fault zone, $\tau = \eta\dot{\varepsilon} = [w(t) + s(t)]\eta_0\dot{\varepsilon}$, becomes greater than the yield stress, τ^y. The yield stress is assumed here to be constant, since gravity (and thus pressure) is disregarded in this study. As shown in Figure 12.8, the combined effects of advective fluid inflow and fracture sealing lead to a progressive increase of fluid pressure within the active fault, from near-hydrostatic values after an earthquake to locally overpressurized zones. Consequently, the effective shear strength of the fault is expected to decrease progressively during the interseismic period (Miller 2002; Sibson and Rowland 2003). Overall weakening of the fault zone due to the formation of gouge within a damage zone should result in a decrease of yield stress, τ^y, with time. We disregard this long-term evolution and focus instead on the seismic cycle computed for a constant assumed value of the yield stress, τ^y.

The competing effects of weakening and strengthening on the evolution of fault strength are illustrated schematically in Figure 12.9b, where strength and shear stress within the fault vary with time. After fracturing, weakening prevails and leads to a sharp decrease of fault strength. Fault strength recovers when strengthening becomes dominant, possibly leading to the seismic nucleation of new fractures.

The kinetics of the weakening and strengthening thus controls the duration of the seismic cycle, as quantified in the following section. Note that the proposed rheological model and thus the predicted duration of the seismic cycle are dimensionless. In the future, quantification of the duration of weakening and strengthening in laboratory experiments and natural slip events (e.g., coseismic and postseismic deformation measurement) could define a characteristic time for cyclical changes in fault zone rheology. This calibration could be achieved by a detailed laboratory study of the kinetics of weakening (pressure solution, reaction softening) and strengthening (healing and sealing) processes.

Boundary Conditions and Numerical Scheme

The model comprises a one-dimensional-layered medium subjected to simple shearing. The bottom of the sheared layers in Figure 12.9 is fixed, while a constant velocity, V, is imposed at the top of the layers. The constant velocity is imposed in order to determine the relative roles of weakening and strengthening in the absence of any change in loading. The average strain rate within the fault zone of width L is $\dot{\varepsilon} = V/L$. Mechanical equilibrium is attained by numerically (finite-element method) using the code SARPP (SARPP 2003). The layered structure is discretized into 100 3-noded Lagrangian elements. The velocity $v(y)$ is the only nodal unknown.

Variation in Shear Stress, Strain Rate, and Velocity

Figure 12.10 presents variations of shear stress at a given depth within the fault zone. The kinetics of weakening and strengthening were arbitrarily set at $\varphi_w = 10$ and $\varphi_s = 1$, respectively. The cyclic variations of shear stress are related to weakening (points 1 to 2) and strengthening (points 2 to 3) that prevailed in the fault zone, as discussed above. Strain rate and velocity as a function of depth within

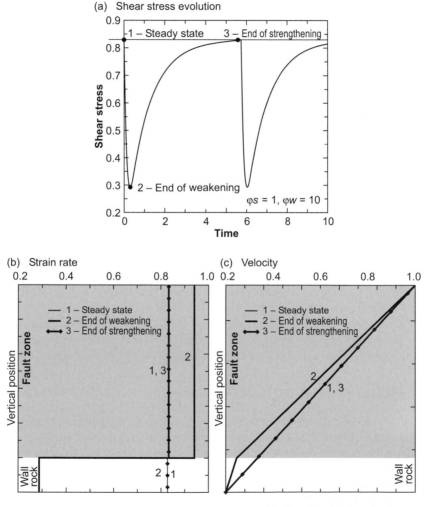

Figure 12.10 (a) Dimensionless shear stress variations with dimensionless time ($\varphi_s = 1$ and $\varphi_w = 10$); (b) dimensionless strain rate profiles and (c) dimensionless velocity profiles of the fault zone and wall rock at the beginning of the earthquake cycle (point 1: steady state in a), at the end of the weakening (point 2 in a) and at the end of strengthening (point 3 in a).

the fault are also given in Figure 12.10b and c, respectively. At the beginning of the cycle, that is, prior to the onset of weakening and strengthening, the strain rate within the fault zone and the wall rock are uniform and identical (points 1 and 3), reflecting the assumed absence of any vertical variation in the material properties of the rheological model and yielding a linear velocity profile. At the end of the weakening, the increase in strain rate of the fault is compensated by a decrease in strain rate of the wall rock in order to maintain mechanical equilibrium (uniform shear stress across the layering during simple shear). This leads to a sharp velocity gradient in the fault zone. At the end of strengthening, the strain rate in the fault zone is almost equal to the strain rate in the wall rock.

Duration of the Seismic Cycle

Figure 12.11 shows the variation of shear stress in the fault zone with time for two sets of strengthening and weakening rate constants ($\varphi_s = 1$, $\varphi_w = 10$ in curve 1 and $\varphi_s = 5$, $\varphi_w = 10$ in curve 2). The width of the fault zone, L, and the displacement velocity, V, are both set to 1, corresponding to an average strain rate of 1 ($\dot{\varepsilon} = V/L$). For $\varphi_s = 1$, $\varphi_w = 10$ (curve 1, Figure 12.11 and also Figure 12.10) a rapid decrease in shear stress is observed after the onset of fracturing,

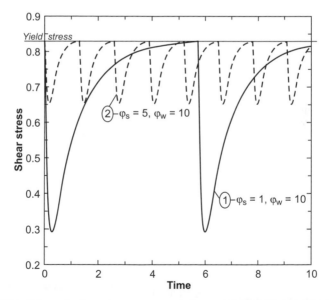

Figure 12.11 Dimensionless shear stress variations with dimensionless time for two rates of strengthening: $\varphi_s = 1$ (curve 1, straight line) and $\varphi_s = 5$ (curve 2, dashed line). The rate of weakening is constant at $\varphi_w = 10$ because $\varphi_w > \varphi_s$, weakening sets in immediately after fracturing at the yield stress, shown as the horizontal line. Subsequently, strengthening dominates, leading to renewed fracturing.

corresponding to post-seismic stress relaxation. The duration and rate of stress decrease are governed by the rate of weakening. For $\varphi_w = 10$, the dimensionless time at the end of weakening is of the order of 0.5 (Eq. 12.12). This estimate is very close to the time when the minimum value of shear stress is attained (~0.4). Note that this time exceeds the time of the end of weakening, reflecting the positive feedback between strain localization and weakening kinetics (Eq. 12.10). During stress relaxation, strain rate increases significantly in the fault zone, leading to faster weakening. For times greater than 0.4, shear stress increases over a time whose duration is controlled by the rate of strengthening. Because this rate is an order of magnitude less than the rate of weakening, the time necessary for shear stress to increase is much greater than the time of stress relaxation. In this case, the time when strengthening ceases is of the order of 5 ($\varphi_s = 1$ and $\dot\varepsilon = 1$, Eq. 12.12). Again, this estimate is consistent with modeling in which strengthening ends at a time of around 6. This difference between estimated and numerical results is attributable to the decrease in bulk strain rate within the fault zone during strengthening (Figure 12.10), leading to a decrease in the rate of strengthening (Eq. 12.11). At the end of strengthening, the increase in shear stress is sufficient to trigger fracturing, which resets the weakening and strengthening variables to their initial values at the beginning of a new cycle of stress relaxation and fault loading. Since the yield stress is constant through time, the time interval between two episodes of fracturing is constant.

The time interval between two episodes of fracturing within the fault zone defines the duration of the earthquake cycle, t_{cycle}. Note again that since the proposed rheological model is dimensionless, the predicted duration of the earthquake cycle is also dimensionless. However, the real-time duration can be estimated for a natural fault if the rates of weakening and strengthening in that fault are known from creep and compaction laws. More specifically, because the earthquake cycle is defined by stress relaxation (during a time close to t_{w_end}, Eq. 12.12) and stress loading (during a time close to t_{s_end}, Eq. 12.12), the duration of the earthquake cycle is simply

$$t_{cycle} = t_{w\_end} + t_{s\_end} \approx \frac{5}{\dot\varepsilon}\left[\frac{1}{\varphi_w} + \frac{1}{\varphi_s}\right]. \tag{12.13}$$

For $\varphi_s = 1$, $\varphi_w = 10$, this earthquake cycle time is about 5.1. This estimate is very close to the numerical result (curve 1, Figure 12.11). Increasing φ_s to 5 at the same φ_w (curve 2, Figure 12.11) yields a smaller earthquake cycle time ($t_{cycle} = 1.5$). Note also that this shorter cycle time coincides with a decrease in the shear stress drop during stress relaxation. This reflects the fact that increasing the strengthening rate causes strengthening to dominate the strength evolution of the fault zone earlier in the earthquake cycle. Weakening prevails during a shorter time interval, leading to less pronounced stress relaxation.

In summary, the rheology of the fault zone controls the timing of postseismic stress relaxation and interseismic stress loading. The duration of the earthquake cycle is thus controlled by the relative rates of weakening and strengthening within the fault zone. Transient weakening and strengthening are activated within the fault zone immediately after fracturing, which opens the system and induces the advective influx of fluid. Weakening is primarily related to fluid-rock interaction (reaction softening, pressure solution), whereas strengthening initiates when free fluids become rare and the fault zone is sealed.

The Role of Fault Zone Width

Figure 12.12 depicts the duration of the seismic cycle as a function of the ratio of the rates of strengthening and weakening for two values of fault zone width, $L = 1$ and $L = 10$. Increasing the width of the fault zone from 1 to 10 leads to a corresponding, order-of-magnitude increased in the cycle duration. Therefore, an increase of L at constant displacement velocity, V, involves a decrease in strain rate and an increase in the duration of the earthquake cycle (Eq. 12.13). If we assume that the width of the fault zone is related to its length, these results indicate that the seismic cycle of a large fault is much longer than that of a minor fault, as observed in nature.

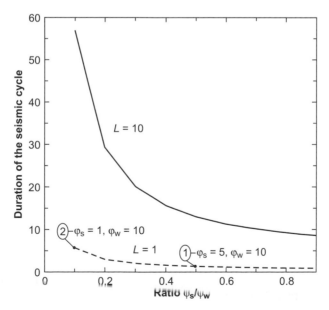

Figure 12.12 Dimensionless duration of the earthquake cycle as a function of the ratio of strengthening to weakening rates for two values of fault zone width, $L = 1$ and 10. Points 1 and 2 refer to numerical simulation of Figure 12.10 for $L = 1$.

CONCLUDING REMARKS

Within upper crustal faults, the interaction of fracturing, pressure solution, and mineral reactions leads to unstable deformation, characterized by fast, episodic slip alternating with slow creep. At the time scale of the earthquake cycle, fracturing speeds up the kinetics of pressure solution and reaction, thereby weakening the fault, whereas healing and sealing of fractures re-strengthen the fault zone. These mechanisms also relax stress during the interseismic period and partly smooth asperities on fault surfaces. The effect of fracturing on the dissolution process is faster than healing and sealing of the fault, such that weakening predominates just after earthquake whereas strengthening occurs more progressively with time. Strengthening associated with fault sealing is sometimes counteracted by weakening related to a decrease in permeability and—given sufficient fluid inflow—to an increase in pore-fluid pressure in the fault zone. Numerical modeling indicates that the relative rates of weakening and strengthening control the temporal evolution of fault strength during the interseismic period, and thus the duration of the earthquake cycle.

Future Research

Microstructural Approach

Numerous parameters control the characteristic times of mechanical and physical responses to seismic events. The kinetics of weakening- and strengthening processes are, of course, key factors in this response. As presented in this chapter, different creep and compaction laws are possible depending on the parameters used (see also Chapter 7). Further experiments are required to constrain these parameters, as well as to establish the rate-limiting steps of the weakening and strengthening processes. A vexing problem is that the processes examined in nature operate on time scales that far exceed the average human life span.

Another major parameter that can only be inferred from the detailed analysis of natural fault zones is the relative contributions of advective and diffusive modes of mass transport. Closely related to this is the mean distance of mass transfer within fluids (see also Chapter 11). Careful microstructural and geochemical analysis of naturally deformed rock is required to establish this distance, which can range from the spacing between fractures in a rock with solution cleavage (diffusive mass transfer) to thousands meters in veined rocks (advective transport).

Multiscaling Approach

We understand how individual asperities on a fault are dissolved or broken, but determining the bulk effect of asperities at all scales on macroscopic fault behavior is a true challenge. A first step will be to evaluate the geometry of the asperities on exhumed fault surfaces and to consider asperities at all scales as deviations from the ideal model of a threaded surface. Also needed for quantitative models are the scaling properties of fractures (their width, length, and spacing), as these are expected to play a key role in weakening and strengthening.

Macrostructural Approach

Testing models requires in-depth studies of real faults. Measurements of how mechanical and physical properties of faults zone evolve during the interseismic period are, of course, very important. In particular, monitoring fluid pressure could be combined with geodetic measurements of fault motion (see Chapter 4) to establish the relationship between fault displacement and fluid pressure evolution along active faults. This would be especially interesting to do immediately after earthquakes, in order to capture the postseismic creep triggered by fracturing, then the expected links between postseismic strengthening and the recovery of high fluid pressure. Finite displacements during aftershocks can be regarded in the broader context of long-term displacements obtained with geodetic measurements in order to constrain the postseismic creep rate and the strain localization along active fault zones. Fault zones with a regular earthquake recurrence time and constant displacement rate (e.g., Parkfield, southern California) lend themselves well to this kind of study, because they are expected to show features that are diagnostic of recurrent, competitive weakening and strengthening during the interseismic period.

Finally, the idea that inherited and induced heterogeneities on several length scales can trigger instabilities introduces the notion of self-organized complexity that prevents us from accurate prediction of earthquakes on useful time and length scales (days to months, meters to kilometers). Earthquake predictions are, and will remain, beset with large uncertainties. Evaluating these uncertainties remains a challenge that can only be met by the careful study of structural heterogeneities and their effects on self-organization.

Numerical Modeling

Future modeling should account not only for changes of fault zone rheology during the earthquake cycle as proposed in this chapter, but also for changes of pore fluid pressure. Following an earthquake, the combined effect of advective fluid inflow and fracture sealing in the fault core could lead to an increase of fluid pressure from near-hydrostatic after the earthquake to locally overpressurized, at least along some parts of the fault (see Chapter 14 for a discussion of postseismic pore-fluid pressure evolution). Consequently, both the yield strength and effective shear strength of a fault could decrease with time during the interseismic period, particularly if a gouge develops. Therefore, an important step will be to incorporate transient yield stress into models of the seismic cycle.

ACKNOWLEDGEMENTS

We thank D. Gapais for his detailed comments on a previous version of the manuscript and C. Spiers, M. Handy, and J. Urai for their very helpful reviews of the final version. We also thank P. Labaume, P. Vialon, and A.-M. Boullier, respectively, for the photographs in Figures 12.3d, 12.5c, and 12.6a.

REFERENCES

Atkinson, B.K. 1982. Subcritical crack propagation in rocks: Theory, experimental results, and applications. *J. Struct. Geol.* **4**:41–56.

Austrheim, H., and T. Boundy. 1994. Pseudotachylites generated during seismic faulting and eclogitization of the deep crust. *Science* **265**:82–83.

Beeler, N.M., and S.H. Hickman. 2004. Stress-induced, time-dependent fracture closure at hydrothermal conditions. *J. Geophys. Res.* **109**:1–16.

Blanpied, M.L., D.A. Lockner, and J.D. Byerlee. 1995. Frictional slip of granite at hydrothermal conditions. *J. Geophys. Res.* **100**:13,045–13,064.

Bokelmann, G.H.R., and R. Kovach. 2003. Long-term creep rate changes and their causes. *Geophys. Res. Lett.* **30**:NIL_9-NIL_12.

Bos, B., C.J. Peach, and C.J. Spiers. 2000. Frictional-viscous flow of simulated fault gouge caused by the combined effects of phyllosilicates and pressure solution. *Tectonophysics* **327**:173–194.

Brantley, S., B. Evans, S.H. Hickman, and D.A. Crerar. 1990. Healing of microcracks in quartz: Implications for fluid flow. *Geology* **18**:136–139.

Cosgrove, J.W. 1976. The formation of crenulation cleavage. *J. Geol. Soc. Lond.* **132**:155–178.

De Boer, R.B., P.J.C. Nagtegaal, and E.M. Duyvus. 1977. Pressure solution experiment on quartz sand. *Geochim. Cosmochim. Acta* **41**:257–264.

de-Meer, S., C.J. Spiers, C.J. Peach, and T. Watanabe. 2002. Diffusive properties of fluid-filled grain boundaries measured electrically during active pressure solution. *Earth Planet. Sci. Lett.* **200**:147–157.

den Brok, S.B. 1998. Effect of microcracking on pressure solution strain rate: The Gratz grain boundary model. *Geology* **26**:915–918.

Dewers, T., and P. Ortoleva. 1990. A coupled reaction/transport/mechanical model for intergranular pressure solution stylolites, and differential compaction and cementation in clean sandstones. *Geochim. Cosmochim. Acta* **54**:1609–1625.

Dixon, J., and G. Williams. 1983. Reaction softening in mylonites from the Arnaboll thrust, Sutherland. *Scott. J. Geol.* **19**:157–168.

Dysthe, D.K., F. Renard, J. Feder et al. 2003. High-resolution measurements of pressure solution creep. *Phys. Rev. E* **6801**:317–329.

Elias, B.P., and A. Hajash. 1992. Changes in quartz solubility and porosity due to effective stress: An experimental investigation of pressure solution. *Geology* **20**:451–454.

Ethcridge, M.A., S.F. Cox, V.J. Wall, and R.H. Vernon. 1984. High fluid pressures during regional metamorphism and deformation: Implications for mass transfer and deformation mechanisms. *J. Geophys. Res.* **89**:4344–4558.

Evans, J.P., and F.M. Chester. 1995. Fluid–rock interaction in faults of the San Andreas system: Inferences from the San Gabriel fault rock geochemistry. *J. Geophys. Res.* **100**:13,007–13,020.

Farver, J.R., and R.A. Yund. 1998. Oxygen grain boundary diffusion in natural and hot-pressed calcite aggregates. *Earth Planet. Sci. Lett.* **161**:189–200.

FitzGerald, J.D., and H. Stünitz. 1993. Deformation of granitoids at low metamorphic grade. I: Reactions and grain size reduction. *Tectonophysics* **221**:269–297.

Fletcher, R.C., and D.D. Pollard. 1981. Anticrack model for pressure solution surfaces. *Geology* **9**:419–424.

Fyfe, W.S., N.J. Price, and A.B. Thomson. 1978. Fluids in the Earth's Crust. Developments in Geochemistry I. Amsterdam: Elsevier.

Gapais, D. 1989. Shear structures within deformed granites: Mechanical and thermal indicators. *Geology* **17**:1144–1147.

Gibbs, J.W. 1877. On the equilibrium of heterogeneous substances. *Trans. Connecticut Acad.* **3**:108–248; 343–524.

Gratier, J.P. 1987. Pressure solution-deposition creep and associated tectonic differentiation in sedimentary rocks. In: Deformation of Sediments and Sedimentary Rocks, ed. M.E. Jones and R.M.F. Preston, Spec. Publ. 29, pp. 25–38. London: Geol. Soc.

Gratier, J.P., P. Favreau, and F. Renard. 2003. Modeling fluid transfer along California faults when integrating pressure solution crack sealing and compaction processes. *J. Geophys. Res.* **108**:B2, 28–52.

Gratier, J.P., and J.F. Gamond. 1990. Transition between seismic and aseismic deformation in the upper crust. In: Deformation Mechanisms, Rheology and Tectonics, ed. R.J. Knipe and E.H. Rutter, Spec. Publ. 54, pp. 461–473. London: Geol. Soc.

Gratier, J.P., F. Renard, and P. Labaume. 1999. How pressure solution and fractures interact in the upper crust to make it behave in both a brittle and viscous manner. *J. Struct. Geol.* **21**:1189–1197.

Gratz, A.J. 1991. Solution-transfer compaction of quartzites: Progress toward a rate law. *Geology* **19**:901–904.

Gueydan, F., Y.M. Leroy, L. Jolivet, and P. Agard. 2003. Analysis of continental midcrustal strain localization induced by reaction softening and microfracturing. *J. Geophys. Res.* **108**:B2, 2064, doi:10.1029/2001JB000611.

Handy, M.R. 1989. Deformation regime and the rheological evolution of fault zones in the lithosphere. The effect of pressure, temperature, grain size and time. *Tectonophysics* **163**:119–159.

Handy, M.R., and H. Stünitz. 2002. Strain localization by fracturing and reaction weakening: A mechanism for initiating exhumation of subcontinental mantle beneath rifted margins. In: Deformation Mechanisms, Rheology and Tectonics: Current Status and Future Perspectives, ed. S. de Meer, M.R. Drury, J.H.P. de Bresser, and J.M. Pennock, Spec. Publ. 200, pp. 387–407. London: Geol. Soc.

Hickman, S.H., and B. Evans. 1991. Experimental pressure solution in halite: The effect of grain–interphase boundary structure. *J. Geol. Soc. Lond.* **148**:549–560.

Hickman, S.H., and B. Evans. 1995. Kinetics of pressure solution at halite–silica interfaces and intergranular clay films. *J. Geophys. Res.* **100**:B7, 13,113–13,132.

Hickman, S., R.H. Sibson, and R. Bruhn. 1995. Introduction to special section: Mechanical involvement of fluids in faulting. *J. Geophys. Res.* **100**:B7, 12,831–12,840.

Hilgers, C., K. Dilg-Gruschinski, and J.L. Urai. 2004. Microstructural evolution of syntaxial veins formed by advective flow. *Geology* **32**:261–264.

Jolivet, L., H. Raimbourg, L. Labrousse et al. 2005. Softening triggered by eclogitization: The first step toward exhumation during continental subduction. *Earth Planet. Sci. Lett.* **237**:532–547.

Kennedy, B.M., Y.K. Kharaka, W.C. Evans et al. 1997. Mantle fluids in the San Andreas fault system, California. *Science* **278**:1278–1280.

Kingery, W.D., H.K. Bowen, and D.R. Uhlmann. 1976. Introduction to Ceramics. New York: Wiley.

Klaper, E.M. 1990. Reaction-enhanced formation of eclogite-facies shear zones in granu-lite-facies anorthosites. In: Deformation Mechanisms, Rheology and Tectonics, ed. R.J. Knipe and E.H. Rutter, Spec. Publ. 54, pp. 167–174. London: Geol. Soc.

Knipe, R.J., and R.P. Wintsch. 1985. Heterogeneous deformation, foliation develop-ment, and metamorphic processes in a polyphase mylonite. In: Metamorphic Reac-tions: Kinetics, Textures, and Deformation, A.B. Thompson and D.C. Rubie, pp. 180–210. Berlin: Springer.

Le Hebel, F., D. Gapais, S. Fourcade, and R. Capdevila. 2002. Fluid-assisted large strains in a crustal-scale décollement (Hercynian Belt of South Brittany, France). In: Deformation Mechanisms, Rheology and Tectonics: Current Status and Future Per-spectives, ed. S. de Meer, M.R. Drury, J.H.P. de Bresser, and J.M. Pennock, Spec. Publ. 200, pp. 85–101. London: Geol. Soc.

Lehner, F.K. 1995. A model for intergranular pressure solution in open systems. *Tectonophysics* **245**:153–170.

Li, Y.G., and J.E. Vidale. 2001. Healing of the shallow fault zone from 1994–1998 after the 1992 *M* 7.5 Landers, California, earthquake. *Geophys. Res. Lett.* **28**: 2999–3002.

Li, Y.G., J.E. Vidale, S.M. Day, D.D. Oglesby, and E. Cochran. 2003. Postseismic fault healing on the rupture zone of the 1999 *M* 7.1 Hector Mine, California, earthquake. *Bull. Seismol. Soc. Am.* **93**:854–869.

McEwen, T.J. 1978. Diffusional mass transfer processes in pitted pebble conglomer-ates. *Contrib. Mineral. Petrol.* **67**:405–415.

Means, W.D., and T. Li. 2001. A laboratory simulation of fibrous veins: Some first observations. *J. Struct. Geol.* **23**: 857–863.

Miller, S.A. 2002. Properties of large ruptures and the dynamical influence of fluids on earthquakes and faulting, *J. Geophys. Res.* **107**:536–548.

Miller, S.A., W. van-der-Zee, D.L. Olgaard, and J.A.D. Connolly. 2003. A fluid pres-sure feedback model of dehydration reactions: Experiments, modeling, and applica-tion to subduction zones. *Tectonophysics* **370**:241–251.

Mitra, G. 1978. Ductile deformation zones and mylonites: The mechanical process involved in the deformation of crystalline basement rocks. *Am. J. Sci.* **278**: 1057–1084.

Muhuri, S.K., T.A. Dewers, G.E. Scott, and Z. Reches. 2003. Interseismic fault strength-ening and earthquake-slip instability: Friction or cohesion? *Geology* **31**:881–884.

Muller, W., D. Aerden, and A.N. Halliday. 2000. Isotopic dating of strain fringe incre-ments: Duration and rates of deformation in shear zones. *Science* **288**:2195–2197.

Nadeau, R.M., and L.R. Johnson. 1998. Seismological studies at Parkfield VI: Moment release rates and estimates of source parameters for small repeating earthquakes. *Bull. Seismol. Soc. Am.* **88**:790–814.

Niemeijer, A.R., and C.J. Spiers. 2002. Compaction creep of quartz sand at 400–600°C: Experimental evidence for dissolution-controlled pressure solution. *Earth Planet. Sci. Lett.* **195**:261–273.

Oelkers, E.H., P.A. Bjørkum, and W.M. Murphy. 1996. A petrographic and computa-tional investigation of quartz cementation and porosity reduction in North Sea sand-stones. *Am. J. Sci.* **296**:420–452.

Paterson, M.S. 1973. Nonhydrostatic thermodynamics and its geologic applications. *Rev. Geophys. & Space Phys.* **11**:355–389.

Paterson, M.S. 2001. Relating experimental and geological rheology. *Intl. J. Earth Sci. (Geol. Rundsch.)* **90**:157–167.

Pfiffner, O.A., and J.G. Ramsay. 1982. Constraints on geological rate: Arguments from finite strain values of naturally deformed rocks. *J. Geophys. Res.* **87**:311–321.

Pili, E.B., B.M. Kennedy, M.S. Conrad, and J.P. Gratier. 1998. Isotope constraints on the involvement of fluids in the San Andreas fault. *EOS Trans. AGU* **79**:S229–S320, Spring meeting

Raj, R. 1982. Creep in polycrystalline aggregates by matter transport through a liquid phase. *J. Geophys. Res.* **87**:4731–4739.

Ramsay, J.G. 1980. The crack-seal mechanism of rock deformation. *Nature* **284**:135–139.

Renard, F., A. Park, J.P. Gratier, and P. Ortoleva. 1997. An integrated model for transitional pressure solution in sandstone. *Tectonophysics* **312**:97–115.

Renard, F., J.P. Gratier, and B. Jamtveit. 2000. Kinetics of crack-sealing, intergranular pressure solution, and compaction around active faults. *J. Struct. Geol.* **22**:1395–1407.

Robin, P.Y. 1978. Pressure solution at grain-to-grain contacts. *Geochim. Cosmochim. Acta* **42**:1383–1389.

Robin, P.Y. 1979. Theory of metamorphic segregation and related processes. *Geochim. Cosmochim. Acta* **43**:1587–1600.

Rutter, E.H. 1976. The kinetics of rock deformation by pressure solution. *Phil. Trans. R. Soc. Lond.* **283**:203–219.

Rutter, E.H. 1983. Pressure solution in nature, theory, and experiment. *J. Geol. Soc. Lond.* **140**:725–740.

Sammis, G.C., and J.R. Rice. 1998. Repeating earthquakes as low-stress-drop events at a border between locked and creeping fault patches. *Bull. Seismol. Soc. Am.* **91**: 532–537.

SARPP. 2003. Structural Analysis and Rock Physics Program, Y.M. Leroy and F. Gueydan, LMS, Ecole Polytechnique, Palaiseau, France.

Shimizu, I. 1995. Kinetics of pressure solution creep in quartz: Theoretical considerations. *Tectonophysics* **245**:121–134.

Sibson, R.H., F. Robert, and H.H.A.F. Poulsen. 1988. High angle faults, fluid pressure cycling, and mesothermal gold-quartz deposits. *Geology* **16**:551–555.

Sibson, R.H., and J.V. Rowland. 2003. Stress, fluid pressure, and structural permeability in seismogenic crust, North Island, New Zealand. *Geophys. J. Intl.* **154**:584–594.

Simpson, C. 1985. Deformation of granitic rocks across the brittle–ductile transition. *J. Struct. Geol.* **7**:503–511.

Sorby, H.C. 1865. On impressed limestone pebbles, as illustrating a new principle in chemical geology. *Proc. West Yorks. Geol. Soc.* **14**:458–461.

Spiers, C.J., and P.M. Schutjens. 1990. Densification of crystalline aggregates by fluid phase diffusional creep. In: Deformation Processes in Minerals, Ceramics, and Rocks, ed. D.J. Barber and P.G. Meredith, pp. 334–353. London: Unwin Hyman.

Stel, H. 1981. Crystal growth in cataclasites; Diagnostic microstructures and implications. *Tectonophysics* **78**:585–600.

Taber, S. 1916. The growth of crystals under external pressure. *Am. J. Sci.* XLI (4th series) **246**:532–556.

Tada, R., and R. Siever. 1986. Experimental knife-edge pressure solution of halite. *Geochim. Cosmochim. Acta* **50**:29–36.

Tenthorey, E., S.F. Cox, and H.F. Todd. 2003. Evolution of strength recovery and permeability during fluid–rock reaction in experimental fault zones. *Earth Planet. Sci. Lett.* **206**:161–172.

Thibaut, M., J.P. Gratier, M. Léger, and J.M. Morvan. 1996. An inverse method for determining three-dimensional fault geometry with thread criterion: Application to strike-slip and thrust faults (Western Alps and California). *J. Struct. Geol.* **18**:1127–1138.

Vissers, R.L.M., J.P. Platt, and D. van der Wal. 1995. Late orogenic extension of the Betic Cordillera and the Alboran Domain: A lithospheric view. *Tectonics* **14**:786–803.

Weyl, P.K. 1959. Pressure solution and the force of crystallization: A phenomenological theory. *J. Geophys. Res.* **64**:2001–2025.

White, S.H., and R.J. Knipe. 1978. Transformation and reaction-enhanced ductility in rocks. *J. Geol. Soc. Lond.* **135**:513–516.

Wibberley, C.A.J. 1999. Are feldspar-to-mica reactions necessarily reaction-softening processes in fault zones? *J. Struct. Geol.* **21**:1219–1227.

Wintsch, R.P., R. Christoffersen, and A.K. Kronenberg. 1995. Fluid–rock reaction weakening in fault zones. *J. Geophys. Res.* **100**:13,021–13,032.

Zhang, X., J. Salemans, C.J. Peach, and C.J. Spizers. 2002. Compaction experiments on wet calcite powder at room temperature: Evidence for operation of intergranular pressure solution. In: Deformation Mechanisms, Rheology and Tectonics: Current Status and Future Perspectives, ed. S. de Meer, M.R. Drury, J.H.P. de Bresser, and J.M. Pennock, Spec. Publ. 200, pp. 29–40. London: Geol. Soc.

Zubtsov, S.F., F. Renard, J.P. Gratier et al. 2004. Experimental pressure solution creep of polymineralic aggregates. *Tectonophysics* **385**:45–47.

13

Effects of Melting on Faulting and Continental Deformation

CLAUDIO L. ROSENBERG, SERGEI MEDVEDEV,
and MARK R. HANDY

Department of Earth Sciences, Freie Universität Berlin, Malteserstr. 74–100,
12249 Berlin, Germany

ABSTRACT

The presence of melt is closely related to the localization of deformation in faults and shear zones in a variety of tectonic settings. This relationship is observed on length scales from the outcrop to plate boundary faults to orogens. However, the question of whether melting induces localization, or localization creates a pathway for melts, can rarely be answered from field observations alone. Experimental studies show that rock strength decreases exponentially with increasing volume percentage of melt. This suggests that melting facilitates strain localization where deformation would be homogeneous in the absence of melt. Yet, the extrapolation of experimental relationships between rock strength and melt content to natural conditions at depth in the lithosphere remains speculative, largely because the grain-scale processes underlying dramatic weakening at small amounts of melt have yet to be investigated in crustal rocks. New geochronological methods for dating minerals that crystallized during deformation in the presence of melt have the potential to constrain the time lag between the onset of melting and deformation in naturally deformed anatectic rocks. An indirect, but clear answer to the question of whether melting induces strain localization on a regional scale comes from numerical models of orogenesis which can be run in the presence or absence of low-viscosity domains that approximate the mechanical behavior of partially melted rock. These models show that melting induces lateral flow of anatectic crust within horizontal channels usually situated at the base of the continental crust. These channels have strong vertical strain gradients, especially at their boundaries where shear zones accommodate lateral extrusion of the anatectic rock in between. Together with their bounding shear zones, these flow channels form a new class of faults, which we term "extrusional faults." Extrusional faults containing long-lived melt (tens of millions of years) can support large, broadly distributed topographic loads such as orogenic plateaus and can exhume deeply buried rocks from beneath orogens. In contrast, strike-slip and oblique-slip faults serve as steep conduits for the rapid ascent, differentiation, and crystallization of melt. The relatively short residence time of melts in such moderately to steeply dipping fault systems can lead to episodic motion, with long periods of creep punctuated by shorter periods of melt veining, magmatic activity, and/or faster slip.

INTRODUCTION

Mechanical coupling within the continental lithosphere is manifested by a wide variety of the first-order structural features, from the geometry of faults and shear zones to the topography of orogens (e.g., Royden 1996) and the architecture of passive margins (e.g., Hopper and Buck 1998; Brun 1999). In the absence of melt, the solid-state creep of minerals governs rock rheology and determines the location of decoupling horizons within lithologically and rheologically stratified lithosphere (Ranalli and Murphy 1987). Melting obviously changes rheology and mechanical coupling for the simple reason that melts have very low viscosities compared to that of rock undergoing solid-state creep (Cruden 1990).

This chapter focuses on how melting affects the structure and the rheology of continental crust. As used below, "melting" refers to the process of partial fusion (anatexis) within crust that is subjected to prolonged temperatures above its solidus. Thus, we consider the effects of regional melting on faulting and shearing in the intermediate and lower crust, rather than any local effects associated with flash-heating and ephemeral melting during coseismic slip on fault surfaces in the upper crust (Chapter 5). We note that regional melting usually occurs well below the depth interval of the brittle-to-ductile transition in melt-free crust (Chapter 1), although we hasten to add that melting can certainly induce fracturing during viscous creep, as previously documented in several studies (e.g., Handy et al. 2001, and references therein).

The dramatic weakening effect of melt in crustal rocks has been known for several decades, both from experiments (Arzi 1978) and field studies (Hollister and Crawford 1986), but the grain-scale mechanisms of melt-induced weakening have been debated to the present day (e.g., Brown and Rushmer 1997; Rosenberg 2001). Renewed interest in synkinematic melting in recent years has stemmed primarily from two discoveries: first, seismic and magnetotelluric campaigns have detected partial melt within active orogens (Nelson et al. 1996; Schilling et al. 1997) usually at or near the base of thickened continental crust over areas of hundreds to thousands of km^2. Second, numerical models of orogenesis show that the geometries of some orogens can only be reproduced if viscosity is reduced by an order of magnitude in at least a part of the lower, orogenic crust (e.g., Beaumont et al. 2001). Experiments on partially melted aggregates have shown that partial melting is the only viable mechanism for inducing such a marked drop in viscosity (e.g., Hirth and Kohlstedt 1995a, b).

These findings support the idea that melt-induced and -assisted flow is fundamentally important for the development of faults, structure, and topography at the orogenic scale.

In this chapter, we assess current knowledge of melt-induced effects on fault rocks and shear zone patterns. After reviewing experimental studies of deforming, melt-bearing rocks on the grain scale, we consider different approaches for obtaining estimates of melt content and residence time on different time and length scales in the continental crust. Numerical models of orogenesis indicate that the topography of mountain belts is inextricably linked to the presence or absence of melt-bearing rocks in the deep crust. We conclude with an outlook on possibly fruitful avenues of future research.

EXPERIMENTAL DEFORMATION OF MELT-BEARING CRUSTAL ROCKS

A long-standing debate centers on the question of whether the reduction of rock strength with increased melt volume is exponential, linear, or is characterized by one or more discontinuities at specific melt volumes (Rosenberg and Handy 2005). The debate began when Arzi (1978) combined experimental strength data and

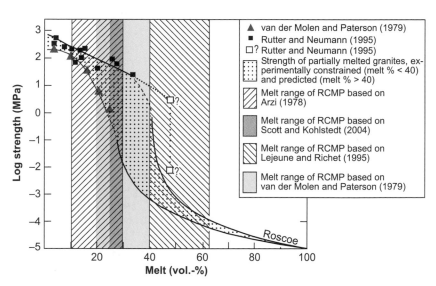

Figure 13.1 Logarithmic strength versus melt volume % (modified after Rosenberg and Handy 2005). Open squares with question marks indicate a possible, but experimentally unconstrained range of strengths for samples containing 40% melt (Rutter and Neumann 1995). Continuous black lines show curves calculated from Roscoe's (1952) equation for a suspension with grain-shape parameters used by Arzi (1978, left curve) and Lejeune and Richet (1995, right curve).

suspension theory to infer the existence of a dramatic strength drop within a critical range of volume percentages of melt (10–40%), termed the rheological critical melt percentage (RCMP, Figure 13.1). This strength drop was interpreted to coincide with a structural transition from a solid framework of crystals with interstitial melt pockets to a suspension of crystals in melt (Arzi 1978).

In Figure 13.2, we show that the available experimental data for deformed, partially melted aggregates can be fit by curves of exponential form, irrespective of the rock composition and starting texture in the experiments (aplite, granite, orthogneiss, amphibolite, experiments cited in the caption). The strength curves in Figure 13.2 are very steep at low melt volumes (<7 vol.-%), indicating that small changes in the amount of melt effect drastic changes of aggregate strength. At melt volumes greater than ~7%, all strength curves flatten, indicating a more moderate dependence of bulk strength on the amount of melt. The maximum change of slope of the exponential curves occurs at melt volumes of ~7%. The strength drop at melt volumes <7% appears to contradict previous work in silicates (Arzi 1978; van der Molen and Paterson 1979; Wickham 1987; Lejeune and Richet 1995; Scott and Kohlstedt 2004) and a variety of nongeological materials (compilation of Vigneresse and Tikoff 1999) claiming that strength drops most markedly at much higher melt volumes (20–50%, Figure 13.1).

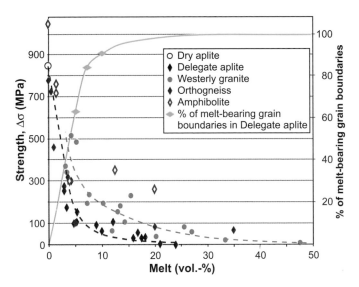

Figure 13.2 Plot of strength versus melt volume % (modified from Rosenberg and Handy 2005). The experimental data are fitted by continuous curves. However, two straight lines intersecting at melt volumes of ~7% could also fit the data (Rosenberg and Handy 2005). Experimental data on delegate aplite is from van der Molen and Paterson (1979); westerly granite: Rutter and Neuman (1995); orthogneiss: Holyoke and Rushmer (2002); amphibolite: Rushmer (1995); dry aplite: Dell'Angelo and Tullis (1988).

It turns out that this discrepancy is not real, but only the result of the different scales used to plot the strength of the experimentally deformed aggregates. All previous authors plotted sample strength on a logarithmic strength scale (Figure 13.1) as a convenient way of depicting a strength decrease of more than four orders of magnitude at melt volumes between 20% and 50%, taken to be the RCMP. As shown in Figure 13.3, however, plotting strength on a linear scale reveals two strength drops: a large drop of ~800 MPa and nearly one order of magnitude at melt volumes between 0 and 6–7%, and a smaller drop of only a few MPa, but nearly 4 orders of magnitude at 20–50 vol.-% melt. This second drop is only visible in linear plots with an expanded lower end of the vertical axis (Figure 13.3). The first, larger strength drop does not correspond to a transition from a solid to a liquid suspension (solid-to-liquid transition of Rosenberg and Handy 2005; Figure 13.3), in the sense of the RCMP defined above. This conclusion relies first, on the relatively high (>100 MPa; Figure 13.2) differential stress that is still supported by the samples at melt volumes of 7%, suggesting the presence of a solid framework, and second, on the fact that solid aggregates collapse to form a suspension only if the liquid attains a minimum volume of 26% (e.g., van der Molen and Paterson 1979).

The more prominent first drop can be attributed to a transition from intragranular deformation of a crystal framework containing melt in isolated or partly connected pockets at 7 vol.-% to intergranular deformation of this framework within an interconnected network of melt film at 7 vol.-% (melt connectivity transition of Rosenberg and Handy 2005). Admittedly, this interpretation is speculative because the melt topology of the samples plotted in Figures 13.2 and 13.3 has not yet been investigated. However, microstructural analysis of the 3D melt network in sheared samples of olivine containing 7 vol.-% of metallic melt revealed interconnected melt films (Bruhn et al. 2000) within a continuous framework of solid grains. In addition, 80% of the grain boundaries of samples of Delegate aplite containing 7 vol.-% of melt were wetted by melt (van der Molen and Paterson 1979; Figure 13.2). At lower melt volumes, the percentage of grain boundaries wetted by melt showed a drastic decrease (Figure 13.2). Hence, the pronounced weakening at 7 vol.-% melt is interpreted to result from the concentration of deformation along interconnected, melt-bearing grain and phase boundaries within a solid aggregate (Rosenberg and Handy 2005). Hirth and Kohlstedt (1995a, b) attributed weakening of olivine aggregates deformed in the presence of basaltic melt primarily to the increased contact area of the melt along the grain boundaries, hence to the change in load-bearing area of the grain contacts. In addition, they showed convincingly that melting leads to an increase in the contribution of grain-boundary sliding during dislocation creep under constant load.

The extrapolation of these laboratory relations to natural rates and temperatures is problematic for several reasons: (a) the curves in Figure 13.2 are only

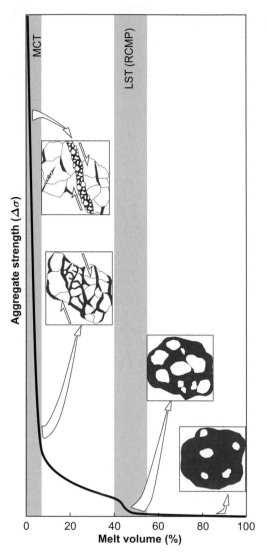

Figure 13.3 Schematic plot of aggregate viscous strength versus melt volume % for silicate rocks between the liquidus and solidus (modified from Figure 4 in Rosenberg and Handy 2005). Note the two strength drops at the melt connectivity transition (MCT) and liquid-to-solid transition (LST); RCMP is rheological critical melt percentage. The vertical scale of the lower part of the ordinate is exaggerated to make the LST visible. The microstructural sketches illustrate deformation at different melt vol.-%. At 3 vol.-%, deformation localizes along a melt-bearing fault. At 7 vol.-%, deformation becomes more distributed, but is localized along the interconnected melt network on the grain scale. At 40–60 vol.-%, the solid crystal framework breaks down, but the grains still interact through the melt. Above 60 vol.-%, the solid particles suspended in the melt do not interact.

valid for the peak strength of samples at very small percentages of shortening (2–5%), not for flow strength at the high shear strains typical of natural deformation; (b) all experiments were performed in a closed system at undrained conditions (Renner et al. 2000), whereas natural deformation of melt-bearing rocks involves melt segregation and migration on a broad range of length scales (mm–km) indicative of drained conditions; (c) no reliable constitutive equation for melt-bearing crustal rock that includes melt volume percentage as an independent variable has been constrained yet on the basis of experimental data. Such flow laws are only available for olivine aggregates in the presence of small melt volume percentages (Hirth and Kohlstedt 2003; Zimmerman and Kohlstedt 2004). The paucity of experimental flow laws for anatectic aggregates reflects the basic difficulty of attaining steady state after only low strains in the laboratory. Moreover, cataclasis pre-empts creep due to the high melt pressures that accrue in undrained samples deformed at unnaturally high strain rates in the laboratory.

We note that the onset of melting may initially result in strengthening rather than weakening if water is partitioned from the crystals into the melt phase (Karato 1986). Water depletion hardens the creeping grains, as described for experimentally deformed olivine aggregates containing small percentages (<4%) of basaltic melt (Hirth and Kohlstedt 1996). However, this process, which is also inferred to be active in oceanic gabbros (Hirth et al. 1998), is limited to the onset of melting.

EXTRAPOLATION OF EXPERIMENTAL DATA TO NATURAL STRAIN RATES

The laboratory experiments discussed above were performed at high strain rates (10^{-5}–10^{-4} s^{-1}), many orders of magnitude greater than natural creep rates. At the outset, we should like to point out a common source of confusion amongst experimentalists and structural geologists when discussing the extrapolation and application of laboratory results to nature: Experimentalists measure either stress at specified strain rate (creep tests) or strain rate as a function of applied stress (constant load tests), and then calculate the effective viscosity of the partially melted material. However, what counts from the perspective of tectonic modeling are viscosities as a function of melt content. For example, knowledge of effective viscosity allows modelers to calculate integrated crustal strengths. Historically, discussion on the mechanical properties of partially melted crust has been based on experimentally derived changes in sample strength as a function of melt content, without explicitly regarding viscosity during the experiments. Confusion has arisen because materials with different viscosities can support similar stresses while deforming at disparate rates in a crustal section. In the following, therefore, we pay special attention to the viscosities of partially melted rocks in the rock-mechanical literature.

Despite the aforementioned difficulties of extrapolating laboratory results, there is evidence that the drastic change in the slope of the strength curves at melt vol. of 5–7% (Figures 13.2, 13.3) also pertains to changes in viscosity during natural deformation. So far, the only experimental flow law with direct application to anatectic continental crust is for quartzite containing very low melt volumes (1–2 vol.-%, Gleason and Tullis 1995). Extrapolating the latter flow laws for both melt-bearing and dry (anhydrous, melt-free) quartzite to a natural strain rate of 10^{-15} s^{-1} indicates that the viscosity ratio of melt-bearing to melt-free quartzite decreases as temperature increases from 700 to 800°C (inset to Figure 13.4). At temperatures inferred to induce melting of pelitic rocks in the lower crust of the Himalayas (750–770°C; Patino Douce and Harris 1998), a melt volume of only 1–2% induces a viscosity drop of 25–30% (Figure 13.4). We note, however, that the differences in viscosities of melt-bearing and melt-free quartzite of Gleason and Tullis (1995) reflect contrasting activation energies with large errors as obtained in a deformation rig with a molten salt cell.

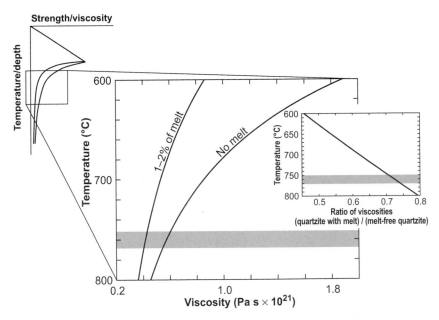

Figure 13.4 Plot of viscosity of quartzite with 1–2 vol.-% of melt versus dry, melt-free quartzite for the temperature range 600–800°C obtained by extrapolating the flow laws of Gleason and Tullis (1995). Note in the inset the decrease in viscosity ratio with increasing temperature, suggesting that melt weakening is more dramatic at lower temperatures. Gray bar indicates the temperature range for crustal anatexis in the footwall of the South Tibetan detachment fault in Tibet (Patino Douce and Harris 1998).

Figure 13.5 shows a theoretical flow law derived by Paterson (2001; inset in Figure 13.5a) for diffusion-accommodated viscous granular flow in a closed system at melt volumes of 0 to ~20% (for a similar formulation see Rutter 1997).

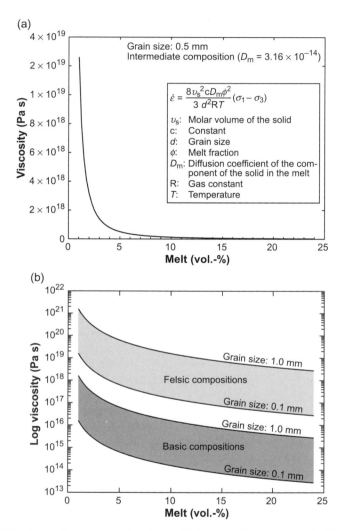

Figure 13.5 Viscosity versus melt-volume % diagrams for the theoretically derived flow law of Paterson (2001). This flow law considers the diffusion of components between grains and melt, and therefore cannot be applied to the solid-state flow of materials (the effective viscosity becomes infinite if the melt vol.-% is 0). Thus, we only calculated the viscosity for melt volumes $\geq 1\%$. (a) Linear plot, for a rock of intermediate composition and grain size of 0.5 mm; see Paterson (2001) for the absolute values of parameters as a function of composition. (b) Log plot, showing the variation in viscosity as a function of grain size and composition.

This curve (Figure 13.5a), which indicates a power law relationship between viscosity and melt-volume percentage, reveals a dramatic change of slope at 3 to 4 vol.-% of melt (Figure 13.5a). The shape of this curve approximates the strength curves of experimentally deformed granite at slightly higher melt-volume percentages (Figures 13.2 and 13.3). Similar power law relationships between melt volume and viscosity apply variously to open systems that allow melt segregation, or to closed systems without melt segregation (Paterson 2001). The viscosity of melt-bearing granite in Figure 13.5a is calculated for a granitoid of intermediate composition and a grain size of 0.5 mm, which is a likely average for migmatitic crustal rocks. The effects of rock composition and grain size on the viscosity of the melt-bearing granitoid are shown on a log diagram in Figure 13.5b.

The available experiments on melt-bearing mantle rocks deformed in the diffusion creep and dislocation creep regimes without cataclasis, and to higher percentages of shortening (15–30%) than the granitoid samples of Figures 13.1 and 13.2, show an exponential decrease in viscosity with increasing melt volume (Hirth and Kohlstedt 2003; Figure 13.6). The viscosity of olivine drops dramatically, similar to the strength of experimentally deformed granite and the viscosity of granite calculated from Paterson's flow law.

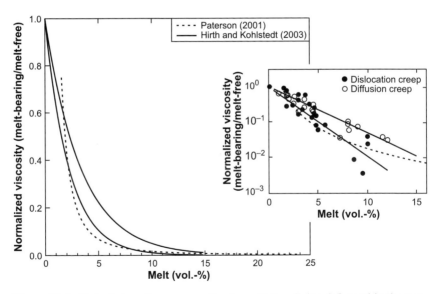

Figure 13.6 Plot of viscosity versus melt volume % for olivine deformed in the presence of basaltic melt, modified from the logarithmic plot of Hirth and Kohlstedt (2003) shown in the inset. The viscosity is normalized to the viscosity of a melt-free aggregate deformed at the same conditions. Continuous curves represent the lower und upper bounds on the experimental data (see inset).

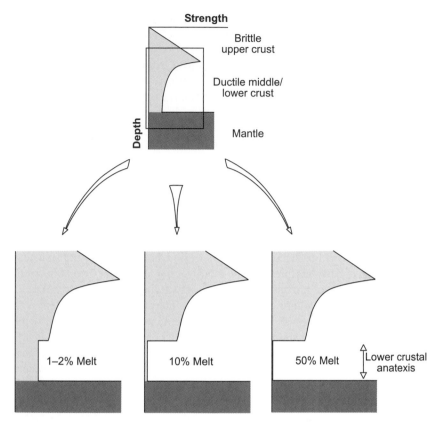

Figure 13.7 Schematic strength profiles of continental crust containing a partially melted layer. The strength drop induced by 10 vol.-% melt is only marginally less than that induced by 50 vol.-%.

The consequence of experimental strength versus melt vol.-% relations for natural melt-bearing systems is visualized in Figure 13.7 in a series of schematic strength profiles for continental lithosphere containing a partially melted, lower crustal layer. These profiles suggest that the upper mantle is easily decoupled from the lower crust once the latter contains more than 7 vol.-% melt. Further melting (e.g., to 50 vol.-% at the solid-to-liquid transition), does not significantly change the structure and integrated strength of the lithosphere (Figure 13.7).

In the following sections of this chapter, we test the validity of the strength profiles in Figure 13.7, first by considering the relationship between estimated present-day melt content and the topography of Tibet, and second by considering the amount of melt-induced weakening necessary to reproduce the geometrical characteristics of orogens in numerical models.

ESTIMATES OF MELT CONTENT AND RHEOLOGICAL TRANSITIONS IN NATURE

Estimates of melt volume in anatectic rocks from exhumed orogens range from 10 to 40 vol.-% (Teyssier and Whitney 2002). However, small amounts of melt (= 5 vol.-%) are probably overlooked in crustal rocks, especially if the melt did not segregate into discrete leucosome lenses. Indeed, several microstructural investigations have reported such small melt percentages (e.g., Sawyer 2001; Marchildon and Brown 2002). In addition, the estimates above cannot tell us whether all of the melt inferred from measurements at the outcrop scale was present in the rocks at the same time.

Real-time distribution of partial melts at depths of more than 20 km beneath the Puna part of the Andean Plateau and Tibetan Plateau have been inferred from anomalies in seismic attenuation (the ratio of P-wave to S-wave velocities, Nelson et al. 1996; Yuan et al. 2000) and electrical conductivity (Schilling et al. 1997; Li et al. 2003; Schmitz et al. 1997; Unsworth et al. 2005). These anomalies have been interpreted to be melt-bearing layers that extend horizontally over hundreds of km and with thicknesses varying from 10 to 40 km (e.g., Nelson et al. 1996; Gaillard et al. 2004). This interpretation is consistent with the inferred temperatures of 700°C at 18 km depth and 800°C at 32 km depth below Tibet, based on the seismically derived depth of the α–β transition in quartz (Mechie et al. 2004). As shown in Figure 13.6, dehydration melting in the Himalayan crust takes place at 750°C (Patino Douce and Harris 1998). Recent magnetotelluric investigations suggest that the melt content below Tibet is not more than 5–14 vol.-%, and may be as small as 2–4 vol.-% below the northwestern part of the Himalayas (Unsworth et al. 2005). By assuming that melt-bearing rocks are porous elastic media on the time scales of the geoelectric measurement, Schilling and Partzsch (2001) used the conductivity results to calculate a melt volume of at least 20% below the Puna and the Tibetan Plateaus.

These estimated melt volumes are consistent with a viscous strength drop of ~15 times below the Puna Plateau and ~10 times below the Tibetan Plateau according to the strength versus melt vol.-% relations compiled in Figure 13.2. As shown below, thermomechanical models of orogenesis require an order-of-magnitude drop in viscous strength within the lower crust to develop orogenic plateaus (Beaumont et al. 2001). In this context, it is interesting to note that the Tibetan Plateau overlies crust containing 5–14 vol.-% melt, whereas only 2–4 vol.-% melt underlies the northwestern part of the adjacent Himalayas (Unsworth et al. 2005). This change in melt content coincides with the range of melt volumes marking the transition from the melt connectivity transition to the more flat-lying part of the strength curve in experimentally deformed, melt-bearing granitic rocks (Figure 13.3). Therefore, we infer that it coincides with a major transition in the integrated strength of the crust. If so, then small amounts of melt may be the prime factor governing variations in plateau topography.

The higher end of the range of melt contents in exhumed anatectic rocks (30–40 vol.-%) corresponds to the liquid-to-solid transition in Figure 13.3 (Rosenberg and Handy 2005; RCMP of Arzi 1978). The liquid-to-solid transition also corresponds to the transition from metatexite to diatexite in partially melted rocks and has been interpreted as the fundamental rheological transition within the ductile part of the continental crust (Vanderhaeghe and Teyssier 2001a). These authors argue that the formation of diatexites at melt volumes as high as the SLT controls the formation of migmatite-bearing domes in the North Canadian Cordillera by weakening the crust to a point that allows the onset of gravitational collapse. We consider this unlikely, however, given that the contact between metatexite and diatexite in these domes does not coincide with any marked structural discontinuity in the sense of a shear zone (Vanderhaeghe and Teyssier 2001a, their Figure 4). These observations reinforce our opinion that the melt connectivity transition is far more important than the liquid-to-solid transition from a rheological standpoint. If indeed melt volumes in natural systems reach 30 to 40 vol.-% at the liquid-to-solid transition, then this results in a comparatively modest drop in strength. To our knowledge, systematic changes of structural style related to variations in viscous strength as a function of melt content have yet to be described in naturally deformed anatectic rocks.

RESIDENCE TIME OF MELT IN OROGENIC CRUST

The rate and time to produce and maintain a rheologically critical amount of melt (= 7 vol.-% at the melt connectivity transition in Figure 13.3) govern the effect of melting on faulting in the lithosphere. Melting times are poorly constrained, partly because this time varies with the volume of melt considered and partly because melting rates are not well known. A minimum melt time of 10^5 years is obtained from studies of plutons (compilation of Petford et al. 2000). An upper limit for the residence time of melt in orogenic crust can be estimated by relating the extensive layers of geophysically imaged, melt-bearing crust beneath the Andean and Tibetan Plateaus to the ages of exposed magmatic bodies inferred to have formed by melting of these layers. Magmatic activity in the northern, Altiplano part of the Andean Plateau started in Miocene time some 23 Ma ago and has continued unabated to the present (de Silva 1989), leading to the formation of a large ignimbritic complex. Dacitic volcanism in the more southerly Puna part of the plateau started at 10 Ma and continued until 2 Ma (Riller et al. 2001). The melt feeding this volcanism is still present in a partially melted layer at 20–25 km depth, as determined by the geophysical studies cited in the previous section (e.g., Yuan et al. 2000). The residence time of melt beneath the Andean Plateau is therefore at least 10 Ma, possibly as much as 23 Ma.

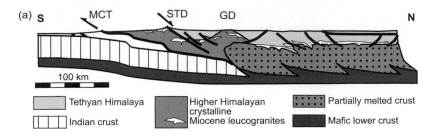

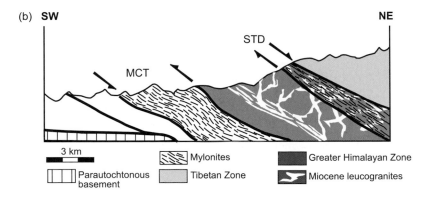

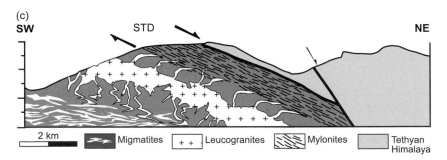

Figure 13.8 Cross sections of the Tibetan-Himalayan orogen. MCT: Main Central Thrust, STD: South Tibetan detachment, GD: gneiss domes. (a) INDEPTH profile, modified from Nelson et al. (1996). The partially melted region is inferred from seismic data. (b) Cross section of the Annapurna area in the Himalayas, modified from Hodges et al. (1996). Note that leucogranites are concentrated in the central part of the High Crystalline Complex, not along the South Tibetan detachment. (c) Cross section of the STD, in the Northwest Himalayan (Zanskar), modified from Dèzes et al. (1999). Migmatites concentrate well below the extensional mylonites of the STD, confirming the observations of the Annapurna section. Leucogranitic plutons are located well below the detachment and are only marginally affected by mylonitic deformation.

The melt imaged below the Tibetan Plateau may be related to a chain of leucogranite bodies situated in the footwall of a large (>1000 km length) low-angle normal fault, the South Tibetan detachment system (Figure 13.8 and discussion in section below on FIELD-BASED MODELS OF THE EFFECTS OF MELT ON LARGE-SCALE FAULT ZONES). These granites range in age from 24 to 10 Ma (e.g., Zhang et al. 2004), most of them from 22 to 19 Ma (review of Searle and Godin 2003). Some authors claim that these leuco-granites are continuous with the partially melted intracrustal layer imaged below the Tibetan Plateau (e.g., Nelson et al. 1996; Wu et al. 1998; Fig. 13.8). If so, the lower crust of Tibet has been partially molten for more than 20 Ma, similar to the maximum residence time of melt in the lower crust of the Andean Plateau. This long time interval covers a significant part of the uplift history, which probably started in latest Cretaceous to Early Tertiary time in the Tibetan region (Yin and Harrison 2000) and in Early Tertiary time in the Andes (e.g., Lamb et al. 1997). Dating of synkinematic leucosomes in anatectic rocks of older orogens (e.g., Variscides: Brown and Dallmeyer 1996) also shows that melt may have been present in the crust for similarly long times.

The long residence times of melts beneath orogenic plateaus contrast with the much shorter times of melts in steeply-dipping shear zones and faults (= 1.5 Ma; e.g., Davidson et al. 1992; Oberli et al. 2004). As dis-cussed below, these short residence times can lead to transient motion (Handy et al. 2001).

FIELD-BASED MODELS OF THE EFFECTS OF MELT ON LARGE-SCALE FAULT ZONES

Melt-induced weakening of the crust is expected within subhorizontal, lower crustal layers for the simple reason that—barring decompression melting dur-ing rapid exhumation—isotherms at or near the solidus are generally subhori-zontal. Shear zones engendered by melt-induced weakening at depth form decoupling horizons within the lithosphere, but their long-term effect on the bulk rheology of the lithosphere is expected to depend on their geometry as well as on the regional stress field. For example, most strike-slip faults are steeply inclined and transect melt-bearing layers at larger angles than thrusts and normal faults. Whereas thrusts and normal faults may root in weak, melt-bearing layers, strike-slip faults are more likely to serve as conduits for the channeled ascent of melts to higher crustal levels (e.g., D'Lemos et al. 1992; Handy et al. 2001) where they crystallize rapidly (e.g., Davidson et al. 1992). In the following, we examine some of the controls on melt-enhanced shearing and lithospheric rheology.

Transpressive Settings

Historical Perspective

Hollister and Crawford (1986) were the first to argue that there is a causal relationship between large-scale deformation and melting in orogenic crust. They proposed that melting weakens the lower crust significantly during orogenesis, thereby increasing strain rates there and augmenting the exhumation rates of crustal blocks confined between melt-bearing shear zones. These enhanced displacements were termed "tectonic surges." Their existence was posited mainly on the following observations: (a) Crustal rocks weaken significantly upon melting, as shown in experimental deformation of partially molten aggregates; (b) There is a close spatial relationship between sites of large deformation (shear zones) and the occurrence of migmatitic or magmatic rocks, for example, in the Coastal Mountains of British Columbia; (c) Rapid decompression (exhumation) in the Coastal Mountains coincided with the thermal peak of metamorphism, which induced partial melting.

Whether or not melting of lower crust is necessary to attain such rapid exhumation rates is questionable. Since Hollister and Crawford's (1986) landmark paper, exhumation rates much higher than those they reported (1 mm yr^{-1}) have been documented in several orogens that lack visible evidence of melted crust (e.g., Milliman and Syvitski 1992). The coincidence of high exhumation rates in crustal blocks with the occurrence of melt in adjacent shear zones does necessarily mean that melt was the main agent of the increased rates of exhumation.

Numerous field-based investigations of large-scale transpressive fault systems have demonstrated the close spatial relationship between migmatites and/or magmatic rocks and mylonitic shear zones (e.g., Davidson et al. 1992; D'Lemos et al. 1992; McCaffrey 1992; Hollister 1993; Ingram and Hutton 1994; Tommasi et al. 1994; Berger et al. 1996; Neves et al. 1996; Vauchez et al. 1997; Tikoff and de Saint-Blanquat 1997; Brown and Solar 1998). All these studies established that mylonitization occurred in the presence of melt. Thus, shear zones were believed to nucleate in the melt-bearing crust and propagate into thermally weakened country rocks (Neves et al. 1996).

Unfortunately, field observations alone are insufficient to discriminate between melt-induced localization of deformation and deformation-induced melt channeling, because the structural evidence of both processes is probably the same, viz., the occurrence of granite in shear zones. The fact remains that the available criteria are equivocal, as pointed out by Vauchez et al. (1997) and discussed below.

Tertiary Plutonism in the Alps

The dextral transpressive Periadriatic fault system in the Alps is closely associated with Late Oligocene plutons (Figure 13.9a) whose source region is inferred to be the base of the thickened Alpine orogenic crust (von Blanckenburg et al. 1998). The Periadriatic fault system extends from the surface down to the top of the lower, mafic crust as shown in geophysical transects (Schmid and Kissling 2000). All plutons exposed adjacent to the Periadriatic fault system crystallized within a restricted time interval of approximately 5 Ma during a broader period of fault activity (review in Rosenberg 2004). The exhumation of crustal levels from the surface down to 25–30 km allows a unique reconstruction of the geometrical relationships between magmatic bodies and shear zones in profile, as shown in Figure 13.9b. This reconstruction shows that magmatic bodies accompany the fault plane almost continuously from the surface to the maximum paleodepth of 25–30 km and possibly beyond, whereas no intrusive bodies occur away from the fault plane (Figure 13.6b). Thus, the plutons ascended along the Periadriatic fault system.

Isotopic ages showed that the base of the Bergell tonalite (marked B in Figure 13.9a) at a paleodepth of ~25 km (Figure 13.9b) remained in a partially molten state for at least 1.5 Ma (Oberli et al. 2004) within the mylonitic belt of the Periadriatic fault system. In contrast, the close similarity of ages obtained by isotopic systems with different closing temperatures on upper crustal (≤10 km depth) plutons such as the Biella Pluton (Western Alps, Italy; Figure 13.9a) and Adamello (Southern Alps, Italy; Figure 13.9a) indicates rapid crystallization. Therefore, the effect of melt on deformation is expected to depend strongly on the level of melt emplacement, and hence on the melt residence time. 1.5 Ma is probably an upper time limit for the existence of melt in a pluton that is ascending as an elongate sheet along a fault plane. This factor represents a fundamental limitation to the process of melt-weakening in transpressive systems. Once the melt crystallizes, the pluton plus its host shear zone are expected to harden (Handy et al. 2001).

We emphasize that the close spatial and temporal relationship between plutons and the Periadriatic fault system (Figure 13.9a) does not result from melt-induced strain localization, but rather from deformation-induced channeling of melts into an active, orogen-scale fault system (Rosenberg 2004). The fault rocks of the Periadriatic fault system overprint first-order Mesozoic paleogeographic and Alpine metamorphic boundaries that have been interpreted as the sites of repeated transform, strike-slip motion in Jurassic and Late Cretaceous times (Schmid et al. 1989, Froitzheim et al. 1996), long before Tertiary intrusive activity and differential exhumation of the plutons affected the retrowedge of the Tertiary Alpine orogen. Moreover, numerical models of the Central Alps indicate that an orogenic retrowedge bounded by a steep backthrust like the Periadriatic fault system in the central to western part of the Alps develops in kinematic response to a subduction singularity irrespective of the presence of melts and of numerous rheological heterogeneities (Schmid et al. 1996). If the present exposure of minor Tertiary dykes that are geochemically related to the plutons (Figure 13.9a) is taken as a first-order proxy for the areal extent of the

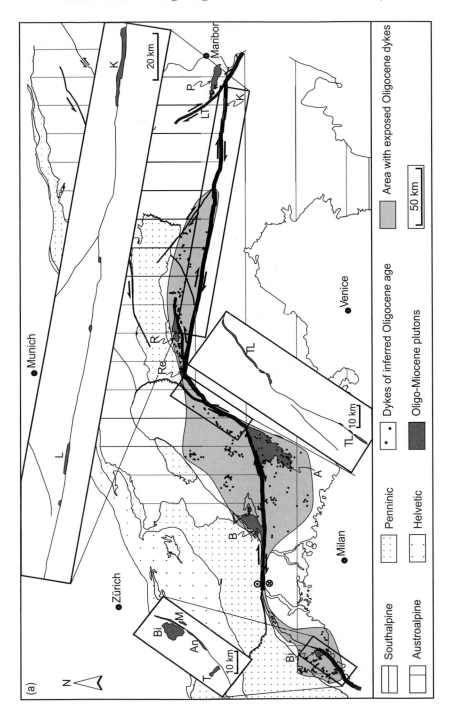

(b)

N   **S**

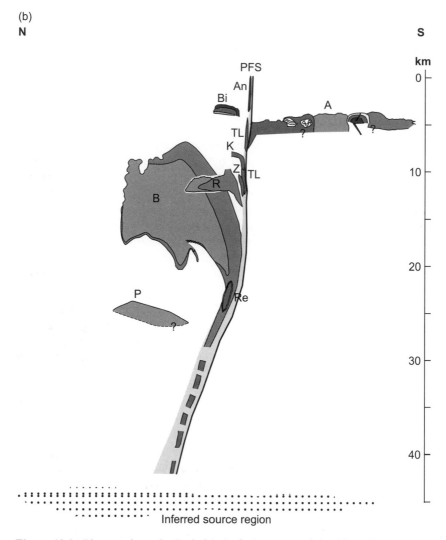

Figure 13.9 Plutons along the Periadriatic fault system of the Alps. *Facing page:* (a) Simplified tectonic map of the Alps showing the major Tertiary faults and shear zones, the Oligo-Miocene plutons, and locations of Tertiary dykes (modified from Rosenberg 2004). Periadriatic fault system is shown with thick black lines. Boxed areas contain detailed maps of thin magmatic sheets along the Periadriatic fault system. *Above:* (b) Synthetic cross section showing the relationship of the Periadriatic fault system (PFS) to the plutons at their original depths of emplacement (modified from Rosenberg 2004). The depth and distance of each pluton from the Periadriatic fault system are constructed from geobarometric and field data. A: Adamello Batholith; B: Bergell Pluton; Bi: Biella Pluton; K: Karawanken Pluton; P: Pohorje Pluton; R: Rensen Pluton; TL: Tonalitic Lamellae; Z: Zinsnock Pluton.

underlying melt source, then the concentration of the synkinematic feeders of the intrusive bodies within the 2–3 km wide mylonites of the Periadriatic fault system gives a good impression of the degree to which upward melt flow was channeled parallel to the steep mylonitic foliation.

Magmatic Arcs

Arc-parallel strike-slip faulting (e.g., Jarrard 1986) is thought to result from strain partitioning of overall oblique convergence into a steep zone of predominantly simple shear flanked by domains of more distributed pure shear (e.g., Teyssier et al. 1995). Besides the obliquity of convergence, crustal strength is an important control on this strain partitioning. Several authors have pointed out the obvious relationship between arc-parallel strike-slip faults and magmatic arcs, which are interpreted as zones of crustal weakness due to the high thermal gradients associated with the advection of heat from melts (e.g., Jarrard 1986; Scheuber and Andriessen 1990). Melt-induced weakening may explain why strike-slip faulting is very common in the upper plate of ocean–continent subduction systems (de Saint-Blanquat et al. 1998), whereas only one fifth of upper oceanic plates in ocean–ocean subduction systems have such structures (Jarrard 1986). This holds true even in subduction systems with high convergence angles (e.g., Sumatra: 50°, Andes: 60–90°), in other words, at angles that do not favor partitioning in the absence of melting. A case in point is South Island, New Zealand, where active magmatism is absent and no strike-slip partitioning occurred, in spite of the highly oblique convergence (16–29°; de Saint-Blanquat et al. 1998, and references therein). Note however, that weakening of the crust, leading to the partitioning of deformation along arc-parallel strike-slip faults may not only result from the presence or absence of melt, but from the occurrence of older anisotropies, as suggested for the Taiwan subduction system (Fitch 1972), or by stronger erosion in the retrowedge of the accreted crust, which may allow the lateral and the convergent components of strain to occur both on the same fault plane, as inferred for South Island (New Zealand; Koons et al. 2003). The geometry of the melt bodies presumed to be responsible for weakening at depth within magmatic arcs is unknown. Several studies have noted a positive correlation between the intensity of magmatism (volume of melt generated per time) and the obliquity of convergence (Western USA: Glazner 1991; Andes: Günther 2001). This was thought to result from the intrusion of melt into secondary extensional structures at releasing bends along the strike-slip faults (Glazner 1991; McNulty et al. 1998). If so, then transcurrent deformation is the cause for the spatial and temporal association of faults and plutons. This contrasts with the idea propounded above that melt induces the partitioning of deformation by reducing the viscous strength of the crust within the arc.

Positive Feedbacks between Melting and Faulting

Different opinions on the cause and effect of strain localization in melt-bearing crust have been reconciled by the idea that deformation and melt interact in a positive feedback loop. De Saint-Blanquat et al. (1998) suggested that magma ascent induces localization of deformation into strike-slip faults, which in turn create the space for melt ascent, which further weakens the crust and hence reinforces localization of deformation into the strike-slip zone. A slightly different process was described for intracontinental settings by Brown and Solar (1999), who pointed out that transpressive deformation on the orogen-scale leads to an upward displacement of the isotherms (e.g., Huerta et al. 1996), which creates an antiformal thermal structure. This structure can be amplified, if heat is advected by ascending melts, thus extending the zone of deformation upward, which in turn favors the upward migration of melts in a positive feedback loop (Brown and Solar 1999).

Positive feedbacks during a single magmatic cycle are depicted in the series of sections through a generic strike-slip fault in Figure 13.10, following Handy et al. (2001, their Figure 11). Incipient melting at depth (juvenile stage, Figure 13.10a, d) thermally weakens the lower crust, increasing the strain rate and loading shallower crustal levels of crust that undergo solid-state creep and frictional sliding. Together with the accumulation of large bodies of segregated melt, this favors melt-induced upward veining which facilitates the rapid, buoyant rise of melt within the fault system (climax stage in Figure 13.10b, e). During this stage, the rise of the isotherms (e.g., Huerta et al. 1996) combined with the presence of low-viscosity melt in veins connecting plutons with their source regions at depth act to accelerate fault movement. For example, Davidson et al. (1992) has estimated that a km-thick syntectonic tonalite within the McLaren Glacier metamorphic belt in Alaska accommodated at least 10 km of displacement within an estimated time to crystallization of only 90,000 years. Mature fault zones (Figure 13.10c, f) are expected to harden as the melts within them crystallize, the isotherms subside, and the geotherm decreases. In fact, mature fault zones can attain an integrated strength greater than their pre-melting strength if the crystallized melts (e.g., mafic melts) have greater solid-state creep strengths than the rocks they displaced during intrusion.

The model in Figure 13.10 suggests that the feedback between deformation and magmatism in oblique-slip fault systems may induce cyclical weakening-then-hardening of the continental crust on time scales of only 10^3–10^5 years (Handy et al. 2001). This is much shorter than the total duration of motion (10^6–10^7 years) along the plate boundaries in which the faults occur. We have pointed out before that episodic melt-induced fault slip may be responsible for repeated, sudden shifts in sedimentary depocenters and volcanic fields along the margins of basins bounded by oblique-slip faults (examples in Biddle and Christie-Blick 1985).

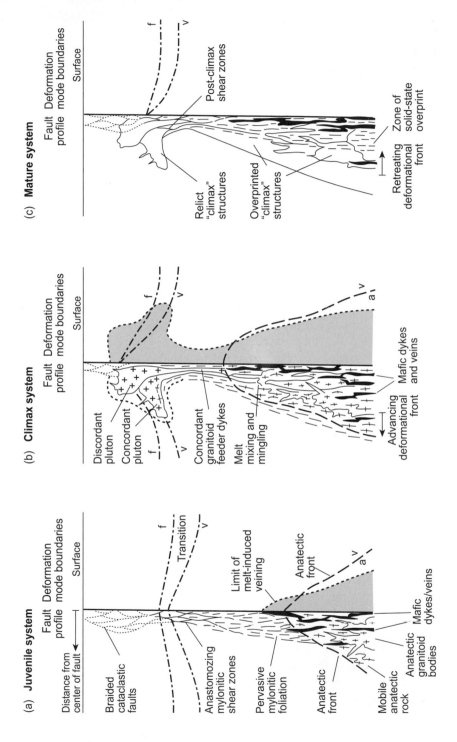

(a) **Juvenile system**

(b) **Climax system**

(c) **Mature system**

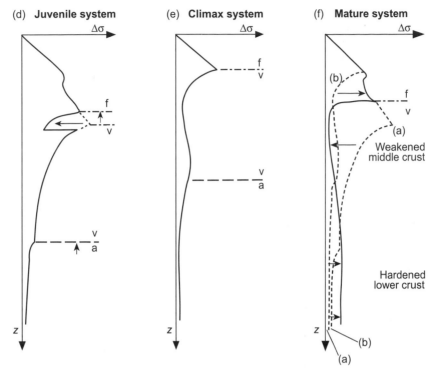

Figure 13.10 *Facing page:* Structure and strength versus depth diagrams for a generic strike-slip fault zone undergoing one cycle of syntectonic magmatism (modified from Fig. 11 of Handy et al. 2001). Structure versus depth diagram is shown for (a) the juvenile, (b) climax, and (c) mature stages. The dashed-dotted curves represent the frictional-viscous (f-v) transition; dashed curve is the transition from solid-state viscous creep (marked v) to melt-assisted viscous granular flow (anatectic flow, marked a); the dotted curve is the limit of melt-induced veining (gray area on right-hand side of the diagram). The figure depicted above shows strength versus depth diagrams for (d) the juvenile stage, (e) the climax stage, and (f) the mature system stage. Bold dotted lines in (f) depict the strength profiles during previous stages; arrows indicate movements of curves since these stages.

Syn-orogenic and Post-orogenic Extensional Settings

Historical Perspective

Wernicke et al. (1987) were the first to note a systematic relationship between the onset of crustal extension and the amount of Late Cretaceous–Early Tertiary plutonism in the North American Cordillera. Areas where extension initiated earliest (55–49 Ma) have the largest volumes of intrusive rock compared to other areas where extension began later (38–20 Ma, minor intrusive rocks), or only ~15 Ma ago (no intrusive rocks). They proposed that extension results from gravitational spreading of the previously thickened lithosphere and that extension initiated in the partially melted lower crust with the attainment there of the rocks' melting temperature, as previously suggested by Coney and Harms (1984). In this interpretation, the low-melt viscosity was not considered to be the direct cause of crustal weakening, but instead melt was interpreted to indicate high temperature, hence thermal weakening of the lithosphere. The close spatial and temporal relationship between magmatism and extension in the North American Cordillera was later corroborated by the synthesis of an immense set of structural and geochronological data covering western North America from Alaska to Mexico (Armstrong and Ward 1991). In contrast to Wernicke et al. (1987), Armstrong and Ward (1991) recognized that melt drastically reduces the viscosity of the crust due its own very low viscosity, in addition to thermal weakening of its solid host rocks. They suggested that melt-induced weakening triggered extension of previously thickened crust. This idea has since been used to explain coeval melting and extensional deformation in Tertiary orogens such as the Tibetan Plateau (e.g., Burchfiel et al. 1992; Hodges 1998), the Hellenides (Vanderhaeghe and Teyssier 2001b), as well as in older orogens like the Variscides (Vanderhaeghe et al. 1999) and the Caledonides (e.g., McLelland and Gilotti 2003).

In the following, we refer to gravitational collapse as "gravity-driven ductile flow that effectively reduces lateral contrasts in gravitational potential energy" (Rey et al. 2001). As shown by Willett and Pope (2004) this process may be transient. Absent any changes in tectonic forces, the critical parameter controlling gravitational collapse is the ratio of the gravitational load to the strength of the crust (Rey et al. 2001). Therefore, weakening the crust by partial melting can trigger such collapse. Structural and geochronological evidence for coeval extensional faulting and magmatism supports, but does not prove, the hypothesis that melting triggers gravitational collapse. Orogenic modeling has shown that both extensional faulting and crustal anatexis may be triggered by other tectonic processes (e.g., loss of a dense, isostatically unstable root or a subducting slab; Houseman et al. 1981). Thus, in some tectonic settings neither process causes its other, but both may have a common cause.

If the onset of extension systematically postdates the onset of melting within a short time interval, then melting may be inferred to trigger gravitational collapse. However, establishing such a causal relationship is beset with the basic problem that extension is commonly dated with crystallization ages of synkinematic migmatites (e.g., Dèzes et al. 1999). Thus derived, the radiometric ages of extension and melting are obviously not independent. Given the duration of magmatic events ($\sim 10^7$ years, see below) during the late stages of orogeny, it is unreasonable to assume that minerals from anatectic leucosome provide anything more than "snapshot" ages of their crystallization during a much longer-lived event.

Likewise, but on a much larger scale, cross-cutting relations between shear zones and intrusive bodies only tell us locally which structure formed first, not whether these structures formed at the very beginning or late in the history of shearing and intrusion, much less which process (shearing or intrusion) started first. Mutually cross-cutting relationships can be interpreted to show that two events were broadly coeval in a given rock volume, but they do not indicate which event began first.

Melting and Syn-orogenic Extension in the Himalayas

The relationship between melting and normal faulting along the South Tibetan detachment system in the Himalayas (STD, Figure 13.8) is the subject of ongoing debate. Some authors infer that crustal melting, which engendered the Miocene leucogranite bodies in the footwall of the STD, resulted from isobaric decompression during extension (Harris and Massey 1994; Dèzes et al. 1999; Harris et al. 2004). Based on thermal modeling, others consider that extensional decompression is unlikely because it requires extremely rapid ($= 20$ mm yr^{-1}) and large-magnitude denudation to produce minor amounts of melt, and because the ages linking slip on the normal faults with melting are not well constrained (Harrison et al. 1999). The latter authors invoke shear heating along the Main Central Thrust (MCT in Figure 13.8; Le Fort 1975; England et al. 1992) as the main heat source for partial melting.

At a very basic level, the debate is fueled by ambiguous field relations. Parts of the STD are truncated by leucogranitic bodies (e.g., Guillot et al. 1994; Edwards et al. 1996) which are themselves deformed by brittle normal faults in the hangingwall of the STD (Brown and Nazarchuk 1993). Some of the granites that intrude the STD have very young ages (12.5 Ma; Edwards and Harrison 1997), which clearly postdate the oldest (late Oligocene) crystallization ages of kyanite-bearing leucosomes in the inferred source region of the leucogranites at the base of the High Himalayan Crystalline Complex (Hodges et al. 1996). In other localities extensional shear zones of the STD overprint the leucogranite (Searle and Godin 2003). Thus, both melting and deformation persisted for several Ma (Searle and Godin 2003; Hodges et al. 1996).

Th-Pb dating of monazites (Kohn et al. 2005) suggests that the MCT initiated as recently as 16 ±1 Ma, earlier than the inferred age of some (but not all) leucogranitic plutons along the STD. Additional complexities arise from recently obtained Oligocene ages (27.5 Ma) of Himalayan plutons north of the STD (Zhang et al. 2004). The composition of these granites indicates that they were derived from the same melt source as the Miocene leucogranites (the High Himalayan Crystalline Complex), but at greater depth. If the Oligocene ages reflect the onset of melting in the High Himalayan Crystalline Complex, then melting definitely initiated before extension along the overlying STD and, indeed, may have triggered this extension. This would rule out shear heating as a cause of melting, a process which we also consider unlikely given the relatively low strain rates (10^{11}–10^{-13} s^{-1}) and low differential stresses (tens of MPa) commonly measured in crustal mylonite (Handy and Streit 1999; see also Chapter 6, this volume), and the extreme paucity of shear zones that experienced higher metamorphic temperatures than their enclosing rocks.

Most of the crystallized melt bodies are not concentrated along the STD, but in its footwall (Figures 13.8b, c). This observation is inconsistent with the idea that melts lubricate mylonitic shear zones (Hollister and Crawford 1986; Hodges 1998) in the sense of a pressurized fluid that reduces effective normal stress and resistance to frictional sliding. Where intrusive bodies do occur within the STD, they show various degrees of overprinting, from mylonitic to undeformed (Dèzes et al. 1999). The intrusive bodies never penetrate to the hangingwall of the STD. Some leucogranites that crosscut the STD were thought to intrude its hangingwall (e.g., Guillot et al. 1994; Edwards et al. 1996), but recent work has shown that these granites are deformed by an extensional shear zone of ~300 m width (Searle and Godin 2003) which is not itself intruded by leucogranites and is part of the STD. Mylonitization on subhorizontal shear zones thus acts as a mechanical barrier to the ascent of melt (Handy et al. 2001).

Perhaps the most important feature in the Himalayan sections is the location of most crystallized melt bodies in the central parts of the High Himalayan Crystalline Complex, between the Southern Himalayan detachment system and MCT (Figures 13.8a, b). Based on independent field evidence this area is inferred to represent a former low-viscosity, melt-bearing channel (Grujic et al. 1996) bounded by shear zones with opposite shear senses. We will return to this below in the context of modeling studies.

Tertiary Melting and Extension in the North American Cordillera

Thin, granitic sills have also been found along the mylonitic tops of metamorphic core complexes and they have been inferred to promote localization of deformation (Whipple Mountains, U.S.A.; Lister and Baldwin 1993). Low viscosity bodies must be included in scaled, analogue, and numerical models of metamorphic core complexes in order to obtain localized extension and core complex formation (Brun et al. 1994; Tirel et al. 2004). Thus, melt intrusion may augment localization during extension. In two other metamorphic core complexes of the North American Cordillera, the onset of melting is inferred to precede the onset of extension by a considerable amount of time (10 Ma in the Shuswap core complex, British Columbia, Canada: Vanderhaeghe et al. 1999; 30 Ma in the Bitterroot core complex, Montana, U.S.A.: Foster and Fanning 1997). These authors therefore conclude that melting triggered gravitational collapse by reducing the strength of the lower crust (also Foster et al. 2001).

Yet, these interpretations are partly based on ambiguous, sometimes contradictory data. Decompression of the Shuswap core complex from 10 to 5 kbar was synchronous with melting (Norlander et al. 2002). However, a temporal distinction between the onset of melting and the onset of extensional faulting is equivocal because both events are dated by the oldest crystallization ages of zircons from synkinematic leucosomes (Vanderhaeghe et al. 1999). In the Bitterroot complex, field evidence points to coeval extensional shearing and magmatism (LaTour and Barnett 1987), but abundant geochronological data indicate prolonged melting and magmatism for at least 30 Ma (beginning before 80 Ma) prior to the onset of extensional deformation at about 53 Ma (Foster and Fanning 1997; Foster et al. 2001). In fact, migmatitic crystallization ages suggest that there were two distinct anatectic events, only the younger of which at ~53 Ma (Foster et al. 2001) was coeval with, or slightly older than the extensional deformation. Thus, gravitational collapse initiated only at the very end of this younger anatectic phase and continued in the absence of magmatism until 43 Ma (Foster and Fanning 1997). If the geochronological data are valid, then anatexis may not have triggered extension, much less the collapse of the Cordilleran orogen. Unfortunately, to our knowledge no structural investigation of the migmatites exists as yet.

The idea that melting triggers syn-orogenic extension is based on the premise that tectonic boundary conditions like the regional convergence rate remained constant for the duration of crustal thickening. This assumption is probably justified for the North American Cordillera, where independent evidence from magnetic anomalies suggests that divergence at the plate boundaries, between the Pacific and the Farallon Plates, began some 10 Ma after the formation of the metamorphic core complexes (Engebretson et al. 1985; Vanderhaeghe and Teyssier 2001a, b). However, the poor constraints on the large-scale plate kinematics of older orogens such as the Variscides (Malavieille et al. 1990; Ledru et al. 2001; Brown 2005) and Caledonides (White and Hodges 2002; McLelland and Gilotti 2003) may not allow one to distinguish between extension induced by changing boundary conditions and by melt-induced changes in the rheology of the thickened crust. For example, the onset of Early Permian extension and magmatism in the Variscides appears to be related to a switch from head-on collision to dextral transpression between Laurussia and Gondwana (Matte 1991).

NUMERICAL MODELING OF FAULTING DURING ANATEXIS

Numerical models allow one to simulate the flow of melt-bearing layers on the scale of the entire crust and to predict the effects of melting and deformation on the surface features of mountain belts (e.g., Beaumont et al. 2001; Babeyko et al. 2002). At the current state of computing power, the resolution of melt-induced flow is practically limited to the scale size of a numerical cell of several kilometers. This resolution is sufficient to predict large-scale flow patterns, but not suited to investigate the nucleation and propagation of individual faults. Melting is modeled by assuming that the crust weakens instantaneously upon reaching the solidus, generally taken to be in the range 700–750°C. Therefore, the segregation of melt during deformation, variations in melt pressure, and changes in energy balance associated with melting and crystallization are usually not considered. An exception to this is the study of Babeyko et al. (2002), in which melt segregation is achieved simply by changing the melt volume percent independent of deformation and pressure.

Influence of Crustal Melting on the Shape of Orogens

Bird (1991) predicted that topographic gradients create pressure gradients within the Earth, which induce flow of a weak lower crust and cause topography to flatten. For example, an initially 2 km high and 300 km wide mountain belt can be reduced to 1 km height within tens of millions of years (Bird 1991, Table 2) due to lower-crustal flow, even in the absence of melt. Weakening the lower crust by a factor of 10 to 20 increases the rate of topographic leveling by at least the same order of magnitude, that is, by 10–20 times. The resulting rate of leveling is of the same order of magnitude as the thickening rate of most orogenic systems (10^{-3} m yr^{-1}). Therefore, orogenesis cannot produce any significant topographic gradients above a partially melted lower crust.

Whereas Bird's model considered local readjustments of topographic gradients, later analytical models investigated the relationship between the shape of

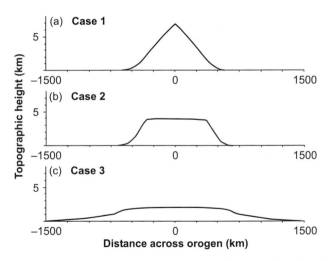

Figure 13.11 Generic cross-sectional shapes of modeled orogens, from Royden (1996). All cross sections show the orogenic geometry after 16 Ma of shortening. Case 1: Uniform viscosity crust; Case 2: Initial uniform viscosity followed by formation of a low-viscosity zone at the base of the crust; Case 3: Presence of low-viscosity zone at the base of the crust prior to the onset of convergence.

orogens and crustal viscosities (Figure 13.11, Royden 1996). Orogens modeled with uniform crustal rheology acquire triangular cross-sectional shapes (back-to-back wedges in Figure 13.11a), whereas significant weakening of the lower part of the thickened crust results in a plateau-like geometry (Figure 13.11b). Royden (1996) did not explicitly mention melting in the lower crust, but the rheology in her preferred plateau model (Figure 13.11b) is based on a viscosity reduction of almost two orders of magnitude per km depth. Such a drop in viscosity can only occur at the transition from solidus to hypersolidus conditions. Vanderhaeghe et al. (2003) presented a series of models which extended Royden's (1996) model and concluded that partial melting in the orogenic crust can change the shape of an orogen from triangular to plateau-like.

To illustrate the effect of lower crustal melting on the height and the width of a generic orogen, we consider the balance of horizontal forces acting on one half of an evolving orogenic crust subjected to continuous compressional basal traction (Figure 13.12), Our approach follows the models outlined in Medvedev (2002) and Vanderhaeghe et al. (2003). The horizontal compressional force, F_c, is inversely proportional to the width of the orogen and therefore decreases progressively during growth of the orogen (Medvedev 2002). This force plays a limited role in the balance of forces for a uniformly linear viscous crustal wedge, so we neglect it for the sake of simplicity. The two remaining forces, the gravitational force (F_g) associated with a difference in elevation between the mountain range and the foreland and the basal tractional force (F_g), must balance: $F_t = F_g$. The gravitational force, F_g, is the difference in potential energy between mountain and foreland. F_g, increases with thickening of the crust. It is proportional to the difference in density between crust and mantle, $\rho_m - \rho_c$, and to the square of the change in crustal thickening, $h_{max}^2 - h_0^2$. The basal tractional force, F_t, is proportional to the viscosity at the base of the crust, η_c, and grows proportionally with the width of the orogen, λ_c (Figure 13.12a). So for a crust with uniform rheology subjected to constant material flux of incoming material, there is no limit to the growth of F_t, and therefore no limit to the growth of F_g. Consequently, there is also no limit to the thickness of the orogenic crust, h_{max} (Vanderhaeghe et al. 2003).

Melting in the crust changes the force balance significantly (Figure 13.12b). In this case, the base of the crust is characterized by two different viscosities: the viscosity of the solid crust, η_c, and the viscosity of the melt-bearing crust, $\eta_b \ll \eta_c$. The basal tractional force, F_t, becomes the sum of the tractions corresponding to different viscosities: F_{tc} and F_{tb}. Assuming that $F_{tc} \gg F_{tb}$ (because $\eta_b \ll \eta_c$), the force balance in the melt-bearing crust becomes $F_{tc} = F_g$. At these conditions the area of the unmelted base of the crust limits the force F_{tc}, and so F_g and, consequently also h_{max} are limited. Thus, melting at the base of the orogenic crust limits the height of mountains and causes the formation of a plateau. Once established, the plateau widens without significant change in its height (Vanderhaeghe et al. 2003; Beaumont et al. 2004).

The most important assumption made in the derivation above is that $F_{tc} \gg F_{tb}$. If partial melting weakens the base of the orogenic crust less significantly (Case 2a in Royden 1996; "double-slope wedge" in Vanderhaeghe

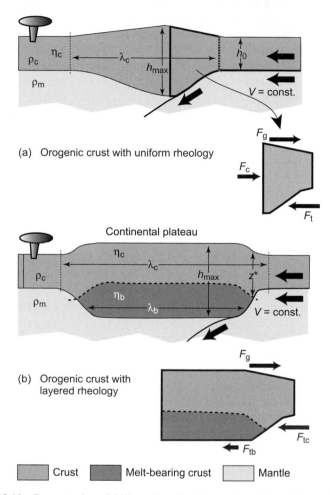

Figure 13.12 Conceptual model illustrating the deformation style and force balance in orogenic crust. Deformation is driven by convergence at a velocity, V, and subduction of the mantle lithosphere. The forces stem from gravity, F_g, compression, F_c, and traction at the base of the crust, F_t. (a) Orogenesis with crust of uniform viscosity (η_c) leads to the formation of "back-to-back wedges" without any limit to the thickness of the crust; (b) Orogenesis with layered crust and rheology leads to the formation of a plateau when $\eta_c \gg \eta_b$ and $F_t = 0$. Basal traction force, F_t, is divided into two parts reflecting the changes at the base of crust due to the formation of a weak (partially melted) basal layer. F_c is neglected in the simplified force balance. ρ_c = density of crust, ρ_m = density of mantle (modified from Vanderhaeghe et al. 2003).

et al. 2003) or if the viscosity of the lower crust decreases monotonously with depth and temperature in the absence of melt-weakening (Model 3 and Figure 6b in Beaumont et al. 2004; Figure 11a in Medvedev and Beaumont 2006), the orogen acquires a shape that is intermediate between wedge-like and plateau-like.

The analysis presented above assumes that the orogenic system is driven at a constant shortening rate by the plate tectonic force, which is theoretically unlimited. However, translation of this force into orogenic deformation is limited by the rheological properties of the crust, hence by melting. A similar conclusion is reached by considering that the compressive forces involved during orogenesis are not unlimited, preventing lithospheric thickening and mountain chains from growing beyond a given height (Molnar and Lyon-Caen 1988, p. 195). The model presented above differs in that it is based on the analysis of viscosity changes due to melting of the crust. These changes are better constrained than the forces driving orogenesis.

Case Studies of Deformation in the Presence of Partially Melted Mid- to Lower Crust

Beaumont and colleagues (Beaumont et al. 2001, 2004; Jamieson et al. 2002, 2004; Vanderhaeghe et al. 2003) developed thermomechanical models in which the effective viscosity of the lower crust is reduced during shortening in order to simulate a melt-bearing crustal layer below Tibet (e.g., Nelson et al. 1996). If viscosity is reduced to 10^{19} Pa s or less at temperatures of 700–750°C, then the weakened part of the lower crust flows laterally within a channel (Figure 13.13). This melt-induced viscosity reduction is actually not very great: only an order of magnitude less than the adjacent, unmelted rock. However, it is important to point out that greater viscosity reductions (to 10^{18} Pa s or less) do not significantly change the result—orogens underlain by enough melt to weaken the crust by at least an order of magnitude always develop plateaus (Beaumont et al. 2001; Vanderhaeghe et al. 2003). We note that these values of viscosity are in line with the theoretically derived flow laws for granitoid rocks of intermediate grain size and composition, containing 2 to 5 vol.-% of melt (Figure 13.4). Models without a low-viscosity layer (i.e., melt-absent) show little if any lower crustal flow and no pronounced development of a plateau-like topography (Beaumont et al. 2004, their Model 3). The melt-bearing layer and crustal channel coincide exactly in all models. In other words, the lateral extent of channel flow always matches the increase in lateral extent of the melt-bearing layer (Figure 13.13a; Beaumont et al. 2004, their Figures 3 and 10). Melting in the lower crust results in instantaneous lateral propagation of the melted domain within a channel.

Models incorporating channel flow successfully explain the formation of the first-order structure of the Himalayan-Tibetan system (Beaumont et al. 2001,

t = 30 Ma   Δ*x* = 1500 km
Deformation

V_P = 5 cm yr^{-1}

Velocity, Temperature

V_P = 5 cm yr^{-1}

700°C

800°C

Moho

t = 48 Ma   Δ*x* = 2400 km

700°C

800°C

Moho

2004) and the exhumation of Miocene migmatitic rocks from beneath the Tibetan Plateau (e.g., Grujic et al. 1996; Jamieson et al. 2004) in response to topographic loading and enhanced erosion (e.g., Wu et al. 1998). If the top of the melt-bearing channel in the model of Beaumont et al. (2004) is taken to represent the STD (Figure 13.8a), then the anatectic rocks of the High Himalayan Crystalline Complex represent the flowing channel of melt-bearing rocks whose movement coincided with the onset of extension along the STD.

We note that modeled and natural orogenic cross sections are only similar if the erosion rate at the southern boundary of the plateau is assumed to be extremely high (1 cm yr^{-1}). If not, then the low-viscosity channel is not drawn to the surface, and anatectic rocks are not exhumed. Only the highest reported erosion rates in the Himalayas are comparable to this value (2–12 mm yr^{-1}; Burbank et al. 1996) which are themselves greater than in the central Andes (e.g., Montgomery et al. 2001) This may explain why, although both plateaus are inferred to overly a melt-bearing crustal layer, no extrusion and exhumation of this layer takes place in the Andes. Comparison of long-term erosion rates in both Tibet and the Andes (methods described in Chapter 9) could constrain the effect of erosion on the geometry of faults in orogens that overlie melt-bearing crust.

Babeyko et al. (2002) demonstrate a different effect of partial melting at the base of the Andean Plateau in their thermomechanical model of the Andean subduction orogen. Petrological data indicate that the orogenic crust in this region became extremely hot (800°C, at 20 km depth) during the formation and evolution of the plateau (20 Ma). To match these data,

---

◄ **Figure 13.13**   Two stages of convergence modeled for a region corresponding to the Himalayas and southern Tibet (from Beaumont et al. 2004). Half arrows indicate the movement of the subducting (Indian) lower plate lithosphere. Deformation of the finite element grid outlines the site of the partially melted, midcrustal channel. For details on thermomechanical parameters, see Beaumont et al. (2004). (a) After 1500 km of shortening: The 750°C isotherm (stippled line interpolated between 700°C and 800°C isotherms) is a proxy for the melt front and outlines the partially melted area in the crust. The site and shape of this isotherm closely matches the margin of the channel as defined by the deformed finite element grid. The coincidence of the melting front and the channel front suggests a nearly instantaneous propagation of channel flow into the newly melted midcrust. This feature can be observed at all stages of the model (Beaumont et al. 2004). An antiformal structure has formed at the margin of the plateau and is caused by localized erosion. (b) After 2400 km of shortening. Lateral propagation of the melt-bearing channel leads to formation of the antiformal structure at the erosional front. Note the upper-crustal antiformal structures, whose position is similar to the gneiss domes north of the STD in Figure 13.8a.

Babeyko et al. (2002) introduced a very high mantle heat flow at the base of the model (60 mW m^{-2}), which leads to melting of the lower crust, melt segregation, and a reduction of lower-crustal viscosity to 10^{17} Pa s. This value is consistent with a melt content of about 20 vol.-% estimated from geoelectric measurements (Schilling and Partzsch 2001) and leads to convection of the lower Andean crust (Babeyko et al. 2002). Models in which crustal convection was inhibited and/or in which the basal heat flow was less than 60 mW m^{-2} failed to match the petrological interpretations (800°C at 20 km depth after 20 Ma of orogenesis).

Interaction of Crustal Melts with Brittle Upper Crust

The effect of lower-crustal channel flow on the brittle upper crust was investigated in a series of numerical experiments in which the crust is weakened by a combination of rapid denudation along the plateau flank and thermal weakening due to heat advected by the channel (Figure 13.13; Beaumont et al. 2001, 2004). The upper crust becomes unstable and slides laterally under the force of gravity. This lateral migration involves thrusting at the plateau margin and normal faulting within the plateau. The low-viscosity (melt-bearing) material flows into the extensional area within the plateau, forming gneiss domes analogous to those exposed north of the STD (GD in Figure 13.8a; Figure 13.13b; Lee et al. 2000). Note that this extensional feature was modeled in the plane containing the direction of convergence between India and Asia. However, similar translation of the upper crust above the weak, lower crust can also effect out-of-section motion, for example, eastward lateral extrusion of Tibet (e.g., Medvedev and Beaumont 2006), as inferred from east–west directed rifting in Southern Tibet (Masek et al. 1994).

In their model of the Altiplano plateau in the Andes, Babeyko et al. (2002) also predicted pronounced faulting of the upper crust above a melt-weakened lower crust during shortening (their Figure 7). However, detailed investigation of the relationship between melting and crustal-scale faulting is limited by the low spatial resolution of the numerical model.

Analytical models provide an alternative approach to investigate the effects of melting on the deformation of the brittle part of the crust. Though based on simple assumptions, analytical models are independent of a given spatial resolution and hence very useful for investigating specific interaction of melts and faults. However, these simplifications may lead to incorrect results. For example, some analytical models employ a purely elastic rheology to approximate the distribution of stresses in the brittle crust (Parsons and Thompson 1993), and predict that extensional faults nucleated near dykes have a low-angle geometry. However, the use of a fully numerical approach and theoretical analysis showed that the latter orientation results from the unrealistic rheology of the boundary conditions (Gerbault et al. 1998).

CONCLUSIONS AND FUTURE OUTLOOK

A small amount of melt has a great effect on the geometry of mountain belts, faults, and shear zones. Melt-bearing crustal rocks deformed at different experimental conditions all show an exponential decay of strength with increasing melt percentage. The greatest strength drop, of about one order of magnitude, takes place between the onset of melting and ~7 vol.-% melt (Figure 13.2). Experimental results on melt-bearing olivine aggregates, and theoretically derived flow laws for melt-bearing granite show that the viscosity decreases most strongly at very low melt volumes (3 to 5%; Figures 13.5 and 13.6), whereas for melt volumes >5% the decrease in viscosity is much less pronounced (Figures 13.5 and 13.6). Several observations from natural and modeling studies suggest that the attainment of this transition is associated with a first-order change in the tectonic and topographic style of orogens. As shown for the Indian–Asian collision, a plateau formed above crust inferred to contain 5–14 vol.-% melt, but not where melt volume is inferred to be 4% (Unsworth et al. 2005). In addition, geodynamic modeling shows that the onset of plateau formation and channel flow in the lower crust is triggered by a melt-induced viscosity reduction of one order of magnitude in the lower crust, i.e., the viscosity reduction corresponding to ~5 vol.-% of melt (Figures 13.5 and 13.6). Modeling also shows that any additional weakening of the lower crust does not significantly change the tectonic style (Beaumont et al. 2001).

The structural and mechanical causes underlying this strength drop have yet to be investigated thoroughly. We suspect that it is controlled by the degree of melt connectivity, in that increasing melt interconnectivity induces grain-boundary sliding and thereby enhances granular flow (Hirth and Kohlstedt 1995a, b; Rosenberg and Handy 2005). To test this hypothesis on crustal rocks, deformation experiments should be combined with detailed microstructural investigation of the evolving 3D melt network at the grain boundaries, especially at melt volumes less than 10%. In addition, new experiments are needed to establish whether mechanical steady state can be achieved for melt-assisted viscous granular flow, and hence derive a constitutive flow law. Determining whether or not the same grain-scale mechanisms operate in the laboratory specimens as in naturally deformed rocks would be an important first step toward establishing at least a phenomenological basis for extrapolating flow laws for melt-bearing rocks.

Large bodies of melt can reside in the crust for up to 20 Ma, possibly longer. Because they localize strain so effectively, such bodies can spread laterally to form weak, subhorizontal channels which drive the lateral topographic growth of the orogen. In fact, numerical modeling suggests that partially melted channels are a requisite for the growth of orogenic plateaus like those presently observed in the central Andes and Tibet. Melt thus plays a major role in shaping orogens. Despite general agreement that melt drastically weakens the crust and fosters vertical decoupling, there is still no consensus on whether melting can trigger gravitational collapse of thickened crust, and if so, what amount of melt is necessary to induce such a large-scale process.

The latter point has rarely been addressed in field investigations. Ideally, future studies would focus on the relationship of strain gradients to gradients in crystallized melt content to evaluate the influence of melt on structural style. Yet, correlating structural style with melt content in natural exposures of anatectic rock is very difficult, if not impossible, for the following reasons: (a) leucosomes taken to represent the melt are generally cumulate or fractionated liquids (Solar and Brown 2001); (b) leucosomes may only represent the small amount of melt remaining at crystallization, rather than the greater volume which originally resided in the rock prior to crystallization; (c) deformation often severely modifies or overprints structures associated with melt and melting.

Establishing whether or not melting triggers gravitational collapse, requires evidence that the onset of melting preceded the onset of crustal extension over the entire area affected by extension. This requires radiometric ages of igneous minerals that grew in the syntectonic melt as well as formational ages of metamorphic minerals that form the dominant schistosity in the extensional shear zones. No studies so far provide such independent ages, relying instead on one or the other (melt and shear zone) ages to date both events.

In the case of a Himalayan-Tibetan-type orogen with a melt-weakened intracrustal channel, numerical modeling (Beaumont et al. 2001) indicates that melting (inferred to attain ~5%) should be nearly contemporaneous with the formation and lateral propagation of the channel and the extrusional flow of melt-bearing rock within it. The subhorizontal channel is bounded along its base and roof by large shear zones with opposite shear senses. At the orogenic scale, these shear zones are recognized as low-angle thrusts and normal faults that accommodate the extrusion of partially melted crust in between. The MCT and the Southern Tibetan detachment system in the Himalayas exemplify such an extrusional system. Together with their bounding shear zones, the channels form a new class of fault which we term "extrusional faults." Exhumation due to extrusional flow is very different from extensional exhumation of anatectic rocks in the footwall of low-angle extensional detachments in the North American Cordillera. However, it is kinematically related to the buoyancy-driven, return motion of subducted crustal slivers that detach from the down-going lithospheric slab, as modeled by Chemenda et al. (1995). Like the extruded anatectites in the Himalayas, these exhuming coherent slivers are bounded above and below by normal faults and thrusts, respectively.

Strike-slip and oblique-slip faults are effective pathways for the rapid, buoyant rise of melts through the crust. These faults may or may not nucleate during melting. The relatively short residence time of melts in these fault systems (<1–2 Ma) can lead to episodic motion, with long periods of creep punctuated by shorter periods of melt veining, magmatic activity and/or faster slip (Handy et al. 2001).

Continental crust subjected to very high heat flux from the asthenosphere may attain melt volumes in excess of the critical 5–7 vol.-% required for localization and decoupling. If the melt content reaches 20–25 vol.-% under these conditions, then crustal shortening can trigger convective overturn of the melt-bearing crust (Babeyko et al. 2002). However, this process, which was proposed to explain "dome and keel" structures of the Archaean crust (Collins et al. 1998), is not yet supported by field studies on Phanerozoic orogens.

Geodynamic models that include the effect of melting on deformation are in their infancy. These models are only valid for 2D plane strain deformation, whereas the interaction between melts and oblique-slip faults is a 3D problem awaiting further investigation. The challenge in using these sophisticated models will be to treat them as controlled experiments and parameter studies rather than as simulations with a large number of interactive, yet poorly constrained variables.

ACKNOWLEDGEMENTS

Discussions with Ulrich Riller and constructive reviews by Mike Brown and Greg Hirth significantly improved our manuscript. Martyn Unsworth kindly provided a preprint of his work. We acknowledge the support of the German Science Foundation (DFG) in the form of grants RO 2177/1-1, HA 2403/3-1, HA 2403/6-1, and project G2 of the SFB-267 "Deformation Processes in the Andes," which provided funding for parts of our work.

REFERENCES

Armstrong, R.L., and P. Ward. 1991. Evolving geographic patterns of Cenozoic magmatism in the Northern American Cordillera: The temporal and spatial association of magmatism and metamorphic core complexes. *J. Geophys. Res.* **96**:13,201–13,224.

Arzi, A. 1978. Critical phenomena in the rheology of partially melted rocks. *Tectonophysics* **44**:173–184.

Babeyko, A.Y., S.V. Sobolev, R.B. Trumbull, O. Oncken, and L.L. Lavier. 2002. Numerical models of crustal scale convection and partial melting beneath the Altiplano-Puna Plateau. *Earth Planet. Sci. Lett.* **199**:373–388.

Beaumont, C., P. Fullsack, and J. Hamilton. 1994. Styles of crustal deformation in compressional orogens caused by subduction of the underlying lithosphere. *Tectonophysics* **232**:119–132.

Beaumont, C., R.A. Jamieson, M.H. Nguyen, and B. Lee. 2001. Himalayan tectonics explained by extrusion of a low-viscosity channel coupled to focused surface denudation. *Nature* **414**:738–742.

Beaumont, C., R.A. Jamieson, M.H. Nguyen, and S. Medvedev. 2004. Crustal channel flows: 1. Numerical models with applications to the tectonics of the Himalayan-Tibetan orogen. *J. Geophys. Res.* **109**:B06406, doi:10.1029/2003JB002809.

Berger, A., C.L. Rosenberg, and S.M. Schmid. 1996. Ascent, emplacement and exhumation of the Bergell pluton within the Southern Steep Belt of the Central Alps. *Schweiz. Mineral. Petrograph. Mitt.* **76**:357–382.

Biddle, K.T., and N. Christie-Blick. 1985. Strike-slip deformation, basin formation, and sedimentation, Spec. Publ. 37, pp. 127–142. Tulsa, OK: Soc. Econ. Paleontologists and Mineralogists.

Bird, P. 1991. Lateral extrusion of lower crust from under high topography, in the isostatic limit. *J. Geophys. Res.* **96**:10,275–10,286, 10.1029/91JB00370.

Brown, M. 2005. Synergistic effects of melting and deformation: An example from the Variscan belt, western France. In: Deformation Mechanisms, Rheology and Tectonics: From Minerals to the Lithosphere, ed. D. Gapais, J.P. Brun, and P.R. Cobbold, Spec. Publ. 243, pp. 205–225. London: Geol. Soc.

Brown, M., and R.D. Dallmeyer. 1996. Rapid Variscan exhumation and role of magma in core complex formation: Southern Brittany metamorphic belt, France. *J. Metamorphic Geol.* **14**:361–379.

Brown, M., and T. Rushmer. 1997. The role of deformation in the movement of granitic melt: Views from the laboratory and the field. In: Deformation-enhanced Fluid Transport in the Earth's Crust and Mantle, ed. M.B. Holness, Mineral. Soc. Ser. 8, pp. 111–144. London: Chapman and Hall.

Brown, M., and G.S. Solar. 1998. Shear zone systems and melts: Feedback relations and self-organization in orogenic belts. *J. Struct. Geol.* **20**:211–227.

Brown, M., and G.S. Solar. 1999. The mechanism of ascent and emplacement of granite magma during transpression: A syntectonic granite paradigm. *Tectonophysics* **312**:1–33.

Brown, R.L., and J.H. Nazarchuk. 1993. Annapurna detachment fault in the Greater Himalaya of central Nepal. In: Himalayan Tectonics, ed. P.J. Treloar, and M.P. Searle, Spec. Publ. 74, pp. 461–473. London: Geol. Soc.

Bruhn, D., N. Groebner, and D.L. Kohlstedt. 2000. An interconnected network of core-forming melts produced by shear deformation. *Nature* **403**:883–886.

Brun, J.-P. 1999. Narrow rifts versus wide rifts: Inferences for the mechanics of rifting from laboratory experiments. *Phil. Trans. R. Soc. Lond. A* **357**:695–712.

Brun, J.-P., D. Sokoutis, and J. Van Den Driesche. 1994. Analogue modelling of detachment fault systems. *Geology* **22**:319–322.

Burbank, D.W., J. Leland†, E. Fielding et al. 1996. Bedrock incision, rock uplift and threshold hillslopes in the northwestern Himalayas. *Nature* **379**:505–510

Burchfiel, B.C., Z. Chen, K.V. Hodges et al. 1992. The South Tibetan Detachment System, Himalayan Orogen: Extension Contemporaneous with and Parallel to Shortening in a Collisional Mountain Belt, Spec. Paper 269, p. 41. Boulder, CO: Geol. Soc. Am.

Chemenda, A.I., M. Mattauer, J. Malavieille, and A.N. Bokum. 1995. A mechanism for syn-collisional rock exhumation and associated normal faulting: Results from physical modeling. *Earth Planet. Sci. Lett.* **132**:225–232.

Collins, W.J., M.J. van Kranendock, and C. Teyssier. 1998. Partial convective overturn of Archaean crust in the eastern Pilbara Craton, Western Australia: Driving mechanisms and tectonic implications. *J. Struct. Geol.* **20**:1405–1424.

Coney, P.J., and T.A. Harms. 1984. Cordilleran metamorphic core complexes: Cenozoic extensional relics of Mesozoic compression. *Geology* **12**:550–554.

Cruden, A.R. 1990. Flow and fabric development during the diapiric rise of magma. *J. Geol.* **98**:681–698.

Davidson, C., L.S. Hollister, and S.M. Schmid. 1992. Role of melt in the formation of a deep-crustal compressive shear zone: The MacLaren glacier metamorphic belt, South Central Alaska. *Tectonics* **11**:348–359.

de Saint-Blanquat, M., B. Tikoff, C. Teyssier, and J.L. Vigneresse. 1998. Transpressional kinematics and magmatic arcs. In: Continental Transpressional and Transtensional Tectonics, ed. R.E. Holdsworth, R.A. Strachan, and J.F. Dewey, Spec. Publ. 135, pp. 327–340. London: Geol. Soc.

de Silva, S.L. 1989. Geochronology and stratigraphy of the ignimbrites from the 21°30' S portion of the Central Andes of N. Chile. *J. Volcanol. & Geotherm. Res.* **37**:93–131.

Dell'Angelo, L.N, and J. Tullis. 1988. Experimental deformation of partially melted granitic aggregates. *J. Metamorphic Geol.* **6**:495–515

Dèzes, P.J., J.-C. Vannay, A. Steck, F. Bussy, and M. Cosca. 1999. Synorogenic extension: Quantitative constraints on the age and displacement of the Zanskar shear zone (northwest Himalaya). *Bull. Geol. Soc. Am.* **111**:364–374.

D'Lemos, R.S., M. Brown, and R.A. Strachan. 1992. Granite magma generation, ascent and emplacement within a transpressional orogen. *J. Geol. Soc. Lond.* **149**:487–490.

Edwards, M.A., W.S.F. Kidd, J. Li, Y. Yue, and M. Clark. 1996. Multi-stage development of the southern Tibet detachment system near Khula Kangri. New data from Gonto La. *Tectonophysics* **260**:1–19.

Edwards, M.A., and M.T. Harrison. 1997. When did the roof collapse? Late Miocene north–south extension in the high Himalaya revealed by Th-Pb monazite dating of the Khula Kangri granite. *Geology* **25**:543–546.

Engebretson, D.C., A. Cox, and R.G. Gordon. 1985. Relative Motions between Oceanic and Continental Plates in the Pacific Basin, Spec. Paper 206. Boulder, CO: Geol. Soc. Am.

England, P., P. Le Fort, P. Molnar, and A. Pecher. 1992. Heat sources for Tertiary metamorphism and anatexis in the Annapurna Manaslu region of Central Nepal. *J. Geophys. Res.* **97**:2107–2128.

Fitch, T.J. 1972. Plate convergence, transcurrent faults, and internal deformation adjacent to southeast Asia and the Western Pacific. *J. Geophys. Res.* **77**:4432–4460.

Foster, D.A., and C.M. Fanning. 1997. Geochronology of the northern Idaho batholith and the Bitterroot metamorphic core complex: Magmatism preceding and contemporaneous with extension. *Bull. Geol. Soc. Am.* **109**:379–394.

Foster, D.A., C. Schafer, C.M. Fanning, and D.W. Hyndmann. 2001. Relationships between crustal partial melting, plutonism, orogeny, and exhumation: Idaho-Bitterroot batholith. *Tectonophysics* **342**:313–350.

Froitzheim, N., S.M. Schmid, and M. Frey. 1996. Mesozoic paleogeography and the timing of eclogite-facies metamorphism in the Alps: A working hypothesis. *Eclogae Geologicae Helvetiae* **89**:81–110.

Gaillard, F., B. Scaillet, and M. Pichavant. 2004. Evidence for present-day leucogranite pluton growth in Tibet. *Geology* **32**:801–804, doi:10.1130/G20577.1

Gerbault, M., A.N.B. Poliakov, and M. Daignieres. 1998. Prediction of faulting from theories of elasticity and plasticity: What are the limits? *J. Struct. Geol.* **20**: 301–320.

Glazner, A.F. 1991. Plutonism, oblique subduction, and continental growth: An example from the Mesozoic of California. *Geology* **19**:784–786.

Gleason, G.C., and J. Tullis. 1995. A flow law for dislocation creep of quartz aggregates determined with the molten salt cell. *Tectonophysics* **247**:1–23.

Grujic, D., M. Casey, C. Davidson et al. 1996. Ductile extrusion of the Higher Himalayan Crystalline in Bhutan: Evidence from quartz microfabrics. *Tectonophysics* **260**:21–44.

Guillot, S., K. Hodges, P. Le Fort, and A. Pecher. 1994. New constraints on the age of the Manaslu granite: Evidence for episodic tectonic denudation in the Central Himalayas. *Geology* **22**:559–562.

Günther, A. 2001. Strukturgeometrie, Kinematik und Deformationsgeschichte des oberkretazisch-alttertiären magmatischen Bogens (Nord-chilenische Präkordillere 21.7–23°S). Ph.D. diss., Freie Universität Berlin, p. 170.

Handy, M.R., A. Mulch, M. Rosenau, and C.L. Rosenberg. 2001. A synthesis of the role of fault zones and melts as agents of weakening, hardening and differentiation of the continental crust. In: The Nature and Tectonic Significance of Fault Zone Weakening, ed. R.E. Holdsworth et al., Spec. Publ. 186, pp. 305–332. London: Geol. Soc.

Harris, N., and J. Massey. 1994. Decompression and anatexis of Himalayan metapelites. *Tectonophysics* **13**:1537–1546.

Harris, N.B.W., M. Caddick, J. Kosler et al. 2004. The pressure-temperature-time path of migmatites from the Sikkim Himalaya. *J. Metamorphic Geol.* **22**:249–264.

Harrison, M.T., M. Grove, K.D. McKeegan et al. 1999. Origin and episodic emplacement of the Manaslu Intrusive Complex, Central Himalaya. *J. Petrol.* **40**:3–19.

Hirth, G., J. Escartin, and J. Lin. 1998. The rheology of the lower oceanic crust: Implications for lithospheric deformation at mid-ocean ridges. In: Faulting and Magmatism at Mid-Ocean Ridges, ed. W. Buck, P. Delaney, J. Karson, and Y. Lagabrielle, Geophys. Monogr. 106, pp. 291–303. Washington, D.C.: Am. Geophys. Union.

Hirth, G., and D.L. Kohlstedt. 1995a. Experimental constraints on the dynamics of the partially molten upper mantle: Deformation in the diffusion creep regime. *J. Geophys. Res.* **100**:1981–2001.

Hirth, G., and D.L. Kohlstedt. 1995b. Experimental constraints on the dynamics of the partially molten upper mantle. 2. Deformation in the dislocation creep regime. *J. Geophys. Res.* **100**:15,441–15,449.

Hirth, G., and D.L. Kohlstedt. 1996. Water in the oceanic upper mantle: Implications for rheology, melt extraction and the evolution of the lithosphere. *Earth Planet. Sci. Lett.* **144**:93–108.

Hirth, G., and D. Kohlstedt. 2003. Rheology of the upper mantle and the mantle wedge: A view from the experimentalists. In: Inside the Subduction Factory, ed. J. Eiler, Geophys. Monogr. 103, pp. 83–105. Washington, D.C.: Am. Geophys. Union.

Hodges, K.V. 1998. The thermodynamics of Himalayan orogenesis. In: What Drives Metamorphism and Metamorphic Reactions? ed. P.J. Treloar and P.J. O'Brien, Spec. Publ. 138, pp. 7–22. London: Geol. Soc.

Hodges, K.V., R.R. Parrish, and M.P. Searle. 1996. Tectonic evolution of the central Annapurna Range, Nepalese Himalayas. *Tectonics* **15**:1264–1291.

Hollister, L.S. 1993. The role of melt in the uplift and exhumation of orogenic belts. *Chem. Geol.* **108**:31–48.

Hollister, L.S, and M.L. Crawford. 1986. Melt-enhanced deformation: A major tectonic process. *Geology* **14**:558–561.

Holyoke, III, C.W., and T. Rushmer. 2002. An experimental study of grain-scale melt segregation mechanisms in two common crustal rock types. *J. Metamorphic Geol.* **20**:493–512.

Hopper, J.R., and R.W. Buck. 1998. Styles of extensional decoupling. *Geology* **26**:699–702.

Houseman, G.A., D.P. McKenzie, and P. Molnar. 1981. Convective instability of a thickened boundary layer and its relevance for the thermal evolution of continental convergent belts. *J. Geophys. Res.* **86**:B7, 6115–6132.

Huerta, A.D., L.H. Royden, and K.V. Hodges. 1996. The interdependence of deformational and thermal processes in mountain belts. *Science* **273**:637–639.

Ingram, G.M., and D.H.W. Hutton. 1994. The Great Tonalite Sill: Emplacement into a contractional shear zone and implications for Late Cretaceous to early Eocene tectonics in southeastern Alaska and British Columbia. *Bull. Geol. Soc. Am.* **106**: 715–728.

Jamieson, R.A., C. Beaumont, S. Medvedev, and M.H. Nguyen. 2004. Crustal channel flows: 2. Numerical models with implications for metamorphism in the Himalayan-Tibetan Orogen. *J. Geophys. Res.* **109**:B06407, doi:10.1029/2003JB002811.

Jamieson, R.A., C. Beaumont, M.H. Nguyen, and B. Lee. 2002. Interaction of metamorphism, deformation and exhumation, in large hot orogens. *J. Metamorphic Geol.* **20**:9–24.

Jarrard, R.D. 1986. Relations among subduction parameters. *Rev. Geophys.* **24**:217–284.

Karato, S. 1986. Does partial melting reduce the creep strength of the upper mantle? *Nature* **319**, 309–310.

Kohn, M.J., M.S. Wieland, C.D. Parkinson, and B.N. Upreti. 2005. Five generations of monazite in Langtang gneisses: Implications for chronology of the Himalayan metamorphic core. *J. Metamorphic Geol.* **23**:399–406. doi:10.1111/j.1525-1314.2005.00584.x

Koons, P.O., R.J. Norris, D. Craw, and A.F. Cooper. 2003. Influence of exhumation on the structural evolution of transpressional plate boundaries: An example from the Southern Alps, New Zealand. *Geology* **31**:3–6.

Lamb, S., L. Hoke, L. Kennan, and J. Dewey. 1997. Cenozoic evolution of the Central Andes in Bolivia and northern Chile. In: Orogeny through time, ed. J.P. Burg, and M. Ford, Spec. Publ. 121, pp. 237–264. London: Geol. Soc.

LaTour, T.E., and R.L. Barnett. 1987. Mineralogical changes accompanying mylonitization in the Bitterroot dome of the Idaho batholith: Implications for timing of deformation. *Bull. Geol. Soc. Am.* **98**:356–373.

Ledru, P., G. Courrioux, C. Dallain et al. 2001. The Velay dome (French Massif Central): Melt generation and granite emplacement during orogenic evolution. *Tectonophysics* **342**:207–237.

Lee, J., B.R. Hacker, W.S. Dinklage et al. 2000. Evolution of the Kangmar Dome, southern Tibet: Structural, petrologic, and thermochronologic constraints. *Tectonics* **19**:872–895. doi1999TC001147.

Le Fort, P. 1975. Himalayas: The collided range. Present knowledge of the continental arc. *Am. J. Sci.* **275A**:1–44.

Lejeune, A. and P. Richet. 1995. Rheology of crystal-bearing silicate melts: An experimental study at high viscosities. *J. Geophys. Res.* **100**:4215–4229.

Li, S., M.J. Unsworth, J.R. Booker, W. Wei, H. Tan, and A.G. Jones. 2003. Partial melt or aqueous fluids in the Tibetan crust: Constraints from INDEPTH magnetotelluric data. *Geophys. J. Intl.* **153**:289–304.

Lister, G.S., and S.L. Baldwin. 1993. Plutonism and the origin of metamorphic core complexes. *Geology* **21**:607–610.

Malavieille, J., P. Guihot, S. Costa, J.M. Lardeaux, and V. Gardien. 1990. Collapse of thickened Variscan crust in the French Massif Central: Mont Pilat extensional shear zone and St. Etienne upper Carboniferous basin. *Tectonophysics* **177**:139–149.

Marchildon, N., and M. Brown. 2002. Grain-scale melt distribution in two contact aureole rocks: Implication for controls on melt localization and deformation. *J. Metamorphic Geol.* **20**:381–396.

Masek, J.G., B.L. Isacks, E.J. Fielding, and J. Browaeys. 1994. Rift flank uplift in Tibet: Evidence for a viscous lower crust. *Tectonics* **13**:659–667.

Matte, P. 1991. Accretionary history and crustal evolution of the Variscan belt in western Europe. *Tectonophysics* **196**:309–337.

McCaffrey, K. 1992. Igneous emplacement in a transpressive shear zone: Ox Mountains igneous complex. *J. Geol. Soc. Lond.* **149**:221–235.

McClelland, W.C., and J.A. Gilotti. 2003. Late-stage extensional exhumation of high-pressure granulites in the Greenland Caledonides. *Geology* **31**:259–262.

McNulty, B.A., D.L. Farber, G.S. Wallace, R. Lopez, and O. Palacios. 1998. Role of plate kinematics and plate-slip-vector partitioning in continental magmatic arcs: Evidence from the Cordillera Blanca, Peru. *Geology* **26**:827–830.

Mechie, J., S.V. Sobolev, L. Ratschbacher et al. 2004. Precise temperature estimation in the Tibetan crust from seismic detection of the α–β quartz transition. *Geology* **32**:601–604.

Medvedev, S. 2002. Mechanics of viscous wedges: Modeling by analytical and numerical approaches. *J. Geophys. Res.* **107**, 10.1029/2001JB000145

Medvedev, S., and C. Beaumont. 2006. Growth of continental plateaus by channel injection: Constraints and thermo-mechanical consistency. In: Channel Flow, Ductile Extrusion and Exhumation in Continental Collision, ed. R.D. Law, M.P. Searle, and L. Godin, Spec. Publ. London: Geol. Soc., in press.

Milliman, J.D., and J.P.M. Syvitski. 1992. Geomorphic/tectonic control of sediment discharge to the ocean: The importance of small mountaineous rivers. *J. Geol.* **100**:525–544.

Molnar, P., and Lyon-Caen, H. 1988. Some Simple Physical Aspects of the Support, Structure, and Evolution of Mountain Belts, Spec. Paper 218, pp. 179–207. Boulder, CO: Geol. Soc. Am.

Montgomery, D.R., G. Balco, and S.D. Willett. 2001. Climate, tectonics, and the morphology of the Andes. *Geology* **29**:579–582.

Nclson, K.D., W. Zhao, L.D. Brown et al. 1996. Partially molten middle crust beneath southern Tibet: Synthesis of Project INDEPTH results. *Science* **274**:1684–1687

Neves, S.P., A. Vauchez, and C.J. Archanjo. 1996. Shear zone-controlled magma emplacement or magma-assisted nucleation of shear zones? Insights from northeast Brazil. *Tectonophysics* **262**:349–364.

Norlander, B.N., D.L. Whitney, C. Teyssier, and O. Vanderhaeghe. 2002. Partial melting and decompression of the Thor-Odin Dome, Shuswap metamorphic core complex, Canadian Cordillera. *Contrib. Mineral. & Petrol.* **61**:103–125.

Oberli, F., M. Meier, A. Berger, C.L. Rosenberg, and R. Gieré. 2004. U-Th-Pb and ^{230}Th/^{238}U Disequilibrium Isotope Systematics:Precise Accessory Mineral Chronology and Melt Evolution Tracing in the Alpine Bergell Intrusion. *Geochim. Cosmochim. Acta* **68**:2543–2560.

Parsons, T., and Thompson, A. 1993. Does magmatism influence low angle faulting? *Geology* **21**:247–250.

Paterson, M.S. 2001. A granular flow theory for the deformation of partially melted rock. *Tectonophysics*, **335**:51–61.

Patino Douce, A.E., and N. Harris. 1998. Experimental constraints on Himalayan anatexis. *J. Petrol.* **39**:689–710.

Petford, N., A.R. Cruden, K.J.W. McCaffrey, and J.-L. Vigneresse. 2000. Granite magma formation, transport and emplacement in the Earth's crust. *Nature* **408**:669–673.

Ranalli, G., and D.C. Murphy. 1987. Rheological stratification of the lithosphere. *Tectonophysics* **132**:281–296.

Renner, J., B. Evans, and G. Hirth. 2000. On the rheologically critical melt fraction. *Earth Planet. Sci. Lett.* **181**:585–594.

Rey, P., O. Vanderhaeghe, and C. Teyssier. 2001. Gravitational collapse of continental lithosphere: Definition, regimes, and modes. *Tectonophysics* **342**:435–444.

Riller, U., I. Petrinovic, J. Ramelow, M. Strecker, and O. Oncken. 2001. Late Cenozoic tectonism, collapse caldera and plateau formation in the central Andes. *Earth Planet. Sci. Lett.* **188**:299–311.

Roscoe, R. 1952. The viscosity of suspensions of rigid spheres. *Brit. J. Appl. Phys.* **3**: 267–269.

Rosenberg, C.L. 2001. Deformation of partially molten granite: A review and comparison of experimental and natural case studies. *Intl. J. Earth Sci.* **90**:60–76.

Rosenberg, C.L. 2004. Shear zones and magma ascent: A model based on a review of the Tertiary magmatism in the Alps. *Tectonics* **23**:TC3002, doi10.1029/2003TC001526.

Rosenberg, C.L., and M.R. Handy. 2005. Experimental deformation of partially melted granite revisited: Implications for the continental crust. *J. Metamorphic Geol.* **23**:19–28. doi:1111/j.1525–1314.2005.005555.x.

Royden, L. 1996. Coupling and decoupling of crust and mantle in convergent orogens: Implications for strain partitioning in the crust. *J. Geophys. Res.* **101**:17,679–17,705.

Rushmer, T. 1995. An experimental deformation study of partially molten amphibolite: Application to low-melt fraction segregation. *J. Geophys. Res.* **100**:15,681–15,695.

Rutter, E. 1997. The influence of deformation on the extraction of crustal melts: A consideration on the role of melt-assisted granular flow. In: Deformation-enhanced Fluid Transport in the Earth's Crust and Mantle, ed. M.B. Holness, Mineral. Soc. Ser. 8, pp. 82–110. London: Chapman & Hall.

Rutter, E., and D.H.K. Neumann. 1995. Experimental deformation of partially molten Westerly granite under fluid-absent conditions, with implications for the extraction of granitic magmas. *J. Geophys. Res.* **100**:15,697–15,715.

Sawyer, E.W. 2001. Grain-scale and outcrop-scale distribution and movement of melt in a crystallizing granite. *Trans. R. Soc. Edinburgh: Earth Sci.* **91**:73–85.

Scheuber, E., and P.A.M. Andriessen. 1990. The kinematic and geodynamic significance of the Atacama Fault Zone, northern Chile. *J. Struct. Geol.* **12**:243–257.

Schilling, F.R., and G.M Partzsch. 2001. Quantifying partial melt fraction in the crust beneath the central Andes and the Tibetan Plateau. *Phys. & Chem. Earth A* **26**: 239–246.

Schilling, F.R., G.M Partzsch, H. Brasse, and G. Schwarz. 1997. Partial melting below the magmatic arc in the Central Andes deduced from geoelectromagnetic field experiments and laboratory data. *Phys. Earth & Planet. Inter.* **103**:17.

Schmid, S.M., H.R. Aebli, F. Heller, and A. Zingg. 1989. The role of the Periadriatic Line in the tectonic evolution of the Alps. In: Alpine Tectonics, ed. M.P. Coward, D. Dietrich, and R. Park, Spec. Publ. 45, pp. 153–171. London: Geol. Soc.

Schmid, S.M., and E. Kissling. 2000. The arc of the western Alps in the light of geophysical data on deep crustal structure. *Tectonics* **19**:62–85.

Schmid, S.M., O.A. Pfiffner, N. Froitzheim, G. Schönborn, and E. Kissling. 1996. Geophysical-geological transect and tectonic evolution of the Swiss Italian Alps *Tectonics* **15**:1036–1064.

Schmitz, M., W.-D. Heinsohn, and F.R. Schilling. 1997. Seismic, gravity, and petrological evidence for partial melt beneath the thickened Central Andean crust (21–23°S). *Tectonophysics* **270**:313–326.

Scott, T.J., and D.L. Kohlstedt. 2004. The effect of large melt fraction on the deformation behaviour of peridotite: Implications for the viscosity of Io's mantle and the rheologically critical melt fraction. *EOS Trans. AGU* **85**, Fall Meeting suppl., Abstract.

Searle, M.P., and L. Godin. 2003. The South Tibetan Detachment and the Manaslu Leucogranite: A structural reinterpretation and restoration of the Annapurna-Manaslu Himalaya, Nepal. *J. Geol.* **111**:505–523.

Solar, G., and M. Brown. 2001. Petrogenesis of migmatites in Maine, USA: Possible source of peraluminous leucogranite in plutons. *J. Petrol.* **42**:789–823.

Teyssier, C., B. Tikoff, and M. Markley. 1995. Oblique plate motion and continental tectonics. *Geology* **23**:447–450.

Teyssier, C., and, D.L. Whitney. 2002. Gneiss domes and orogeny. *Geology* **30**: 1139–1142.

Tikoff, B., and M. de Saint Blanquat. 1997. Transpressional shearing and strike-slip partitioning in the late Cretaceous Sierra Nevada magmatic arc, California. *Tectonics* **16**:442–459.

Tirel, C., J.-P. Brun, and E. Burov. 2004. Thermomechanical modeling of extensional gneiss domes. In: Gneiss Domes in Orogeny, ed. D. Whitney, C. Teyssier, and S. Siddoway, Spec. Paper 380, pp. 67–78. Boulder, CO: Geol. Soc. Am.

Tommasi, A., A. Vauchez, L.A.D. Fernandes, and C. Porcher. 1994. Magma-assisted strain localization in an orogen-parallel transcurrent shear zone of southern Brazil. *Tectonics* **13**:421–437.

Unsworth, M.J., A.G. Jones, W. Wei et al. 2005. Crustal rheology of the Himalaya and Southern Tibet inferred from magnetotelluric data. *Nature* **438**:78–81.

Vanderhaeghe, O., J.-P. Burg, and C. Teyssier. 1999. Exhumation of migmatites in two collapsed orogens: Canadian Cordillera and French Variscides. In: Exhumation Processes: Normal Faulting, Ductile Thinning and Erosion, ed. U. Ring, M.T. Brandon, G.S. Lister, and S.D. Willet, Spec. Publ. 154, pp. 181–204. London: Geol. Soc.

Vanderhaeghe, O., S. Medvedev, C. Beaumont, P. Fullsack, and R.A. Jamieson. 2003. Evolution of orogenic wedges and continental plateaus: Insights from crustal thermal-mechanical models overlying subducting mantle lithosphere. *Geophys. J. Intl.* **153**:27–51.

Vanderhaeghe, O., and C. Teyssier. 2001a. Crustal-scale rheological transitions during late-orogenic collapse. *Tectonophysics* **335**:211–228.

Vanderhaeghe, O., and C. Teyssier. 2001b. Partial melting and flow of orogens. *Tectonophysics* **342**:451–472.

van der Molen, I., and M.S. Paterson. 1979. Experimental deformation of partially melted granite. *Contrib. Mineral. & Petrol.* **70**:299–318.

Vauchez, A., S. Pacheco Neves, and A. Tommasi. 1997. Transcurrent shear zones and magma emplacement in Neoproterozoic belts of Brazil. In: Granite: From Segregation of Melt to Emplacement Fabrics, ed. J.-L. Bouchez, D.H.W. Hutton, and W.E. Stephens, pp. 275–293. Dordrecht: Kluwer Academic.

Vigneresse, J.-L., and B. Tikoff. 1999. Strain partitioning during partial melting and crystallizing of felsic magmas. *Tectonophysics* **312**:117–132.

von Blanckenburg, F., H. Kagami, A. Deutsch et al. 1998. The origin of Alpine plutons along the Periadriatic Lineament. *Schweiz. Mineral. Petrograph. Mitt.* **78**:57–68.

Wernicke, B.P., R.L. Christiansen, P.C. England, and L.J. Sonder. 1987. Tectonomagmatic evolution of Cenozoic extension in the North American Cordillera. In: Continental Extensional Tectonics, ed. M.P. Coward, J.F. Dewey, and P.L. Hancock, Spec. Publ. 28, pp. 203–221. London: Geol. Soc.

White, A.P., and K.V. Hodges. 2002. Multistage extensional evolution of the central East Greenland Caledonides. *Tectonics* **21**:1048, doi:10.1029/2001/C001308.

Wickham, S.M. 1987. The segregation and emplacement of granitic magmas. *J. Geol. Soc. Lond.* **144**:281–297.

Willett, S.D., and D.C. Pope. 2004. Thermo-mechanical models of convergent orogenesis: Thermal and rheologic dependence of crustal deformation. In: Rheology and deformation of the lithosphere at continental margins, ed. G.D. Karner, B. Taylor, N.W. Driscoll, and D.L. Kohlstedt, pp. 179–222. New York: Columbia Univ. Press.

Wu, C., K.D. Nelson, G. Wortman et al. 1998. Yadong cross-structure and South Tibetan Detachment in the east central Himalaya (89–90°E). *Tectonics* **17**:28–45, 97TC03386.

Yin, A., and M.T. Harrison. 2000. Geologic evolution of the Himalayan-Tibetan orogen. *Ann. Rev. Earth & Planet. Sci.* **28**:211–280.

Yuan, X., S.V. Sobolev, R. Kind et al. 2000. Subduction and collision processes in the Central Andes constrained by converted seismic phases. *Nature* **408**:958–961.

Zhang, H., N. Harris, R. Parrish et al. 2004. Causes and consequences of protracted melting of the mid-crust exposed in the North Himalayan antiform. *Earth Planet. Sci. Lett.* **228**:195–212.

Zimmerman, M.E., and D.L. Kohlstedt. 2004. Rheological properties of partially molten lherzolite. *J. Petrol.* **45**:275–298.

From left to right: Jean-Pierre Gratier, Frédéric Gueydan, Bart Bos, Claudio Rosenberg, Lukas Baumgartner, Mark Person, Steve Miller, Bruce Yardley, James Connolly, Janos Urai

14

Group Report:
Fluids, Geochemical Cycles, and
Mass Transport in Fault Zones

MARK PERSON, Rapporteur

LUKAS P. BAUMGARTNER, BART BOS, JAMES A. D. CONNOLLY,
JEAN-PIERRE GRATIER, FRÉDÉRIC GUEYDAN,
STEPHEN A. MILLER, CLAUDIO L. ROSENBERG, JANOS L. URAI,
and BRUCE W. D. YARDLEY

OVERVIEW

This report considers the role of fluids and fluid–rock geochemical reactions that potentially influence earthquake behavior and fault zone strength throughout the lithosphere. Our discussions highlighted the distinctions between processes that occur within brittle, upper crustal fault zones (0 to ~10 km) where fluids can be pressured hydrostatically to lithostatically and deeper parts of fault zones (>10 km) where fluids can be either near-lithostatically pressured or where a free fluid phase is absent. We recognized that fluid flow and geochemical processes vary with the style of faulting. We conclude that the comprehensive understanding of fluid–rock geochemical interactions in seismogenic faults is in its infancy. The major areas of future research can be grouped according to three thematic questions:

- What are the controls on fluid–rock chemical reactions in and adjacent to fault zones?
- How do fluid-flow processes change before, during, and, after earthquakes?
- What are the magnitudes of fluid fluxes in fault zones from the surface down to the asthenosphere and how do these fluxes vary with tectonic environment?

THEME 1:
WHAT ARE THE CONTROLS ON FLUID–ROCK CHEMICAL INTERACTION IN AND ADJACENT TO FAULT ZONES?

Introduction

Several statements can be made regarding fluid–rock chemical interactions within fault zones: Faults and shear zones can be sites of both upward and downward flow of fluids, but they can also act as barriers for fluid flow (Bense and Person 2006). The effect of fluids on rock strength is one of the main reasons for strain localization. Shallower parts of faults tend to show more extensive hydration than deep parts. In the shallow parts, fluids are usually introduced from the surface. At deeper levels faults tend to separate blocks with different hydraulic heads (i.e., pressure compartments). This heterogeneous fluid-pressure structure may change as faulting progresses and the rocks move through different pressure regimes. Thus, fluid pressure within fault zones can fluctuate with time. At yet deeper levels, faults in sedimentary basins and hydrous crystalline rocks are potential sites of fluid escape, at least while the faults are active. Faults and shear zones undergoing retrograde metamorphism (e.g., involving chloritization of crystalline rocks) in nominally dry rocks require the introduction of a fluid phase. The cores of these faults become water-saturated and are lined with quartz or fully hydrated minerals. Rates of fluid production in metamorphism are much too low to account for such highly mineralized/altered fault zones. Transcurrent faulting, overthrusting, and subduction may result in the release and upward migration of fluids from deep levels of fault zones (Screaton and Saffer 2005; Saffer and McKiernan 2005). Aseismic or silent earthquakes may be driven by metamorphic fluid production. Different patterns of metasomatic alteration are associated with down-temperature and up-temperature flow systems in subduction zones.

Chemical Processes

There are a number of geochemical processes by which fluids affect fault-zone strength. Precipitation of phyllosilicate minerals, iron oxides, quartz/calcite cements, and other minerals along fault zones can either increase or decrease fault rock strength while reducing fault zone permeability. Precipitation reactions are potentially driven by a number of mechanisms, including temperature and pressure changes, fluid mixing, and fluid–rock reaction. However, little is known about how these fluids originate, and how they migrate to and from fault zones during the earthquake cycle. Integrated fluid volumes and fluid-flow rates in fault zones are poorly constrained. We know very little about the chemical saturation state of fluids and their pressure evolution in seismogenic zones. We know even less about the background permeability structure of deep crustal fault zones.

In rock masses exhumed from great depth where little or no fluid is initially present, vein minerals crystallized in fractures usually include chlorite which forms at <250°C. This indicates that minerals do not begin to precipitate in cracks until rocks are exhumed to a depth of about 10 km due to slow reaction kinetics. There are some exceptions to this, for example, in shear zones where fracture-filling minerals appear to have formed in equilibrium at higher temperature conditions.

Chemical reactions (e.g., dissolution–precipitation and pressure solution) can cause the fault core and the damaged zone around this core to heal (Chapter 12). Within the fault core, hydrothermal alteration tends to be fast due to the relatively large surface area of fine, comminuted grains in cataclastic fault rock. Time scales of geochemical processes in the damage zone and the core vary widely due to differences in temperature history, grain surface area, and magnitude of fluid flow. Temperature changes affect reaction rate-controlled processes more than diffusion-controlled processes (pressure-solution creep and compaction) due to the lesser temperature dependence of the latter. Frictional heating at temperatures in excess of 500°C can occur locally in the fault core (see Chapter 7) leading to flash boiling close to the fault surface followed by rapid cooling to 100°C shortly thereafter. Fission-track analysis of zircon and apatite grains collected adjacent to fault cores could be used to determine whether frictional heating cycles occur and perhaps to establish the duration of these cycles (Murakami et al. 2002). Faulting can juxtapose mineral phases that are geochemically incompatible and lead to rapid reactions, (e.g., ultramafic rocks like serpentinites emplaced next to quartz-bearing pelites that react to form biotite, chlorite, and amphibole-rich zones).

Faults are usually open geochemical systems, at least on the time scale of a large earthquake cycle. Open system reactions entail the flux of water and dissolved chemical components to the reactive surfaces. On the other hand, pressure solution can occur in a closed system and only requires a small amount of fluid to progress.

For deep parts of faults undergoing viscous creep, the fluid-flow regime depends strongly on whether the rocks within and adjacent to the fault are undergoing prograde or retrograde metamorphism (Chapter 11). Prograde reactions create fluid and raise fluid pressure, potentially even leading to fluid overpressure in the case of rapid fluid production in low-permeability rocks. Evidence for lithostatic pore-fluid pressures is indirect and comes from fluid inclusions, mineral stability relations, and structures (e.g., Yardley 1989; Peacock 1990). In contrast, retrograde reactions generally consume water and therefore reduce fluid volume and mineralization rates. A common example for this is the formation of phyllosilicates at the expense of stronger solid mineral phases (e.g., feldspar in granitic rocks). Reactions like these have a profound effect on crustal rheology because the hydrous products are much weaker than the anhydrous (or less hydrous) reactant phases. In the following, our group focused its discussion on several mechanisms including dissolution–precipitation, pressure solution, stress corrosion, phase transformations, and melting.

Dissolution–Precipitation

Dissolution–precipitation reactions are driven by chemical potential gradients. These can be induced by cooling or heating of fluids, or by fluids in equilibrium with one mineral assemblage in the crust coming into contact with another mineral assemblage in the fault zone. Both dissolution and precipitation occur on surfaces of mineral grains that are in contact with a fluid. Surface kinetics control the rate of the reaction. Dissolution–precipitation has been investigated in the laboratory (e.g., Rimstidt and Barnes 1980), but the influence of deformation on reaction rates is not well known. Likewise, the effect of the force of crystallization (Fletcher and Merino 2001; Hilgers and Urai 2005) on fault zone strength has not been considered to date. The rate of advective transport of reaction products away from reactive surfaces places an important limit on reaction rate, which in turn often governs the precipitation rate. Mineral growth requires new chemical components to be transferred in a fluid. Surface kinetics are easier to investigate than fluid transport processes. The solubility of aluminum- and silica-bearing minerals increases with pressure and depth. However, increasing chloride concentrations slightly decreases the solubility of aluminum and silicate and therefore hinders their transportability in a fluid phase. Hofstra and Cline (2000) have shown that calcite dissolution can enhance the permeability of faults in carbonate-hosted ore deposits. There, carbonate rocks contain calcite vugs replaced by silicates. Yet, as carbonate rocks dissolve, CO_2 enters the fluid phase and reduces the solubility of all minerals. Many in our group felt that dissolution and precipitation reactions are by far the most important mechanisms for mineral transformation, an opinion that is largely based on isotopic evidence for large scale isotopic exchange (Mulch et al. 2004). However, our understanding of how these geochemical reactions are coupled with deformation is incomplete. The following hypothesis can be tested in well-exposed, exhumed (fossil) fault zones, and may lead to better constraints on the role of fluids in fluid–rock chemical reactions in fault zones.

Hypothesis 1: Downward-flowing fluids in fault zones enhance dissolution reactions, whereas upward-flowing fluids favor precipitation reactions.

Test: Collect geochemical, isotopic and thermochronometric data from exhumed fault zones to determine whether there are monotonic cooling and heating trends. Establish the degree of fluid–rock isotopic exchange within and away from the fault zone. Develop numerical models of fluid–rock isotopic exchange to constrain the depth of fluid circulation, the temperature history, and the duration of the flow systems.

Pressure Solution

Pressure solution is driven by stress-induced gradients in chemical potential and by diffusion-controlled reaction rates that make new chemical components available along the fault zone. There is a fundamental difference between pressure solution at monomineralic grain boundaries and at polymineralic boundaries. In rocks containing both mono- and polymineralic layers, the monomineralic layers that underwent pressure solution always appear to be less deformed (boudinaged) and therefore stronger than the adjacent polymineralic layers (Chapter 12). An explanation for this is that stressed boundaries of like mineral phases (e.g., quartz-quartz, calcite-calcite, feldspar-feldspar) behave as single, large crystals, especially if their crystallographic orientation is similar. In contrast, stressed boundaries of different mineral phases (and other less-soluble species like mica and illite) are likely to remain open to diffusive mass transfer, allowing pressure solution to progress (Hickman and Evans 1995).

There is clear petrographic evidence of pressure-solution reactions in active fault zones. In the post-seismic interval immediately after an earthquake, pressure solution is important for several reasons: First, coseismic micro-fracturing within the fault zone reduces grain size, enhancing diffusion and pressure-solution creep, and in turn relaxing any residual stress in the fault system. Pressure solution does this in part by producing mineral fabrics whose preferred orientation is parallel to the fault plane, thus reducing the shear strength of rocks. Second, pressure solution can produce chemical components that seal coseismic fractures, leading to hardening. Third, pressure solution is an efficient compaction mechanism, especially in fine-grained fault gouge. Compaction can lead to hardening, but also to a decrease in permeability and therefore to weakening if fluid is produced within the fault system. The weakening due to fracturing and dissolution is faster than the strengthening associated with healing and sealing of the fault, and the associated increase of fluid pressure (Chapter 12).

To be effective, pressure solution requires the presence of a fluid phase. Small amounts of trapped fluid forming thin layers on the scale of nanometers are sufficient for pressure solution to be rate competitive with other deformation mechanisms. Pressure solution is affected by the following factors: (a) kinetics of mineral dissolution at highly stressed grain contacts; (b) precipitation rates at fluid–mineral interfaces; (c) catalytic effects of minerals (e.g., phyllosilicates); (d) mass transfer properties of trapped fluid paths under stress; (e) the nature of fluids and minerals present; (f) presence and interconnectedness of the fluid phase; (g) the size of the system in which the transfer process (diffusion, nm to dm; advection, hm to km) operates; and (h) temperature.

The thermodynamics of pressure solution are reasonably well known (e.g., Gibbs 1877; Paterson 1973; Lehner 1995; Shimizu 1995), however, the thermodynamics of pressure solution in general 3D stress states needs further development. Laboratory measurements of pressure-solution rates (e.g., Rutter 1976; Gratier and Guiguet 1986; Hickman and Evans 1995; de Meer et al. 2002) allow us to make order-of-magnitude predictions of these rates. A major problem is that the characteristic time of deformation by stress-induced mass transfer is several orders of magnitude less than the average human life-span. Acceleration of pressure-solution rates in experiments renders some geological interpretations of extrapolated experimental results uncertain. Another major factor that can only be inferred from the analysis of natural fault zones is the relative contributions of advection and diffusive modes of fluid flow. Related to that is the mean distance of mass transfer within fluid-flow zones (see also Chapter 11). Evaluating the nature of the fluid and the minerals at depth along the fault is also an important issue.

Mineral Reactions (Hydration and Dehydration)

Mineral reactions affect rheology principally through the variations that they induce in fluid content, fluid pressure, and lithology in a rock. Yardley and Baumgartner (Chapter 11) point out that rocks generate fluid and become water-saturated throughout their heating cycle (prograde metamorphism), but dry out rapidly through retrograde absorbtion of water once they begin to cool (retrograde metamorphism). Seismic low-velocity zones detected in geophysical surveys have been interpreted as water-rich zones in some regions currently inferred to be undergoing metamorphism, for example off the coast of Vancouver Island, as imaged by the Lithoprobe traverse (Hyndman 1995) and beneath the Southern Alps, New Zealand (Koons et al. 1998; Wannamaker et al. 2002; Upton et al. 2003; see section below on *Key Observations of Fluid Flux in Convergent Settings*).

Prograde mineral reactions commonly involve the release of fluid and are accompanied by changes in volume and grain size. A major rheological consequence of prograde reactions is likely to be local embrittlement, because an increase in fluid pressure through volatile release reduces effective normal stress. Most fluid production occurs stepwise during a series of devolitalization reactions, so it is likely that deformation mechanisms facilitated by the presence of a water-rich fluid (pressure solution or grain-boundary slip) are possible throughout the prograde cycle. Exceptional behavior is possible in carbonate rocks, where the solid volume changes accompanying the breakdown of carbonate minerals may lead to transient porosity, leading to focused fluid flow and the production of skarns.

Retrograde reactions can have two opposing effects. Water consumption can drive water fugacity to just a few bars, potentially inhibiting material transport along grain boundaries even if individual grains are hydrolytically weak by remaining dissociated water in defect centers. As a result, dry rocks are very strong, retaining old textures and fabrics where wet rocks in the vicinity undergo extensive deformation in response to the same stress field. However, the presence of water can lead to the production of intrinsically weak, fine grained reaction products, especially phyllosilicates, which focus deformation and even favor superplastic behavior.

Some large-scale shear zones contain abundant microstructural evidence for reaction softening assisted by microfracturing in the presence of water. Fracturing and dilatation opens the rock system to episodic fluid infiltration (e.g., Sibson 1988; Bauer et al. 2000) whereas reaction softening obviously localizes strain (Chapter 12, Figure 12.4).

Based on studies of pressure solution and mineral reactions, we advance the following hypotheses to better constrain the role of fluids in fluid–rock chemical reactions in fault zones.

Hypothesis 2: Fluid–rock and mineral interactions affect interseismic permeability and strength, and thus control the recurrence time of earthquakes.

Tests: Determine the reaction rates both in the lab and in naturally deformed, exhumed fault rocks as a basis for formulating macroscopic numerical models that include strain-dependent transport properties and rheologies. Then, test this model with geophysical and geochemical measurements of fluid flux during interseismic periods. Such measurements (involving GPS, inclinometry, seismology, microgravimetry, chemical analysis of springs) may yield insight into how fluid transfer, porosity, permeability, and rock strength evolve at depth. The aim is to incorporate permeability and fluid pressure in empirical rate- and state-dependent constitutive laws. This would yield insight on how fluid–rock interaction "restores" rheological and transport properties during interseismic periods, setting the stage for renewed seismicity.

Stress Corrosion

Research on stress corrosion in rocks was active in the 1970s and 1980s (reviews of Atkinson 1982, 1984, 1987), but has stagnated since. Stress corrosion is controlled by reaction rates, cracking, and rate- and state friction. Subcritical crack growth at grain contacts in fault zones can lead to a time-dependent increase in contact area. Experiments on the rate- and state-friction of olivine reveal a component of crystal plasticity at grain contacts, even in the absence of water. Still missing in many investigations of experimentally and naturally deformed fault rocks are detailed studies of the interplay of fracturing and intracrystalline plasticity at grain boundaries. The search for stress corrosion should involve looking at the leading edges of cracks, for example, with TEM or laser interferometric microscopy.

One of several possible explanations for the origin of rate and state friction (discussed at length in Chapters 5 and 7) is that the real area of contact during frictional sliding increases due to subcritical crack growth. The idea is that if static contacts enlarge with time, or equivalently in the sliding case, enlarge if the slip velocity is lowered, the frictional coefficient will be larger because there is more contact area over which the resistance occurs. Thus, the total amount of bonding across the surface will be larger, thereby increasing resistance. This explanation is plausible because an increase in the area of contact with an increase in normal stress is the accepted explanation for friction, namely the ratio of normal stress to shear stress. One way the contacts might enlarge is by limited dislocation motion at highly stressed grain contacts. Another way is that contacting grains fracture slowly, involving subcritical crack growth. Although fracturing of the contacts might seem to weaken rocks, it could also increase contact area and cause increased friction, leading to strengthening. This explanation is consistent with observations that both subcritical crack growth and changes in the coefficient of friction occur in wet rather than dry environments.

Melting

From a mechanical standpoint, melt is often considered to be a fluid insomuch as it exerts a pore-fluid pressure that counteracts normal stress and therefore reduces effective pressure. An important distinction between water-rich or CO_2-rich fluids and melts is that melts crystallize completely. When fluids cool, the volume (mass) of the dissolved phase removed by precipitation is small compared to that of the fluid volume (mass). The effect of melt on the deformation mechanisms of solid aggregates is probably similar to compositional effects on fluid viscosity observed in many less-viscous fluids. For example, a transition from dislocation creep to diffusion creep has been observed in experiments by adding a few percent (3 to 5%) melt to fine-grained (<10 μm) granitic samples (e.g., Dell'Angelo et al. 1987) and by adding ~0.8% water to feldspathic aggregates (Tullis et al. 1996). This transition is also inferred to occur in natural fault zones, even for large grain sizes (>300 μm) typical of natural mylonitic fault zones containing a melt (Rosenberg and Berger 2001). Strength reduction due to an increase in pore-fluid (-melt) pressure during melting has also been observed in aplite samples experimentally deformed in the presence of 5 to 10% melt (Dell'Angelo and Tullis 1988).

An additional similarity in the way fluids and melts may affect the deformation of crustal rocks concerns the localization of deformation within viscously deforming aggregates. The presence of a few volume percentages of fluids and/or melts in a viscously deforming aggregate can induce the transition from homogeneous- to highly localized deformation along interconnected networks of fluid- or melt-bearing shear bands (Bauer et al. 2000; Rosenberg and Handy 2000). Analogue experiments indicate that localization coincides instantly in space and time with the introduction of fluids into the deforming aggregate. More rheological data is needed to estimate the effect of this transition on the strength of real rock.

The residence time of melt in a rock can be much longer (10^7 yr) than for metamorphic fluids. This is probably due to the higher density and viscosity of melts with respect to aqueous fluids, reducing the mobility of melts compared to other fluids. The long residence time of melts in the lower crust may not only affect fault nucleation, but the geometry of entire faulted domains, even orogens (e.g., Beaumont et al. 2001; Chapter 13, this volume). When present, melt volumes in the crust are generally much higher than fluid volumes. This conclusion is based on the debatable interpretation of seismic low-velocity zones as representative of high melt fractions. From a rheological standpoint, however, the critical question is not just "How much melt?", but "Is the melt interconnected?" (see Chapters 4 and 7). Inferred melt percentages at a depth >20 km beneath the Tibetan Plateau in the Himalayas and southern Altiplano (Puna) Plateau in the Andes are ~20% (Schilling and Partzsch 2001). In this case, it is not certain what length scale of melt interconnection is represented by this low-velocity zone. A rheologically critical amount of melt (3–7 % by volume) could exist in much smaller amounts, e.g., on the grain scale, without its being observable with existing imaging methods. Within exposed fossil orogenic migmatites, the inferred melt volumes commonly range from 10 to 40% (Teyssier and Whitney 2002). The strength of a shear zone is nonlinearly proportional to the melt volume it contains and the strength versus melt fraction relationship probably follows an exponential law (Chapter 13). However, no experimental data exist yet to reliably constrain the mechanism and amount of melt-induced weakening in the viscous creep regime. The presence of water in the melt phase can also change the rheology of the melt-rock system. Water enters the melt phase above 650°C in granitic rocks. The melting temperature can be as low as 500°C when water is introduced. Formation of silica gels can also occur in fault zones at temperatures in excess of 200°C. An important question that needs to be resolved regarding the role of melts on fault dynamics is addressed with the following hypothesis:

Hypothesis 3: The relationship between melt content and viscous strength of shear zones follows an exponential law.

Tests: Deformation experiments on partially melted crustal rocks in the viscous deformation regime (dislocation and diffusion creep regimes) must be conducted with different melt contents (especially at very low melt fractions, <3%) in order to obtain a relationship relating strength to melt fraction.

THEME 2:
HOW DOES FLUID FLOW CHANGE
BEFORE, DURING, AND AFTER EARTHQUAKES?

Introduction

Fluid pressure has significant effects on the mechanical and chemical behavior of the crust and fault zones, because it can change dramatically on different time- and length scales throughout the earthquake cycle. The earthquake cycle can last from hundreds of years along major plate boundary faults to thousands of years in intraplate environments. During interseismic intervals, diffusive processes in surrounding stable crust relax any overpressures, leading to hydrostatic fluid pressures. In low-permeability fault zones, however, pore pressures can rise through a variety of relatively sluggish mechanisms such as pressure solution, shear-enhanced compaction, chemical osmosis, or dehydration of water-bearing clays. Quantification of pore-pressure changes due to these processes is difficult because the conditions at depth that control them are still not sufficiently constrained. Crack healing and mineral precipitation can also increase pore pressure by drastically reducing the fault and damage zone permeability between successive earthquakes.

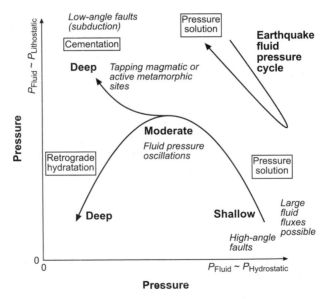

Figure 14.1 Schematic diagram showing the range of fluid pressures and fluid–rock geochemical interactions at different depths (shallow, moderate, and deep). Dry conditions are represented at the origin, water-saturated conditions at the top and right, with hydrostatic and lithostatic fluid-pressure environments represented by the *x*- and *y*-axes, respectively.

There is a wide range of possible fluid-pressure regimes within seismogenic zones of the shallow and deep crust (Figure 14.1). We argue here that each of these regimes should produce characteristic seismogenic features such as pseudotachylites, low-angle normal faults, and postseismic, limited-magnitude earthquakes, known as aftershocks. If deep crustal rocks are exhumed and undergo retrograde mineral reactions, rocks dry out and become strong. This is conducive to the formation of pseudotachylite (region labeled 'Deep' in Figure 14.1). Supracrustal rocks undergoing burial attain near-lithostatic pressures if permeabilities are sufficiently low (10^{-20} m^2; upper left portion of Figure 14.1). Examples of this include both the deep parts of oilfields, as well as the cases from New Zealand and Vancouver Island noted above. However, if rocks are more permeable ($>10^{-16}$ m^2), they are hydrostatically pressured (region labeled 'Shallow' in Figure 14.1) and may be flushed by surface fluids, depending on the permeability structure of crustal rocks with depth and within the fault zone. We would expect that these environments would host high-angle faults. Examples of this include the Rio Grande Rift, U.S.A. (Mailloux et al. 1999), the Rhine Graben, central Europe (Smith et al. 1998), and the Basin and Range, U.S.A. (McKenna and Blackwell 2004). These different environments give rise to different styles of fluid flow (escape, infiltration, or circulation) and hence of mineral reaction. It is clear that fluid-rock geochemical interactions need to be integrated into hydrothermal and mechanical descriptions of earthquake cycles.

There is evidence for hydrostatically pressured saline fluids in some deep boreholes. For example, at the German KTB borehole hydrostatically pressured fluids have been observed at 9 km depth (Moller at al. 1997). This is quite old water and raises questions about its chemical evolution and salinization mechanisms. One possible mechanism involves hydration reactions, whereby water is removed from the system and incorporated within retrograde minerals (e.g., phyllosilicates). This is augmented if Cl-bearing minerals, such as high-temperature biotite, are replaced by chloride-free phases such as chlorite. However, a number of other mechanisms, including the ingress of brines from former overlying sedimentary basins and the release of brine from fluid inclusions, have been proposed (Fritz and Frape 1987).

Various mechanisms for generating anomalous pore-fluid pressures have different characteristic time- and length scales of operation. For example, Miller et al. (2004) recently proposed that the coseismic release of high-pressure fluids at depth can initiate a pressure pulse that may propagate at a rate of up to a kilometer per day. This pressure pulse can trigger aftershocks by reducing the effective normal stress of incipient slip planes. Normally pressured fluids flowing within high permeability environments may flow at rates of up to about 10 m yr^{-1}.

Fluid flow is controlled predominantly by permeability, and permeability may change drastically at the onset of seismic rupture (e.g., Rojstaczer and Wolfe 1994). Just prior to an earthquake, the fault zone is most likely sealed

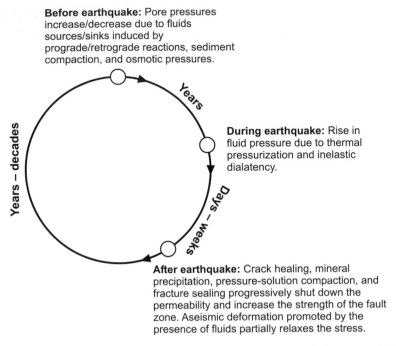

Before earthquake: Pore pressures increase/decrease due to fluids sources/sinks induced by prograde/retrograde reactions, sediment compaction, and osmotic pressures.

During earthquake: Rise in fluid pressure due to thermal pressurization and inelastic dialatency.

After earthquake: Crack healing, mineral precipitation, pressure-solution compaction, and fracture sealing progressively shut down the permeability and increase the strength of the fault zone. Aseismic deformation promoted by the presence of fluids partially relaxes the stress.

Figure 14.2 Schematic diagram indicating processes affecting pore-fluid pressure during the earthquake cycle.

from the damaged surroundings by crack healing and other long-term, time-dependent processes that operated during the previous earthquake cycle. The temporal evolution of pore-fluid pressure in fault zones can be thought of in terms of the time cycle of earthquakes (Figure 14.2). Prior to an earthquake, fluid pressure can rise due to inflow of fluids from depth, prograde metamorphic reactions, sediment compaction (in sedimentary basins), and possibly also osmotic pressurization provided there is a build up of salinity within the clay-rich core zone (salinity could increase due to hydration reactions in the fault gouge). Increases in osmotic pressure (Neuzil 2000) in shallow fault zones (<5 km) have not received much attention and we therefore propose the following hypothesis to study this mechanism.

Hypothesis 4: Relatively shallow earthquakes are triggered by the buildup of high pore-fluid pressures in a clay-rich fault core due to the generation of osmotic pressure.

Tests: Conduct ring-shear experiments in which salts are introduced to the clay-rich core of an experimental fault zone. Determine if this procedure reduces the frictional coefficient during shearing. In the field, drill a seismogenic fault zone and sample pore fluids within the clay-rich core of the fault zone. If the core has elevated salinities and low porosity, this is consistent with osmotic pressure generation. Conduct numerical experiments using realistic hydration kinetic rates and osmotic pressure terms included into the groundwater flow equation (e.g., Neuzil 2000).

During an earthquake, fluid pressure can change dramatically due to inelastic dilatancy and thermal pressurization. Once a rupture begins, frictional heating of trapped pore fluids can raise the pore-fluid pressure to near-lithostatic values, thus weakening the fault. However, this process is self-limiting because lithostatic fluid pressures imply near-frictionless faulting, which in turn reduces frictional heating (Mase and Smith 1987a, b). Dynamic rupture across a region of fault overpressure is not yet understood in the context of the rate- and state formulation of friction, so this needs much additional study. Dilatant rupture causes pore-pressure dissipation both parallel and perpendicular to the fault zone. Based on known kinetic rates of precipitation/dissolution reactions, the rates of chemical reaction during earthquakes are probably small. However, frictional heating can generate silica gels rapidly, as discussed in Chapters 5 and 7.

When considering the dissipation of pore-fluid pressure, it is important to consider how pore-pressure changes both within the low-permeability, clay-rich core of a fault, and the higher permeability damage zone within adjacent basement rocks. Seismic rupture can reset the hydraulic properties of the system and any postseismic flow is controlled by the new permeability structure. At the crack tip during dynamic rupture and at fault bends, significant tensile stresses are developed just adjacent to the fault, providing a mechanism for the creation of a highly fractured damage zone surrounding the slip surface. Calculations and geophysical imaging indicate that the width of the damage zone diminishes with depth, from about 1–100 m near the surface to much less at midcrustal levels (~10 km, Chapter 2). The damage zone plays an important role in the postseismic redistribution of fluid pressure because it represents a highly permeable crack network that can transport fluid and mitigate fluid pressure over short time periods. Once earthquake rupture ceases, the heavily damaged and fractured region near the slip plane controls both the rate at which rupture-induced overpressures subside and the rate at which trapped pockets of overpressure dissipate. If there exist locally overpressured compartments with high permeability and earthquake-related fractures access these pockets, then the magnitude of the fluid-pressure pulse can be significant. The ensuing pressure pulse will exceed any local changes in shear and normal stress, thus providing a viable mechanism for the generation of aftershocks.

The build-up of osmotic pressures could be an important mechanism for generating earthquakes at shallow depths (~5 km) if porosity is low (<10%) and there is an order-of-magnitude increase in salinity within the fault zone and the clay-rich core (Neuzil 2000). Elevated fluid pressures can clearly occur within the fault core zone. An important issue centers on how fluids feed into faults at great depth during the interseismic period. Depending on the fluid source, it may be necessary to break some sort of seal at midcrustal levels (10 km). This seismic zone is expected to widen toward the surface (Chapter 2).

Earthquakes can be triggered by changes in the Coulomb Failure Stress (ΔCFS):

$$\Delta\text{CFS} = \Delta\tau + \mu(\Delta\sigma_n + \Delta P_f) \qquad (14.1)$$

where $\Delta\tau$ is the shear stress change on incipient fault planes, μ depicts the friction coefficient, $\Delta\sigma_n$ is the normal stress change on incipient fault planes, and ΔP_f represents pore pressure change.

If dynamic rupture of an earthquake taps into a source of high-pressure fluids at depth, then the propagating pressure pulse can have a significant effect on ΔP_f, potentially much more than on $\Delta\tau$ or $\Delta\sigma_n$. This hypothesis can be tested by accurate measurements of the space-time evolution of aftershocks. If this mechanism is important for aftershock generation, then a correlation should be found between aftershock evolution and predictions from a model of this process (Miller et al. 2004). Some data is already available for analysis (e.g., San Andreas fault, North Anatolian fault), but much more data is needed from a variety of tectonic environments.

In the postseismic interval immediately after a large earthquake, chemical reactions driven by residual stresses or by other earthquake-related processes (fractures, fluid transfers, temperature and pressure changes, etc.) progressively restore the strength and mass-transfer properties (i.e., permeability) of the gouge and damage zone to values that prevailed in the interseismic period up to the earthquake (Chapter 12).

Time and Length Scales of Fluid Flow and Geochemical Reactions in Fault Zones

Processes like pressure solution operate at the grain scale, whereas precipitation/dissolution reactions can occur at the km scale, linking areas of mineral dissolution and precipitation. Dehydration reactions occur over a very specific temperature range. At temperatures below about 300°C, reaction kinetics can be very important. At higher temperatures, thermodynamic equilibrium can be maintained provided there is ample fluid to facilitate material transport between crystals. Grinding up mineral grains within the fault core drives chemical reactions faster, primarily due to the increase of surface area with respect to volume. Kinetics may play an important role in the history of the earthquake cycle. Retrograde reactions can consume water and increase salinity, but the kinetics of these reactions suggest that they continue long after the earthquake. Reaction rates are strongly temperature-dependent and reaction times on the order of days to many tens of years seem probable (Wood and Walther 1983).

Hydroseismicity and Surfacial Expressions of Fluid Flow

It is well known that natural and anthropogenic changes in pore-fluid pressure can induce seismic events (e.g., Saar and Manga 2003). Following an earthquake, pore space collapses (Muir-Wood and King 1993) and the resulting increase in matrix permeability (Rojstaczer and Wolf 1994) has been proposed as a mechanism to explain changes in surface water discharge as the rocks adjust to changes in the stress field. Changes in surface water chemistry have also been observed following earthquakes (Curry et al. 1994). While fluid flow is known to occur during the earthquake cycle, the mechanisms driving fluid flow during seismic events remain poorly understood. Below we propose the following hypothesis which may shed light on this issue.

Hypothesis 5: Seismicity is triggered by fluid movement within and around fault zones. Earthquakes can tap mid- to lower-crustal reservoirs of high-pressure fluid, generating a fluid-pressure pulse and triggering aftershocks by suddenly reducing the effective normal stress on incipient slip planes.

Test: Analyze the space-time evolution of aftershocks to determine precisely the hypocenter locations. The spatio-temporal patterns of hypocenters will have predictable patterns that depend on tectonic setting. These should be correlated to surface flow features, including changes in aquifer geochemistry, subtle land deformation features associated with pressurization of the upper crust and depressurization of the lower crust as revealed through geodetic (InSAR and GPS) surveys, changes in seismic velocity associated with high pore-pressure zones, and possible changes in electrical conductivity.

Porosity and Permeability within Fault Zones

Studies within petroleum basins have generated a large body of knowledge regarding pore-fluid pressure distribution and the hydraulic properties of faults in the shallow crust. In soft shales or clays, fault zones act as barriers to fluid flow (e.g., Knipe 1997; Caine et al. 1996; Bense and Van Balen 2004). This is evidenced by pressure changes or compartments that have been observed to develop in fault-bounded oil reservoirs in the North Sea and Gulf of Mexico. On the other hand, petroleum basins are replete with examples of vertical migration of petroleum along fault zones (Rinkbeiner et al. 2001). Evidently, the relative permeability of fault rocks can evolve through time, especially during and after movement. In some situations, the main effect of faulting on fluid flow is to juxtapose high and low-permeability units (e.g., Mailloux et al. 1999).

Despite several attempts to quantify the permeability of the upper crust (e.g., Manning and Ingebritsen 1999; Saar and Manga 2003), little is known about the hydraulic properties of fault zones beneath the sedimentary pile. There, the only suggestion of fluid flow comes from indirect measures such as the isotopic alteration of fault rocks (Mulch et al. 2004), the genesis of fault-hosted ore deposits (Sibson 1989), and thermochronological and numerical modeling (Mailloux et al. 1999). To date, studies of fault mineralization have rarely been undertaken with seismogenesis in mind. Due to the relatively low permeability of wall rocks, faults in basement rocks are much more likely to act as fluid conduits rather than barriers to flow, as evidenced by the common occurrence of ore deposits along and adjacent to faults. However, Sibson (1989) pointed out that large ore deposits are commonly associated with relatively low-displacement faults, perhaps because the roughness of the fault plane permits it to act as a valve. Fluid flow may take place through tubes opened at jogs from the time of displacement until the permeability has been resealed by mineral precipitates. This should be tested by careful monitoring of fluid flow in the vicinity of jogs along active faults, for example, in the Walker Lane area of the Great Basin, Nevada, U.S.A.

THEME 3:
WHAT ARE THE MAGNITUDES OF FLUID FLUX THROUGHOUT THE LITHOSPHERE IN DIFFERENT TECTONIC ENVIRONMENTS?

Introduction

Some information regarding fluid fluxes in fault zones is already available from drilling campaigns in European and North America. For example, the Corinth drilling project sponsored by the European Union (www.corinth-rift-lab.org/index_en.html) has drilled an active fault system to a depth of about 2 km in an extensional setting in Greece. Pressure solution was found to be an important mechanism controlling deformation (compaction and sealing) in this carbonate-hosted fault zone. Other sites offer additional data: Kobe (Japan), Chichi (Taiwan), and Parkfield (California, U.S.A.). At Parkfield, boreholes have been emplaced to monitor active processes on the San Andreas fault.

Key Observations of Fluid Flow in Rift Zones

Fluid inclusions from fault-bound ore deposits in extensional tectonic settings (e.g., Irish Pb-Zn) indicate that fluids were at near-hydrostatic pressures. The ore deposits formed in the damage zone where there is a source of sulfide-rich waters, suggesting that the damage zone is very permeable. Interestingly, big ore deposits do not normally occur directly along the slip planes of large faults (Sibson et al. 1988). Biological processes may have enhanced the formation of ore precipitation in the Irish deposits, providing a source of reduced S to precipitate metals from hydrothermal solutions (Blakeman et al. 2002). Given the likely importance of damage zones for fluid mixing in the upper continental crust, it is probable that detailed fluid inclusion studies could further constrain flow models by providing information about fluid pressure fluctuations.

Orogenic gold deposits at midcrustal levels are more controversial (Chapter 11). Perhaps the fluids associated with these gold deposits come from prograde metamorphic reactions. Alternatively magmatic fluids or even high level waters may be responsible.

Key Observations of Fluid Flux in Convergent Settings

Fluid flow associated with active, prograde metamorphism has been imaged as a zone of anomalously high conductivity beneath both the Southern Alps of New Zealand (Figure 14.3) and Vancouver Island, Canada. High-conductivity zones in the vicinity of Nanga Parbat (Himalayas) can also be interpreted as a zone of melts (see discussion above). The Southern Alps is a remarkable system in that rapid convergence and erosion has brought rocks to the surface that are inferred to have passed through the shallowest of the metamorphic zones within only the past 1–2 Ma.

The U-shaped region of high conductivity within the lower crust in Figure 14.3 is interpreted to be a region of fluid production and strain-induced permeability (Upton et al. 2003). Several fluid-flow paths connect this U-shaped region both with the surface and with sources at depth. Each path may have a different cause: topographically driven meteoric flow in the upper crust around the Main Divide, fluid expulsion from the lower crust, interconnected fluid including meteoric and basinal components corresponding to the eastern vertical conductor which may be undergoing buoyancy-driven flow through the upper crust and episodic upward flow in the Main Divide region.

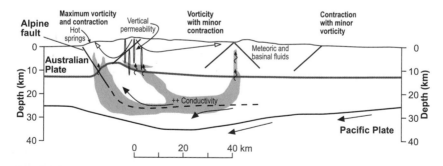

Figure 14.3 Schematic cross section of the Southern Alps and South Island, New Zealand, showing inferred paths of fluid flow based on conductivity studies (Upton et al. 2003). West is to the left, and the Alpine Fault is the prominent steep fault to the west of the topographic culmination (Main Divide). Strain regions (bold arrows) are taken from Koons et al. (1998).

SUMMARY

What Have We Learned?

To date, very little is understood about the role of fluid–rock chemical interactions, the mechanisms that drive hydrochemical fluid flow, and the magnitude of fluid flux in seismogenic zones. Even less is known about how these factors and processes interact during the earthquake cycle. Fluid–rock geochemical reactions have received considerable attention by metamorphic petrologists and geoscientists studying sedimentary basins, but not from a seismogenic perspective. Given the complexity of fluid–rock geochemical interactions (e.g., pressure solution, dissolution–precipitation reactions), the mineralogy in different fault zones, and the thermobarometric history during faulting (burial versus exhumation), few generalizations can be made regarding this topic. It is critical to develop well-posed, testable hypotheses that can better constrain the role of fluid–rock chemical reactions within the framework of the seismogenic cycle.

Our understanding of the role of pore fluids and associated geochemical interactions in earthquake dynamics would be greatly improved with the realization of advances in isotope geochemistry, hydrogeology, reaction kinetics, and numerical modeling. In addition, integrated field campaigns are needed in different tectonic settings. These would include working out the kinetics of deuterium isotope exchange for different minerals at elevated temperature and pressure, constraining the effective surface areas of fault zones, and developing a testable model of fault permeability and permeability anisotropy for faults in crystalline and sedimentary rocks.

In rock-dominated systems, geochemical markers are overwhelmed by the local wall–rock signature. Whereas oxygen isotopes are often useless as fluid tracers (because fluid compositions are dominated by wall–rock interactions in a fluid-poor system), hydrogen isotopes may work reasonably well to determine whether fluids in fault zone originate from fluid-producing metamorphic reactions in the crust or at the surface. Nevertheless, the effects of hydrogen (in adsorbed water in defects as well as in free aqueous fluid) on mineral-fluid fractionation of hydrogen isotopes are not well understood. The presence of other fluid species, notably methane, may have a major influence on hydrogen isotopic composition. Cl-stable isotopes, and Cl/Br ratios are very conservative (i.e., are not modified by mineral interactions), but they may not be very distinctive (Banks et al. 2000). He^3/He^4 isotope ratios give good indications of inputs of mantle volatiles but must be carefully linked to fluid flows, because He is very insoluble in saline water. Improvements to analytical techniques for conservative element and isotope ratios are needed to assess the origin of fluids in orogenic belts.

With the development of hot dry rock projects, the seismicity observed during fluid injection demonstrates the potential for using seismicity to track fluid migration. The Camborne School of Mines HDR project, the German KTB deep drilling project, and the Soultz-sous-Forets HDR project are promising examples of this. Microseismicity is a marker of where fluid-induced crack propagation leads to higher permeability. Thus, more high-resolution instrumentation of seismogenic zones is needed.

New analytical techniques are needed to date mineralization events and associated mineral fabrics, as well as fault zone fluids. Where it can be established that fluid inclusions are conserved chemically and physically during exhumation of the fault zone, detailed fluid inclusion studies can reveal changes from hydrostatic to lithostatic pressures. Fault permeability and permeability anisotropy within fault zones is poorly understood. Additional measurements of these parameters are needed, both with depth and in different tectonic settings.

Integrative community models such as those developed by atmospheric scientists to study climate change (e.g., COHMAP, Hadley Center) are required to quantify coupled fluid flow and geochemical-thermomechanical processes within fault zones. Processes to be considered include changes in porosity and permeability due to variations in effective stress, thermal expansion, and mineral precipitation, fluid flow induced by fluid source/sink terms, multicomponent fluid transport (H_2O, CO_2, NaCl), and elastic-plastic-viscous deformation. The model should also include advective-diffusive heat and solute mass transport, fluid–rock chemical interactions by pressure solution, and precipitation–dissolution reactions. In addition, we need a suite of models at the pore- or grain-scale to compliment those at the continuum scale (e.g., Discrete Elements Models). Grain-scale models may help to constrain field observations at the hand-specimen scale by determining effective parameters for continuum-scale models. Such models may help us understand the rheological and geochemical transitions which occur along fault zones connecting Earth's surface with the asthenosphere. The problem remains of testing of the numerical codes, an exercise that is too often neglected by the geoscience community.

Our group recognized the need for monitoring both modern flow systems and for reconstructing fossil flow systems in exhumed fault zones. The best field sites for modern flow systems are characterized by extensive internal brittle deformation and are covered with an extensive seismic network to obtain continuous, 3D images of the microseismicity. Further drilling of active fault zones are required to measure fluid characteristics (pressure, chemistry, isotopic) in and around the damage zone and fault core. Possible candidate field areas are the Apennines in Italy and the Southern Alps of New Zealand. These areas would compliment existing or planned sites in the U.S.A., Japan, Greece, and Taiwan where long-term observation of active faults is scheduled.

Unresolved Questions

Where do fluids in fault zones originate and at what rate are they available to fault zones?

How do different fluids from different sources mix in fault zones, if at all?

What is the nature of fluid–chemical–mechanical–thermal interactions within fault systems?

How can fluids be pumped down into dry basement rocks? The transient fluid pressure gradient is there, but what is the mechanism?

More generally, can we reconcile the petrological prediction of a dry, strong lower crust with seismic data?

How much time is required for the fluid to achieve local geochemical equilibrium with a rock?

What criteria can be used to distinguish pressure solution from precipitation–dissolution reactions?

Which fluid–rock geochemical reactions are most relevant to fault strength?

Why do aftershocks on most fault zones have the same frequency?

Why do oceanic transform faults not have many aftershocks?

These questions can only be addressed with an integrated effort among different specialists in the Earth Science community. They need the support of governmental funding agencies that are willing to make long-term investments in basic research.

REFERENCES

Atkinson, B.K. 1982. Subcritical crack propagation in rocks: Theory, experimental results, and applications. *J. Struct. Geol.* **4**:41–56.

Atkinson, B.K. 1984. Subcritical crack growth in geological materials. *J. Geophys. Res.* **89**:4077–4114.

Atkinson, B.K. 1987. Fracture Mechanics of Rock. London: Academic.

Banks, D.A., R. Green, R.A. Cliff, and B.W.D. Yardley. 2000. Chlorine isotopes in fluid inclusions: Determination of the origins of salinity in magmatic fluids. *Geochim. Cosmochim. Acta* **64**:1785–1789.

Bauer, P., S. Palm, and M.R. Handy. 2000. Strain localization and fluid pathways in mylonite: Inferences from *in situ* deformation of a water-bearing quartz analogue (norcamphor). *Tectonophysics* **320**:141–165.

Beaumont, C., R.A. Jamieson, M.H. Nguyen, and B. Lee. 2001. Himalayan tectonics explained by extrusion of a low-viscosity channel coupled to focused surface denudation. *Nature* **414**:738–742.

Bense, V.F., and M.A. Person. 2006. Faults as conduit-barrier systems to fluid flow in siliciclastic sedimentary aquifers. *Water Resour. Res.* **42**:W05421; doi:10.1029/2005WR004480.

Bense, V.F., and R.T. Van Balen. 2004. The impact of clay-smearing and fault relay on groundwater flow patterns in the Lower Rhine Embayment. *Basin Res.* **16**:397–411.

Blakeman, R.J., J.H. Ashton, A.J. Boyce, A.E. Fallick, and M.J. Russell. 2002. Timing of interplay between hydrothermal and surface fluids in the Navan Zn plus Pb orebody, Ireland: Evidence from metal distribution trends, mineral textures, and δ^{34}S analyses. *Econ. Geol.* **97**:73–91.

Caine, J.S., J.P. Evans, and C.B. Forster. 1996. Fault zone architecture and permeability structure *Geology* **24**:1025–1028.

Curry, R.R., B.A. Emery, and T.G. Kidwell. 1994. Sources and magnitudes of increased stream flow in the Santa Cruz Mountains for the 1990 water year after the earthquake. USGS Professional Paper 1551-E. p. E31f.

Dell'Angelo, L.N., and J. Tullis. 1988. Experimental deformation of partially melted granitic aggregates. *J. Metamorphic Geol.* **6**:495–515.

Dell'Angelo, L.N., J. Tullis, and R.A. Yund. 1987. Transition from dislocation creep to melt-enhanced diffusion creep in fine-grained granitic aggregates. *Tectonophysics* **139**:325–332

de Meer, S., C.J. Spiers, C.J. Peach, and T. Watanabe. 2002. Diffusive properties of fluid-filled grain boundaries measured electrically during active pressure solution. *Earth Planet. Sci. Lett.* **200**:147–157.

Fletcher, R.C., and E. Merino. 2001. Mineral growth in rocks: Kinetic-rheological models of replacement, vein formation, and syntectonic crystallization. *Geochim. Cosmochim. Acta* **65**:3733–3748.

Fritz, P., and S.K. Frape, eds. 1987. Saline water and gases in crystalline rocks. Geol. Assoc. Canada Spec. Paper vol. 33. Toronto: Geol. Assoc. Canada.

Gibbs, J.W. 1877. On the equilibrium of heterogeneous substances. *Trans. Connecticut Acad.* **3**:108–248; 343–524.

Gratier, J.P., and R. Guiguet. 1986. Experimental pressure solution-deposition on quartz grains: The crucial effect of the nature of the fluid. *J. Struct. Geol.* **21**:1189–1197.

Hickman, S.H., and B. Evans. 1995. Kinetics of pressure solution at halite–silica interfaces and intergranular clay films. *J. Geophys. Res.* **100**:B7, 13,113–13,132.

Hilgers, C., and J.L. Urai. 2005. On the arrangment of solid inclusions in fibrous veins and the role of the crack-seal mechanism. *J. Struct. Geol.* **27**:481–494.

Hofstra, A.H., and J. Cline. 2000. Characteristics and models for carlin-type gold deposits. *Rev. Econ. Geol.* **13**:163–220.

Hyndman, R.D. 1995. The Lithoprobe corridor across the Vancouver Island continental margin: The structural and tectonic consequences of subduction. *Canad. J. Earth Sci.* **32**:1777–1802.

Knipe, R.J. 1997. Juxtaposition and seal diagrams to help analyze fault seals in hydrocarbon reservoirs. *AAPG Bull.* **81**:187–195.

Koons, P.O., D. Craw, S.C. Cox, P. Upton, and C.P. Chamberlain. 1998. Fluid flow during active oblique convergence: A Southern Alps model from mechanical and geochemical observations. *Geology* **26**:159–162.

Lehner, F.K. 1995. A model for intergranular pressure solution in open systems. *Tectonophysics* **245**:153–170.

Mailloux, B., M. Person, P. Strayer et al. 1999. Tectonic and stratigraphic controls on the hydrothermal evolution of the Rio Grande Rift. *Water Resour. Res.* **35**: 2641–2659.

Manning C., and S. Ingebritsen. 1999. Permeability of the continental crust: Implications of geothermal and metamorphic systems. *Rev. Geophys.* **37**:127–150.

Mase, C.W., and L. Smith. 1987a. Effects of frictional heating on the thermal, hydrologic, and mechanical response of a fault. *J. Geophys. Res. B* **92**:6249–6272.

Mase, C.W., and L. Smith. 1987b. The role of pore fluids in tectonic processes. In: U.S. National Report to International Union of Geodesy and Geophysics 1983–1986: Seismology; Tectonophysics; Oceanography (chemical). *Rev. Geophys.* **25**:1348–1358.

McKenna, R.R., and D. Blackwell. 2004. Numerical modeling of transient Basin and Range extensional geothermal systems. Selected papers from the TOUGH symposium 2003, ed. K. Pruess. *Geothermics* **33**:457–476.

Miller, S.A., C. Collettini, L. Chiaraluce et al. 2004. Aftershocks driven by a high pressure CO_2 source at depth. *Nature* **724**:724–727.

Moller, P. et al., 1997. Palaeofluids and recent fluids from the upper continental crust: Results from the German continental deep-drilling program (KTB). *J. Geophys. Res.* **102**:18,233–18,254.

Muir-Wood, R., and G. King. 1993. Hydrologic signatures of earthquake strain. *J. Geophys. Res.* **98**:22,035–22,068.

Mulch, A., C. Teyssier, M. Cosca, O. Vanderhaeghe, and T. Vennemann. 2004. Reconstructing paleoevevation in eroded orogens. *Geology* **32**:525–528.

Murakami, M., R. Yamada, and T. Tagami. 2002. Detection of frictional heating of fault motion by zircon fission track thermochronology. *Geochim. Cosmochim. Acta* **66**:537.

Neuzil, C.E. 2000. Osmotic generation of "anomalous" subsurface fluid pressures in geological environments. *Nature* **403**:182–184.

Paterson, M.S., 1973. Nonhydrostatic thermodynamics and its geologic applications. *Rev. Geophys. & Space Phys.* **11**:355–389.

Peacock, S.M. 1990. Numerical simulation of metamorphic pressure-temperature-time paths and fluid production in subducting slabs. *Tectonics* **9**:1197–1211.

Rimstidt, J.D., and H.L. Barnes. 1980. The kinetics of silica–water reactions. *Geochim. Cosmochim. Acta* **44**:1683–1699.

Rinkbeiner, T., M. Zoback, P. Flemings, and B. Stump. 2001. Stress, pore pressure, and dynamically constrained hydrocarbon columns in the South Eugene Island 330 field, north Gulf of Mexico. *AAPG Bull.* **85**:1007–1031.

Rojstaczer, S., and S. Wolfe. 1994. Hydrologic changes associated with the earthquake in the San Lorenzo and Pescadero Drainage Basins. USGS Professional Paper 1551–E, p. E51–. Reston, VA: U.S. Geol. Survey

Rosenberg, C.L., and A. Berger. 2001. Syntectonic melt pathways in granite, and melt-induced transition in deformation mechanisms. *Phys. & Chem. Earth A* **26**:287–293.

Rosenberg, C.L., and M.R. Handy. 2000. Syntectonic melt pathways during simple shearing of a partially molten rock analogue (norcamphor-benzamide). *J. Geophys. Res.* **105**:3135–3149.

Rutter, E.H. 1976. The kinetics of rock deformation by pressure solution. *Phil. Trans. R. Soc. Lond.* **283**:203–219.

Saar, M., and M. Manga. 2003. Seismicity induced by seasonal groundwater recharge at Mt. Hood, Oregon. *Earth Planet. Sci. Lett.* **214**:605–618.

Saffer, D.M., and A.W. McKiernan. 2005. Permeability of underthrust sediments at the Costa Rican margin: Scale dependence and implications for dewatering. *Geophys. Res. Lett.* **32**:L02302, doi:10.1029/2004GL021388.

Schilling, F.R., and G.M. Partzsch. 2001. Quantifying partial melt fraction in the crust beneath the central Andes and the Tibetan Plateau. *Phys. & Chem. Earth A* **26**: 239–246.

Screaton, E.J., and D.M. Saffer. 2005. Fluid expulsion and overpressure development during initial subduction at the Costa Rica convergent margin. *Earth Planet Sci. Lett.* **233**:361–374.

Shimizu, I. 1995. Kinetics of pressure solution creep in quartz: Theoretical consideration. *Tectonophysics* **245**:121–134.

Sibson, R.H. 1989. Earthquake faulting, induced fluid flow, and fault-hosted gold-quartz mineralization. In: Characterization and Comparison of Ancient and Mesozoic Continental Margins, publ. 8, pp. 603–614. Salt Lake City: Intl. Basement Tectonics Assoc.

Sibson, R.H., F. Robert, and K.H. Poulsen. 1988. High-angle reverse faults, fluid-pressure cycling, and mesothermal gold-quartz deposits. *Geology* **16**:551–555.

Smith, M.P., V. Savary, B.W.D. Yardley et al. 1998. The evolution of the deep flow regime at Soultz-sous-Forets, Rhine Graben, eastern France: Evidence from a composite quartz vein. *J. Geophys. Res.* **103**:B11, 27,223–23,237.

Teyssier, C., and D.L. Whitney. 2002. Gneiss domes and orogeny. *Geology* **30**: 1139–1142.

Tullis, J., R. Yund, and J. Farver. 1996. Deformation-enhanced fluid distribution in feldspar aggregates and implications for ductile shear zones. *Geology* **24**:63–66.

Upton, P., P.O. Koons, and D. Eberhart-Phillips. 2003. Extension and partitioning in an oblique subduction zone, New Zealand: Constraints from three-dimensional numerical modeling. *Tectonics* **22**:1068, doi:10.1029/2002TC001431.

Wannamaker, P.E., G.R. Jiracek., J.A. Stodt et al. 2002. Fluid generation and pathways beneath an active compressional orogen, the New Zealand Southern Alps, inferred from magnetotelluric data. *J. Geophys. Res.* **107**:B6, doi:10.1029/2001JB000186.

Wood, B.J., and J.V. Walther. 1983. Rates of hydrothermal reactions. *Science* **222**: 413–415.

Yardley, B.W.D. 1989. An Introduction to Metamorphic Petrology. Harlow: Longmans.

Author Index

Baumgartner, L. P. 295–318, 403–425
Beroza, G. C. 9–46, 79–98
Bos, B. 403–425
Brun, J.-P. 79–98
Buck, W. R. 273–294
Bürgmann, R. 139–181, 183–204

Cocco, M. 99–138, 183–204
Connolly, J. A. D. 403–425
Cowie, P. A. 47–77, 79–98

Densmore, A. L. 273–294

Friedrich, A. M. 273–294
Furlong, K. 79–98

Gratier, J.-P. 319–356, 403–425
Gueydan, F. 319–356, 403–425

Handy, M. R. 1–8, 79–98, 139–181,
              357–401
Hirth, G. 1–8, 139–181, 183–204
Hovius, N. 1–8, 231–271, 273–294

Kind, R. 9–46
King, G. C. P. 183–204
Kirby, E. 205–230, 273–294
Koons, P. O. 205–230, 273–294

Medvedev, S. 357–401
Miller, S. A. 403–425
Mooney, W. D. 9–46, 79–98

Mortimer, E. 47–77

Nagel, T. J. 273–294

Oncken, O. 183–204
Otsuki, K. 183–204

Person, M. 403–425

Rice, J. R. 99–138, 183–204
Roberts, G. P. 47–77
Rosenberg, C. L. 357–401, 403–425
Rubin, A. 183–204

Schlunegger, F. 273–294
Segall, P. 183–204
Shapiro, S. A. 183–204
Strecker, M. R. 273–294

Taymaz, T. 79–98
Teyssier, C. 79–98
Tullis, T. E. 183–204

Urai, J. L. 403–425

Vauchez, A. 79–98
von Blanckenburg, F. 231–271, 273–294

Wernicke, B. 79–98
Wibberley, C. A. J. 183–204

Yardley, B. W. D. 295–318, 403–425

Subject Index

Aegean Sea (*see also* Hellenic trench-
arc system) 6, 93
aggregate
  anatectic 363
  polymineralic 335
Alps, European
  deep structure 18
  denudation rate 251
  erosion rate 250
  orogenic crust 373
  Periadriatic fault system 17
  seismic profile 18
  tectonic map 374, 375
  Tertiary plutonism 372
  Western 312, 373
Altiplano *see* Andean Plateau
anatexis 358, 383
Andean Plateau 5, 9, 368, 369, 389, 390
Andes 32, 276, 371, 376, 389, 390
anisotropy, seismic 5, 10, 29
Appalachians 251, 253
Asia 390
  continental plate 254
  crust 21
  lithospheric mantle 34
asperity 112, 144, 333, 334, 339
asthenosphere 2, 30, 393
atmosphere
  circulation pattern 232
  dynamics 274

basement
  Caledonian 301
  crystalline 300, 301
  faulting 300
  rock, Precambrian 301
basin (*see also* sediment, sedimentary)
  –basement interface 301
  extensional 61

  fault-bounded 4
Basin
  Caledonian Foreland 301
  East Shetland 60, 61
  Fucino 57
  Galicia Interior 61
  Inner Moray Firth Rift 50, 51
  Loreto 62–66
  Los Angeles 21
  North Sea 61
  Ruili (China) 23
Basin and Range Province 93, 277,
  281, 413
Bay Area Seismic Experiment 19
BDT *see* brittle-ductile transition
bedrock
  channel gradient 278
  denudation rate 241
  fluvial erosion 234
  fluvial incision 211, 234, 243, 245
  wear 246
brine 300
  magmatic 297
  sedimentary 297
brittle (*see also* cataclasis, failure,
  fracture/fracturing, frictional,
  rupture)
  behavior 275
  damage 62
  field, crustal deformation 48
brittle-ductile transition (BDT) (*see
  also* frictional-to-viscous transition,
  FVT) 17, 26–28, 139–152,
  308–310
  criteria for recognition 149–152
  dependence on physical conditions
  139–149
  depth and time-dependence of 26,
  28, 141, 144

brittle-ductile transition (BDT) *cont.*
    effects of fluid on 139, 141, 142,
        144–147, 295, 301, 308–310
    relationship to the earthquake cycle
        139, 141–145, 148, 163–167, 310
Byerlee's Law 147, 308

Caledonides 380
    basement 301
    eclogite 313
    thrust 15
carbon, global cycle 233
carbon dioxide 231, 296, 306
    activity diagram 307
    consumption 231
    drawdown 232, 233, 279, 288
    sink 279
cataclasis 4, 320, 327, 334, 366
catchment 236, 248, 258
CDF *see* chemical depletion fraction
channel
    erosion 234
    flow, lower-crustal 390
    fluvial 222
    steepness index 224
    topography 223
chemical depletion fraction (CDF)
    258, 259, 306, 322
chemical potential 306, 322, 406, 407
clay minerals 5, 191, 260, 300, 301,
    319, 335, 412
climate (*see also* ENSO) 233, 243,
    259, 279, 292
    change 2, 262
        relationship to faulting 280–282
    Quaternary 247
        oscillation 281
cohesion 207, 342
collapse, gravitational 380, 383, 392
compaction 301, 341, 405
conductivity/resistivity, electrical 12,
    23, 299
continental crust 330, 358
Cordilleran orogen 6, 383
    North American 369, 379, 382, 383
cosmogenic nuclide 4, 233, 240, 241,
    259, 261, 279, 280
coupling/decoupling 17, 167–171

displacement 167
    relationship to strain partitioning 168
    seismic 168
crack/cracking *see* fracture/fracturing
crack-sealing 339, 341
cross sections
    electrical conductivity
        Chile 22
        Himalaya 21
        San Andreas Fault 22, 23
        Southern Alps, New Zealand 419
        Western Fissure Zone 21
    geological
        Alps 18, 171, 375
        Annapurna area, Himalayas 370
        Fucino Basin 57
        Himalaya 370
        Loreto Basin 64
        Periadriatic fault system 171, 375
        South Tibetan detachment 370
        Tibetan-Himalayan orogen 370
    seismic 141, 150, 219
        Alpine Fault, New Zealand 20
        Alps 18, 151
        British Isles 16
        California Coast Ranges 19
        East Shetland Basin 60, 61
        Himalaya 21
        InnerMoray Firth Rift Basin 51
        Puna High Plateau 33
        San Andreas Fault 20
        Tenda Massif, Corsica 331
        Tibet 34, 35
crust
    Andean 390
    Asian 21, 36
    continental 5, 300, 302, 306, 310,
        358, 390, 393
    Himalayan 368
    Indian 36
    melt-bearing 369, 377, 385
    seismogenic 342
    Tibetan 36
cycle
    geochemical 403
    hydrological 296
    metamorphic 308, 312
    seismic 4, 188, 190, 284

damage zone 10–13, 405
Dead Sea Transform fault 9, 14, 21, 24, 30
denudation 3, 4, 205, 231, 232, 240, 243, 381
  definition 265, 266
  flux 253
  history reconstruction 261
  quantification 234
  rate 222, 233, 240, 242, 244, 245, 250–253, 259, 280
diagenesis 297, 302
diffusion 323–325
  creep 366
diffusivity 287
dislocation
  creep 84, 330, 366
  glide 327
displacement 12, 332
  accumulation 280
  coseismic 286
  gradients 53
  length, scaling 67, 70
  postseismic 165
  rate 328, 329
dissolution 322–325, 328, 332–336, 405, 406
dissolved load 238, 265
drilling sites
  Corinth 418
  deep 296
  KTB 300, 413
  Saatly 300
dynamic rupture 190, 192
dynamic weakening 111, 190

earthquake 216, 232, 255, 283, 287, 320, 337
  cluster 287
  cycle 4, 14, 141, 188, 281, 284, 310, 319, 346, 347–351, 412
  deep continental 299
  depth
    distribution 24
    time dependence 29
  effect on fluid flow 403, 412
  energy, prediction 117
  erosional response 254

frequency 281
hazard risk mitigation 290
location 24, 47
nucleation 5, 107
prediction 28, 117, 351
predictive model 124, 127
relocation 25, 26, 28
sedimentary record of 264
size-frequency distribution 283
triggering 154, 414–417
earthquake locations/names
  Big Bear (1992) 25
  Borah Peak (1983) 284
  Chi-Chi (1999) 12, 255, 256, 263, 289–291
  Duzce (1999) 13, 14
  Fort Tejon (1857) 26
  Hector Mine (1999) 15, 71, 90, 164–166, 187, 337
  Joshua Tree (1992) 25
  Kobe (1995) 418
  Landers (1992) 14, 24, 25, 29, 71, 90, 144, 164–166, 187
  Morgan Hill (1984) 29
  San Francisco (1906) 26
  Sumatra-Andaman (2004) 7, 288, 376
eclogite/eclogitization 308, 312, 313, 330, 331
education, with respect to seismic hazard and risk 291
El Niño Southern Oscillations (ENSO) 275, 289
ELA *see* equilibrium line altitude
elastic behavior 275
embrittlement 301, 309, 331
ENSO *see* El Niño Southern Oscillations
equilibrium line altitude (ELA) 225
erosion (*see also* erosion rate, erosional unloading) 206, 207, 234, 240, 243, 274, 278, 283, 285, 389
  climatic control 282
  continental 235
  definition 265
  footwall 277
  measurement 6, 231, 233–236, 238, 240, 243–259, 264
  pattern 220

erosional unloading 275
erosion rate 221, 223, 234, 236, 250, 254
    British Columbia 250, 372, 382
    catchment scale 234
    cosmogenic nuclide-derived 248
    definition 266
    estimation 235
    Germany (Bayerischer Wald) 251
    global 261
    Himalayas 208, 248, 250
    measurement 231–264, 279
    relationship to topography 221, 237, 243, 287
    Taiwan 250
Europe
    Ardennes 251, 262, 263
    Eifel 262, 263
    France 250, 252
    Germany 251, 252
    North Sea 50, 51, 54, 60, 61
    Norway 251, 301, 313, 330
    Scotland 15, 16
    Switzerland 253
European Alps *see* Alps, European
eustatic sea-level change 248
exhumation 4, 211–215, 299, 309
    rate 247
experiment 327, 363
    laboratory deformation 359, 363
    magnetotelluric 21
    numerical deformation 125
extension 47–51, 55, 56, 66, 73, 148, 150
    ductile 9, 15, 62, 72, 84, 96, 140
    faulting 16, 48, 51, 54–56, 58–64, 84, 85, 88, 148, 151, 379, 380
    lateral 71, 93
    syn-orogenic 381

failure 206, 320
fault (*see also* fault system, fault zone, rheology, strength, strengthening, strike-slip fault, weakening)
    activity and population centers 1, 289–292
    array 47, 56
    cementation 320

core 9–12, 15, 405
    damage zone 9, 38
    evolution 205, 240, 285
    friction 120, 342
    geometry 3, 5, 9, 10, 198
    gouge 337, 339, 341
    healing 9, 15, 337, 342, 345
    imaging 9, 15, 23
    initiation 277
    length 12, 277
    linkage 48–55, 58, 59, 66, 69–73
    loading 348
    localization 79, 85
    migration 59
    motion 3, 4, 83, 168, 184, 351
    network 47, 163
    nucleation 79, 188, 325, 330, 383
    processes, scale of 80, 186
    propagation 52
    response to stressing 107
    scaling 122, 158–160
    sealing 338, 345
    segmentation 14, 69
    strengthening 120, 335
    upper mantle, in 5, 15, 16, 29, 30, 62, 84, 91, 93, 97
    viscous creep 405
    weakening 3, 5, 7, 62, 67, 99–101, 103, 106, 111–119, 124–129, 153, 154, 157, 159, 162, 190–194, 319–322, 330, 334, 337, 342–351, 357, 358, 361, 371, 373, 376, 407, 411, 415
faulting (*see also* extension, subduction, strike-slip fault, oblique-slip fault) 231, 240, 243, 248, 251, 260, 273, 282, 292, 301, 357, 390
    climatic and surficial controls 280–282
    denudational response 222, 223, 231, 232, 243, 253, 254, 256, 263, 276, 278, 279, 283, 284
    geomorphic record 223, 233, 283, 285–287
    hazard associated with 289–292

impact on human environment 274, 288, 289
numerical modeling 357, 383
positive feedback with melting 377
surface environmental effect 231, 263, 273, 274, 277, 289
fault localities/names (*see also* shear zone localities/names)
    Abruzzo 56, 58
    Alpine (New Zealand) 20, 21, 31, 419
    Altun 67
    Barlett Spring 19
    Blackwater 68–70, 72
    Brent-Statfjord 59
    Calaveras 19, 24, 27, 28
    Calico 68–70
    Chelungpu 12, 255, 289
    Dead Sea Transform 14, 24, 30
    Eastern Structural High 65
    Emerson 24
    Farallon Ridge 19
    Fucino 55, 57
    Garlock 71
    Great Glen 21
    Harper 71
    Hayward 12, 19, 28
    Helendale 71
    Homestead Valley 24
    Hosgri 19
    Imperial 25, 26
    Johnson Valley 24
    Kun-Lun 34
    Lenwood 71
    Little Pine 332
    Lockhart 71
    Loreto 62, 64, 65
    Los Angeles basin 21
    Lost River 284
    Main Himalayan 246
    Marlborough (New Zealand) 31
    Nacimiento 19
    North Anatolian 6, 88, 89
    Periadriatic 86, 87, 171
    Periadriatic f. system (PFS) 17, 87, 171, 372, 373
    Precordilleran 23
    Punchbowl 10, 102

    Raikhot 208–211, 244
    Rangitaiki 55
    San Andreas 6, 19, 26, 38
    San Gabriel 10
    San Gregorio 19
    Santa Lucia Bank 19
    Salt Lake 282
    Sierra Madre 21
    Stak 244
    Strathspey-Brent-Statfjord 54
    Tre Monte 55
    Wasatch 58, 282
    West Fissure Zone (Chile) 22
fault scarp 287
fault slip
    geomorphic response 283
    melt-induced 371, 376–379
    temporal variation 62
fault structure 9, 10, 13, 24
    continental 139
fault system
    active 6, 283
    evolution 233
    growth 79
    localization 79
    nucleation 79
    transpressive 372
fault zone 9, 279, 311, 320, 322, 346, 348, 377
    deep 295
    fluid fluxes 418
    fluids 9
    geometry 10, 24
    head wave 14
    healing 14
    interaction with layering 92
    permeability 404, 417
    porosity 417
    postseismic sealing 338
    pressure solution 407
    properties 9, 12
    relation to rupture dynamics 101
    restrengthening 194
    rheology 337, 345, 349, 351
    seismic properties 14
    strength 342
    structure 102
    system 373

fault zone *cont.*
  viscosity 344
  width 349
feedback 6
  climate and weathering 264
  climate, erosion and tectonics 6, 273
  climatic 221, 260
  crustal deformation, seismicity, and
    surface erosion 276
  deformation and magmatism 377
  faulting 2
  fluid–rock 306
  fluid systems 91
  loop, positive 377
  mechanism, record 276
  melting 191
  melting and faulting 377
  rock uplift and topography 4
  strain localization and thermal
    structure of lithosphere 59
  stresses 226
  surface and tectonic processes 208,
    277
  surface erosion and crustal
    deformation 277, 278
  surface erosion and strain
    accumulation 277
  surface processes 225
  thermo-mechanical 92
  topography and precipitation 222
  weakening and strain localization
    344, 348
Flannan reflector 15, 16
flash heating 112, 117, 358
flow
  granular, frictional (cataclasis) 4
  law 363–365
  lower-crustal 384
  systems, temperature 404
  viscoelastic 275
  viscous 206
    granular 365
fluid (*see also* fluid flow, fluid
  pressure, fluid–rock interaction) 3,
  4, 12, 296, 319, 320, 335, 341, 403
  advection 1, 332
  amount in crust 300
  aqueous 314, 410

chemical processes 404
cooling 410
crustal 295–297, 299
distribution
  in crust 303, 407
  in rocks 304, 311
downward- and upwardflowing 406
evidence for 300
generation 312
geochemical equilibrium 422
inclusion 214, 301, 306
infiltration 308
in fractures 300
in upper crust 340, 341
isotopic composition of 320
metamorphic 312, 404
mineral wetting angle 303
mixing 404, 422
origin 312, 313, 422
phase, stressed, model of 324
process kinetics in crust 304
pumping 422
release 408
residence time in rocks 411
–rock reaction 403, 404
saline 297, 413
segregation 303
transfer 342
transient flux 337
viscosity 105
fluid flow 296, 311, 404, 412
  advective 324
  controls 413
  in convergent settings 419
  in rift zones 418
  magnitude 403, 418
  processes 403
  scales 416
  surfacial expressions 417
fluid pressure 4, 12, 301, 302, 311,
  339, 412
  change 415
  control 306
  diagram 412
  earthquake triggering 417
  effect on earthquakes 414
  evolution 342
  fluctuation 302

fluid–rock interaction 9, 14, 306, 312, 322, 349, 403, 412
   controls 404
   effect on recurrence time of earthquakes 409
   effect on seismic cycle 319
flux
   chemical 260
   denudational 253
   erosional 208, 221
   sediment 225
force
   basal tractional 385
   gravitational 385
forcing, tectonic 221, 250, 259
fracture/fracturing 14, 31, 322, 328, 330–333, 337–341, 409
   effect on pressure-solution creep 326, 327
   effect on seismic cycle 319
friction 409, 410
   coefficient 117
   constitutive law 108
   laws 5, 127
   localized 108
   rate-state 188, 409, 410
frictional sliding 4, 308, 334
frictional-to-viscous transition (FVT) (*see also* brittle-ductile transition) 149, 150

geochemical
   cycles 403
   processes, timescales 405
   tracer 312
geochronology 4, 152, 233, 261, 262, 273, 276, 277, 282
geodetic data analysis 47
geodynamic
   evolution 273
   model 232, 393
geomorphic
   record 232, 233, 235, 236, 241, 256, 261, 262, 264, 274, 278, 279, 281, 283–288
   response 283
   transport law 220
Gibbs free energy 322

glacial valley 225
glacier 225, 277
Global Position System (GPS) 55, 163
   Basin and Range 83
   EGAN 83
   measurement 10
   network 58, 82, 84
   sites 83, 165, 332
   use for estimating rheology 163–167
Goetze's criterion 147
GPS *see* Global Position System
Graben
   North Viking (North Sea) 60, 61
   Rhine 251, 413
   Whakatane (New Zealand) 55
gradient
   stress-induced 407
   thermal 376
   topographic 220, 253, 278, 384
grain
   boundaries, pressure solution 407
   dissolution 14
   fractured 329
   growth 148, 154, 166
granite, granitoid 118, 330, 360, 371
   decompression melt 214
granulite 309, 310, 330, 331
   Precambrian basement 313
   shield terrane 310
   terrane 310
gravitational collapse 380, 383
gravity
   method 10
   sliding 275
greenhouse effect 279
Greenland
   LAB depth 36
   S-wave receiver functions 37
ground acceleration 255, 289
groundwater
   infiltration rate 281
   pathway 279
Gulf
   Coast Extensional Province 66
   extensional faults 54
   of California 63–65, 287
   of Corinth 59
   of Suez 59

hazard, seismic 1, 7, 292
heat flux 208, 393
heating 117, 301
   frictional 17, 405
   radiogenic 17
   rate 305
   shear 113
Hellenic trench-arc system (*see also*
   Aegean Sea) 88
Hellenides 380
Helmholtz free energy 322
hillslope 221, 234
   denudation, measurement 280
   gradient 221
   mass wasting 238,
Himalaya, Himalayan, Himalayas 221,
   244, 246, 368, 381, 389
   Annapurna 370
   convergence model 388, 389
   crust 368
   Crystalline Complex 370, 389
   erosion rate 250
   front 208, 216, 218
      stress state 216, 217
   leucogranite 371, 381
   Main Central Thrust (MCT) 246,
      370, 381
   Main Frontal Thrust (MFT) 223
   Nepal 246
   orogen 6, 9, 21, 329, 387
   rivers 218
   sutures 33, 34
   Pakistani 245
human environment
   hominid evolution 2, 288
   impact of faulting 288
hydration 308, 408
   reaction 305
hydrocarbon 296
hydrofracture 306, 311
hydrological cycle 296
hydrology, subglacial 225
hydrometric station 235
hydroseismicity 417

IASP91 global reference model 32
Iceland
   LAB depth 36

S-wave receiver functions 37
ice sheet melting 275
imaging
   deep structures, of 31
   geo-electrical 23, 24
   geophysical 4, 95
   methods 4, 79
   seismic 7, 17
indenter, orogenic 327, 328
INDEPTH
   profile 370
   Project 21
India 21, 390
InSar (synthetic aperture radar
   interferometry) 15, 68, 69, 71, 163,
   417
instabilities, mechanical 152
interaction
   fluid–rock 9, 14
   tectonic–surface processes 275
interseismic
   creep 332
   period 320
   stressing 107
intracrystalline plasticity 327
isotopic geochronology
   Be-10 244, 246
   exposure age 249

Juan de Fuca, oceanic slab 18

Kontinentales Tiefbohrprogramm der
   Bundesrepublik Deutschland (KTB)
   300
   borehole 300
   hydrostatically pressured fluids
      413

LAB *see* lithosphere–asthenosphere
   boundary
laboratory
   experimental 7, 363
   natural 6
lake localities
   Bonneville 167, 281
Landers (*see also* earthquake localities/
   names)
   aftershock sequence 28

earthquake (1992) 14, 24, 25, 29, 71, 90, 144, 164, 166, 187
   postseismic displacement 165
   fault zone 15
landscape
   change 240
   erosional 237, 243
   evolution
      model 232, 279
      tectonic control on 231
   form 278
   geomorphic components 242
   perturbation 243
   response 221, 222
   scale 221
landslide 240
   earthquake-triggered 290
   size distributions 239
LARSE *see* Los Angeles Area Regional Seismic Experiment
Last Glacial Maximum 248
law
   Effective Stress, of *see* Terzaghi's Law
   geomorphic transport 220
   rate-and-state frictional constitutive 108
leucosome 381
   synkinematic 371, 383
lithosphere 232, 275, 358, 380
   continental 358
   contracting 1
   definition 37
   extending 1
   fluid flux 418
   long-term rheology 17
   seismicity patterns 17
   viscous 152
      depth 36
lithosphere–asthenosphere boundary (LAB) 10, 31, 36
   delay time 36
   mapping 36
lithospheric
   strength 17
   structure 17
   weakness 225

load
   lithostatic 206
   topographic 232, 275, 389
   transient 280
Los Angeles Area Regional Seismic Experiment (LARSE) 18
lower crust, effective viscosity 387
low-velocity damaged zone 13, 14, 38
low-viscosity channel 389

magma, ascent 17, 377
magmatic
   arc 376
   brine 297
   cycle 377
Main Central Thrust (MCT) 246, 370, 381
Main Frontal Thrust (MFT) 223
mantle
   deformation 29
   reflector 15
   seismic anisotropy of 29
mass
   flux, reconstruction 278
   redistribution 273, 280, 282
   transfer 322, 324, 334, 403
      diffusive 320, 328, 330
   wasting, process 238
Massif
   Central 251
   Nanga Parbat 308
   Rhenish 262
McLaren Glacier metamorphic belt 377
MCT *see* Main Central Thrust
MFT *see* Main Frontal Thrust
measurement of
   dissolved load 238
   hillslope mass wasting 238
   river sediment load 234
   water discharge 235
melt 5, 410
   connectivity transition 361
   content
      estimate 368
      relationship with viscous strength 411

melt *cont.*
    crustal, interaction with brittle
        upper crust 390
    effect on fault zone 371
    fraction, critical 304
    reduction of crustal viscosity 380
    residence time 371
        in orogenic crust 369
        in rocks 411
    segregation 303
    upward migration 377
    volume 359, 360, 362, 365, 366
    weakening 358, 364
melting 119, 358, 381, 384, 410
    dehydration 368
    effect on continental deformation 357
    effect on faulting 357
    influence on orogen shape 384
    modeling 384
    partial 386
    positive feedback with faulting 377
    relationship to large-scale
        deformation 372
    synkinematic 358
Mendocino Triple Junction 19
metamorphic core complex 254
    Bitterroot 382
    frictional-viscous transition in 149
    model 382
    Shuswap 382
metamorphic cycle 308, 312
metamorphism 5, 205, 208, 302, 305,
    309, 312, 372, 404, 408
    burial 5
    fluid flow 413
    fluid production 404
    low-grade 305
    prograde 5, 88, 295, 302, 305, 306,
        405, 408, 413
    retrograde 305, 404, 405, 408
metatexite 369
method
    borehole geophysical 15
    double-difference location 24, 25
    geo-electrical 10
    gravity 10
    magnetic 10
    magnetotelluric studies 23

    receiver function 31, 32
    seismic 15
    thermochronometric 233
microcrack growth 328
microcracking (*see also* microfacture)
    337
    model 324
microearthquake 10, 14
    activity 27
    locations 25
    relocation 25, 27
microfracture 332, 329
microseismicity 4
microstructural approach 350, 361
mineral 256
    catalytic effects 407
    dissolution 407
        rate 256
    precipitation 404
    reactions 319, 408
    solution–precipitation 276, 404
    weathering rate 257
model
    Altiplano 390
    convergence 218, 219, 388, 389
    decoupling zones 171
    fluid distribution 299
    IASP91 global reference 32
    island-channel 324
    landscape evolution 232, 279
    metamorphic core complexes 382
    microcracking 324
    orogen 384
    orogenesis 368
    orogenic crust 386
    plateau 385
    progressive strain localization 50
    rheological 342, 348
    strain localization 49
    strengthening 343
    thermomechanical 387, 389
    thin-fluid film 324
    topographic 208
    weakening 343
modeling 341
    earthquake 127
    finite-element 345
    numerical 7, 351, 383

orogenic 380
rupture nucleation 127
thermodynamic 302
Moho
  California 21
  central Andes 32
  continental, mapping 32
  delay time 36
  fault 19
  offset 21
    San Andreas fault 17
  temperature 93
  Tibet 21, 34
Mohr Coulomb failure, criteria 210
Mojave desert 68, 69, 71, 164
mountain belt 278
  active 222, 249, 253, 276
  front 287
  range 225, 273
mountain localities/names
  Alaska Range 250, 312, 377, 380
  Alps 6, 17, 18, 32, 87, 251–253,
    372–275, 278
  Andes 5, 9, 32, 276, 368–371, 389,
    390
  Appalachians, 251, 253
    Great Smoky 251
  Ardennes 262, 263
  Finisterre 287
  Himalaya 221, 244, 246, 368, 381,
    389
  Juneau belt 312, 313
  King Range 223
  San Gabriel 21
  Sierra Nevada 253
  Whipple 382
  Yucca 12
mylonite, mylonitization 4, 83, 87, 94,
  95, 140, 142, 143–146, 149, 150,
  153–158, 173, 330, 370, 372, 382

NAFZ *see* North Anatolian fault zone
New Zealand 30, 208, 376, 413
  SAPSE network 31
North Anatolian fault zone (NAFZ) *(see
  also* fault localities/names) 6, 88, 89
nucleation, fault 79, 188, 189, 325,
  330, 345, 383

oblique convergence 376
oblique-slip fault 392
olivine 330
  aggregate 391
  transient creep 167
  viscosity 366
ore, fault-bounded deposits 297, 418
orogen 206, 380, 391
  active 221, 358
  boundary 207
  exhumed 368
  modeled 384
  shape, influence of crustal melting
    on 384
orogenic
  crust 386
  modeling 380
  plateau 368
  wedge 275
orogen names
  Alps (European) 17, 87, 151, 373
  Andes 9, 369, 389
  Appalachians 253
  Caledonides 15, 380, 383
  Cordilleran 6, 383
  Himalayas 6, 9, 21, 250, 369, 370,
    380, 392
  Southern Alps (New Zealand)
    250
  Taiwan 247, 250, 253–255
  Variscides 253, 371, 380, 383
outreach, public 7, 291, 292
oxygen, isotope signature 312

paleodenudation
  estimate 261
  rate 261, 262
  sequence 261
Péclet number 214
Periadriatic fault system (PFS) 87,
  171, 372, 373
  map 86
  plutons 374, 375
permeability 3, 306, 339, 417
  as an intrinsic rock property 306
PFS *see* fault localities/names,
  Periadriatic fault system
plasticity, intracrystalline 327

plate
  Asian continental 254
  Australian 30
  boundary 206
  Eurasian–Indian boundary 216
  Indian 21
  motion 1, 31
  Pacific 30
  Philippine Sea 254
  structure 211
plateau, orogenic
  Andean 5, 9, 368, 369, 389, 390
  margin 390
  model 385
  Tibetan 38, 208–216, 250, 368–371,
    380, 389, 411
pluton localities/names
  Himalayan 381
  Periadriatic 373–375
pollen record 262
pore fluid
  impurity 338
  pressure 5, 276, 281, 342, 415, 414
  thermal pressurization 113, 115,
    117
pore inflation 306
porosity 3, 417
postseismic 332
power law
  relationship 366
  scaling 240
precipitation 275, 281, 322, 325
  rates 407
  reactions 404, 406
pressure (*see also* pressure solution)
  change 404
  lithostatic 342
  osmotic 415
  pore fluid 5, 414
pressure solution 324, 329, 332, 341,
    345, 349, 405, 407
  creep 323, 326, 327, 332, 342,
    405
  drivers 407
  factors 407
  grain boundaries, at 407
  sliding 334
  thermodynamics 407

pressurization, thermal 113, 115
profile (*see also* section)
  deep reflection 15
  NRP-20 17
  receiver function 33–35
  reflection 10, 17, 20
  refraction 17, 20
  seismic 15, 16, 18, 370
prograde metamorphism 302, 305, 306
progressive dehydration 306
pseudotachylite 308, 330
Puna, central Andes 32, 368, 369
  crust 32
  Plateau 32, 33, 368
    receiver function profile 33
P-wave 12, 32
  receiver function 32, 33
  velocity 337, 368

quartz 297, 330, 335, 337
  alpha-beta transition 368
  cement 404
  dissolution 338
  hydrous 299
  solubility 297, 337
  viscosity 364

rate
  friction 188, 409, 410
  frictional constitutive law 108
  -limiting process 324
RCMP *see* rheological critical melt
  percentage
reaction
  dehydration 305
  geochemical 403
    scales 416
  metamorphic 5
  rates 409
  softening 342, 345, 349
  weakening 330
receiver function 31, 37
  profile 32–35
  technique 31, 32
record
  geologic 283, 286
  geomorphic 262, 283, 284, 287
  hydrometric 235

stratigraphic 286
recrystallization 9
    dynamic 153, 305
    eclogite facies 313
    grain-boundary migration 144,
        145, 153
    grain-size reduction 67, 73
    static 191
reflection
    profile 10, 17, 20
    seismic 15
        imaging 16
        profile 32
reflectivity, crustal 15
reflector, seismic 299
refraction 15
    active-source 10
    imaging 15
    profile 17, 20
relief (*see* topographic relief)
resistivity, electrical 12, 23, 24
response
    geomorphic 283
    mechanical 275
    topographic 278
retrograde metamorphism 299, 303,
    305
retrowedge, orogenic 373
rheological
    critical melt percentage (RCMP)
        360, 361, 369
    evolution 211
    model 342, 348
    modification 226
    stratification 93
    transition 368, 369
rheology 3, 12, 139, 163, 220, 232,
    275, 299, 314, 337, 342, 349, 390
    continental crust 358
    crustal 206, 213, 216, 385
        from satellite-based data
            163–167
    experiment 361
    fault rock 183
Rift
    Inner Moray Firth 50, 51
    Rio Grande 413
    Southeast Rift Zone (Hawaii) 28

risk *see* seismic risk
river 235
    anticline 218
    carrying capacity 237
    catchment 206
    incision, climatic control 246
    incision rate 222, 247, 248
    sediment load 234
    stage 236
    terrace 247, 287
River
    Astor 211
    Burhi Gandaki 206, 237, 246
    Choshui 255, 256
    Indus 244, 245
    Jinsha 34
    Leine 235
    Liwu 246, 247
    Meuse 262
    Nsimi 260
    Peinan 237
    Potomac 248, 249
    Susquehanna 248, 249
    Wendebach 235
rock
    anatectic 368, 369, 371, 389
    deformation, effect of fluid 410
    fluid content 304, 309, 405
    -fluid interaction 304
    melt-bearing 391
    permeability 306
    polymineralic 322, 332
    seismic anisotropy 29
    strengthening 5
    uplift 4, 253, 287
    viscosity, evolution of 335
runoff, definition 265
rupture 255
    dynamic 101, 190, 192
    nucleation 127
    parameter 129

SAFOD *see* San Andreas fault deep
    drill hole site
salt, salinity 238, 296, 297, 415
San Andreas fault 6, 9, 19, 26, 38, 243,
    342, 418
    deep drill hole site (SAFOD) 24

San Andreas fault *cont.*
  microearthquake locations 27
  Parkfield segment 28
  profile 20
  resistivity 22, 24
  seismic anisotropy of upper mantle
    29
  seismic surveys 17
  streaks 28
scale, tectonic processes 285
scaling, multiscaling 122, 158–160, 350
sea-level change 248, 287
sediment
  concentration 235, 236
  discharge 254, 255
  fluvial transport 233
  flux 225
  homogenization during transport
    243
  load 235, 236, 247, 265
  suspended, concentration 235, 247,
    256
  transport 220, 278, 282
  trap 238
  yield 261, 266
sedimentary basin *see* basin
sedimentary brine 297
seismic anisotropy 29
seismic cycle *see* earthquake cycle
seismic event 264, 283, 286
seismic hazard 289–292
seismic moment release 254
seismic period 194
seismic reflection
  imaging 15, 16
  profile 32
seismic risk 7, 289–292
seismic velocity 12, 19
seismicity 1, 24, 231, 232, 264, 283
  crustal 17
  dating 286
  depth distribution 29
  pattern 10
  rate 108
  streaks 10, 28, 29
seismogenesis, temperature limit for
  depth 108
seismogenic zone 14, 38

setting
  extensional 379
  transpressive 372
shear (*see also* shear zone) 376
  heating 113, 381
  localized 121
  strength 12, 115, 334
  wave, splitting 31
shear band 154–158
  relationship to seismicity 157, 158
  strain 155
  stress on 156
shear zone 1, 5, 15, 31, 301, 303, 312,
  330, 381
  dislocation creep 84
  ductile 322
  eclogitic 331
  formation 330
  geometrical scaling properties
    158
  lithospheric 31
  melt-bearing 372
  midcrustal 330
  pattern 359
  reactivated 15
  stress state 160
  subhorizontal 382
shear zone localities/names (*see also*
  fault zone localities/names)
  East California Shear Zone 71
  Pogallo 143
  Tenda 330
shield names
  Brazilian 32
  Scandinavian 32
short slip, duration 128
silica 297
silica gel 118
silicate melt 296
silicate weathering 256, 261, 279
  influence on climate 279
  rate 259
  tectonic control on 259
slip 115, 117, 124, 350
  coseismic 358
  distributions 192
  duration 128
  evolution 125

inversion 129
localization 12
propagation 192
rapid 120
rate 4, 117
seismic 111
temperature increase 103
temporal variation 280
unstable 188
weakening 124
slip-line theory 170
slip-parallel streaks 38
slope, topographic 206
SLT *see* solid-to-liquid transition
soil
  denudation rate 241
  mechanics 207
  production 243, 244, 280
solid-to-liquid transition (SLT) 369
solidus 358, 362, 385
solubility 325
solution cleavage 341
Southern Alps, New Zealand 6, 373,
  408
  cross section 419
  erosion rate 250
  landslide size distribution 239
South Island, New Zealand 376
  cross section 419
South Tibetan detachment system
  (STD) 370, 381, 389, 390
splitting, shear wave 31
spreading, gravitational 380
STD *see* South Tibetan detachment
  system
strain (*see also* strain localization, strain
  partitioning, strain rate) 1, 335
  crustal 278, 279
  field 279
  model 50
  Nanga Parbat 212, 213
  release 282
  tectonic 211, 226
  weakening 11
strain localization 5, 47, 51, 54, 154,
  155, 159, 274, 277, 280, 330, 377
  evolution 48
  model 50

rapid 49
strike-slip settings, in 67
strain partitioning 167, 168, 171, 220,
  301, 376
strain rate 329, 346, 348, 363
  axial 324
  extrapolation 363
  profile 346
stratigraphic record 286
streaks
  of seismicity 28
  slip-parallel 38
strength 359, 360
  crustal 160–163, 299, 309,
    141–149, 367
  diagram 378, 379
  drop 361, 391
  fault 276, 334, 342, 345
  lithospheric 17
  postseismic 337
  profile 367
  recovery 337
  reduction 214, 215
  tensile strength 306
  viscous 362
strengthening 336, 344–346, 349
  effect on strain delocalization 344
  model 343
  pervasive 335
  process kinetics 345
  rate 344, 348, 349
  restrengthening 194–197
  shear stress variation 347
stress 1, 124, 156, 201
  compressive horizontal 207
  corrosion 409
  crustal 206, 280
  distribution 390
  drop, coseismic 144, 147, 162
  jump 109
  loading, interseismic 349
  relaxation 348, 349
  shear zones, in 160
  topographically generated 206,
    207, 210–212, 216, 223
strike-slip (*see also* strike-slip fault)
  deformation 30
  environment 14

strike-slip *cont.*
    segment boundary 69
    settings, strain localization 67
    shear zone 29
strike-slip fault 10, 12, 14, 17, 30, 66,
    141, 184, 185, 243, 376, 378, 379,
    392
    arc parallel 376
    array evolution 70
    damage zone 12
    displacement 141
    earthquake frequency 141
    pore-fluid pressure 141
    seismic anisotropy 10
stylolite 341
subduction 17
    Japan 13
    thrust 199
    zones, flow systems 404
Sumatra-Andaman earthquake (2004)
    7, 288, 376
surface (*see also* surface deformation,
    surface processes)
    chemical potential 322
    dating 4
    displacement field 287
    environmental effect 273
    erosion 277
    load 280
    rupture 286
    state 206
    velocity field 279
surface deformation 283
    climate impact on 280
    measurement 163
surface processes 205, 206, 213, 220,
    225, 273, 274, 292
    scales 80
    tectonic response 275
S-wave 12, 29, 36
    receiver functions 32, 36, 37
    velocity 337, 368

Taiwan 237, 244, 246, 254, 263, 289, 290
    Central Range 254
    erosion rate 250
    orogen 247, 250, 253–255
    subduction system 376

tectonic
    aneurysm 208, 214
    denudation 265
    surge 372
temperature (*see also* temperature
    effect) 3, 297
    frictional-driven change in 103
    increase by slip 103
    lithospheric distribution 96
    Moho 93
    stress jump 109
temperature effect on
    brittle–ductile transition 17,
       142
    chemical weathering rate 258
    deformation zone 93
    dynamic slip 105
    failure in the lower crust 52
    friction 139
    grain growth 166
    pressure solution 404, 407
    rheology 91, 142, 364
    shear strength 112
    silicate weathering 261
    slip 107
tephrachronology 262
Terzaghi's Law 276
thermal pressurization 113, 115
thrust
    Caledonian 15
    Chelungpu 289
    Main Central (Himalayan) 246,
       370, 381
    Main Frontal (Himalayan) 223
    subduction 199
Tibet 5, 32, 371, 389
    convergence model 388, 389
    deep structure 33
    earthquakes 34
    Moho offset 21
    plateau 38, 368, 369, 371, 380,
       389
    receiver function profile 34, 35
tomography, seismic 10
topographic
    gradient 220, 253, 278, 384
    index (TSI) 210
    load 232, 275, 389

model 208
relief 248
    reduction, timescale 250
response 278
slope 206
stress 206, 207, 223
topography 2, 205–207, 279
    correlation length 279
    effect on regional strain 210
    erosional 287
    faulted 280
tracer 312
traction evolution 124, 128
transient load 280
transition
    aseismic to seismic 107
    brittle-to-ductile 17, 308–310
    frictional-to-viscous 139, 142, 149,
        150, 154, 163, 308
        aseismic 144, 145
        coseismic 143
        fossil 149
    liquid-to-solid 362
    rheological 368, 369
transport
    advective 339
    geomorphic 220
tree-ring record 238
TSI *see* topographic stress index
tsunami 7, 80, 289
turbidity current 256
typhoon, flood 256

United States 244, 253, 277
uplift 251, 302
    rate 223, 224
upper mantle 330
    decoupling from lower crust 367
up-scaling 159

valley
    formation 243
    glacial 225
Valley
    Astor River 209–213
    Big Lost 284
    Death 93
    Indus River 208, 209, 211–213

Meuse 261
Owen's 71
Tennessee 243, 244
Vancouver Island 408, 413
Variscan
    granite 312
    orogen 380
    orogenic event 253
vein
    cementation 341
    diagenetic 297
    gold-bearing 311, 312
    metamorphic 297
    quartz
        oil-bearing 313
        system 311, 312
    sealing of 342
velocity
    profile 346
    seismic 12, 19
Very Long Baseline Interferometry
    (VLBI) 68, 69, 163
viscoelastic flow 275
viscosity 275, 329, 343, 364–366, 380,
    385, 387, 390
    decrease 344
    effective 167, 363–365
    evolution of 332
    lithospheric 36, 163, 165–167,
        275, 330, 335, 343, 344, 380,
        384–391
    postseismic 332
viscous behavior 163–167, 206, 275
VLBI *see* Very Long Baseline
    Interferometry
volcanic island, flexure pattern 274
volcanism 355, 369
    in Eifel area (Germany) 262, 263
    on Hawaii 28, 29

water 296, 306
    activity diagram 307
    discharge 235, 247
        definition 265
        measurement 235
    partial pressure 300
wave
    fault zone-guided 10, 13

wave *cont.*
  fault zone head 14
  Love 13
  P- 12, 32
  S- 12, 29
  seismic propagation 13
  teleseismic converted 10
waveform cross-correlation (*see* also
  method, double-difference) 24, 25
weakening 112, 344–347, 349, 384
  advection 216
  dynamic 111, 190
  kinetics of 345
  melt-induced 5, 358, 371, 376, 380
  model 343
  rate 344, 348, 349
weakness
  crustal 376
  lithospheric 225

weathering 232, 234, 240, 243, 274,
  278, 279, 285
  chemical 259
    catchment scale 256
    flux 260
    rate 258
  climate dependence of 259
  definition 265
  index 280
  rate 256, 257
  silicate 261
  transport control on 257
wedge, orogenic
  critical equilibrium 207
  crustal 275, 385
Western Gneiss Region 330
wetting angle, fluid 303, 304

xenoliths, seismic imaging 94